# Clay Mineral Transformations after Bentonite/Clayrocks and Heater/Water Interactions from Lab and Large-Scale Tests

# Clay Mineral Transformations after Bentonite/Clayrocks and Heater/Water Interactions from Lab and Large-Scale Tests

Editors

**Ana María Fernández**
**Stephan Kaufhold**
**Markus Olin**
**Lian-Ge Zheng**
**Paul Wersin**
**James Wilson**

MDPI • Basel • Beijing • Wuhan • Barcelona • Belgrade • Manchester • Tokyo • Cluj • Tianjin

*Editors*

Ana María Fernández
Department of
Enviroment Ciemat
Madrid, Spain

Stephan Kaufhold
Geophysical Exploration—
Technical Mineralogy BGR
Hannover, Germany

Markus Olin
VTT Technical Research
Centre of Finland Espoo,
Otaniemi, Finland

Lian-Ge Zheng
Energy Geoscience Division
Lawrence Berkeley National
Laboratory (LBNL)
Berkeley, CA, USA

Paul Wersin
Institute of Geological Sciences,
University of Bern
Bern, Switzerland

James Wilson
Wilson Scientific Ltd.,
Wilson Scientific
Warrington, UK

*Editorial Office*
MDPI
St. Alban-Anlage 66
4052 Basel, Switzerland

This is a reprint of articles from the Special Issue published online in the open access journal *Minerals* (ISSN 2075-163X) (available at: https://www.mdpi.com/journal/minerals/special_issues/CMTB).

For citation purposes, cite each article independently as indicated on the article page online and as indicated below:

LastName, A.A.; LastName, B.B.; LastName, C.C. Article Title. *Journal Name* **Year**, *Volume Number*, Page Range.

ISBN 978-3-0365-4430-4 (Hbk)
ISBN 978-3-0365-4429-8 (PDF)

# Contents

*Editorial*

# Editorial for Special Issue "Clay Mineral Transformations after Bentonite/Clayrocks and Heater/Water Interactions from Lab and Large-Scale Tests"

Ana María Fernández [1,*], Stephan Kaufhold [2], Markus Olin [3], Lian-Ge Zheng [4], Paul Wersin [5] and James Wilson [6]

[1]  Centro de Investigaciones Energéticas Medioambientales y Tecnológicas, 28040 Madrid, Spain
[2]  Bundesanstalt für Geowissenschaften und Rohstoffe, 30655 Hannover, Germany; stephan.kaufhold@bgr.de
[3]  VTT Technical Research Centre of Finland Ltd., 02044 Espoo, Finland; markus.olin@vtt.fi
[4]  Energy Geoscience Division, Lawrence Berkeley National Laboratory (LBNL), Berkeley, CA 94720, USA; lzheng@lbl.gov
[5]  Institute of Geological Sciences, University of Bern, CH-3012 Bern, Switzerland; paul.wersin@geo.unibe.ch
[6]  Wilson Scientific Ltd., Warrington WA3 6TR, UK; jim@wilsonscientific.co.uk
*   Correspondence: anamaria.fernandez@ciemat.es

**Citation:** Fernández, A.M.; Kaufhold, S.; Olin, M.; Zheng, L.-G.;Wersin, P.; Wilson, J. Editorial for Special Issue "Clay Mineral Transformations after Bentonite/Clayrocks and Heater/Water Interactions from Lab and Large-Scale Tests". *Minerals* **2022**, *12*, 569. https://doi.org/10.3390/min12050569

Received: 20 April 2022
Accepted: 26 April 2022
Published: 30 April 2022

**Publisher's Note:** MDPI stays neutral with regard to jurisdictional claims in published maps and institutional affiliations.



This Special Issue "Clay Mineral Transformations after Bentonite/Clayrocks and Heater/Water Interactions from Lab and Large-Scale Tests" covers a broad range of relevant and interesting topics related to deep geological disposal of nuclear fuels and radioactive waste.

Most countries that generate nuclear power have developed radioactive waste management programmes during the last 50 years to emplace long-lived and/or high-level radioactive wastes in a deep underground repository in a suitably chosen host rock formation. The aim is to remove these wastes from the human environment, considering the ethical undertaking that this geological disposal should be pursued now and not left to future generations [1,2]. Because radioactive waste retains potentially harmful levels of radioactivity for hundreds of thousands of years, safely disposing of high-level nuclear waste is not a temporary problem, but it involves a million-year solution.

The earliest discussions of solutions for the disposal of nuclear waste date from the mid-1950s and the first US National Academy of Sciences (NAS) Committee on Waste Disposal report [3], where it was proposed that the waste could be disposed of hundreds of meters underground in specially constructed mined cavities. If a site were properly chosen, a repository system comprising both natural and engineered barriers would provide a high level of protection from the toxic effects of the waste. Since then, all countries having a long-term management strategy have accepted this approach [4].

The search for technically suitable and socially acceptable repository sites for high-level radioactive waste (HLRW) and spent fuel (SF) began in the mid-1960s. Nuclear nations in Western Europe and elsewhere began to plan for geological disposal during the late 1970s and 1980s. Since the mid-1980s, France and Belgium initiated projects for analysing the disposal of HLRW in clayrocks, whereas Finland, Sweden, Canada, Switzerland and Spain did the same in crystalline granitic rocks. In 1998, the European Community recommended to its members to "continue activities on siting, construction, operation and closure of HLRW repositories". For this purpose, most countries have interacted through the International Atomic Energy Authority, which has established safety requirements for the disposal of radioactive waste [5] and guidance on how geological disposal facilities should be developed [6].

Nowadays, the site-selection process for HLRW has been reached in only three countries: Finland, Sweden and France. The first construction license for a geological disposal facility was presented by Posiva at Olkiluoto in Finland in 2012, which was accepted in 2015. Final disposal activities are expected to start in 2025. On 16 March 2011, SKB applied for a license to build a repository for spent nuclear fuel at the Forsmark Site (Sweden, north of Stockholm), which was finally approved in January 2022. In France, the National Radioactive Waste Management Agency (ANDRA) is currently preparing its license application for the Industrial Centre for Geological Disposal (Cigeo project, Meuse/Haute-Marne Site). In Canada and Switzerland, national waste management agencies (NWMO and NAGRA, respectively) are still investigating appropriate sites through site characterisation programmes. In other countries, such as the United States, Germany and Spain, the process appears to have stalled [2,7]. In the case of Russia, the research into the searching of geological formations and sites for building an underground facility for HLRW disposal started in 1993, later than in the other countries. In 2015, it approved the programme "Maintaining Nuclear and Radiation Safety, 2016–2030", for the construction of a high-level and intermediate-level radioactive waste isolation facility in the Nizhnekansk Granitic Gneiss Crystalline Massif (Krasnoyarsk Territory, Central Siberia) [8].

Significant research efforts world-wide have increased our knowledge and understanding of how underground disposal systems will function over very long periods of time. Characterisation works have been performed, not only at laboratory scale but also by large scale in situ tests in underground research laboratories (URLs). These URLs, conducted for experimental and demonstration programmes, have played an important role in the assessment of the geological and the engineered barrier systems, the development of research methodologies and techniques, the development of construction techniques, the demonstration of repository operations, confidence building and international co-operation [1,7].

The first URLs were already developed in the ninety-sixties and ninety-seventies. Since then, the main URLs constructed were [9]: Asse Mine (salt formation, Germany, 1965–1997), Nevada Test Site (tuff, USA, 1979–1990), Sellafield (tuff, UK, stopped in 1997), HADES-URF (clay, Belgium, since 1980), Konrad (limestone, Fe-mine, Germany, since 1980), URC Josef (volcanic and sedimentary rock, Czech Republic, since 1981), Grimsel Test Site (granite, Switzerland, since 1983), Lac du Bonnet (granite, Canada, since 1984), Gorleben (salt dome, Germany, 1985–2020), Tono (sandstone, Japan, since 1986), Äspö (granite, Sweden, since 1990), Tournemire (clay, France, since 1990), Olkiluoto (granite, Finland, since 1993), Mont Terri (clay, Switzerland, since 1995), Yucca Mountain (tuff, 1996–2010), Meuse/Haute-Marne (clay, Bure, France, since 2000), ONKALO (granite, Okiluoto, Finland, since 2003), Mizunami URL (granite, Japan, since 2004), Horonobe (sedimentary rock, Japan, since 2005), KURT (granite, Korea, since 2006) and Beishan URL (igneous rock, Gobi Desert, China, since 2021). An underground research laboratory will be built in the Nizhnekansk Granitic Massif in 2025–2030, as the first phase in constructing a HLRW final isolation facility in Russia [8].

The 17 papers published in this Special Issue show that bentonites and clayrocks are an essential component of the multi-barrier system ensuring the long-term safety of the final disposal of nuclear waste. The efficiency of such engineered and natural clay barriers relies on their physical and chemical confinement properties, which should be preserved in the long-term. From a geochemical point of view, the clayey mineral's function to isolate the canisters from water and retard the migration of radionuclides also means maintaining a suitable chemical and mineralogical environment for canister integrity, radionuclide retention and mechanical stability over time, buffering possible alteration/deterioration processes of the nanoporous clay materials [10,11].

The selected papers can be grouped in four main investigating issues: (a) characterisation of the barrier materials and their properties [12,13], (b) impact of transient variables and external perturbations on clay minerals' properties: cement, temperature, organics and microbial activity [14–18], (c) long-term behaviour of bentonite and clayrock barriers from large-scale tests inside URLs [19–25] and (d) long-term predictions of clay behaviour by means of reactive-transport modelling after different perturbations linked to interactions between the clay materials and the allochthonous engineered solid materials (ground-/pore-water, heat, cement/concrete, iron corrosion, organics, etc.) [26–28]. Most of these studies are currently being performed in the framework of various international co-operation programmes, such as the Mont Terri Project and the European Commission BEACON and EURAD (tasks ACED, HITEC, CONCORD) Projects, which are a base for improving technical and scientific knowledge in the context of nuclear waste disposal.

The Guest Editors would like to acknowledge the authors and reviewers for their foremost contributions and the *Minerals* journal for the opportunity to share this valuable knowledge.

**Conflicts of Interest:** The authors declare no conflict of interest.


## References

1. OECD. *Geological Disposal of Radioactive Waste. Review of Developments in the Last Decade*; OECD Nuclear Energy Agency (NEA): Paris, France, 2000; 106p.
2. Ewing, R.C.; Whittleston, R.A.; Yardley, B.W.D. Geological Disposal of Nuclear Waste: A Primer. *Elements* **2016**, *12*, 233–237. [CrossRef]
3. NAS–NRC. *Disposal of Radioactive Waste on Land*; National Academy of Sciences–National Research Council Report; National Academies Press: Washington, DC, USA, 1957; 146p.
4. Metlay, D.S. Selecting a Site for a Radioactive Waste Repository: A Historical Analysis. *Elements* **2016**, *12*, 269–274. [CrossRef]
5. IAEA (International Atomic Energy Agency). *Disposal of Radioactive Waste. Specific Safety Requirements*; International Atomic Energy Agency Safety Standards Series No. SSR-5; IAEA: Vienna, Austria, 2011; 62p.
6. IAEA (International Atomic Energy Agency). *Geological Disposal Facilities for Radioactive Waste. Specific Safety Guide*; International Atomic Energy Agency Safety Standards Series No. SSG-14; IAEA: Vienna, Austria, 2011; 104p.
7. Delay, J.; Bossart, P.; Ling, L.X.; Blechschmidt, I.; Ohlsson, M.; Vinsot, A.; Nussbaum, C.; Maes, M. Three decades of underground research laboratories: What have we learned? *Geol. Soc. Lond. Spec. Publ.* **2014**, *400*, 7–32. [CrossRef]
8. Nikitin, A. *The Underground Research Laboratory in the Deep Geological Repository in the Nizhnekansk Massif, Krasnoyarsk Territory*; A Bellona Working Paper; Bellona: Oslo, Norway, 2018; 23p.
9. OECD. *Underground Research Laboratories (URL)*; OECD Nuclear Energy Agency (NEA): Paris, France, 2013; 52p.
10. Altman, S. Geochemical research: A key building block for nuclear waste disposal safety cases. *J. Contam. Hydrol.* **2008**, *112*, 174–179. [CrossRef] [PubMed]
11. Apted, M.J.; Ahn, J. *Geological Repository Systems for Safe Disposal of Spent Nuclear Fuels and Radioactive Waste*, 2nd ed.; Woodhead Publishing Series in Energy; Elsevier: Duxford, UK, 2017.
12. Daniels, K.A.; Harrington, J.F.; Milodowski, A.E.; Kemp, S.J.; Mounteney, I.; Sellin, P. Gel Formation at the Front of Expanding Calcium Bentonites. *Minerals* **2021**, *11*, 215. [CrossRef]
13. Meleshyn, A.Y.; Zakusin, S.V.; Krupskaya, V.V. Swelling Pressure and Permeability of Compacted Bentonite from 10th Khutor Deposit (Russia). *Minerals* **2021**, *11*, 742. [CrossRef]
14. Manzel, T.; Podlech, C.; Grathoff, G.; Kaufhold, S.; Warr, L.N. In Situ Measurements of the Hydration Behavior of Compacted Milos (SD80) Bentonite by Wet-Cell X-ray Diffraction in an Opalinus Clay Pore Water and a Diluted Cap Rock Brine. *Minerals* **2021**, *11*, 1082. [CrossRef]
15. Bateman, K.; Amano, Y.; Kubota, M.; Ohuchi, Y.; Tachi, Y. Reaction and Alteration of Mudstone with Ordinary Portland Cement and Low Alkali Cement Pore Fluids. *Minerals* **2021**, *11*, 588. [CrossRef]
16. Kašpar, V.; Šachlová, Š.; Hofmanová, E.; Komárková, B.; Havlová, V.; Aparicio, C.; Černá, K.; Bartak, D.; Hlaváčková, V. Geochemical, Geotechnical, and Microbiological Changes in Mg/Ca Bentonite after Thermal Loading at 150 °C. *Minerals* **2021**, *11*, 965. [CrossRef]
17. Laufek, F.; Hanusová, I.; Svoboda, J.; Vašíček, R.; Najser, J.; Koubová, M.; Čurda, M.; Pticen, F.; Vaculíková, L.; Sun, H.; et al. Mineralogical, Geochemical and Geotechnical Study of BCV 2017 Bentonite—The Initial State and the State following Thermal Treatment at 200 °C. *Minerals* **2021**, *11*, 871. [CrossRef]
18. Podlech, C.; Matschiavelli, N.; Peltz, M.; Kluge, S.; Arnold, T.; Cherkouk, A.; Meleshyn, A.; Grathoff, G.; Warr, L.N. Bentonite Alteration in Batch Reactor Experiments with and without Organic Supplements: Implications for the Disposal of Radioactive Waste. *Minerals* **2021**, *11*, 932. [CrossRef]
19. Bleyen, N.; Small, J.S.; Mijnendonckx, K.; Hendrix, K.; Albrecht, A.; De Cannière, P.; Surkova, M.; Wittebroodt, C.; Valcke, E. Ex and In Situ Reactivity and Sorption of Selenium in Opalinus Clay in the Presence of a Selenium Reducing Microbial Community. *Minerals* **2021**, *11*, 757. [CrossRef]
20. Emmerich, K.; Bakker, E.; Königer, F.; Rölke, C.; Popp, T.; Häußer, S.; Diedel, R.; Schuhmann, R. A MiniSandwich Experiment with Blended Ca-Bentonite and Pearson Water—Hydration, Swelling, Solute Transport and Cation Exchange. *Minerals* **2021**, *11*, 1061. [CrossRef]
21. Wersin, P.; Hadi, J.; Jenni, A.; Svensson, D.; Grenèche, J.-M.; Sellin, P.; Leupin, O.X. Interaction of Corroding Iron with Eight Bentonites in the Alternative Buffer Materials Field Experiment (ABM2). *Minerals* **2021**, *11*, 907. [CrossRef]
22. Sudheer Kumar, R.; Podlech, C.; Grathoff, G.; Warr, L.N.; Svensson, D. Thermally Induced Bentonite Alterations in the SKB ABM5 Hot Bentonite Experiment. *Minerals* **2021**, *11*, 1017. [CrossRef]
23. Kaufhold, S.; Dohrmann, R.; Ufer, K.; Svensson, D.; Sellin, P. Mineralogical Analysis of Bentonite from the ABM5 Heater Experiment at Äspö Hard Rock Laboratory, Sweden. *Minerals* **2021**, *11*, 669. [CrossRef]

24. Fernández, A.M.; Marco, J.F.; Nieto, N.; León, F.J.; Robredo, L.M.; Clavero, M.A.; Fernández, S.; Svensson, D.; Sellin, P. Characterization of Bentonites from the in situ ABM5 Heater Experiment at Äspö Hard Rock Laboratory, Sweden. *Minerals* **2022**, *11*, 471. [CrossRef]
25. Yokoyama, S.; Shimbashi, M.; Minato, D.; Watanabe, Y.; Jenni, A.; Mäder, U. Alteration of Bentonite Reacted with Cementitious Materials for 5 and 10 years in the Mont Terri Rock Laboratory (CI Experiment). *Minerals* **2021**, *11*, 251. [CrossRef]
26. Bateman, K.; Murayama, S.; Hanamachi, Y.; Wilson, J.; Seta, T.; Amano, Y.; Kubota, M.; Ohuchi, Y.; Tachi, Y. Evolution of the Reaction and Alteration of Mudstone with Ordinary Portland Cement Leachates: Sequential Flow Experiments and Reactive-Transport Modelling. *Minerals* **2021**, *11*, 1026. [CrossRef]
27. Jenni, A.; Mäder, U. Reactive Transport Simulation of Low-pH Cement Interacting with Opalinus Clay Using a Dual Porosity Electrostatic Model. *Minerals* **2021**, *11*, 664. [CrossRef]
28. Chaparro, M.C.; Finck, N.; Metz, V.; Geckeis, H. Reactive Transport Modelling of the Long-Term Interaction between Carbon Steel and MX-80 Bentonite at 25 °C. *Minerals* **2021**, *11*, 1272. [CrossRef]

*Article*

# Gel Formation at the Front of Expanding Calcium Bentonites

Katherine A. Daniels [1,*], Jon F. Harrington [1], Antoni E. Milodowski [1], Simon J. Kemp [1], Ian Mounteney [1] and Patrik Sellin [2]

[1]  British Geological Survey, Keyworth, Nottinghamshire NG12 5GG, UK; jfha@bgs.ac.uk (J.F.H.); aem@bgs.ac.uk (A.E.M.); sjk@bgs.ac.uk (S.J.K.); iaian1@bgs.ac.uk (I.M.)

[2]  Svensk Kärnbränslehantering AB (SKB), Brahegatan 47, 10240 Stockholm, Sweden; patrik.sellin@skb.se

*  Correspondence: katdan@bgs.ac.uk; Tel.: +44-(0)115-936-3370

**Abstract:** The removal of potentially harmful radioactive waste from the anthroposphere will require disposal in geological repositories, the designs of which often favour the inclusion of a clay backfill or engineered barrier around the waste. Bentonite is often proposed as this engineered barrier and understanding its long-term performance and behaviour is vital in establishing the safety case for its usage. There are many different compositions of bentonite that exist and much research has focussed on the properties and behaviour of both sodium (Na) and calcium (Ca) bentonites. This study focusses on the results of a swelling test on Bulgarian Ca bentonite that showed an unusual gel formation at the expanding front, unobserved in previous tests of this type using the sodium bentonite MX80. The Bulgarian Ca bentonite was able to swell to completely fill an internal void space over the duration of the test, with a thin gel layer present on one end of the sample. The properties of the gel, along with the rest of the bulk sample, have been investigated using ESEM, EXDA and XRD analyses and the formation mechanism has been attributed to the migration of nanoparticulate smectite through a more silica-rich matrix of the bentonite substrate. The migration of smectite clay out of the bulk of the sample has important implications for bentonite erosion where this engineered barrier interacts with flowing groundwater in repository host rocks.

**Keywords:** calcium bentonite; gel; swelling; water uptake; ESEM; EDXA; surface area; XRD; radioactive waste disposal

**Citation:** Daniels, K.A.; Harrington, J.F.; Milodowski, A.E.; Kemp, S.J.; Mounteney, I.; Sellin, P. Gel Formation at the Front of Expanding Calcium Bentonites. *Minerals* **2021**, *11*, 215. https://doi.org/10.3390/min11020215

Academic Editor: Ana María Fernández

Received: 20 November 2020
Accepted: 8 February 2021
Published: 20 February 2021

**Publisher's Note:** MDPI stays neutral with regard to jurisdictional claims in published maps and institutional affiliations.

## 1. Introduction

Geological disposal of intermediate and high level hazardous waste in an especially constructed repository is the favoured choice for the long-term removal of radioactive material from the anthroposphere [1–3]. The repository designs often include a number of different natural and engineered barriers to prevent the waste from contaminating the environment (e.g., [1,4,5]). Bentonite is commonly included as the clay backfill or engineered barrier in these designs [6–8] because of its low permeability, high swelling capacity and self-sealing properties [9–12]. Many different compositions of bentonite exist, and much research has been focussed on the properties and behaviour of sodium (Na) and calcium (Ca) bentonites [13–19]. Within Europe to date, more focus has been turned towards the sodium bentonites for example the MX80 [20] and Turkish Resadiye [21] bentonites. However, numerous nations [22–24] are still considering the use of a Ca bentonite at different dry densities in their radioactive waste disposal concepts, because of a natural availability (e.g., Fourges Clay, Černý Vrch deposit).

Studies have looked at the differences in performance between these two types of bentonite. Marcial et al. [25] have investigated the water retention, permeability and compression characteristics, and Ben Rhaïem et al. [26] have looked at pore size distribution, hydration and swelling capacity. The studies demonstrated that the Ca-bentonite had fewer small pores, and a larger void ratio at higher compressional stresses than the Na-bentonite, with Marcial et al. [25] attributing the higher void ratios of the Ca-bentonite to the larger size of the $Ca^{2+}$ cation. Their results [25] were consistent with the observations of Tessier [27] and Ben Rhaïem et al. [26], who documented the removal of water molecules from calcium smectites at high suction pressures (>5 MPa). Ben Rhaïem et al. [26] also suggested that the reduced number of smaller pores in the Ca-bentonite led to a decrease in both the hydration and swelling capacity compared to Na-bentonites. This is congruous with the slightly higher water permeability values derived from consolidation curves for the Ca-bentonite, and observed using a constant head procedure [25]. However, for void ratios ($e$) of 1, the permeabilities of the Ca-bentonites tested (the natural Fourges bentonite, and the non-Na-activated equivalent of the FoCa7 clay) were on the order of $10^{-12}$ to $10^{-13}$ and close to the values obtained for the other Na-bentonites in their study (Na-Kunigel and Na-Ca MX80).

The formation of bentonite gels from the suspension of low clay concentrations (of a few weight %) in water has long had useful applications in drilling, pharmaceuticals and paper manufacture, amongst other industries [28]. These gels are held together by electrostatic and van der Waals forces and have been found to have a pseudoplastic rheological behaviour. When the gels have clay particles in a sufficient concentration, the interactions between the particles gives the gel a yield stress which causes thixotropic behaviour, whilst concentrations below a threshold will result in soft-elastic behaviour, both of which can be represented by the Bingham Model [28,29]. In radioactive waste repository design, bentonite gels have not been considered because the dry densities of the clay backfill and engineered barriers are intended to be significantly higher than the low clay percentages required for gel formation. However, the disposal concepts frequently incorporate the engineered barrier as a series of stacked individual blocks [4,30–32], which will introduce void spaces between the blocks and at their edge next to the host rock [33]. The high swelling capacity of the bentonite barrier and the availability of water will cause the clay to swell to fill these spaces, removing their ability to transmit fluids and effectively sealing the waste away from the environment. Studies have looked at the material behaviour at ambient and elevated temperatures, examining the swelling capacity [9–11,34] and swelling (or suction) pressure, both at steady-state and over time [33,35–48]. The most recent studies [47,48] have investigated the development and evolution of swelling pressure within the barrier, looking at a case where the void space is nearly equivalent in size to that of the clay material. These large void spaces are especially likely to occur at the edges of the engineered barrier [33], where the availability of water is also likely to be high as it migrates away from the waste canister and heat source (e.g., [49]). Moreover, at the outer edge of the barrier where the bentonite abuts the host rock formation, the presence of fractures and fluid-filled pathways may provide an additional opportunity for the barrier to expand. Should these fractures contain flowing groundwater, there is a significant concern that the bentonite may erode, gradually reducing the effectiveness of the barrier [1,50,51].

In this paper we present the results of an experimental swelling pressure development test conducted on Bulgarian Ca bentonite where the formation of a gel was observed at the swelling front of the clay. The results of this test are compared against analyses made on an MX80 bentonite sample that has undergone a similar test history but does not show the gel formation. The formation of a gel is an unusual observation and this segregation of the sample into a hydrated bentonite with a gel layer has implications for the mechanical properties and behaviour of the material within the repository, and for the erosion of bentonites into fluid-filled fractures in the host rock.

## 2. Materials and Methods

A Bulgarian Ca bentonite was used for the laboratory experimentation presented in this study. The material was supplied as a granular powder by Imerys (sourced by Svensk Kärnbränslehantering AB) and was compacted to form test samples. The powder had been previously analysed and found to contain 82.6 wt.% montmorillonite, 6.4 wt.% cristobalite, 6.3 wt.% mica/illite and 4.0 wt.% calcite, as well as <1 wt.% gypsum and tridymite, and the bulk material had a cation exchange capacity (CEC) of 79.4 (cmol(+)/kg) [52]. In the study by Svenson et al. [52], the material is termed Bulgaria 2018 20 tonnes batch. The powder was mixed with a specified quantity of water using a dispersing spray bottle and pallete knife to ensure that the mixture was as homogenous as possible. The mixture was then compacted axially with a ram at 80 MPa to produce a 100% saturated cylindrical sample with a target dry density of 1700 kg/m$^3$ and a diameter of 60 mm (Table 1). The sample was turned down using a machine lathe to precisely fit the testing apparatus. A figure (Figure S1) and description of the testing apparatus are included in the Supplementary Materials.

**Table 1.** Sample geotechnical parameters. The sample pre-test dimensions were measured using digital calipers and can be considered accurate to the nearest 0.1 mm. The sample weight is accurate to the nearest 0.01 g.

| Composition | Sample Length (mm) | Void Length (mm) | Sample Diameter (mm) | Sample Weight (g) | Bulk Density (kg/m$^3$) | Moisture Content | Fluid Pressure (kPa) | Test Duration (Days) |
|---|---|---|---|---|---|---|---|---|
| Bulgarian Ca bentonite | 64.97 | 51.03 | 59.76 | 368.32 | 2021.16 | 0.1525 | 4500 | 101.7 |

The apparatus was instrumented with 5 load cells (3 radial and 2 axial) to measure the stress distribution in the sample and 12 radial pore pressure transducers. The apparatus tubework was sequentially flushed with distilled water (the test fluid) through each available port to remove any residual air before being connected together and the apparatus calibrated. The calibration for each of the sensors was compared against the laboratory standard, providing a linear least squares regression which was used to apply a correction to the measured data from that sensor. This calibration, in increasing and decreasing increments, also allowed the hysteresis in the apparatus to be quantified. The sample installation took place immediately after the apparatus had been flushed and calibrated to minimise any drying of the sample. Distilled water was used as the test fluid in two Teledyne ISCO 260 D Series high-precision syringe pumps that supplied a porewater pressure of 4500 kPa to each end of the apparatus and were used to control and monitor the flow into and out of the sample. Distilled water was also used to fill the remaining void space inside the vessel after the sample had been inserted and pushed to the left-hand end of the vessel. After the sample had been inserted and porewater pressure had been applied to both ends of the sample, the porewater pressure was held constant and no hydraulic gradient was applied to the sample. The position of the syringe pumps was determined by an optically encoded disc with increments equivalent to a volume change of 16.63 nL. Both were controlled by a single digital control unit connected to a microprocessor within the pump piston that could monitor and adjust the rate of rotation of the encoded disc using a geared worm drive and a DC motor connected to the piston assembly. As such, the pumps could run in either continuous flow or constant pressure modes. The testing procedure was designed to minimise possible pump errors, because although the volumetric control system for each pump was factory calibrated, laboratory calibration of the volumetric data was not practical with the available equipment. FieldPoint™ and cRIO logging hardware combined with the LabVIEW™ data acquisition software (National Instruments Corporation, Austin, TX, USA) were used to record the total stress, pore pressures and pump flow rates at intervals of 2 min, providing detailed time series information. The total testing duration was 101.7 days, which was long enough to ensure that the swelling response of the clay was observed [47,48].

Once the swelling response had reached an asymptote, the test was terminated and the sample was carefully and incrementally removed from the vessel using a hydraulic jack. As the sample was removed from the vessel, it was sliced into ~10 mm thick slices using a sharp blade; 11 slices were obtained. Each of these 11 slices was then cut in half along its horizontal midplane to provide a top half (pieces A to K) and bottom half (pieces AA to KK) that could be separately analysed to quantify differences in the post-test sample that might be attributable to gravitational effects. Slicing the sample in this manner also allowed the moisture content of the sample to be quantified at intervals along the final sample length. The slices were weighed immediately after removal and were then dried in an oven at 105 °C for 48 h to provide the moisture content information.

After drying, two half slices of the final sample were taken for detailed petrographic and microchemical characterisation by scanning electron microscopy (SEM) (piece AA from slice 1 and piece BB from slice 2). These slices were taken from the end of the sample that was originally sited next to the internal void space. High-resolution secondary electron (SE) images were recorded using an FEI Quanta 600 environmental scanning electron microscope (ESEM). This was equipped with an Oxford Instruments INCA Energy 450 energy-dispersive X-ray microanalysis (EDXA) system, with a 50 mm$^2$ Peltier-cooled silicon drift detector (SSD) calibrated using pure metal, metal oxide and mineral standards. EDXA spectra were processed and interpreted using the Oxford Energy INCA Suite software package (Version 4.15 Issue 18d + SP3, 2009), and used to provide microchemical characterisation and aid mineralogical identification during SEM observation. Freshly broken pieces ($\sim$5–10 mm in size) were taken from each sample slice, representing slice surfaces and a cross-section surface, and mounted onto 10 mm diameter aluminium SEM stubs using Leit-C conductive carbon cement. To limit further drying and damage to the microfabric during sample preparation and SEM observation, the uncoated sample surfaces were examined directly in the ESEM instrument under a water vapour atmosphere of 130 Pa (variable pressure mode). SE images were recorded using the bespoke FEI large field secondary electron detector, using electron beam accelerating voltages of 10–20 kV and beam probe currents of 0.25–1.25 nA, at a working distance of 10 mm.

In addition, half of each sample slice was hand-crushed in a pestle and mortar and half of the crushate was ball-milled to a fine powder for whole-rock X-ray diffraction (XRD) and 2-ethoxyethanol (EGME) surface area analyses. The same set of tests was also run on a sample of MX80 bentonite that had experienced the same experimental history (the Test 4 sample from Harrington et al. [47]), to provide a reference for comparison with the Bulgarian Ca bentonite. To provide a finer and uniform particle-size for powder XRD analysis, a 2.7 g portion of each sample was micronised under acetone for 10 min with 10% (0.3 g) corundum (American Elements—PN:AL-OY-03-P). The addition of an internal standard allowed the validation and quantification of the results and also the detection of any amorphous species present in the samples. Corundum was selected because its principal XRD peaks are suitably remote from those produced by most of the phases present in the samples. The micronised samples were then dried at 55 °C, disaggregated and back-loaded into standard stainless steel sample holders for analysis. To separate a fine fraction for clay mineral XRD analysis, portions of the crushed samples were dispersed in distilled water using a reciprocal shaker combined with ultrasound treatment. The suspensions were then sieved on 63 μm and the <63 μm material placed in a measuring cylinder and allowed to stand. No deflocculant was added to ensure preservation of the interlayer cation chemistry of the samples. After a time period determined from Stokes' Law, a nominal <2 μm fraction was removed and dried at 55 °C. Approximately 100 mg of the <2 μm material was then re-suspended in a minimum of distilled water and pipetted onto a ceramic tile in a vacuum apparatus to produce an oriented mount, and allowed to air-dry overnight.

XRD analysis was carried out using a PANalytical X'Pert Pro series diffractometer equipped with a cobalt-target tube, X'Celerator detector and operated at 45 kV and 40 mA. The micronised powder samples were scanned from 4.5–85° 2θ at 2.06° 2θ/minute. Diffraction data were initially analysed using PANalytical X'Pert HighScore Plus version 4.9 software coupled to the latest version of the International Centre for Diffraction Data (ICDD) database. Following identification of the mineral species present in the samples, mineral quantification was achieved using the Rietveld refinement technique (e.g., [53]) using SiroQuant v4 software. This method avoids the need to produce synthetic mixtures and involves the least squares fitting of measured to calculated XRD profiles using a crystal structure databank. Errors for the quoted mineral concentrations calculated from synthetic mixtures of minerals, are better than ±1% for concentrations >50 wt.%, ±5% for concentrations between 50 and 20 wt.% and ±10% for concentrations <10 wt.% [54]. Where a phase was detected but its concentration was indicated to be below 0.5%, it was assigned a value of <0.5%, since the error associated with quantification at such low levels becomes too large. The <2 μm oriented mounts were scanned from 2–40° 2θ at 1.02° 2θ/minute after air-drying, after glycol-solvation and after heating to 550 °C for 2 h. Total surface area determinations were carried out using the 2-ethoxyethanol (ethylene glycol monoethyl ether, EGME) adsorption method [55]. Briefly, the method is based on the formation of a monolayer of EGME molecules on the clay surface under vacuum. Errors for the surface area values are ±3%. The Bulgarian Ca bentonite sample data were compared against the same set of analyses conducted on the sliced sample of MX80 bentonite used in Test 4 presented in Harrington et al. [47]. This sample was analysed for comparison because it was manufactured to the same sample size and dry density, had undergone the same experimental procedure and was removed from the test vessel and dried in the same manner as the Bulgarian Ca bentonite sample.

## 3. Results

### 3.1. Development of Swelling and Pore Pressure

Over the duration of the test period, the test sample swelled from its original length (Table 1) to completely fill the internal dimensions of the test vessel (48.3% increase in size). The swelling pressure recorded over this period showed an evolving picture of material swelling occurring at different locations within the test vessel at different times during the test (Figure 1). The swelling pressure development was both spatially and temporally complicated, but there were patterns to the behaviour observed. At early times, a rapid rise in the swelling pressure was recorded on the axial A1 sensor, where the sample was originally sited at the start of the test (Figure 1A). The swelling pressure recorded by radial load cells R2 and R3 also quickly increased at the start of the test, both of which were sited next to the clay at the start of the test. Load cell R3 was located on the sample midplane (for precise sensor locations, please see Figure S1), and as such, although it was in contact with the sample, it was close to the starting internal void space. As soon as the test was started, the water was able to infiltrate into the sample from both ends. As a result that the radial R3 load cell on the vessel midplane was closer to the right-hand end of the sample than the radial R2 sensor was to the left-hand end, swelling pressures were seen on the R3 sensor before the R2 sensor (Figure 1B). However, because the clay then expanded rightwards (i.e., towards the R4 sensor) filling the internal void space, the maximum swelling pressure observed on sensor R3 was lower than the pressure seen on sensor R2. This is because the density of clay situated next to the R2 sensor remained higher for longer as the sample could not expand equally towards the left. Load cell R4 at the start of the test was located adjacent to the distilled water filling the internal void space. This was the same for the axial A5 load cell, which was located in the right hand end-closure of the test vessel. With time, as the sample started to swell to fill the internal fluid-filled void, the swelling pressure recorded by these last two sensors gradually increased, although the swelling pressures remained low (below 1 MPa) (Figure 1). By the end of the test, the swelling pressures measured by sensors A1, R2 and R3 had decreased to an approximate asymptote, whilst the swelling pressures measured by sensors R4 and A5 had increased to give values in a similar range. By the time the test was terminated at 101.7 days, a variance in the range of recorded swelling pressures from the different sensors still existed (∼600 kPa), although given how slowly they were evolving at this stage in the test, it is unclear how long it would have taken for the pressures to completely converge.

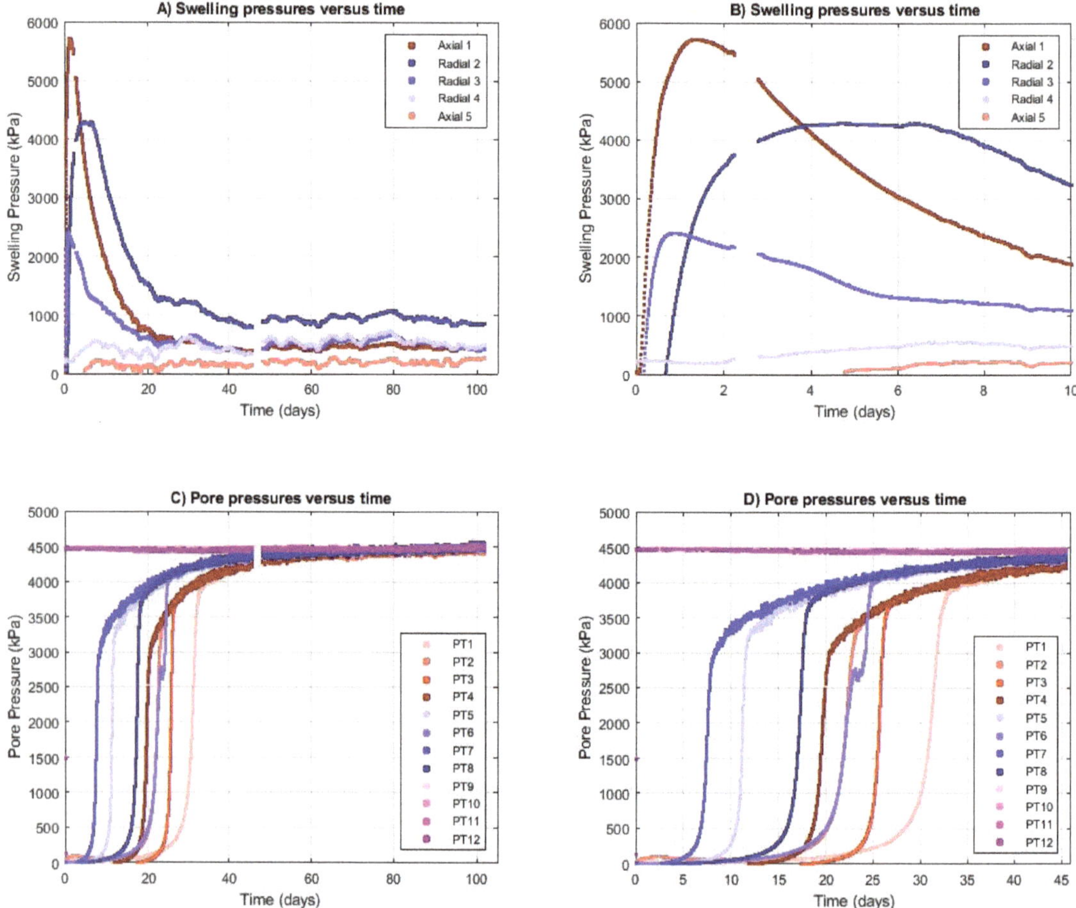

**Figure 1.** (**A**) Swelling pressures recorded on all sensors over the duration of the test, and (**B**) over the first 10 days. (**C**) Pore pressures recorded on all transducers over the duration of the test, and (**D**) over the first 45 days.

This pattern of sample expansion corresponding with the swelling pressure changes was consistent with the data recorded by the pore pressure transducers over the three arrays along the length of the test vessel (Figure 1). The pore pressure recorded by the 4 transducers (PT13-16) that were sited in line with the internal void immediately recorded the applied pore pressure of 4500 kPa. As the water percolated into the sample both from the voidage and the filter at the left-hand end of the sample, a pore pressure increase was seen. This occurred first on the transducers (PT5, PT10-12), which were on the midplane of the vessel. The pore pressure data showed that the water did not percolate into the sample in an even manner and did not arrive at each of the transducers in a given array at the same time. There was as much as 15 days difference in the arrival times, observed as a change in the recorded pore pressure (Figure 1). An overview of the swelling and pore pressure data is also available from Harrington et al. [47], where the swelling and pore pressure data from this test have been compared to tests conducted using a different composition of bentonite and in a vertical orientation.

### 3.2. Post-Test Moisture Contents

The moisture content of the post-test sample was determined at increments along its final length by slicing the sample across its long-axis into ~10 mm pieces (slices 1–11). The slices were also divided into a top (pieces A to K) and bottom (pieces AA to KK) half to ascertain whether gravity altered the moisture content through the sample. Samples were weighed initially, then placed in an oven to dry at 105° for 48 h; the samples were then reweighed to calculate the moisture content (Figure 2). The moisture content changed along the length of the sample, with the lowest moisture contents recorded at the end of the sample furthest away from the starting internal void space. The moisture contents at the right hand end of the final sample were significantly higher (4–6 times larger). Here, the sample took in the distilled water filling the initial void space and expanded to fill the entire cavity. The moisture content data supported the pore pressure observations and showed that the distilled water did not penetrate evenly through the test sample (Figure 1C,D). That the moisture contents were different along the length of the sample at the end of the test also showed that the sample had not fully homogenised in the 102 days of testing.

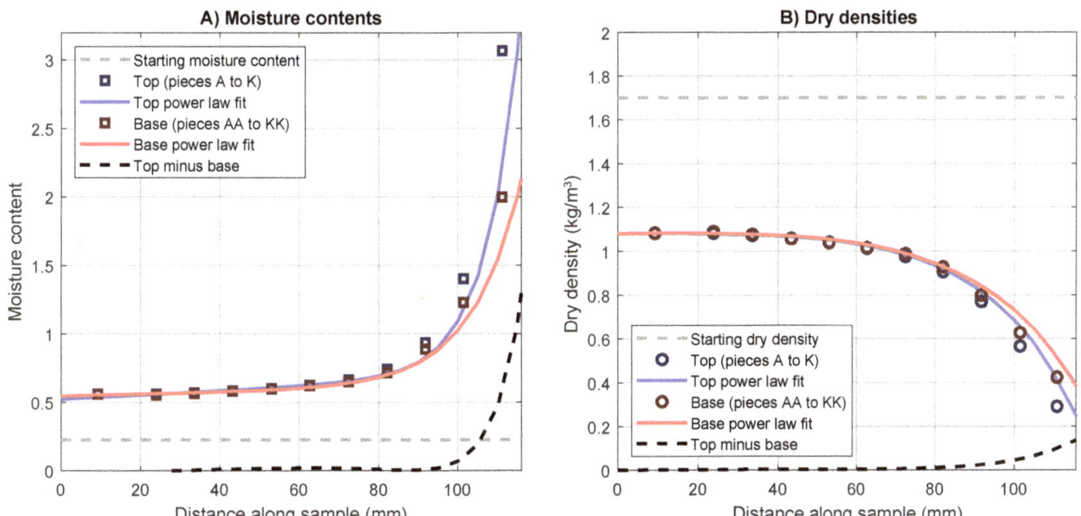


**Figure 2.** (**A**) Moisture contents and (**B**) dry densities along the length of the post-test sample. The moisture content has increased and the dry density has decreased at all points along the length of the sample from the starting values.

Minor differences were observed between the moisture content of the top and bottom halves of the oven-dried slices (Figure 2A). These differences were most notable at the right-hand end of the sample, where the sample had the highest moisture contents. The measurement of moisture content made at the far right-hand end showed that the top half of the slice had a moisture content that was 150% of the moisture content of the bottom half. This provided some evidence of gravitational settling of clay particles at the right-hand end of the sample, presumably made possible by the high moisture contents and the lack of redistribution of water evenly through the sample during the test period. The dry density of each slice was calculated using the moisture content value, a specific gravity of 2.77 and a saturation of 1. The dry density profiles along the sample length for the two halves of each slice showed the same trend as the moisture contents (Figure 2B), and were calculated in the range 0.29–1.08 kg/m³. This indicated that, despite placing a small swelling pressure on the entire surface area of the interior of the vessel, the post-test sample was very pastelike and only semi-solid. Using an exponential fit to the moisture content data from each half of the sample, the continuous difference in moisture content and calculated dry density between the two halves along the length of the sample could be plotted (Figure 2, black dashed line). This clearly showed that the only significant difference in the moisture content and dry density between the two halves occurred at the extreme right-hand end of the sample.

### 3.3. Gel Formation

As well as having a much higher moisture content, the right-most slice of the post-test sample was interesting because it showed that a layer of extremely fine clay particles formed on the advancing face of the right-hand end of the sample. As the sample swelled sideways through the internal fluid-filled void space, this collection of much smaller clay particles formed a gel ∼5 mm thick at the very front of the sample (Figure 3). Whilst it is not possible to draw firm conclusions from a single test study, this gel formation appears to be a unique feature of using a Ca bentonite in this type of swelling test, as it has not been observed in any of the same tests using sodium bentonites [47,48]. Clearly, further work is required to establish the conditions required for its formation and understand why it has not been seen in the same type of test using MX80.

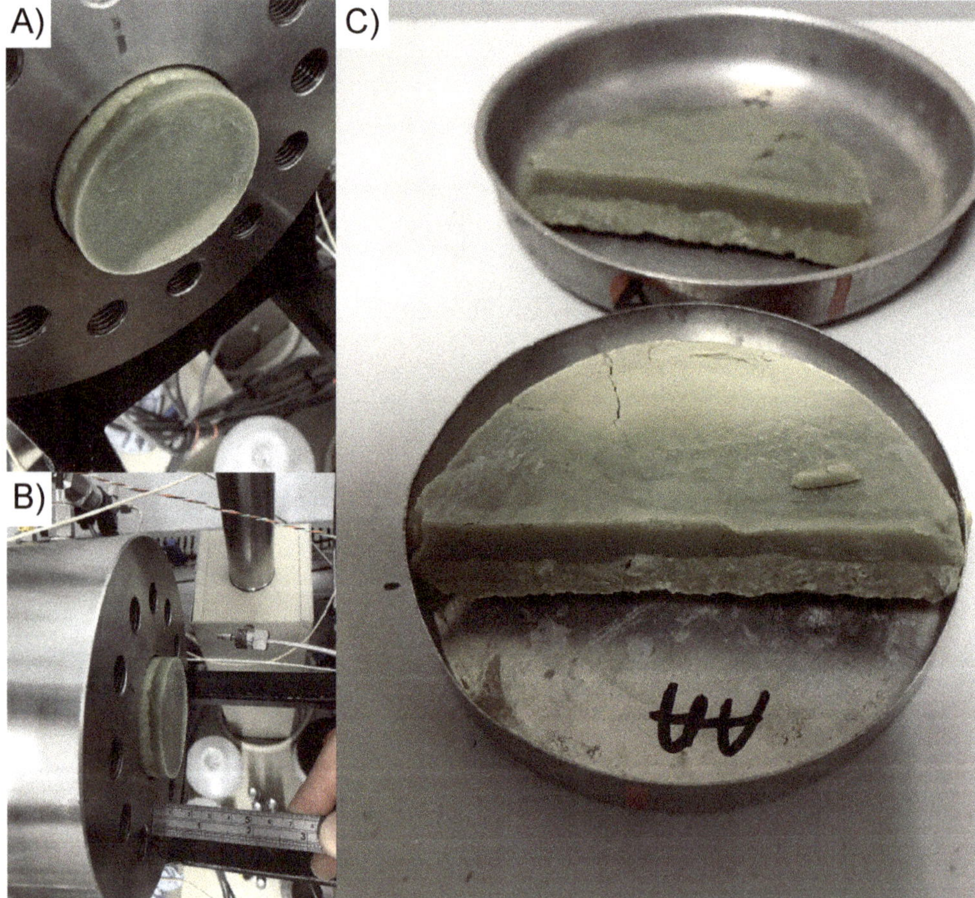


**Figure 3.** (**A,B**) The right-hand end of the sample being extruded from the test vessel. (**C**) Cross-section through the gel showing the top and bottom halves of the first sample slice.

### 3.3.1. Petrographic and Microchemical Analysis

The dried post-test bentonite sample was imaged under an ESEM to provide a more detailed picture of the processes that had occurred at the front face of the swelling material. ESEM imaging was carried out on the bottom halves of the first two slices removed from the apparatus (Figure 4); this was the material closest to the right hand end of the vessel at the end of the test period. The bottom halves of two slices (AA the rightmost end of the sample, and BB the slice behind the rightmost piece) were chosen for imaging under ESEM because both of these slices had a slight paler discolouration of the bentonite within the bottom section of the slice. The discolouration was only evident in the first two slices and was more pronounced in the first slice (slice 1 piece AA) than the second (slice 2 piece BB). The slices had already been dried to provide post-test moisture content information before examining under the ESEM, so there were some initial differences between the two slices associated with the oven drying of the material (Figure 4). Slice 1 piece AA had a moisture content of nearly 200% compared to the ~125% of slice 2 piece BB, which affected the amount of contraction of the sample as it dried. Visual inspection of the front face of slice 1 piece AA showed a smooth surface with very little cracking except for two large cracks in the bottom edge of the slice, making the surface appear as though it was covered by a skin (Figure 4). The rear side however showed deep cracks from top to bottom of the slice that had opened by up to 5 mm (Figure 4). Both sides of slice 1 piece AA showed that the material was extremely fine grained. The surface of both sides of slice 2 piece BB looked much more grainy with larger particles of the original clay matrix clearly visible on both sides of the slice (Figure 4). The cracking on the front face of slice 2 piece BB was more prevalent than slice 1 piece AA, but individually the cracks had opened a smaller amount than those evident on the rear of slice 1 piece AA.

**Figure 4.** Front and rear views of the dried post-test sample slices 1 (piece AA) and 2 (piece BB) (bottom halves).

Detailed SEM observations of the gel formed on the external end of slice 1 piece AA revealed that although the outer surface of the gel was smooth and featureless, fractured cross-sections through the gel displayed a very finely-laminated fabric (Figure 5A). The laminae were extremely thin (typically <10 µm thick), and showed significant distortion, parting and spalling between the layers, as a result of shrinkage, caused by water-loss during drying. In addition, the gel laminae displayed a highly wrinkled skin texture (Figure 5B–D), again indicative of shrinkage due to loss of a significant volume of water consistent with the very high measured moisture content (described earlier). Although the gel displayed a wrinkled texture, the material was extremely fine grained and very uniform (Figure 5D); the bulk of the material was clearly nanoparticulate with a particle size well below the resolution of the SE imaging (i.e., much finer than 50 nm). In contrast, the compacted bentonite substrate surface immediately beneath the gel layer was notably much coarser, with clearly discernable close-packed grains of aggregate clasts of partially expanded bentonite, and minor quartz, feldspar, calcite and gypsum, within a clay-gel matrix (Figure 5E). These particles ranged in size from 5 to 200 µm in diameter. Irregular sigmoidal shrinkage cracks were common around the grain boundaries. Similarly, the bentonite in slice 2 piece BB was much coarser than the gel layer on the outer surface of slice 1. However, whilst aggregate particles of bentonite and minor quartz, feldspar, gypsum and calcite could be discerned, they were less obvious than in the bentonite substrate immediately beneath the gel in slice 1 piece AA (Figure 5F). Grain boundary shrinkage cracks were also relatively uncommon in slice 2 piece BB compared to the bentonite substrate in slice 1 piece AA.

EDXA spectra recorded from the gel layer (slice 1) were consistent with that expected from a montmorillonite, comprising: major Si, Al, O and subordinate Mg, Na, Ca, Fe (Figure 6A). Traces of Ti and S were also detected in most analyses. Rare microcrystals (<2 µm) and microcrystalline aggregates (up to 50 µm) were occasionally observed within the gel under high magnification SE imaging (Figure 5B,C). EDXA spectra (processed to strip the background X-ray signal from the host gel) clearly identified this trace phase to be Ca-S-rich (Figure 6B), and it therefore most likely represented microcrystalline gypsum. It is unclear whether this gypsum nucleated within the gel phase during the experiment, or precipitated from the porewater solution during the post-experimental drying of the sample.

Although fully-quantitative EDXA was not possible, semi-quantitative estimates of the composition of the smectitic gel and clay matrix in the background bentonite were obtained from suitably orientated surfaces. EDXA analytical totals were very low for the gel (typically 20–30 wt.%) indicative of a very high degree of microporosity (void space) included within the electron probe volume of this material, and which was consistent with the very high water content determined from slice 1 piece AA. Higher and more variable analytical totals (25–80 wt.%) were obtained from the bentonite clay matrix but again were low due to the high water content and/or nanoporosity. Semi-quantitative EDXA data obtained from the gel and bentonite substrate from slice 1 piece AA, and the bentonite from slice 2 piece BB are summarised in Figures 7 and 8, and more detailed background data are provided in Tables S1 and S2. The EDXA data processing assumes all Fe was present as $Fe^{3+}$ (the most common valence state in smectite), and O was calculated on the basis of the oxide stoichiometry (note: $H_2O$ and OH are not determined by EDXA). As a result of the very low and variable analytical totals, the analyses have been normalised to 100 wt.% to enable comparison of the data.

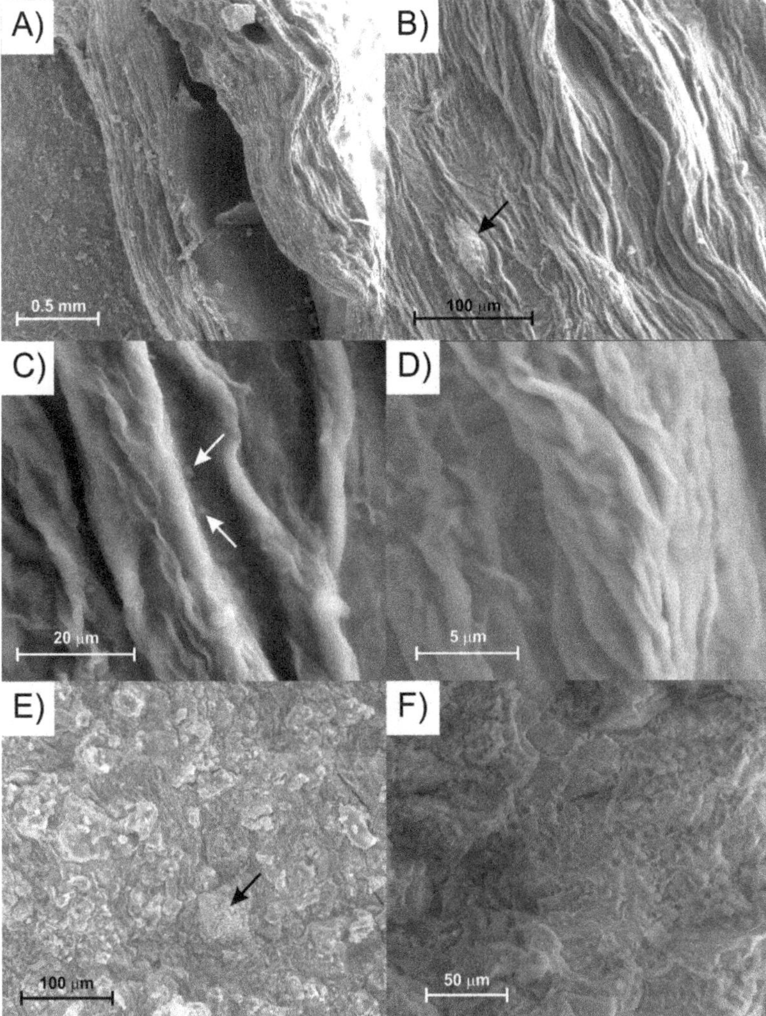

**Figure 5.** Secondary electron (SE) images of the bottom half of the post-test sample (slice 1 piece AA and slice 2 piece BB). (**A**) Finely-laminated smectitic gel layer resting on bentonite substrate in slice 1 piece AA. Parting of the lamination is visible in the gel layer. (**B**) "Wrinkled skin" texture caused by drying and shrinkage of the smectitic gel from slice 1 piece AA. A lenticular aggregate of microcrystalline gypsum (arrowed) can be seen to nucleate within the gel. (**C**) SE image of the smectitic gel layer (slice 1 piece AA), showing higher-resolution detail of the "wrinkled skin" texture. Tiny microcrystals of gypsum (arrowed) can be seen within the gel (slice 1 piece AA). (**D**) Higher magnification SE image of the extremely fine and uniform nature of the gel on slice 1 piece AA, within which no discrete clay particles can be resolved under the SEM. (**E**) SE image illustrating the relatively coarse, granular fabric of the bentonite substrate immediately beneath the gel layer (slice 1 piece AA). Subrounded, partially-hydrated bentonite aggregate grains within a fine, smectite-rich clay matrix are clearly discernable. A discrete grain of gypsum (arrowed) is also shown. Fine, sigmoidal grain boundary microfractures are evident. (**F**) Typical compacted bentonite fabric from slice 2 piece BB, showing close-packed bentonite aggregate grains within tight clay matrix.

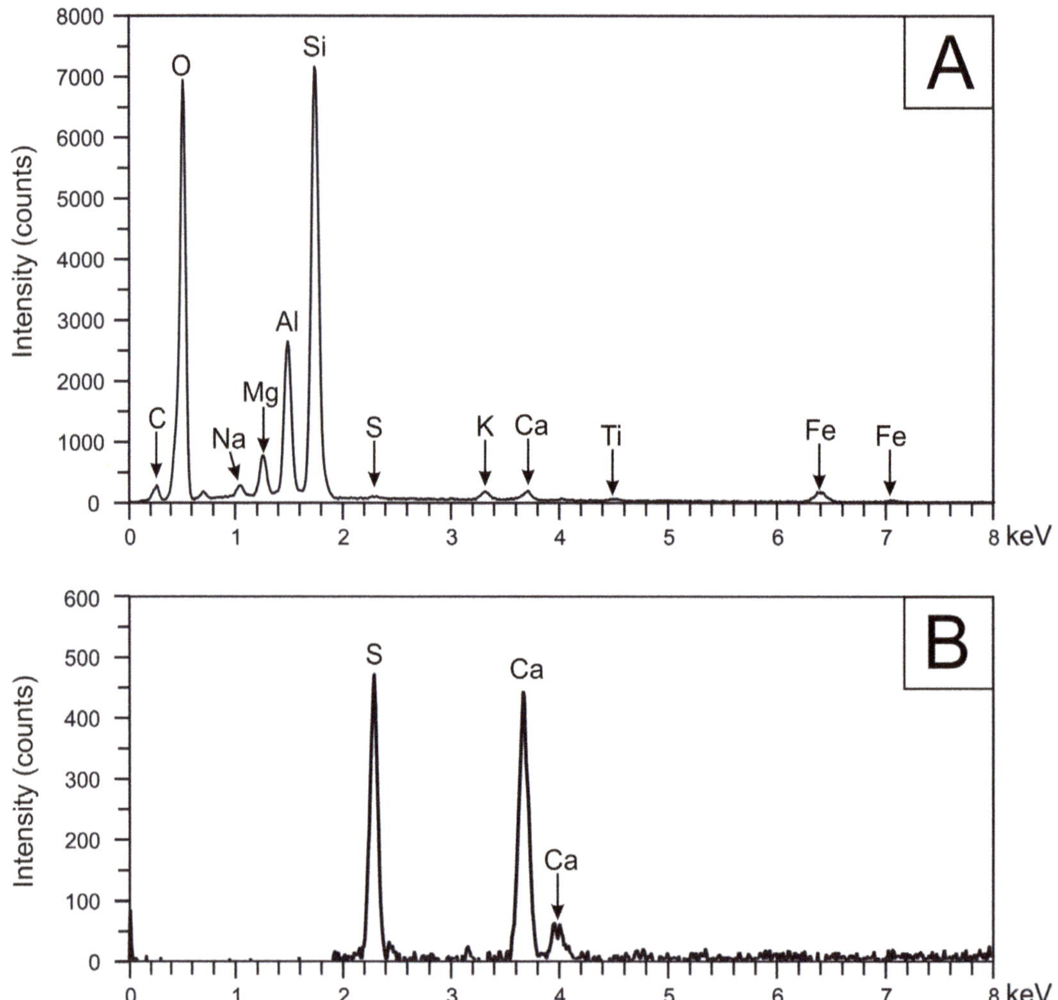

**Figure 6.** (**A**) Typical energy-dispersive X-ray microanalysis (EDXA) spectrum of the hydrous gel formed on the external surface of slice 1 piece AA. (**B**) Residual processed EDXA spectrum (after subtraction of background from the host gel matrix), indicative of calcium sulphate, recorded from the microcrystalline secondary phase that nucleated within the hydrous gel on the external surface of slice 1 piece AA.

Figure 7 illustrates the chemical variation between the gel and the immediately underlying bentonite, presented as a triplot of the molar ratios of [Si]: [total Al, Fe, Mg, Ti]: [total N, K, Ca]. The data clearly showed that there was a difference between the composition of the gel formed on the external surface of slice 1 piece AA and the clay matrix of the background bentonite in slice 2 piece BB. The EDXA data for the gel clustered tightly with little variation, and lay close to a composition compatible with a dioctahedral smectite composition, suggesting that it corresponded to a virtually pure smectite gel. In contrast, the EDXA data from the background bentonite (slice 2 piece BB) and bentonite substrate (slice 1 piece AA) varied significantly, lying on a mixing line between the smectite composition and a much more silica-rich material. Published X-ray diffraction data [52] showed that the original bentonite contained a significant amount of amorphous or poorly crystalline silica minerals (including >6 wt.% cristobalite and tridymite). Together with the ESEM observations, the EDXA data in Figure 6 suggest the gel in slice 1 piece AA may have represented very fine smectite that has been physically mobilised and has migrated from the bentonite matrix exposed in the end surface of the core, leaving a more silica-rich residual matrix in the substrate beneath.

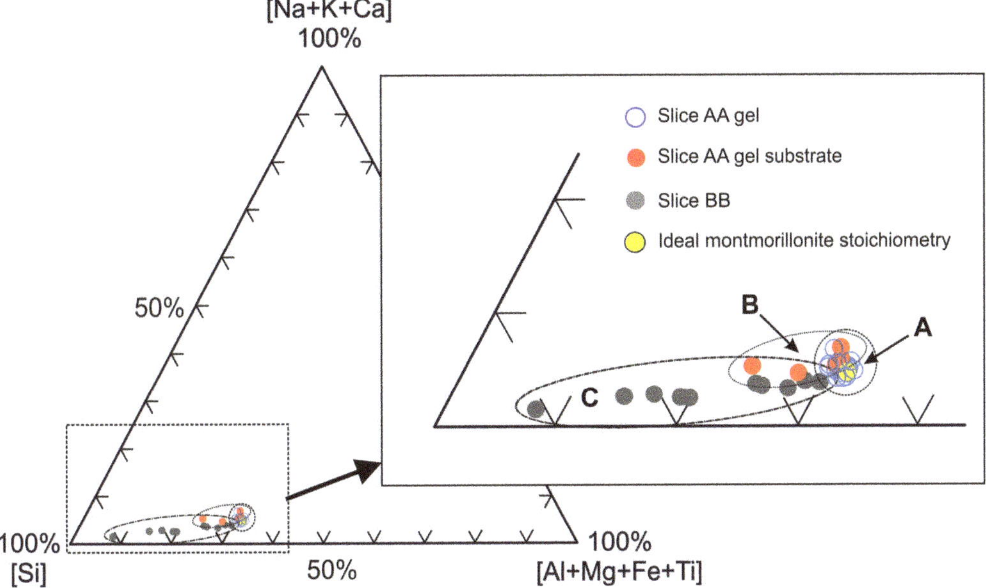


**Figure 7.** Molar ratio triplot of [Si] : [total Al, Fe, Mg, Ti] : [total N, K, Ca] from EDXA, illustrating the compositional variations between the hydrous gel (field A) and gel substrate (field B) from slice 1 piece AA, and background bentonite matrix (field C) from slice 2 piece BB. 'Ideal' montmorillonite stoichiometry is shown for comparison.

Figure 8 shows the variation in the composition of the gel (slice 1 piece AA) and background bentonite (slice 2 piece BB) determined by EDXA presented as a molar ratio plot of Ca:Na:K, which will be largely attributable to the interlayer cation component in the smectite. These data have been corrected to allow for any contribution of Ca from the potentially small amounts of $CaSO_4$ (gypsum) detected by EDXA (seen in the Supplementary Materials Tables S1 and S2). The analyses of the gel and clay matrix of the background bentonite (Figure 8) showed a large scatter of data points, with some degree of overlap. Overall, however, the data appeared to suggest that the smectite gel was lower in Ca and higher in Na/K than the smectite in the background bentonite. The precipitation of Ca as gypsum within the gel may have played a part in this by reducing the Ca:(Na,K) ratio in solution in the pore fluid within the gel.

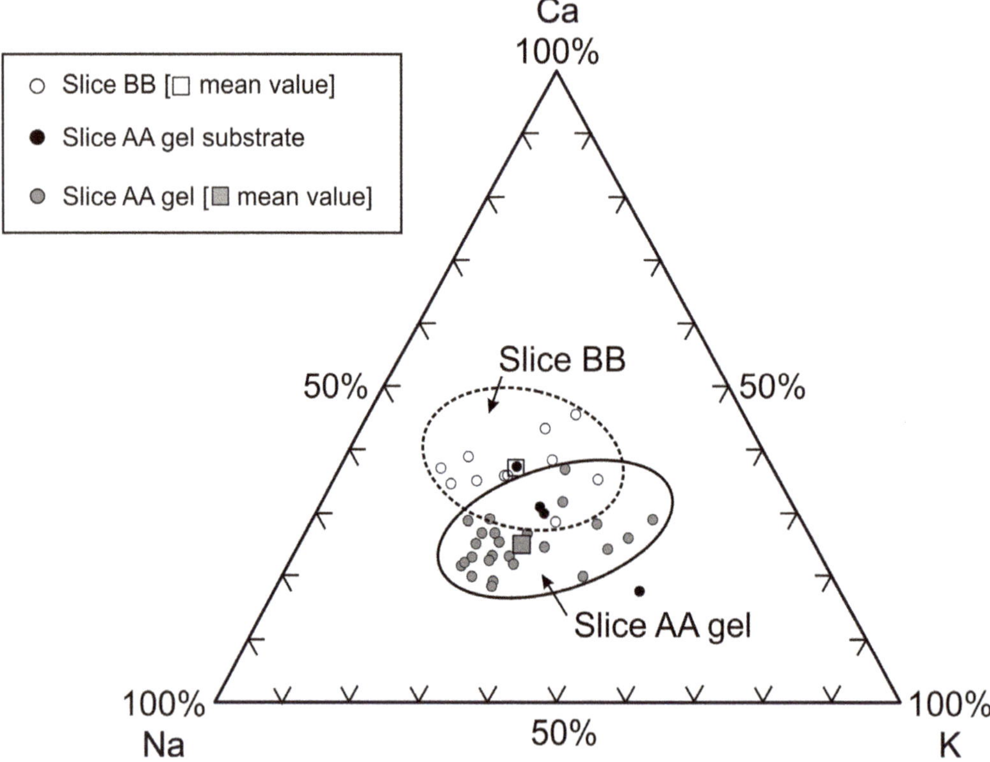



**Figure 8.** Molar ratio triplot for Ca-Na-K from EDXA. The EDXA illustrates the compositional variations of the hydrous gel and gel substrate (slice 1 piece AA) and background bentonite matrix (slice 2 piece BB). Note: these data have been corrected to allow for the potential contribution of Ca from the presence of minor amounts of gypsum detected during EDXA.

### 3.3.2. Surface Area Determinations

The surface area analysis was conducted on slices 1–11 pieces A-K (top halves of the 11 slices) (Table 2). Pieces B-K had a mean surface area of 590 m$^2$/g (Figure 9A). Slice 1 piece A had a relatively higher surface area of 629 m$^2$/g. Slice 9 piece I had an anomalously low surface area of 530 m$^2$/g. In isolation, this suggested that the smectite concentration in the gel (slice 1 piece A) was higher (by ∼5%) than in the rest of the Ca-bentonite. Pure smectite has a surface area of ∼800 m$^2$/g. Unsurprisingly, therefore these samples with their high smectite content produced high surface areas in the range 530–704 m$^2$/g. In the equivalent experiment using MX80 rather than Bulgarian Ca bentonite (Test 4 from [47]), slices 4–13 (the majority of the sample) had a mean surface area of 695 m$^2$/g (Figure 9B). Slices 3 to 1 had sequentially lower surface areas of 663 m$^2$/g, 659 m$^2$/g and 631 m$^2$/g, respectively. In isolation, this suggested that the smectite concentration in MX80 slices 1–3 were between ∼4–8% lower than in the rest of the MX80 slices.

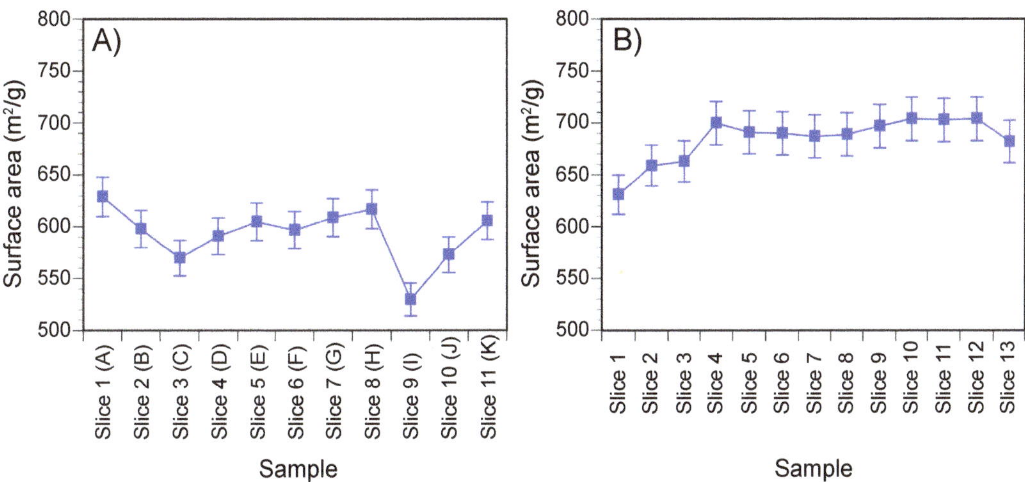

**Figure 9.** Surface area data for (**A**) the Bulgarian Ca bentonite and (**B**) the MX80 slices of the post-test samples. For both materials, the post-test samples were sliced into ~10 mm slices, and each of these was subdivided into a top and bottom half. The top-half data are presented for both samples.

### 3.3.3. <2 μm and Whole-Rock XRD Analysis

Four of the Bulgarian Ca bentonite and MX80 slices were used for <2 μm oriented-mount XRD analysis (Table 2, Figure 10). These were from the top halves of the post-test samples. The analyses suggest that the clay mineral assemblages of both the Bulgarian Ca bentonite and MX80 are principally composed of a smectite-group mineral, dioctahedral (on the basis of its d060 spacing from whole-rock analysis) and most likely to be a montmorillonite. Small quantities of illite were also detected in all of the Bulgarian Ca bentonite slices analysed. The d001 spacing provided an indication of the interlayer cation chemistry of the smectite mineral. In general terms, an air-dry d001 spacing of ~14.5–15 Å would indicate a predominance of divalent (Ca/Mg) interlayer cations and a double layer of water molecules, while a ~12.5 Å would indicate a predominance of monovalent (Na/K) interlayer cations and a single water layer. The air-dry d001 spacing for the montmorillonite in the Bulgarian Ca bentonite pieces B, F and K were in the range ~14.5–15.4 Å, confirming a predominance of divalent interlayer cations (Ca/Mg). However, the significantly smaller d001 spacing measured for piece A of 13.3 Å would suggest an approximately equal mix of divalent (Ca/Mg) and monovalent (Na/K) interlayer cations. The air-dry d001 spacing for the montmorillonite in the MX80 bentonite was consistently 12.7 ± 0.2 Å suggesting an approximately uniform monovalent (Na, K) interlayer cation chemistry.

The width of XRD peaks (e.g., full width at half maximum height, FWHM) can be used to assess the crystallite size distribution of mineral species, with broader peaks indicating smaller crystallite sizes and sharper peaks indicating larger crystallite sizes. For the Bulgarian Ca bentonite, the mean FWHM of the smectite air-dry 001 peak was ~2.23° 2θ but slice A had a broader FWHM of 2.95° 2θ. This suggested that the smectite in piece A had a smaller crystallite-size distribution compared to the general Bulgarian Ca bentonite. The MX80 slices showed a mean FWHM of ~2.24 ± 0.1° 2θ for the smectite air-dry 001 peak which suggested a generally similar crystallite size distribution for all the subsamples. It should be noted that the measurements in the present study were sub optimal since they were made on potentially mixed interlayer cation smectite species, with no control on the ambient humidity.

Like the <2 μm oriented-mount XRD analysis, whole-rock XRD analysis was carried out on 4 of the Bulgarian Ca bentonite and MX80 slices (Table 2). The analyses showed that both the Bulgarian Ca bentonite and MX80 test samples were essentially homogeneous, with no significant variation in smectite content along their length. For the Bulgarian Ca bentonite, there was a small increase in the amount of quartz observed in slice 1 piece A compared to the measured amounts throughout the remainder of the sample. In comparison, the MX80 sample showed a very slight decrease in the quartz content at the equivalent end of the test sample (slice 1). However, the compositional differences between the slices were small. In addition, whilst the EDXA analyses were conducted separately on the gel and the substrate as discrete entities beneath within slice 1 piece AA, the surface area and XRD measurements were made on the whole of slice 1 piece A together, and therefore represent the average values across the slice.

**Table 2.** Surface area, <2 μm oriented and whole-rock powder XRD data. Analyses have been carried out on the Bulgarian Ca bentonite sample as well as a comparable MX80 sample (Test 4) from Harrington et al. [47]. The piece of each slice used is given in brackets in column 2. 'Mica' indicates undifferentiated mica species possibly including muscovite, biotite, illite, illite/smectite.

| Bentonite Type | Slice No. | Surface Area $(m^2/g)$ | Oriented XRD d001 (Å) | FWHM | Smectite | 'Mica' | Quartz | Calcite | Cristo-Balite | Total |
|---|---|---|---|---|---|---|---|---|---|---|
| | 1 (A) | 629 | 13.28 | 2.95 | 77.8 | 15.1 | 1.7 | 4.1 | 1.2 | 100.0 |
| | 2 (B) | 598 | 14.50 | 2.35 | 77.2 | 15.4 | 0.8 | 4.8 | 1.8 | 100.0 |
| | 3 (C) | 570 | — | — | — | — | — | — | — | — |
| | 4 (D) | 591 | — | — | — | — | — | — | — | — |
| | 5 (E) | 605 | — | — | — | — | — | — | — | — |
| Ca | 6 (F) | 597 | 15.38 | 1.98 | 77.4 | 14.7 | <0.5 | 5.8 | 1.7 | 100.0 |
| | 7 (G) | 609 | — | — | — | — | — | — | — | — |
| | 8 (H) | 617 | — | — | — | — | — | — | — | — |
| | 9 (I) | 530 | — | — | — | — | — | — | — | — |
| | 10 (J) | 573 | — | — | — | — | — | — | — | — |
| | 11 (K) | 606 | 14.47 | 2.36 | 77.9 | 15.0 | <0.5 | 5.0 | 1.7 | 100.0 |
| | 1 | 631 | 12.61 | 2.13 | 77.0 | 17.2 | 4.0 | 1.5 | <0.5 | 100.0 |
| | 2 | 659 | 12.62 | 2.29 | 78.4 | 16.8 | 3.6 | 1.3 | <0.5 | 100.0 |
| | 3 | 663 | — | — | — | — | — | — | — | — |
| | 4 | 700 | — | — | — | — | — | — | — | — |
| | 5 | 691 | — | — | — | — | — | — | — | — |
| | 6 | 690 | — | — | — | — | — | — | — | — |
| MX80 | 7 | 687 | 12.91 | 2.26 | 77.7 | 15.8 | 4.8 | 1.2 | <0.5 | 100.0 |
| | 8 | 689 | — | — | — | — | — | — | — | — |
| | 9 | 697 | — | — | — | — | — | — | — | — |
| | 10 | 704 | — | — | — | — | — | — | — | — |
| | 11 | 703 | — | — | — | — | — | — | — | — |
| | 12 | 704 | — | — | — | — | — | — | — | — |
| | 13 | 682 | 12.87 | 2.27 | 77.4 | 16.5 | 4.5 | 1.1 | 0.5 | 100.0 |

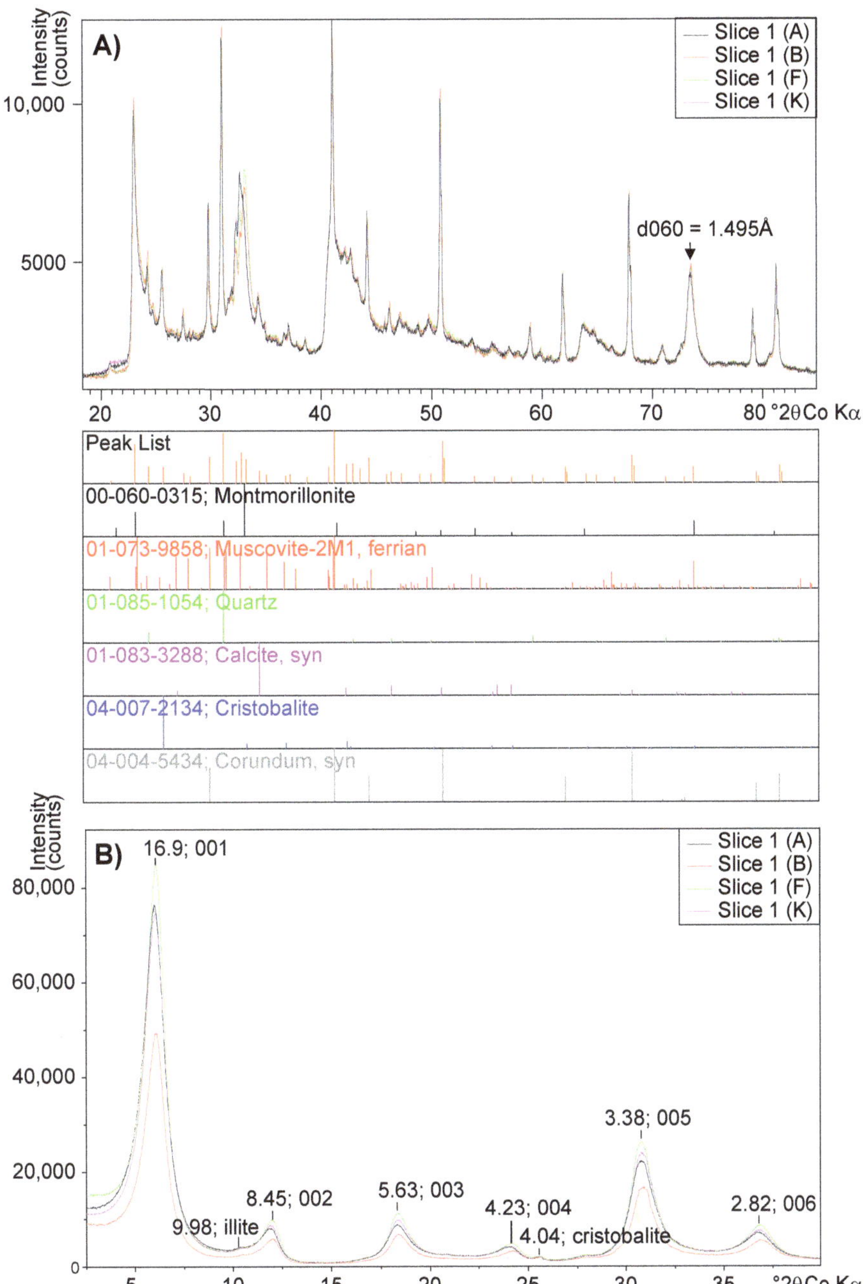

**Figure 10.** (**A**) Whole-rock powder XRD traces for selected Ca-bentonite subsamples showing the similarity of the traces and the diagnostic dioctahedral smectite d060 peak spacing at 1.495 Å. The lower part of the figure shows the extracted peak positions and intensities (shown orange) compared to similar data for the identified International Centre for Diffraction Data (ICDD) mineral standards. (**B**) <2 μm ethylene glycol-solvated oriented mount XRD traces for the same selected Ca-bentonite subsamples showing the very similar positions of the smectite peak positions (labelled with their appropriate 00l and d spacing in Å).

## 4. Discussion

The test results presented in this study show that the formation of a gel at the front of an expanding Ca bentonite can occur under the specific conditions of an ambient temperature, zero hydraulic gradient, large sample to void ratio and free access to water at the expanding bentonite face. In this study, the sample swelling took place slowly; the pore pressure data (Figure 1C,D) showed that the sample took at least 30 days to swell to fill the initial void space in the experimental set up. This was consistent with the swelling pressure data (Figure 1A); these data indicated that the peaks in swelling pressure that occurred on the axial and radial load cells sited next to the sample at the start of the test (Axial 1 and Radial 2), had dropped and asymptoted to constant values that were consistent with all of the sensors on the vessel, between days 30 and 45 of the test. At the end of the test period, the material had expanded to fully fill the initial void space, with a $\sim$5 mm thick very fine grained gel at the front end of the sample. Observations made by the ESEM showed that the gel that formed on the external surface of slice 1 (i.e., the very front of the expanding sample) was extremely fine grained (<50 nm) and uniform, with the particle size well below the resolution of the SEM instrument (Figure 5D). This is consistent with the high surface area value (629 m$^2$/g) obtained for slice 1 piece A, which was significantly larger ($\sim$7%) than the average background Bulgarian Ca bentonite. In isolation this suggests a higher smectite content; this is consistent with the ESEM-EDXA observations, which indicate the gel layer had a composition that was very close to the composition expected for a pure smectite (Figure 6). However, a higher surface area might also be the result of a finer overall particle size, which is consistent with the significantly broader d001 diffraction peak for smectite (FWHM = 2.95° 2θ in slice 1 piece A compared to FWHM = $\sim$2.23° 2θ in the background Bulgarian Ca bentonite) suggesting that gel smectite has a smaller crystallite size than the background bentonite. Both the increased surface area and the indication of a finer crystallite size concurs with the ESEM-EDXA observations that the gel is very fine grained and nanoparticulate, whereas the smectite in the substrate and background bentonite is much coarser grained and granular (slice 2 piece BB; Figure 5E,F). The ESEM-EDXA observations also showed that the bentonite substrate composition was much richer in silica (Figure 7), and demonstrated highly variable silica contents further away from the gel (slice 2 piece BB). Furthermore, the air-dried XRD d001 diffraction peak spacing difference between the gel and the smectite in the background bentonite suggests that the smectite in the gel has significantly more monovalent (Na/K) interlayer cations than the background bentonite, which is predominantly divalent (Ca/Mg) interlayer cations. This confirms the observations from EDXA that show that smectite in the gel and the residual substrate in slice 1 piece AA have significantly higher Na/K than that expected from previous characterisation of the Bulgarian Ca bentonites [52]. A finer sampling discrimination was conducted for the ESEM-EDXA, whereby discrimination was made in slice 1 piece A between the outer gel coating and the residual silica-rich substrate; these data further suggest that the outer gel is even more Na/K rich than the smectite remaining in the residual substrate. This is an interesting finding, which implies that there is a difference in the exchangeable cation component between the two parts of slice 1. The reason for this remains unclear but possibly the precipitation of Ca as gypsum within the gel may have played a part in this by reducing the Ca:(Na,K) ratio in solution in the pore fluid within the gel, and thereby enhancing the exchange of Na/K for Ca.

The presence of higher concentrations of monovalent Na and K cations in the gel is consistent with the results of Missana et al. [56] who found that colloid detachment was preferentially favoured by Na compared to Ca. It should be noted however, that the absolute quantities of Na and K measured here are low and close to the detection limits, indicating that there could be additional noise in these data. The Missana et al. [56] experiments suspended 1 g/L of a bulk material (fifteen different clays including the MX80, FEBEX and three Czech bentonites were used) in deionised water and centrifuged the suspension to obtain particle sizes <1 μm. These colloids were dried and the particle quantities measured by gravimetry. The results showed that the Ca bentonites produced a significantly smaller colloid fraction compared to the Na bentonites (e.g., Ca-MX80 $53 \pm 7$ ppm, Na-MX80 $600 \pm 50$ ppm), and that whilst the Na and Na-Ca mixed clays showed a positive relationship between colloid formation and starting smectite content, the Ca clays colloid formation was invariant with smectite content [56]. The smectite content of the Ca bentonite used in this study and in the MX80 of the bentonites tested in the same way in previous studies [47,48] was very similar ($\sim$83 wt.%), and below that of the MX80 tested by Missana et al. [56] (89 wt.%). The lower smectite contents of the materials tested in these studies (this study and [47,48]) are closer to the values measured for the Czech bentonite samples (e.g., Rokle-S65: 78 wt.% smectite), which produced a much smaller quantity of colloidal material ($153 \pm 19$ ppm), possibly explaining why a gel was not observed in the tests using MX80 [47,48].

The formation of a gel is unusual and has not been previously observed in swelling tests of this type incorporating a large initial void space [47,48], although the tests presented in these studies were conducted using an MX80 bentonite rather than a Ca bentonite composition. Together, the ESEM and EDXA observations suggest that during hydration and expansion of the bentonite, very fine (nanoparticulate) smectite effectively migrated from the bentonite matrix towards the front of the expanding sample, forming a layer between the sample and the void space; here it formed an extremely hydrous gel. This flushing of the smectite minerals towards the front has acted as a mechanical sieve and purified the smectite into this layer. The loss of the smectite from the bentonite matrix left a more silica-rich residual substrate, which appeared to be less mobile. The ESEM and EDXA observations also indicated that the fine grained silica and other components in the bentonite that formed the substrate did not migrate during the bentonite expansion. However, the whole-rock (i.e., whole-sample) XRD results contradict the ESEM-EDXA and surface area results by indicating that there is slightly more quartz in slice 1 piece A than in the background Bulgarian Ca bentonite. One possible reconciliation may lie in the degree of the subsampling undertaken during analysis of slice 1 between the two approaches. In the ESEM-EDXA case, discrimination is drawn between the actual nanoparticulate gel and the substrate immediately beneath, on which the gel was observed to rest. The ESEM-EDXA analysed these separately, showing that the outer gel layer was chemically consistent with a nearly pure smectite, whereas in contrast, the immediately-underlying residual substrate was notably less smectitic and very silica-rich. The XRD analysis by contrast represents the whole of slice 1 piece A and does not discriminate between the two fabric layers within the slice (Figure 3); the apparent difference between the XRD and ESEM-EDXA results may simply arise from this.

When compacted bentonite is hydrated, it swells and expands into available void spaces. In the case of this experiment, the bentonite was able to expand into the water-filled void space in the test cell. The chemistry of the system has a significant effect on the swelling and dispersion behaviour; porewaters with low ionic strength and a low concentration of divalent cations (e.g., $Ca^{2+}$) promote peptisation (deflocculation) of smectite and encourage the separation of individual smectite layers, which can lead to spontaneous colloid formation [57–59]. This experiment used deionised water as the saturating fluid, and this would be expected to encourage deflocculation of the smectite during hydration and swelling. As no hydraulic gradient was applied during the experiment, diffusion must have been the driving force for any colloid release and migration (cf. Kallay et al. [60], as cited by Alonso et al. [51]) from the bentonite matrix to give rise to the formation of the gel at the end of the test sample. Previous studies [51,56] have shown that the nature of interlayer cation strongly influences the smectite colloid generation. The predominance of more hydrated ions in the interlayer cation site (such as $Na^+$) leads to the disaggregation of clay platelets and formation of smaller tactoids (liquid-crystal colloidal phase) with higher electrophoretic mobilities. In contrast, $Ca^{2+}$ is a more strongly binding cation producing a tighter, more compact smectite structure, which favours the formation of larger aggregates and lower colloid detachment and mobility. Interestingly, both the XRD analyses and the ESEM-EDXA observations of the gel formed from the bentonite during this experiment suggested that it was relatively Na/K-rich compared to the smectite in the background matrix of the bentonite. Although the reason for this apparent compositional difference is not clear, the extremely fine nanoparticulate (<50 nm) nature of this material is consistent with formation of a Na-rich colloidal smectite. Whilst the ESEM observations indicated that the nanoparticulate material imaged in slice 1 piece A was extremely fine-grained and uniform, because the resolution of the ESEM was 50 nm, it is not possible to discount both micro- and mesoporosity existing in the sample that could take up water in addition to the interlayer water.

There was very little change in the moisture content along the majority of the length of the sample (Figure 2A), except in the three slices (1, 2 and 3) that were from the end of the sample that had been adjacent to the void space at the start of the test. In these slices, especially slice 1 where the gel was located, the moisture contents were markedly higher; this corresponded with a very low dry density measurement for the sample in this same region (Figure 2B). It should also be noted that the moisture content measurement for slice 1 incorporated both the gel layer (approximately half the thickness of the slice) and the more silica-rich residual substrate underneath (Figure 3). The specific moisture content of the gel layer alone is not known. However, given the fine-grained nature of the gel and the large cracks and wrinkled texture of the material post-drying, it is likely that the moisture content of the gel layer alone would have been much higher; the residual substrate is likely to have had a moisture content closer to the value obtained for slice 2. The averaged moisture content of the gel and the residual substrate together are therefore likely to be lower than the moisture content of the gel alone. The liquid limit for Ca bentonites is reported to be lower than for Na bentonites, but within the range 150–600% (determined using the Casagrande method for a variety of commercial Ca bentonites) [61]. The measured moisture content for slice 1 was 307% indicating that the moisture content could have been much higher whilst the sample still demonstrated plastic behaviour.

As would be expected in a repository, the void space in the experiment was finite; it is not expected therefore that this mechanical separation of the smectite could continue indefinitely because, whilst it continues to contain smectite minerals, the matrix would also expand filling the void space and competing against the migration of the smectite to the expanding front of the material. In addition, the measured stresses (Figure 1A) reached an asymptote indicating that the gel layer was stable and further homogenisation or segregation would not occur mechanically. Whether chemical processes could lead to homogenisation or continued formation of the gel layer, it is not possible to say from the analysis presented. It is possible that previous studies using MX80 bentonite [47,48] have not seen this gel layer formation because the expansion of the matrix took place more quickly than it did in the Bulgarian Ca bentonite in this study. Figure 7 in Harrington et al. [47] suggests that the swelling pressures have asymptoted by approximately day 35 for the MX80 (Test 4) and day 40 for the Bulgarian Ca bentonite (Test 5), although the data are noisy and the other tests presented (in a vertical rather than horizontal orientation) show asymptotes in the stress data much later. The cumulative inflow of distilled water into the Bulgarian Ca bentonite however is much larger than in any of the MX80 tests presented by Harrington et al. [47]. Without the presence of a hydraulic gradient across the sample, this could be attributed to a lower starting saturation of the Bulgarian Ca bentonite compared to the MX80, although the saturations are reported as close to 100% and the dry densities all 1.7 g/cm$^3$. For bentonites that contain a lower smectite content and where material expansion upon contact with water can be expected to take place more slowly, a larger gel layer may be able to form, subject to the smectite content also being high enough to supply the nanoparticulate material to migrate. A decrease of only 5 wt.% in the smectite content has been seen to substantially increase the hydraulic conductivity and decrease the swelling pressures of bentonites [52]. Previous XRD analysis of the Bulgarian Ca bentonite has found that it has a smectite (montmorillonite) content of 82.6 wt.% [52], whilst the MX80 Wyoming bentonite used in the swelling studies [47,48] has been analysed to contain 85.43% montmorillonite [62], which may in part explain the lack of gel formation seen in the tests using MX80 bentonite, due to the differences between these materials in terms of their swelling capacity.

This mechanism of gel formation has important implications for bentonites being used as barriers where the material has free access to flowing water, as might occur within fractures in the host rock. This gel formation shows that bentonite in contact with an open void containing water will expand and the fine grained and swelling smectite minerals may migrate towards the expanding face of the bentonite. If the water in the voidage is moving, the smecite leaving the silica matrix could erode and eventually all the smectite would be lost from the matrix. The high swelling capacity and low permeability of the smectite minerals provide the sealing ability of the bentonite as a barrier to flow, and thus the loss of the smectite content will severely impact the performance of the barrier. For how long this process may continue in the presence of flowing water cannot be predicted from the results presented here, however they may be showing a snapshot of the process that is initiating bentonite erosion in voidages. Bentonite erosion is considered to be one of the main processes leading to mass loss, which could deleteriously affect the ability of the bentonite to perform as a barrier [1,50,51]. It is also thought that the erosion of colloids (<1 µm fluid-suspended clay particles) could provide a mechanism for radionuclide transport [63–65]. The data presented in this study seem to suggest that smectite particles can be chemically flushed through the bentonite matrix to form a gel with a particle size below 50 nm, and that this could occur with free access to water. It is important to note that the test was conducted using distilled water as the fluid in the void space and the test fluid. Distilled water is likely to have been more aggressive in terms of cation dissolution than a groundwater composition might have been expected to be on the test sample, and additional cations may have dissolved into the test fluid. Indeed, bentonite erosion is considered to be more likely to occur when non-saline groundwaters, and waters with a low ionic strength, interact with the bentonite barrier around the wasteform [58,66]. In addition, clay colloid dispersion is likely to be maximised by the use of a test fluid that does not contain any salts [56], suggesting that the test fluid will have exacerbated the formation of the gel layer. Regardless of the testing fluid, some exchange with the hydroxyl ions in the water that interacted with the bentonite would have been expected, and in situ, the extent would be dictated by the type of groundwater present. In the UK, where Ca bentonites are being considered for use as the engineered barrier for the geological disposal of high heat generating radioactive waste, there is a variety of different groundwaters that the engineered barrier might encounter. Groundwater chemistry is dependent on the reactions that have taken place between the groundwater and the host rock it is flowing through [67]. It is also controlled by the soils and unsaturated substrate that it passes through on the way to the saturated zone beneath. Thus, depending on the location and geology, the groundwater hardness and salinity will be different [67]. In addition, the depth of a repository and the proximity to the coastline will also control the groundwater chemistry present. The type of groundwater present and the composition of the bentonite chosen in radioactive waste repository designs will thus have an effect on the performance and behaviour of the barrier. Bentonite erosion is an important topic of research for the safety assessment of radioactive waste repositories. Previous work has investigated both experimentally (e.g., [58,59,66,68–70]) and numerically ( e.g., [71]) the expansion of a range of bentonites [51,72] into fractures filled with flowing groundwater. These studies have found that whilst accessory minerals may prevent the loss of material, bentonite barriers with high smectite concentrations are likely to erode given the expected fluid compositions and flowrates [58]. Whilst only one test has been undertaken here, the data suggest that a process of physical separation and mobilisation of smectite from the bentonite matrix occurs during hydration and expansion; further work using relevant fluid and bentonite compositions is required to elucidate the mechanisms behind these processes more clearly, as these results may have important implications for understanding the erosion of bentonite in contact with moving groundwater.

## 5. Conclusions

The results of this study have shown that a Ca bentonite in contact with a fluid-filled void space is able to form a gel layer at the expanding front, which comprises smectite particles with diameters <50 nm. This gel layer also has a significantly higher moisture content than the rest of the test sample. This is an unusual observation, and has not previously been observed in experimental tests of this type [47,48]. ESEM and EDXA analyses have shown that the gel layer itself is composed of pure smectite, whilst the residual substrate beneath the gel is enriched in silica. The surface area analysis supports the ESEM-EXDA data showing a higher surface area value for the slice containing the gel. The <2 μm XRD analysis and EDXA data also both show an increase in monovalent Na and K cations in the slice containing the gel. The whole-rock XRD data do not show a change in the amount of smectite in slice 1 piece A, but this is likely due to the averaging of the slice 1 piece A data. Together the observations are supportive of a chemically driven flushing mechanism occurring towards the front of the bentonite sample during expansion, whereby the very fine grained smectite is carried through a silica matrix to be deposited in a layer at the front of the sample. This finding has important implications for the erosion of bentonite in radioactive waste repositories, where the bentonite barrier material between the cannister and the host rock encounters fractures filled with flowing groundwater. Under these circumstances, expansion of the bentonite into the fracture would occur and the fine grained smectite particles would be expected to be removed from the front of the expanding bentonite by the flowing groundwater. This process could gradually deplete the barrier of the smectite content, adversely affecting its performance by reducing the swelling capacity and increasing the permeability.

**Supplementary Materials:** The following are available online at https://www.mdpi.com/2075-163X/11/2/215/s1, Figure S1: Testing apparatus. Table S1: EDXA analyses (weight % element). Table S2: EDXA analyses (molar % ions).

**Author Contributions:** The individual contributions to this paper are as follows. Conceptualization and methodology, K.A.D. and J.F.H.; validation, K.A.D., J.F.H. and P.S.; investigation and formal analysis, K.A.D., J.F.H., A.E.M., S.J.K. and I.M.; resources, P.S.; data curation, K.A.D., J.F.H., S.J.K. and I.M.; writing—original draft preparation, K.A.D., J.F.H., A.E.M. and S.J.K.; writing—review and editing, all authors; visualization, K.A.D. and A.E.M.; supervision, J.F.H. and P.S.; project administration, J.F.H.; funding acquisition, J.F.H. and P.S. All authors have read and agreed to the published version of the manuscript.

**Funding:** This research was funded by the European Commission Bentonite Mechanical Evolution (BEACON) project, with grant number 745942. The APC was funded by Svensk Kärnbränslehantering AB.

**Institutional Review Board Statement:** Not applicable.

**Informed Consent Statement:** Not applicable.

**Data Availability Statement:** The data used in this study is available from the authors upon request.

**Acknowledgments:** The authors acknowledge SKB for the supply of the material used in this study. Simon Holyoake is thanked for his assistance with the instrumentation of the experimental apparatus. Humphrey Wallis and Wayne Leman are acknowledged for the design and manufacture of the experimental apparatus. The authors thank Andrew Wiseall of the British Geological Survey for his helpful reviews of an earlier version of this manuscript. This paper is published with the permission of the Director of the British Geological Survey, part of United Kingdom Research and Innovation (UKRI).

**Conflicts of Interest:** The authors declare no conflict of interest. The funders had no role in the design of the study; in the collection, analyses, or interpretation of data; in the writing of the manuscript, or in the decision to publish the results.

**Sample Availability:** The test sample used in this study was specifically manufactured for testing, and was destructively tested to obtain information about the post-test moisture content and for the SEM, EDXA, surface area and XRD analyses. However, the powder used for the sample manufacture is available from the authors upon request.

## References


1. Sellin, P.; Leupin, O.X. The use of clay as an engineered barrier in radioactive waste management—A review. *Clays Clay Mineral.* **2013**, *61*, 477–498. [CrossRef]
2. Johansson, E.; Siren, T.; Kemppainen, K. Onkalo—Underground Rock Characterization Facility for In-Situ Testing for Nuclear Waste Disposal. In Proceedings of the 13th ISRM International Congress of Rock Mechanics, Montreal, QC, Canada, 10–13 May 2015.
3. WIPP. *Waste Isolation Pilot Plant*; U.S. Department of Energy, 2019; Available online: https://wipp.energy.gov/wipp-site.asp (accessed on 15 February 2021).
4. Andra. *Référentiel des Matériaux d'un Stockage de Déchets à Haute Activité et à Vie Longue—Tome 4: Les Matériaux à Base D'argilites Excavées et Remaniées*; Technical Report CRPASCM040015B; Agence nationale pour la gestion des déchets radioactifs (ANDRA): Châtenay-Malabry, France, 2005.
5. Liu, L. Permeability and expansibility of sodium bentonite in dilute solutions. *Colloids Surfaces A Physicochem. Eng. Asp.* **2010**, *358*, 68–78. [CrossRef]
6. Pusch, R. *The Buffer and Backfill Handbook, Part 1: Definitions, Basic Relationships, and Laboratory Methods*; Technical Report TR-02-20; Svensk Kärnbränslehantering AB (SKB): Stockholm, Sweden, 2002.
7. Wersin, P.; Johnson, L.H.; McKinley, I.G. Performance of the bentonite barrier at temperatures beyond 100: A critical review. *Phys. Chem. Earth* **2007**, *32*, 780–788. [CrossRef]
8. Gens, A.; Vallején, B.; Zandarín, M.T.; Sánchez, M. Homogenization in clay barriers and seals: Two case studies. *J. Rock Mech. Geotech. Eng.* **2013**, *5*, 191–199. [CrossRef]
9. Deniau, I.; Devol-Brown, I.; Derenne, S.; Behar, F.; Largeau, C. Comparison of the bulk geochemical features and thermal reactivity of kerogens from Mol (Boom Clay), Bure (Callovo-Oxfordian argillite) and Tournemire (Toarcian shales) underground research laboratories. *Sci. Total Environ.* **2008**, *389*, 475–485. [CrossRef] [PubMed]
10. Villar, M.V.; Lloret, A. Influence of dry density and water content on the swelling of a compacted bentonite. *Appl. Clay Sci.* **2008**, *39*, 38–49. [CrossRef]
11. Zheng, L.; Rutqvist, J.; Birkholzer, J.T.; Liu, H.H. On the impact of temperatures up to 200 °C in clay repositories with bentonite engineer barrier systems: A study with coupled thermal, hydrological, chemical, and mechanical modelling. *Eng. Geol.* **2015**, *197*, 278–295. [CrossRef]
12. Daniels, K.A.; Harrington, J.F.; Zihms, S.G.; Wiseall, A.C. Bentonite Permeability at Elevated Temperature. *Geosciences* **2017**, *7*. [CrossRef]
13. SKB. *Final Storage of Spent Nuclear Fuel—KBS-3*; Technical Report Art716-1; Svensk Kärnbränslehantering AB (SKB): Stockholm, Sweden, 1983.
14. NAGRA. *Project Gewahr 1985, Nuclear Waste Management in Switzerland: Feasibility Studies and Safety Analysis*; Technical Report NAGRA-NTB-85-09; Nationale Genossenschaft für die Lagerung Radioaktiver Abfälle (NAGRA): Baden, Switzerland, 1985.
15. Coulon, H.; Lajudie, A.; Debrabant, P.; Atabek, R.; Jordia, M.; Jehan, R. Choice of French clays as engineered barrier components for waste disposal. *Sci. Basis Nucl. Waste Manag.* **1987**, *10*, 813–824. [CrossRef]
16. Linares, J. Spanish Research Activities in Thefield of Backfilling and Sealing: A Pre-Liminary Study of Some Spanish Sedimentary (Madrid Basin) and Hydrothermal (Almeria) Bentonite; Technical Report. In Proceedings of the NEA/CEC Workshop on Sealing of RadioactiveWaste Repositories, Paris, France, 9–19 June 1989.
17. Vieno, T.; Hautojarvi, A.; Koskinen, L.; Nordman, H. *TVO-92 Safety Analysis of Spent Fuel Disposal*; Technical Report VTJ-92-33E; VTT Technical Research Centre of Finland: Espoo, Finland, 1992.
18. Villar, M.V. *Thermo-Hydro-Mechanical Characterisation of a Bentonite from Cabo de Gata: A Study Applied to the Use of Bentonite as Sealing Material in High Level Radioactive Waste Repositories*; Technical Report 01/2002; ENRESA: Madrid, Spain, 2002.
19. Villar, M.V.; Iglesias, R.J.; García-Siñeriz, J.L. State of the in situ Febex test (GTS, Switzerland) after 18 years: A heterogeneous bentonite barrier. *Environ. Geotech.* **2020**, *7*, 147–159. [CrossRef]
20. Pusch, R. *The Microstructure of MX-80 Clay with Respect to Its Bulk Physical Properties under Different Environmental Conditions*; Technical Report TR-01-08; Svensk Kärnbränslehantering AB (SKB): Stockholm, Sweden, 2001.
21. Tabak, A.; Afsin, B.; Caglar, B.; Koksal, E. Characterization and pillaring of a Turkish bentonite (Resadiye). *J. Colloid Interface Sci.* **2007**, *313*, 5–11. [CrossRef] [PubMed]
22. Coulon, H. Proprietes Physico-Chimiques des Sediments Argileux Francais: Application au Stockage des Dechets Radioactifs. Ph.D. Thesis, University of Lille, Lille, France, 1987.
23. Sun, H.; Mašín, D.; Boháč, J. Experimental characterization of retention properties and microstructure of the Czech bentonite B75. In Proceedings of the 19th International Conference on Soil Mechanics and Geotechnical Engineering, Seoul, Korea, 17–21 September 2017.

24. Tessier, D.; Dardaine, M.; Beaumont, A.; Jaumet, A.M. Swelling pressure and microstructure of an activated swelling clay with temperature. *Clay Miner.* **1998**, *33*, 255–267. [CrossRef]
25. Marcial, D.; Delage, P.; Cui, Y.J. On the high stress compression of bentonites. *Can. Geotech. J.* **2002**, *39*, 812–820. [CrossRef]
26. Ben Rhaïem, H.; Pons, C.H.; Tessier, D. *Factors Affecting the Microstructure of Smectites. Role of Cations and History of Applied Stresses*; Technical Report; The Clay Mineral Society: Denver, CO, USA, 1987, pp. 292-297.
27. Tessier, D. Etude Expérimentale de L'organisation des Matériaux Argileux. Ph.D. Thesis, Institut National de la Recherche Agronomique (INRA), Versailles, France, 1984.
28. Chafe, N.P.; de Bruyn, J.R. Drag and relaxation in a bentonite clay suspension. *J. Non Newton. Fluid Mech.* **2005**, *131*, 44–52. [CrossRef]
29. Whorlow, R.W. *Rheological Techniques*; Ellis Horwood: New York, NY, USA, 1992.
30. Martin, P.L.; Barcala, J.M.; Huertas, F. Large-scale and long-term coupled thermo-hydro-mechanic experiments with bentonite: The febex mock-up test. *J. Iber. Geol.* **2006**, *32*, 259–282.
31. Juvankoski, M. *Description of Basic Design for Buffer*; Posiva Technical Report 2009-131; Posiva: Eurajoki, Finland, 2010.
32. Wang, Q.; Tang, A.M.; Cui, Y.J.; Delage, P.; Barnichon, J.D.; Ye, W.M. The effects of technological voids on the hydro-mechanical behaviour of compacted bentonite–sand mixture. *Soils Found.* **2013**, *53*, 232–245. [CrossRef]
33. Komine, H. Simplified evaluation for swelling characteristics of bentonites. *Eng. Geol.* **2004**, *71*, 265–279. [CrossRef]
34. Cuevas, J.; Villar, M.V.; Martın, M.; Cobena, J.; Leguey, S. Thermo-hydraulic gradients on bentonite: Distribution of soluble salts, microstructure and modification of the hydraulic and mechanical behaviour. *Appl. Clay Sci.* **2002**, *22*, 25–38. [CrossRef]
35. Thomson, S.; Ali, P. A Laboratory Study of the Swelling Properties of Sodium and Calcium Modifications of Lake Edmon Clay. In *Proceedings of the 2nd International Conference on Expensive Soils*; Texas A and M University: College Station, TX, USA, 1969; pp. 256–262.
36. Westsik, J.H.; Bray, L.A.; Hodges, F.N.; Wheelwright, E.J. Permeability, swelling and radionuclide retardation properties of candidate backfill materials. In *Scientific Basis for Nuclear Waste Management IV, Symposium D*; Cambridge University Press: Cambridge, UK, 1981; Volume 6, p. 329. [CrossRef]
37. Börgesson, L.; Pusch, R. *Rheological Properties of a Calcium Smectite*; Technical Report TR-87-31; Svensk Kärnbränslehantering AB (SKB): Stockholm, Sweden, 1987.
38. Börgesson, L.; Hokmark, H.; Karnland, O. *Rheological Properties of a Sodium Smectite Clay*; Technical Report TR-88-30; Svensk Kärnbränslehantering AB (SKB): Stockholm, Sweden, 1988.
39. Gens, A.; Alonso, E. A framework for the behaviour of unsaturated clay. *Can. Geotech. J.* **1992**, *29*, 1013–1032. [CrossRef]
40. Komine, H.; Ogata, N. Simplified evaluation for swelling characteristics of bentonites. *Can. Geotech. J.* **1994**, *31*, 478–490. [CrossRef]
41. Delage, P.; Howat, M.D.; Cui, Y.J. The relationship between suction and swelling properties in a heavily compacted unsaturated clay. *Eng. Geol.* **1998**, *50*, 31–48. [CrossRef]
42. Cui, Y.J.; Yahia-Aissa, M.; Delage, P. A model for the volume change behaviour of heavily compacted swelling clays. *Eng. Geol.* **2002**, *64*, 233–250. [CrossRef]
43. Karnland, O.; Muurinen, A.; Karlsson, F. Bentonite Swelling Pressure in NaCl Solutions: Experimentally Determined Data and Model Calculations. In *Advances in Understanding Engineered Clay Barriers, Proceedings of the International Symposium on Large Scale Field Tests in Granite, Sitges, Barcelona, Spain, 12–14 November 2003*; Eduardo E.A., Alberto, L., Eds.; CRC Press: Boca Raton, FL, USA, 2005.
44. Xie, M.; Moog, H.C.; Kolditz, O. Geochemical Effects on Swelling Pressure of Highly Compacted Bentonite: Experiments and Model Analysis; In *Theoretical and Numerical Unsaturated Soil Mechanics*; Springer Proceedings in Physics 113 Series; Springer: Berlin/Heidelberg, Germany, 2007.
45. Jönsson, B.; Kesson, T.; Jönsson, B.; Meehdi, S.; Janiak, J.; Wallenberg, R. Structure and Forces in Bentonite MX-80; Technical Report TR-09-06; Svensk Kärnbränslehantering AB (SKB): Stockholm, Sweden, 2009.
46. Lee, J.O.; Lim, J.G.; Kang, I.M.; Kwon, S. Swelling pressures of compacted Ca-bentonite. *Eng. Geol.* **2012**, *129–130*, 20–26. [CrossRef]
47. Harrington, J.F.; Daniels, K.A.; Wiseall, A.C.; Sellin, P. Bentonite homogenisation during the closure of repository voids. *Int. J. Rock Mech. Min. Sci.* **2020**, *136*, 104535. [CrossRef]
48. Daniels, K.A.; Harrington, J.F.; Sellin, P.; Norris, S. Closing repository void spaces using bentonite: Does heat make a difference? *Appl. Clay Sci.* **2021**.
49. Tripathy, S.; Thomas, H.R.; Stratos, P. Response of Compacted Bentonites to Thermal and Thermo-Hydraulic Loadings at High Temperatures. *Geosciences* **2017**, *7*. [CrossRef]
50. Missana, T.; Alonso, U.; Albarran, N.; Garcia-Gutierrez, M.; Cormenzana, J.L. Analysis of colloids erosion from the bentonite barrier of a high level radioactive waste repository and implications in safety assessment. *Phys. Chem. Earth* **2011**, *36*, 1607–1615. [CrossRef]
51. Alonso, U.; Missana, T.; Fernandez, A.M.; Garcia-Gutierrez, M. Erosion behaviour of raw bentonites under compacted and confined conditions: Relevance of smectite content and clay/water interactions. *Appl. Geochem.* **2018**, *94*, 11–20. [CrossRef]

52. Svensson, D.; Eriksson, P.; Johannesson, L.E.; Lundgren, C.; Bladström, T. *Developing Strategies and Methods for Material Control of Bentonite for a High Level Radioactive Waste Repository*; Technical Report TR-19-25; Svensk Kärnbränslehantering AB (SKB), Stockholm, Sweden, 2019.

53. Snyder, R.L.; Bish, D.L. Quantitative analysis. In *Modern Powder Diffraction, Reviews in Mineralogy*; Bish, D.L., Post, J.E., Eds.; Mineralogical Society of America: Chantilly, VA, USA, 1989; Volume 20, Chapter 5, pp. 101–144.

54. Kemp, S.; Smith, F.; Wagner, D.; Mounteney, I.; Bell, C.; Milne, C.; Gowing, C.; Pottas, T. An improved approach to characterise potash-bearing evaporite deposits, evidenced in North Yorkshire, U.K. *Eng. Geol.* **2016**, *111*, 719–742. [CrossRef]

55. Carter, D.L.; Heilman, M.D.; Gonzalez, F.L. Ethylene glycol monoethyl ether for determining surface area of silicate minerals. *Soil Sci.* **1965**, *100*, 356–360. [CrossRef]

56. Missana, T.; Alonso, U.; Fernandez, A.M.; García-Gutiérrez, M. Colloidal properties of different smectite clays: Significance for the bentonite barrier erosion and radionuclide transport in radioactive waste repositories. *Appl. Geochem.* **2018**, *97*, 157–166. [CrossRef]

57. Sen, T.; Khilar, K. Review on subsurface colloids and colloid-associated contaminant transport in saturated porous media. *Adv. Colloid Interface Sci.* **2006**, *119*, 71–96. [CrossRef]

58. Neretnieks, I.; Liu, L.; Moreno, L. *Mechanisms and Models for Bentonite Erosion*; Technical Report TR-09-35; Svensk Kärnbränslehantering AB (SKB): Stockholm, Sweden, 2009.

59. Birgersson, M.; Hedström, M.; Karnland, O. Sol formation ability of Ca/Na-montmorillonite at low ionic strength. *Phys. Chem. Earth* **2011**, *36*, 1572–1579. [CrossRef]

60. Kallay, N.; Barouch, E.; Matijević, E. Diffusional detachment of colloidal particles from solid/solution interfaces. *Adv. Colloid Interface Sci.* **1987**, *27*, 1–42. [CrossRef]

61. Inglethorpe, S.D.J.; Morgan, D.J.; Highley, D.E.; Bloodworth, A.J. *Industrial Minerals Laboratory Manual: Bentonite*; Technical Report WG-93-20; British Geological Survey: Nottingham, UK, 1993.

62. Svensson, D.; Lundgren, C.; Johannesson, L.E.; Norrfors, K. *Developing Strategies for Acquisition and Control of Bentonite for a High Level Radioactive Waste Repository*; Technical Report TR-16-14; Svensk Kärnbränslehantering AB (SKB): Stockholm, Sweden, 2017.

63. Ryan, J.N.; Elimelech, M. Review: Colloid mobilization and transport in groundwater. *Colloids Surf. A Physicochem. Eng. Asp.* **1996**, *107*, 1–56. [CrossRef]

64. Missana, T.; Alonso, U.; García-Gutiérrez, M.; Mingarro, M. Role of bentonite colloids on europium and plutonium migration in a granite fracture. *Appl. Geochem.* **2008**, *23*, 1484–1497. [CrossRef]

65. Schaefer, T.; Huber, F.; Seher, H.; Missana, T.; Alonso, U.; Kumke, M.; Eidner, S.; Claret, F.; Enzmann, F. Nanoparticles and their influence on radionuclide mobility in deep geological formations. *Appl. Geochem.* **2012**, *27*, 390–403. [CrossRef]

66. Baik, M.H.; Cho, W.J.; Hahn, P.S. Erosion of bentonite particles at the interface of a compacted bentonite and a fractured granite. *Eng. Geol.* **2007**, *91*, 229–239. [CrossRef]

67. Edmunds, W.M.; Shand, P.; Hart, P.; Ward, R.S. The natural (baseline) quality of groundwater: A UK pilot study. *Sci. Total Environ.* **2003**, *310*, 25–35. [CrossRef]

68. Kurosawa, S.; Kato, H.; Ueta, S.; Yokoyama, K.; Fujihara, H. *Erosion Properties and Dispersion-Flocculation Behavior of Bentonite Particles*; (Symposium QQ—Scientific Basis for Nuclear Waste Management XXII); Cambridge University Press: Cambridge, UK, 1999; Volume 556, p. 679. [CrossRef]

69. Jansson, M. *Bentonite Erosion Laboratory Studies*; Technical Report TR-09-33; Svensk Kärnbränslehantering AB (SKB): Stockholm, Sweden, 2009.

70. Albarran, N.; Degueldre, C.; Missana, T.; Alonso, U.; Garcia-Gutierrez, M.; Lopez, T. Size distribution analysis of colloid generated from compacted bentonite in low ionic strength aqueous solutions. *Appl. Clay Sci.* **2014**, *95*, 284–293. [CrossRef]

71. Birgersson, M.; Börgesson, L.; Hedström, M.; Karnland, O.; Nilsson, U. *Bentonite Erosion Final Report*; Technical Report TR-09-34; Svensk Kärnbränslehantering AB (SKB): Stockholm, Sweden, 2009.

72. Fernández, A.M.; Missana, T.; Alonso, U.; Rey, J.J.; Sánchez-Ledesma, D.M.; Melón, A.; Robredo, L.M. *Characterisation of Different Bentonites in the Context of BELBAR Project (Bentonite Erosion: Effects on the Long Term Performance of the Engineered Barrier and Radionuclide Transport)*; Technical Report; CIEMAT: Madrid, Spain, 2017.

*Article*

# Swelling Pressure and Permeability of Compacted Bentonite from 10th Khutor Deposit (Russia)

Artur Yu. Meleshyn [1,*], Sergey V. Zakusin [2,3] and Victoria V. Krupskaya [2,4,*]

[1] Gesellschaft für Anlagen- und Reaktorsicherheit (GRS) gGmbH, 38122 Braunschweig, Germany; artur.meleshyn@grs.de

[2] Institute of Ore Geology, Petrography, Mineralogy and Geochemistry, Russian Academy of Science (IGEM RAS), 119017 Moscow, Russia; zakusinsergey@gmail.com

[3] Department of Engineering and Environmental Geology, Lomonosov Moscow State University, 119991 Moscow, Russia

[4] Nuclear Safety Institute, Russian Academy of Science (IBRAE RAS), 115191 Moscow, Russia

* Correspondence: krupskaya@ruclay.com; Tel.: +007-926-819-6398

**Citation:** Meleshyn, A.Y.; Zakusin, S.V.; Krupskaya, V.V. Swelling Pressure and Permeability of Compacted Bentonite from 10th Khutor Deposit (Russia). *Minerals* **2021**, *11*, 742. https://doi.org/10.3390/min11070742

Academic Editors: Ana María Fernández, Stephan Kaufhold, Markus Olin, Lian-Ge Zheng, Paul Wersin and James Wilson

Received: 5 May 2021
Accepted: 28 June 2021
Published: 7 July 2021

**Publisher's Note:** MDPI stays neutral with regard to jurisdictional claims in published maps and institutional affiliations.

**Abstract:** Bentonites from the 10th Khutor deposit (Republic of Khakassia, Russia) are considered a potential buffer material for isolation of radioactive waste in the crystalline rocks of Yeniseyskiy site (Krasnoyarskiy region). This study presents the results of a series of permeameter experiments with bentonite compacted to dry densities of 1.4, 1.6, and 1.8 g/cm$^3$, saturated and permeated by the artificial groundwater from Yeniseyskiy Site. Permeation was conducted at hydraulic gradients of 180–80,000 m/m to simulate potential hydraulic conditions in the early post-closure phase of a deep geological repository (DGR). The respective swelling pressures of $0.8 \pm 0.3$, $2.2 \pm 0.6$, and $6.3 \pm 0.3$ MPa and permeabilities of $(27 \pm 15) \times 10^{-20}$, $(3.4 \pm 0.8) \times 10^{-20}$, and $(0.96 \pm 0.26) \times 10^{-20}$ m$^2$ were observed for the hydraulic gradient of 2000 m/m, which is recommended for the determination of undisturbed swelling pressures and permeabilities in permeameter experiments. Upon incremental increases in the hydraulic gradient, swelling pressures at all densities and permeability at the density of 1.8 g/cm$^3$ remained unchanged, whereas permeabilities at 1.4 and 1.6 g/cm$^3$ decreased overall by a factor of approximately 5 and 1.7, respectively. Seepage-induced consolidation and/or reorganisation of bentonite microstructure are considered possible reasons for these decreases.

**Keywords:** radioactive waste; bentonite; swelling pressure; permeability; hydraulic gradient; engineered barriers; geological repository

## 1. Introduction

In the Russian Federation, a concept is being developed to isolate class 1 and 2 high-level radioactive waste and spent nuclear fuel in crystalline rocks (Yeniseyskiy Site, Krasnoyarskiy region) at depths of about 500–700 m [1–4]. In this case, the safety function of radioactive waste disposal is provided by a passive multi-barrier system that ensures the isolation of radioactive waste from the biosphere [5–7].

The current Russian sealing concept of a deep geological radioactive waste disposal facility (DGR) in crystalline rocks foresees the use of compacted bentonite as a buffer material enclosing containers with radioactive waste [1–4] due to the stability of its properties and ability to prevent the mass transfer of radionuclides from advection and diffusion mechanisms [2–7]. The 10th Khutor bentonite deposit is located relatively close to the area of the deep geological disposal site and considered as a source for the potential buffer material in study [4] based on its favourable sealing properties and deposit resources. Swelling pressure and permeability (or proportional to that hydraulic conductivity) of compacted bentonite are key parameters in determining the sealing performance of bentonite-based geotechnical barriers. A direct determination of these parameters for a realistic buffer configuration under in situ conditions of DGR would require conducting sophisticated, long-term, and expensive experiments in an Underground Research Laboratory. Laboratory experiments with small specimens of compacted bentonite provide a feasible way to obtain estimates of the parameters necessary to conduct preliminary performance assessments of bentonite-based geotechnical barriers and facilitate the planning and optimisation in situ studies, possibly reducing their extent.

During the construction of geotechnical barriers, unsaturated bentonites will experience high fluid pressure differences—up to several MPa—which may result across the buffer in the course of establishing hydraulic conditions characteristic of the host formation. Therefore, fluid pressure differences of up to 8 MPa across the compacted specimens were imposed on the present study in order to test the sealing performance of a bentonite sample from the 10th Khutor deposit (described as "10X" or "10X bentonite"). Swelling pressure and hydraulic conductivity of 10X were estimated in a series of preliminary laboratory experiments with distilled water [4]. The present study provides an independent laboratory estimation of swelling pressure and permeability for 10X upon saturation and permeation with the synthetic groundwater of crystalline rocks at the Yeniseyskiy site.

## 2. Materials and Methods

### 2.1. Geological Site and Composition of 10X

The 10th Khutor deposit is located in the steppe area 8 km southwest of Chernogorsk city (Republic of Khakassia). It has reserves of about 3–4 million tons and belongs to a group of deposits with total reserves of approximately 13 million [8]. The deposit has a Carboniferous age ($C_2$) expressed in several bentonite layers interbedded with clay rocks and carbonaceous layers [8–10]. Its genesis is considered volcanogenic-sedimentary.

In the process of studying bentonite powder with specified technical conditions, obtained characteristics were required to justify the use of clay material as a component of engineering safety barriers. This material was assigned a commercial name, 10X.

The study of particle size distribution was conducted using laser deflation on an Analysette-22 (Fritsch GmbH, Idar-Oberstein, Germany) particle analyser in suspension for the measurement range of 0.1–300 μm in five different positions of the measuring cell and 100 measurements in each position. The results are shown in Table 1.

**Table 1.** Basic characteristics of bentonite powder 10X.

**a—Particle Size Distribution**

| Mods, μm | Median, μm | Particle Size Distribution, % | | |
| --- | --- | --- | --- | --- |
| | | 0.1–1 μm | 1–10 μm | 10–25 μm |
| 1.79 | 2.90 | 15.6 | 72.6 | 11.8 |

**b—Mineral Composition (%)**

| Smectite (Montmorillonite) | Chlorite | Quartz | Feldspars (Microcline and Albite) | Calcite | Anatase | Gypsum |
| --- | --- | --- | --- | --- | --- | --- |
| 73.0 | 0.9 | 13.9 | 10.8 | 0.7 | 0.3 | 0.4 |

**c—Chemical Composition**

| $Na_2O$ | MgO | $Al_2O_3$ | $SiO_2$ | $K_2O$ | CaO | $TiO_2$ | $Fe_2O_3$ | FeO | MnO | LOI (105–1000 °C) |
| --- | --- | --- | --- | --- | --- | --- | --- | --- | --- | --- |
| 0.107 | 2.67 | 18.84 | 51.87 | 0.48 | 6.58 | 0.70 | 3.37 | 0.33 | 0.01 | 7.64 |

The mineral composition of the 10X bentonite [10,11] was estimated by X-ray diffraction using an Ultima-IV (Rigaku, Tokyo, Japan) X-ray diffractometer (Cu-$K_\alpha$ radiation, detector—D/Tex-Ultra, scan range 3–65° 2θ). The quantitative mineral analysis (Table 1) was conducted using the Rietveld method [12] with PROFEX GUI ver. 3.1.3, (Nicola Doeblin, Solothurn, Switzerland) software for BGMN [13].

The chemical composition (Table 1) was determined using X-ray fluorescence analysis using an Axios mAX (PANalytical, Almelo, the Netherlands) according to standard methods [11]. The hygroscopic water content of 8.94% was determined upon drying at 105 °C. The suction of the 10X bentonite was not determined.

The density of the solid phase was measured using the pycnometer method with kerosene following [14,15] and amounted to 2.8 g/cm$^3$.

The value of the cation exchange capacity was determined from the adsorption of the Cu(II)-trien-complex using the spectrophotometric method [16–18] and amounted to 73.5 meq/100g. The composition of the exchangeable complex is mainly Na-Mg, and comprises Na—23.6, Ca—16.6, Mg—30.3, and K—1.2 meq/100g, measured with atomic absorption spectrometry (AAS) after repeated extraction with ammonium acetate solution.

### 2.2. Experiments and Conditions

For the research, triplicate experiments of 10X bentonite at the dry densities of 1.4, 1.6, and 1.8 g/cm$^3$ were conducted using nine constant-head, rigid-wall permeameter cells made of titanium (Figure 1a). Compacted specimens were prepared by pressing directly into permeameter cells. Bentonite powder (Table 1a) was filled into a permeameter between two sintered steel filters (GKN SIKA-R 20 AX, a porosity of 43%, an average pore diameter of 23 μm, with an average permeability—measured in triplicate using three permeameter cells—of (3.4 ± 0.5) × 10$^{-14}$ m$^2$) and, after installing a load-transfer ram and a press flange above the top sintered filter, it was statically compacted (without adding any water) to a pellet specimen with a diameter of 5 cm, a height of 1 cm, and a dry density of 1.4, 1.6, and 1.8 g/cm$^3$. The cap nut of the press flange was then tightened without adding axial stress. Based on the density of soil solids (2.80 g/cm$^3$), porosity of specimens can be estimated to equal 0.50, 0.43, and 0.36, pore volumes to 9.8, 8.4, and 7.0 cm$^3$, and saturation to 25%, 33%, and 45%, respectively. Respective bulk densities equal 2.02, 2.17, and 2.32 g/cm$^3$.

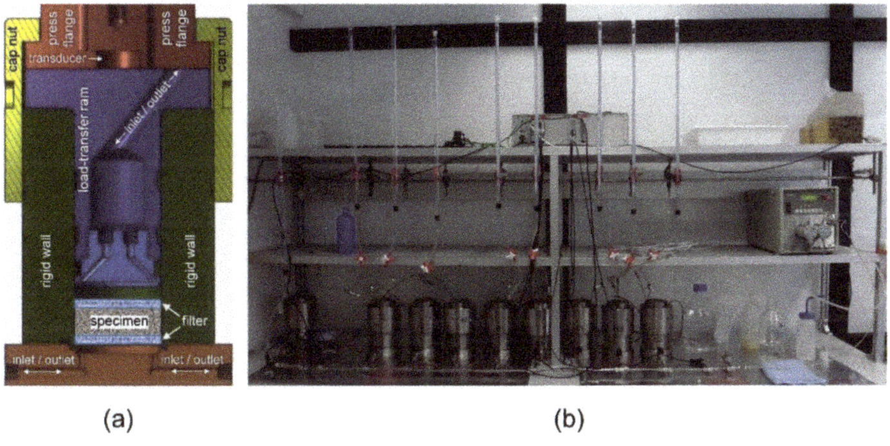


**Figure 1.** A construction drawing of experiments; (**a**); supply of SGW to inlets at cell bases by a piston pump; (**b**) permeability experiments (note that burettes connected to outlets at cell tops to collect percolating solutions).

The swelling pressure of the specimens in the permeameter cells was measured continuously by force transducers FKA613 (Ahlborn Mess- und Regelungstechnik GmbH, Holzkirchen, Germany). It should be noted that for the specimen in the ninth cell (at 1.8 g/cm$^3$, the leftmost cell in Figure 1b), a behaviour strongly deviating from the other two specimens at 1.8 g/cm$^3$ was observed, as its swelling pressure exceeded the measuring capacity of force transducers at 10.2 MPa within two days of starting the experiment. Two restarts and additional tests suggested a permeameter malfunction as the reason for this deviating behaviour; therefore, the experiment with the ninth cell was discontinued.

Since no reliable data currently exists on the composition of the groundwater of the Yeniseyskiy Site at the target depth, synthetic groundwater (hereafter referred to as SGW) was used in the experiments and developed based on available hydrogeological data and data in [19], which is characterized by a pH of 8.1, ionic strength of 6.5 mmol/L, and ionic composition: $Na^+$—0.760, $Mg^{2+}$—0.494, $Ca^{2+}$—1.217, $K^+$—0.115, $Cl^-$—2.549, $SO_4^{2-}$—0.494, and $HCO_3^-$—0.76 mmol/L.

Compacted bentonite specimens were allowed to saturate by absorbing SGW from burettes connected to inlets at cell bases to let trapped air escape through outlets at cell tops for 28 days, then to inlets at cell tops for 87 days to reach the most homogeneous saturation of bentonite possible (Figure 1b). The duration of 28 days for the first saturation stage was chosen based on a preceding study involving 15 different bentonites [20,21]. According to the experimental protocol established in that study, the second saturation stage was continued until the swelling pressure of the specimens was judged to level off. The fluid pressure difference across the specimens was equal to 0.018 MPa during this stage.

After disconnecting the burettes, permeation tests were started on the 133rd day (after a delay due to a malfunction and temporary unavailability of the pump) by injecting SGW through inlets at cell bases using a piston pump (BESTA HD 2-200, BESTA-Technik GmbH, Wilhelmsfeld, Germany) and a bypass (Figure 1b) to prevent uncontrolled fluid injection pressure surges observed in a preceding study discussed in [20,21]. Outlets at cell tops were connected to burettes for measuring volumes of percolating SGW used to calculate permeability. Swelling of compacted bentonites upon saturation was assumed to seal off possible flow pathways along the bentonite-cell interface with the solution percolating only through bentonite pores during permeability measurements, which was confirmed by the observed low permeabilities of specimens. The fluid pressure difference of 0.018 MPa maintained across the specimens for 132 days by the water column in the burettes and was then incrementally increased to 0.2, 0.5, 2.0, and 8.0 MPa using the pump. These fluid pressure differences correspond to hydraulic gradients of 180 up to 80,000 m/m across the compacted bentonite specimens. For the first stage of the permeation tests at a hydraulic gradient of 2000 m/m, levelling off of permeabilities while observing for the levelling off of swelling pressures, if reasonably possible, was used as a termination criterion. During this stage, the pore volume of specimens was replaced on average 13.7, 2.2, and 0.5 times at a dry density of 1.4, 1.6, and 1.8 $g/cm^3$, respectively. For the following stages, a sufficient number of measurement points to capture the permeability trend was aimed, and the pore volume of specimens was replaced at a respective dry density of 1.4, 1.6, and 1.8 $g/cm^3$ on average 0.9, 0.1, and 0.02 times at a hydraulic gradient of 5000 m/m up to 6.9, 4.0, and 1.8 times at a hydraulic gradient of 80,000 m/m. Room temperature and relative humidity during the entire experiment were 18.1 ± 0.4 °C and 48% ± 7%, respectively. A dynamic viscosity of 1.014 ± 0.009 μPa·s was determined for SGW from eight replicate measurements at room temperature.

Apart from the specimens for the swelling and permeation experiments described above, but from the same batch of 10X, further specimens were compacted at hygroscopic humidity for studies using the method of computed microtomography (μCT, using Yamato TDM1000H-2) and qualitative analyses of the microstructure by scanning electron microscopy (SEM) using LEO 1450 VP (Zeiss, Jena, Germany) with a guaranteed resolution of 5 nm). These specimens were prepared in the same manner as the swelling and permeation experiments with distilled water described in [4]. To study changes in the microstructure of 10X during swelling, SEM analyses were conducted with specimens after permeation with distilled water [4].

## 3. Results and Discussion

The μCT study of the structure of compacted 10X showed that the specimen at a density of 1.4 $g/cm^3$ is characterized by an uneven distribution of moisture (Figure 2), and areas of non-compacted material are no more than 0.25 mm in size. The specimen at a density of 1.6 $g/cm^3$ is characterized by greater homogeneity, although a small amount of a non-compacted material (up to 0.1 mm on average, some areas with 0.5 mm) and some moisture inhomogeneity are observed. When compacted to a density of 1.8 $g/cm^3$, the obtained specimen does not contain any non-compacted parts, and no inhomogeneous distribution of moisture between the formed aggregates was observed. The appearance of cracks at the edges of the specimen at a density of 1.8 g/cm$^3$ (Figure 2c) was attributed to a relaxation of the 10X's fabric upon taking the specimen out of the compaction mold for the µCT study.

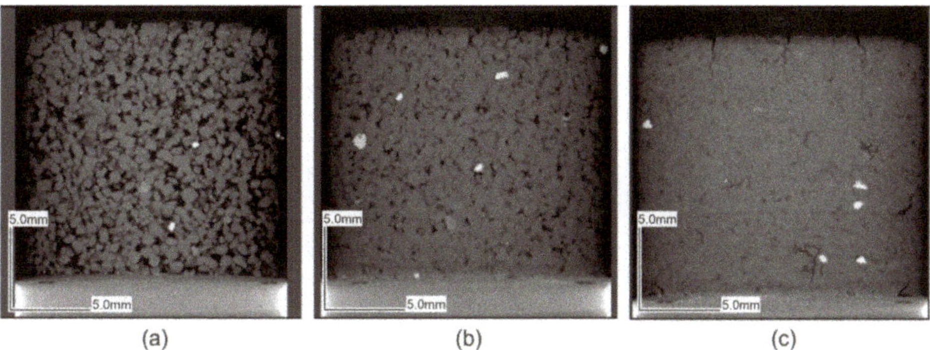


**Figure 2.** µCT images of 10X specimens compacted to dry densities of (**a**) 1.4 g/cm$^3$; (**b**) 1.6 g/cm$^3$; (**c**) 1.8 g/cm$^3$.

The initial stage of bentonite swelling upon saturation with solutions of low salinity [22,23] and—for some bentonites—with brines [20,21] is characterized by a "double-peak" pattern as a result of (i) swelling of aggregates (the first peak), (ii) their subsequent partial decomposition and a collapse of the inter-aggregate pores (the depression), and (iii) swelling of the reorganised aggregates (the second peak). This pattern is also observed for 10X specimens at the studied densities (graphs on left-side in Figure 3). The variation of this pattern with changing density further exemplifies that the collapse of the inter-aggregate pores is more strongly pronounced at the higher porosity so that the second peak is much lower than the first peak at 1.4 g/cm$^3$. This effect diminishes with decreasing porosity so that the second peak becomes higher than the first peak within 1 to 4 days at 1.8 g/cm$^3$. The higher second peak was observed previously for low- and high-salinity solutions ([20–25] and references therein), whereas the lower second peak was observed for high-salinity solutions [20,21]. By combining the latter observation with the present one, it can be suggested that the collapse of the inter-aggregate pores may depend on the electrostatic repulsion between diffuse-double layers (DDL) of single montmorillonite particles or their aggregates, as high salinity is known to depress their DDL [26] and increase pore volume available for their reorganisation accordingly.

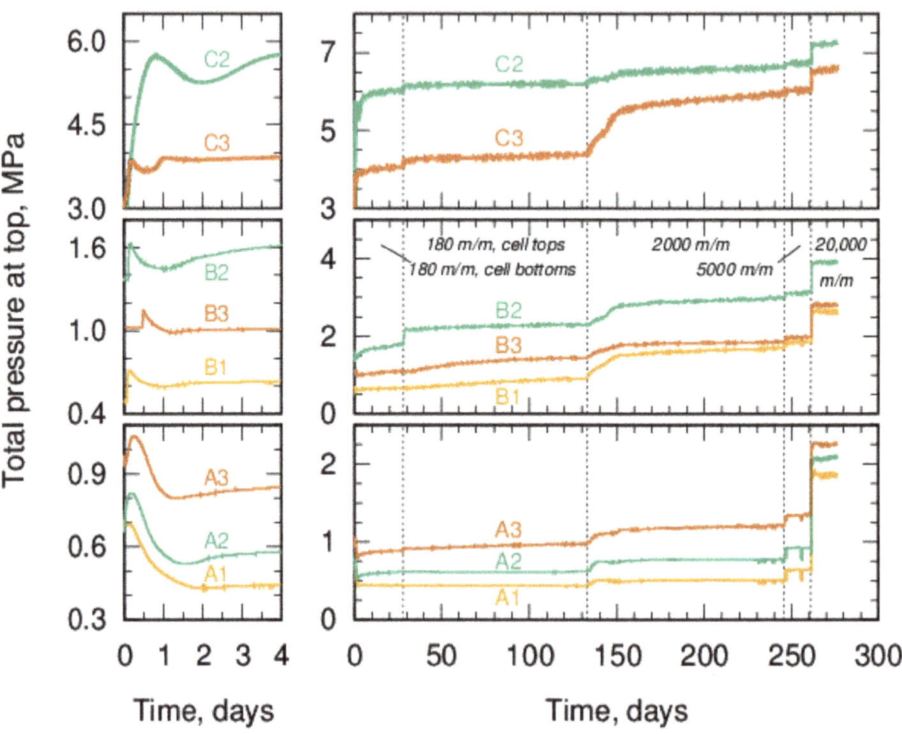


**Figure 3.** Total pressure at tops of permeameter cells for 10X specimens compacted to dry densities of 1.4 g/cm$^3$ (cells A1–A3); 1.6 g/cm$^3$ (cells B1–B3); and 1.8 g/cm$^3$ (cells C2, C3), saturated and permeated by SGW. Graphs on left-side represent x- and y-axis close-ups of the data measured within the first four days of saturation. Vertical dotted lines mark onsets of solution supply by burettes through cell tops (28th day) and by the pump at hydraulic gradients of 2000 m/m (133rd day), 5000 m/m (246th day), and 20,000 m/m (261st day) (note that corresponding labels in the 2nd graph on right-side).

It can be further seen in Figure 3 that despite the identical preparation protocol of specimens, the method used did not eliminate the inherent inhomogeneity of the microstructure of compacted bentonites. As a result of this inhomogeneity, measured swelling pressures and the time span of development for the second peak of the swelling pattern show significant variation at a given density and are otherwise similar to initial swelling patterns. This variation would not disappear (and even became larger at the densities of 1.4 and 1.6 g/cm$^3$) upon switching the solution supply from cell bases to cell tops on the 28th day. Possible reasons for this behaviour can be better understood considering the solution volumes absorbed by specimens and channels of permeameter cells from burettes within the first 115 days of the experiment (Figure 4). Firstly, these data reveal that the bentonite saturation was not accomplished within this stage, as the solution uptake did not level off. This occurrence is a well-known drawback of permeameter tests performed without a back pressure high enough to ensure full saturation of compacted specimens [27]. Secondly, for specimens with estimated pore volumes of 7.0–9.8 cm$^3$ at a given density, a variation in the absorbed solution volumes of up to 10.7 cm$^3$ was observed, which clarifies that the air was still present in the channels of some permeameters or even in the specimens. Hence, a reliable determination of the solution uptake by specimens upon saturation from burettes in the experimental setup used seems only possible through a test termination, dismantling of permeameters, and weighing the specimens.

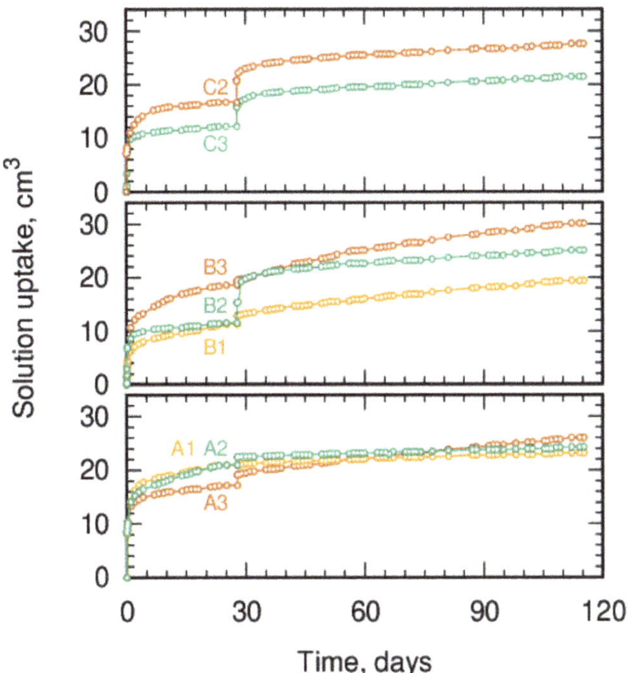

**Figure 4.** Uptake of SGW by permeameter cells with 10X specimens compacted to dry densities of 1.4 g/cm$^3$ (A1–A3); 1.6 g/cm$^3$ (B1–B3); and 1.8 g/cm$^3$ (C2, C3), upon saturation from burettes connected to inlets at cell bases for 27 days and cell tops afterwards.

The subsequent application of the hydraulic gradient of 2000 m/m led to a significant gradual increase of swelling pressure in all specimens, presumably, as a result of displacement of residual air from specimens by the injected solution (as discussed below to permeabilities) and/or microstructural reorganisation (as discussed below). During this stage and the preceding saturation stage using burettes, the total pressure measured at cell tops was equal to the swelling pressure of specimens. Starting from the 246th day, however, it was equal to the sum of the swelling pressure of the specimens and the fluid pressure established at cell tops upon imposing the hydraulic gradient of $\geq$5000 m/m across the specimens. This can be seen in steep increases of the total pressure at top of all cells on the 246th, 261st (Figure 3), and 281st day (Figure 5) upon increasing the hydraulic gradient to 5000; 20,000, and 80,000 m/m, respectively. It also results from its steep decrease to the values measured before the 246th day upon the intermediate and final switching off the pump on the 294th and 297th day, respectively (Figure 5). Therefore, the swelling pressure of 10X compacted to the densities of 1.4, 1.6, and 1.8 g/cm$^3$ can be concluded to equal to $0.8 \pm 0.3$, $2.2 \pm 0.6$, and $6.3 \pm 0.3$ MPa, respectively, measured at the hydraulic gradient of 2000 m/m (Table 2).

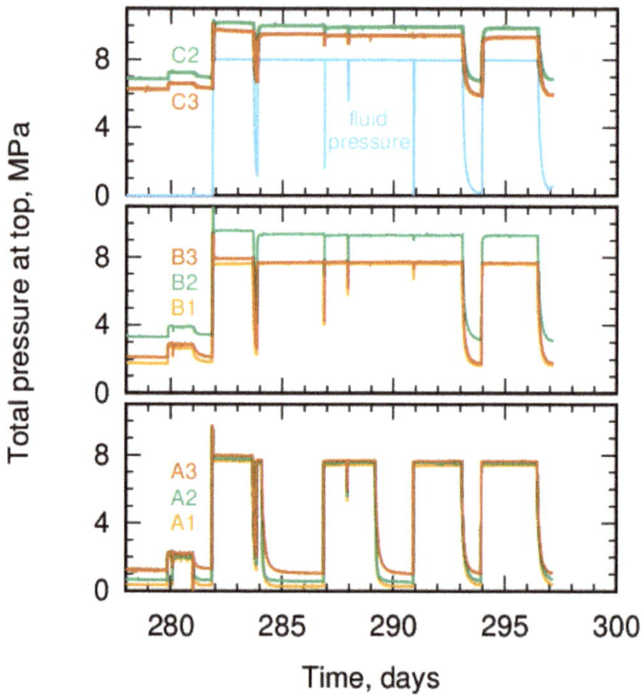



**Figure 5.** Total pressure at tops of permeameter cells for 10X specimens compacted to dry densities of 1.4 g/cm$^3$ (A1–A3), 1.6 g/cm$^3$ (B1–B3), and 1.8 g/cm$^3$ (C2, C3) and permeated by SGW at a hydraulic gradient of 80,000 m/m. Real fluid injection pressure imposed by the pump is shown in the upper graph (note that the abrupt fluid injection pressure changes were caused by short switching off the pump before weekends to disconnect cells A1–A3, preventing an overflow of burettes connected to them and after weekends, reconnecting them).

**Table 2.** Total pressures at the top ($\sigma$) and permeabilities ($\kappa$) averaged over the last two weeks of measurement at a given hydraulic gradient ($i_H$) of 10X specimens compacted to dry densities of 1.4, 1.6, and 1.8 g/cm$^3$; saturated and permeated by SGW.

| $i_H$, m/m | $\sigma$, MPa | | | $\kappa$, $10^{-20}$ m$^2$ | | |
|---|---|---|---|---|---|---|
| | 1.4 g/cm$^3$ | 1.6 g/cm$^3$ | 1.8 g/cm$^3$ | 1.4 g/cm$^3$ | 1.6 g/cm$^3$ | 1.8 g/cm$^3$ |
| 180 | $0.7 \pm 0.2$ | $1.5 \pm 0.6$ | $5.3 \pm 0.9$ | - | - | - |
| 2000 | $0.8 \pm 0.3$ | $2.2 \pm 0.6$ | $6.3 \pm 0.3$ | $27 \pm 15$ | $3.4 \pm 0.8$ | $0.96 \pm 0.26$ |
| 5000 | $0.9 \pm 0.3$ | $2.3 \pm 0.6$ | $6.4 \pm 0.3$ | $25 \pm 14$ | $3.2 \pm 0.4$ | $0.54 \pm 0.10$ |
| 20,000 | $2.1 \pm 0.2$ | $3.1 \pm 0.6$ | $6.9 \pm 0.3$ | $11 \pm 4$ | $2.9 \pm 0.4$ | $0.68 \pm 0.02$ |
| 80,000 | $7.6 \pm 0.1$ | $8.3 \pm 0.8$ | $9.7 \pm 0.3$ | $5.4 \pm 1.5$ | $2.0 \pm 0.2$ | $0.67 \pm 0.07$ |

Based on these values, the fluid pressure at cell tops can be estimated to equal 1.3, 0.9, and 0.6 MPa for the fluid pressure of 2.0 MPa at cell base (hydraulic gradient of 20,000 m/m) and 6.8, 6.1, and 3.4 MPa for the fluid pressure of 8.0 MPa at cell base (hydraulic gradient of 80,000 m/m) at dry densities of 1.4, 1.6, and 1.8 g/cm$^3$, respectively. The decrease of fluid pressure from cell bases to cell tops is in line with previous observations of the non-linear variation of the excess fluid pressure across specimens in constant-head experiments with compacted clays [28]. A decrease in these values with increasing density further indicates that the degree of non-linearity in this variation increases with the decreasing porosity of the specimens (0.50, 0.43, and 0.36, respectively).

In the preliminary 7-day experiments [4] with 10X and distilled water using permeameters similar to those of the present study, for dry densities in the range of 1.44–1.82 g/cm$^3$, swelling pressures in the range of 0.52–9.39 MPa and a relationship of the form $P = \exp(6.4\rho_d - 9.403)$ between swelling pressure $P$ (MPa) and dry density $\rho_d$ (g/cm$^3$) were obtained. At the hydraulic gradient of 2000 m/m in the present study, a somewhat different relationship $P = \exp(5.16\rho_d - 7.445)$ was obtained (Figure 6). The difference might be due to the use of only three data points for obtaining the relationship in the present study, or the difference between SGW and distilled water, or for other reasons, which cannot be identified in view of the limited information on parameters of the experiments and statistical weight of fitted data in the study [4]. In the study by Wang et al. [29], relationships of the form $P = \exp(6.75\rho_d - 8.634)$, $P = \exp(6.85\rho_d - 9.231)$, and $P = \exp(3.32\rho_d - 5.608)$ were provided for MX-80 (75–90% smectite), FEBEX (92% smectite), and Kunigel V1 (48% smectite) bentonites, respectively. Based on the smectite content of 73 mass% of 10X [4], it may be speculated that the value of the coefficient preceding the density ought to be in a range from 3.3 to 6.8. Although the values obtained in [4] and in the present study agree with this range, establishing values that are characteristic of 10X would require additional experimental data.

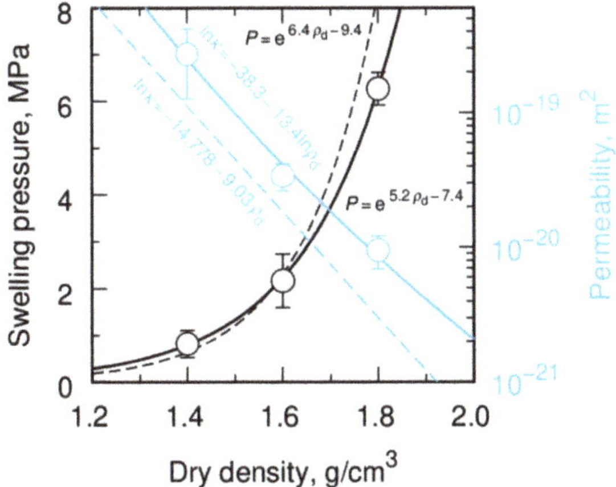


**Figure 6.** Swelling pressure and permeability (m$^2$) of 10X specimens permeated by SGW as a function of dry density $\rho_d$. Circles and solid lines represent measured values and fitted curves of the present study, whereas dashed lines represent fitted curves of the study [4] (hydraulic conductivities reported in [4] were converted to permeabilities for the purpose of comparison).

Permeabilities measured for 10X in the present study show a gradual decrease at all densities during the initial experimental stage at the hydraulic gradient of 2000 m/m (Figure 7). The gradual decrease of permeabilities suggests that the residual air, if any, was largely displaced from the specimens at the very beginning of this stage since air removal would otherwise lead to a non-transient increase of permeability [27], which was not observed later on. At densities of 1.4 and 1.6 g/cm$^3$, a double linear decrease was observed with a steeper initial decrease during the first 10 days at 1.4 g/cm$^3$ and 40 days at 1.6 g/cm$^3$ followed by a flatter one. Shortly after further increasing the hydraulic gradients at these densities, considerably higher transient permeabilities were measured (data points on the vertical dotted lines in Figure 7, which should not be treated as outliers but rather attributed to the additional amount of water squeezed out of specimens by seepage-induced (local) consolidation under the pressure gradient [28,30]. According to [30], the initial transient response preceding the steady-state flow condition can take up to a few hours for clay specimens due to time-dependent changes in the volume or distribution of pore space in a specimen.

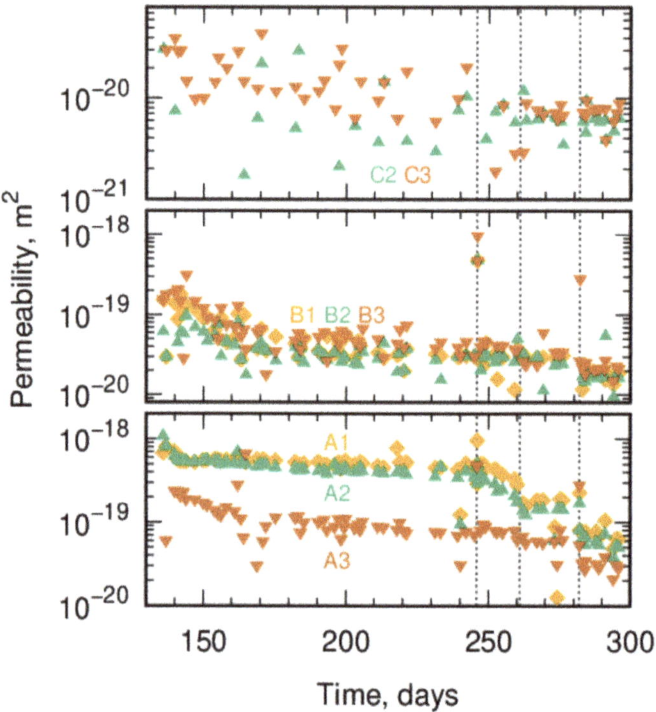



**Figure 7.** Permeability ($m^2$) of 10X specimens compacted to dry densities of 1.4 g/cm$^3$ (A1–A3); 1.6 g/cm$^3$ (B1–B3); 1.8 g/cm$^3$ (C2, C3), and permeated by SGW. Vertical dotted lines mark increases of hydraulic gradient to 5000 m/m (246th day); 20,000 m/m (261st day), and 80,000 m/m (281st day).

As a result of hydraulic gradient increases above 2000 m/m, steady-state permeability showed two major decreases by a factor of ~2–2.5 at a density of 1.4 g/cm$^3$ and one major decrease by a factor of ~1.5 at 1.6 g/cm$^3$, whereas it stabilized at ~6 × 10$^{-21}$ m$^2$ at 1.8 g/cm$^3$ (Figure 7, Table 2). The seepage-induced consolidation, which reduces the specimen's pore volume, was proposed as a reason for permeability decrease with increasing pressure gradient [28,30]. This reason cannot be discarded for the present experiment, as the observed dependence of the extent of permeability decrease on the density would be in line with a larger seepage-induced consolidation for less dense specimens. If this proposal is valid, a larger hydraulic gradient than those applied in the present study can be expected to cause a major decrease of steady-state permeability of 10X specimen at 1.8 g/cm$^3$ as well.

Alternatively, the hydraulic conductivity of expansive materials was observed to decrease after full saturation due to reorganisation of microstructure in the course of a water potential re-equilibration and corresponding water redistribution between the inter-aggregate and intra-aggregate pores [31]. Accordingly, it can be assumed that such microstructural reorganisation may have occurred to a larger extent for 10X specimen at a density of 1.4 g/cm$^3$ and to a lesser extent at 1.6 g/cm$^3$, whereas it did not take place at 1.8 g/cm$^3$, at least for hydraulic gradients ≥5000 m/m. This assumption would be in line with the decreasing pore volume available for a microstructural reorganisation with increasing density.

A change in the pore space during the SGW saturation was detected by scanning electron microscopy. The microstructure of the original compacted 10X is characterized by the predominance of fine interparticle micropores and large inter-micro-aggregate pores. During saturation with the SGW, the number and size of pores are decreased (Figure 8), inter-micro-aggregate and small inter-micro-aggregate micropores predominate, and an increase in the size and isometricity of large inter-micro-aggregate pores is also noted. There is also a decrease in the degree of orientation of structural elements (clay aggregates of different sizes) in the plane orthogonal to the pressing direction, acquired during the process of compacting the specimen, in which the microstructure becomes more uniform.

(a)    (b)

**Figure 8.** SEM images of the compacted 10X, fraction <0.25 mm, density 1.65 g/cm$^3$, at 32,000×: (**a**) natural micro-aggregates of clay particles after an air-drying; (**b**) partially swelled micro-aggregate of clay particles. Sidebars on the top corners: structural model of clay swelling with different moisture (W) content: (**a**) W ≤ Wmg; (**b**) Wp < W < Wsw (Wmg—maximal hygroscopic moisture content; Wp—moisture of the plastic limit; Wsw—moisture of swelling). 1—clay particle; 2—adsorbed water; 3—osmotic water.

This uniformity in microstructure is explained by the processes of hydration of highly hydrophilic clay particles upon saturation of the specimens and accompanying swelling of microaggregates, closure of fine pores, and opening of inter-ultra-micro-aggregate (inter-micro-aggregate) micropores. As a result, the microstructure of the specimens after permeation becomes more homogeneous, intra- and inter-micro-aggregate pores become predominant since they open significantly during the swelling stage but are not affected by the fluid pressure during the permeation stage. This effect can also be associated with the swelling of the specimens during saturation and with the so-called "hinge-joint effect" of clay particles in a micro-aggregate during hydration [32].

The swelling of clays is often characterized by the effect of disordering the microstructure. According to the results of experiments and calculations given in the article [33], this effect is associated with the distribution of forces acting on particles and microaggregates, the result of which leads to the appearance of the hinge-joint effect. Thus, as a result of the wedging pressure of the hydration shells around the particles, their mutual repulsion occurs, while the forces of attraction at their edges prevent complete separation, resulting in the formation of a honeycomb structure with higher porosity and a lower degree of microstructure uniformity (Figure 8, sidebars).

A displacement of particles, which are not directly involved in the load-carrying of specimen's fabric downward the flow direction, resulting in clogging of specimen's pores was proposed as a further possible reason for seepage-induced changes in the permeability [30]. In the study by Al-Taie et al. [34], clogging of sintered filters by this mechanism was proposed as an alternative explanation for permeability decreases observed upon increases of hydraulic gradients above 1000 m/m in experiments with a smectite-rich clay and distilled water. However, in the infiltration tests of hydraulic gradients up to ~64,000 m/m [29], no density changes along the profile of a bentonite specimen, which would provide evidence of such displacement, were found. Furthermore, the initial permeability of compacted 10X was ~$10^{-18}$ m$^2$ at 1.4 g/cm$^3$ (Figure 7), whereas that of sintered filters was $(3.4 \pm 0.5) \times 10^{-14}$ m$^2$, so that clogging of pore space would be more probable than that of sintered filters at 1.4 g/cm$^3$ and even more so at 1.8 g/cm$^3$. Since, no major decreases of steady-state permeability occurred at this density, clogging of pore space or sintered filters due to displacement of particles within the studied specimens was not a factor contributing significantly to the observed permeability decreases. Indeed, the permeability of uncleaned sintered filters after the experiment was measured to equal $(1.9 \pm 1.2) \times 10^{-14}$, $(1.5 \pm 0.6) \times 10^{-14}$, and $(1.2 \pm 0.2) \times 10^{-14}$ m$^2$ at a dry density of 1.4, 1.6, and 1.8 g/cm$^3$, respectively, or a factor of 2 to 3 lower than that before the experiment. Thus, though some clogging of sintered filters occurred during the experiment, its contribution to the observed permeability decreases can be neglected.

In view of remaining uncertainty about the mechanism behind the permeability decreases observed at higher hydraulic gradients, permeability values of $(27 \pm 15) \times 10^{-20}$, $(3.4 \pm 0.8) \times 10^{-20}$, and $(0.96 \pm 0.26) \times 10^{-20}$ m$^2$ (Table 2) obtained at a hydraulic gradient of 2000 m/m were suggested as conservative estimates of permeabilities of 10X for SGW at a temperature of $18.1 \pm 0.4$ °C and dry densities of 1.4, 1.6, and 1.8 g/cm$^3$, respectively. When assessing the performance of the bentonite in the early post-closure phase of DGR at hydraulic gradients higher than 2000 m/m, corresponding permeability estimates given in Table 2 may be used.

In the experiments with seawater, permeability $\kappa$ (expressed in millidarcies, $9.869233 \times 10^{-19}$ m$^2$) of pure smectite and kaolinite was found to be related to bulk density $\rho_{bulk}$ (g/cm$^3$) by respective equations $\ln\rho_{bulk} = -0.037\ln\kappa + 0.27$ and $\ln\rho_{bulk} = -0.074\ln\kappa + 0.40$ [35]. Notably, as can be seen from the equations derived in the present study (Figure 9), the coefficient preceding $\ln \kappa$ at hydraulic gradients of 2000 and 5000 m/m ($0.040 \pm 0.005$ and $0.035 \pm 0.001$, respectively) is close to the value obtained in study [35] for pure smectite, whereas at 80,000 m/m it is close to that for a mixture of 20% smectite and 80% kaolinite.

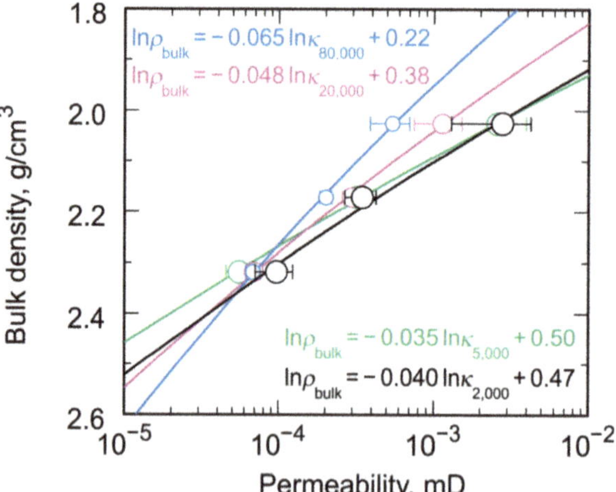


**Figure 9.** Permeability (mD) of 10X permeated by SGW at hydraulic gradients of 2000–80,000 m/m as a function of bulk density. Circles and solid lines represent measured values and fitted curves.

The equation $\ln \rho_{bulk} = -0.040\ln\kappa_{(mD)} + 0.47$ (Figure 9), obtained in the present study at the hydraulic gradient of 2000 m/m, corresponds to $\ln \kappa_{(m^2)} = -13.4\ln\rho_{dry} - 38.3$ (Figure 6), with $\kappa$ being a power function of $\rho_d$. In this form it can be compared with the relationship $\ln \kappa_{(m^2)} = -9.03\rho_d - 14.778$ obtained in the experiments with distilled water [4] where $\kappa$ is an exponential function of $\rho_d$. In the semi-logarithmic plot (Figure 6), the latter relationship has the linear form whereas that used in the present study and the study by Mondol et al. [35] does not. In view of this difference, it appears that establishing the form of the relationship between permeability and density that is characteristic of 10X requires additional experimental data.

The comparison of the fitted curves in Figure 6 further reveals that permeabilities obtained in [4] at hydraulic gradients of 22,000 up to 92,000 m/m are about 2.5 times smaller than those in the present study obtained at a hydraulic gradient of 2000 m/m. Upon increasing the hydraulic gradient to 20,000 m/m in the present study, the permeability decreased by a factor of 2.5 at 1.4 g/cm$^3$ and of 1.5 at 1.8 g/cm$^3$ (Table 2), whereas at 1.6 g/cm$^3$ it decreased by a factor of 1.7 only upon an increase of hydraulic gradient to 80,000 m/m. Therefore, the difference between the permeabilities measured in the present study and in [4] may be due to hydraulic gradients higher than 2000 m/m in the latter study. A contribution of other reasons to this difference cannot be excluded and needs to be verified in additional experiments.

## 4. Conclusions

Differences in the swelling pressures and permeabilities observed in the present study for 10X bentonite at given dry densities suggests that because of inherent inhomogeneity in the microstructure of compacted bentonites, at least duplicate, preferably triplicate, experiments should be conducted when using the present method to estimate swelling pressures or permeabilities. In this regard, a saturation of compacted bentonite specimens and measurements of their swelling pressure and permeability are recommended to be conducted at a hydraulic gradient of 2000 m/m. As higher hydraulic gradients cause modifications of compacted bentonite microstructure, which are still not well understood, the estimates obtained with those should be used with care, and their applicability may be restricted to the early post-closure phase of DGR.

To obtain a reliable estimate of the dependence of swelling pressure and permeability of compacted bentonite on its density, a series of triplicate experiments for at least 7–8 densities, preferably with an even distribution over a broader density range than that of 1.4–1.8 g/cm$^3$ studied in work [4] and the present study, is needed. The dataset for 10X bentonite should be accordingly extended.

**Author Contributions:** Conceptualization, data curation, investigation, methodology, visualization, supervision, and writing—review and editing, A.Y.M.; investigation, methodology, visualization, S.V.Z. and V.V.K. All authors have read and agreed to the published version of the manuscript.

**Funding:** The study was funded by the Federal Ministry of Economic Affairs and Energy (BMWi), represented by the Project Management Agency Karlsruhe (PTKA-WTE) within the frames of the project, "Sicherheitsanalytische Untersuchungen zu Endlagersystemen im Kristallin" (SUSE) with contract no. 02E11577B. Initial data, as well as some additional properties' tests of bentonites and microstructure of compacted specimens, were supported by the Russian Science Foundation (project 16-17-10270).

**Data Availability Statement:** Not Applicable.

**Acknowledgments:** Swelling pressure and permeability results were obtained owing to the skillful lab work of Julia Gose of the geoscientific laboratory of the GRS. The authors thank Michail Chernov and Ruslan Kuznetsov (MSU, Geological Faculty) for providing data by the SEM and μCT. Experimental studies were partially performed using equipment acquired with the funding of the Moscow State University Development Program (X-ray Diffractometer Ultima-IV, Rigaku, and Scanning Electron Microscope LEO 1450VP, Carl Zeiss, X-Ray tomograph Yamato TDM1000, Yamato Scientific). We would like to thank two anonymous reviewers and Marvin Middelhoff (GRS) for constructive comments that helped to improve the original manuscript.

**Conflicts of Interest:** The authors declare no conflict of interest.

## References


1. Abramov, A.A.; Bolshov, L.A.; Dorofeeev, A.N.; Igin, I.M.; Kazakov, K.S.; Krasilnikov, V.Y.; Linge, I.I.; Trokhov, N.N.; Utkin, S.S. Underground research laboratory in the Nizhnekanskiy massif: Evolutionary design study. *Radioact. Waste* **2020**, *1*, 9–21. (In Russian) [CrossRef]
2. Dorofeev, A.N.; Bolshov, L.A.; Linge, I.I.; Utkin, S.S.; Saveleva, E.A. Strategic master-plan for research demonstrating the safety of construction, operation and closure of a deep geological repository for radioactive waste. *Radioact. Waste* **2017**, *1*, 19–26.
3. Krupskaya, V.V.; Biryukov, D.V.; Belousov, P.E.; Lekhov, V.A.; Romanchuk, A.Y.; Kalmykov, S.N. The use of natural clay materials to increase the nuclear and radiation safety level of nuclear legacy facilities. *Radioact. Waste* **2018**, *2*, 30–43.
4. Krupskaya, V.V.; Zakusin, S.V.; Lekhov, V.A.; Dorzhieva, O.V.; Belousov, P.E.; Tyupina, E.A. Buffer properties of bentonite barrier systems for radioactive waste isolation in geological repository in the Nizhnekanskiy Massif. *Radioact. Waste* **2020**, *1*, 35–55. [CrossRef]
5. Birgersson, M.; Hedström, M.; Karnland, O.; Sjöland, A. Bentonite buffer: Macroscopic performance from nanoscale properties. In *Geological Repository Systems for Safe Disposal of Spent Nuclear Fuels and Radioactive Waste*, 2nd ed.; Michael, J.A., Joonhong, A., Eds.; Woodhead Publishing: Duxford, UK, 2017; Volume 12, pp. 319–364. [CrossRef]
6. Kaufhold, S.; Dohrmann, R. Distinguishing between more and less suitable bentonites for storage of high-level radioactive waste. *Clay Miner.* **2016**, *51*, 289–302. [CrossRef]
7. Sellin, P.; Leupin, O.X. The use of clay as an engineered barrier in radioactive-waste management—A review. *Clays Clay Miner.* **2013**, *61*, 477–498. [CrossRef]
8. Belousov, P.E.; Krupskaya, V.V. Bentonite clays of Russia and neighboring countries. *Georesursy* **2019**, *21*, 79–90. [CrossRef]
9. Belousov, P.E.; Krupskaya, V.V.; Zakusin, S.V.; Zhigarev, V.V. Bentonite clays from 10th Khutor deposite: Features of genesis, composition and adsorption properties. *RUDN J. Eng. Res.* **2017**, *18*, 135–143. [CrossRef]
10. Belousov, P.; Chupalenkov, N.; Zakusin, S.; Morozov, I.; Dorzhieva, O.; Chernov, M.; Krupskaya, V. 10th Khutor bentonite deposit (Russia): Geology, mineralogy and industrial application. *Appl. Clay Sci.* **2021**, in press.
11. Belousov, P.E.; Pokidko, B.V.; Zakusin, S.V.; Krupskaya, V.V. Quantitative methods for quantification of montmorillonite content in bentonite clays. *Georesursy* **2020**, *22*, 38–47. [CrossRef]
12. Post, J.E.; Bish, D.L. Rietveld refinement of crystal structures using powder X-ray diffraction data. *Rev. Miner. Geochem.* **1989**, *20*, 277–308.
13. Doebelin, N.; Kleeberg, R. Profex: A graphical user interface for the Rietveld refinement program BGMN. *J. Appl. Crystallogr.* **2015**, *48*, 1573–1580. [CrossRef] [PubMed]
14. GOST 5180-2015 Soils. *Laboratory Methods for Determination of Physical Characteristics*; Standartinform: Moscow, Russia, 2019.
15. ASTM D854-14. *Standard Test Methods for Specific Gravity of Soil Solids by Water Pycnometer*; ASTM International: West Conshohocken, PA, USA, 2014.
16. Lorenz, P.; Meier, L.; Kahr, G. Determination of the cation exchange capacity (CEC) of clays minerals using the complexes of copper (II) ion with triethylenetetramine and tetraethylenepentamine. *Clays Clay Miner.* **1999**, *47*, 386–388.
17. Dohrmann, R.; Genske, D.; Karnland, O.; Kaufhold, S.; Kiviranta, L.; Olsson, S.; Plötze, M.; Sandén, T.; Sellin, P.; Svensson, D.; et al. Interlaboratory CEC and exchangeable cation study of bentonite buffer materials: I. Cu(II)-triethylenetetramine method. *Clays Clay Miner.* **2012**, *60*, 162–175. [CrossRef]
18. Stanjek, H.; Künkel, D. CEC determination with Cu-triethylenetetramine: Recommendations for improving reproducibility and accuracy. *Clay Miner.* **2016**, *51*, 1–17. [CrossRef]
19. Rozov, K.B.; Rumynin, V.G.; Nikulenkov, A.M.; Leskova, P.G. Sorption of 137Cs, 90Sr, Se, 99Tc, 152(154)Eu, 239(240)Pu on fractured rocks of the Yeniseysky site (Nizhne-Kansky massif, Krasnoyarsk region, Russia). *J. Environ. Radioact.* **2018**, *192*, 513–523. [CrossRef]
20. Meleshyn, A.Y. Swelling behaviour and permeability of compacted bentonites at high salinity as influenced by fluid pressure changes. *Radioact. Waste* **2020**, *2*, 109–119, (In Russian; the English version is in press). [CrossRef]
21. Meleshyn, A. *Mechanisms of Transformation of Bentonite Barriers (UMB)*; Report GRS-390 BMWi-FKZ 02 E 11344A.; GRS: Braunschweig, Germany, 2019; Available online: https://www.grs.de/publikationen/grs-390 (accessed on 30 June 2021).
22. Yigzaw, Z.G.; Cuisinier, O.; Massat, L.; Masrouri, F. Role of different suction components on swelling behavior of compacted bentonites. *Appl. Clay Sci.* **2016**, *120*, 81–90. [CrossRef]
23. Massat, L.; Cuisinier, O.; Bihannic, I.; Claret, F.; Pelletier, M.; Masrouri, F.; Gaboreau, S. Swelling pressure development and inter-aggregate porosity evolution upon hydration of a compacted swelling clay. *Appl. Clay Sci.* **2016**, *124*, 197–210. [CrossRef]
24. Zhu, C.M.; Ye, W.M.; Chen, Y.G.; Chen, B.; Cui, Y.J. Influence of salt solutions on the swelling pressure and hydraulic conductivity of compacted GMZ01 bentonite. *Eng. Geol.* **2013**, *166*, 74–80. [CrossRef]
25. Ye, W.M.; Wan, M.; Chen, B.; Chen, Y.G.; Cui, Y.J.; Wang, J. Temperature effects on the swelling pressure and saturated hydraulic conductivity of the compacted GMZ01 bentonite. *Environ. Earth Sci.* **2013**, *68*, 281–288. [CrossRef]
26. Zhang, F.; Ye, W.M.; Wang, Q.; Chen, Y.G.; Chen, B. An insight into the swelling pressure of GMZ01 bentonite with consideration of salt solution effects. *Eng. Geol.* **2019**, *251*, 190–196. [CrossRef]
27. Chapuis, R.P. Sand–bentonite liners: Predicting permeability from laboratory tests. *Can. Geotech. J.* **1990**, *27*, 47–57. [CrossRef]
28. Pane, V.; Croce, P.; Znidarcic, D.; Ko, H.Y.; Olsen, H.W.; Schiffman, R.L. Effects of consolidation on permeability measurements for soft clay. *Geotechnique* **1983**, *33*, 67–72. [CrossRef]

29. Wang, Q.; Tang, A.M.; Cui, Y.J.; Delage, P.; Gatmiri, B. Experimental study on the swelling behaviour of bentonite/claystone mixture. *Eng. Geol.* **2012**, *124*, 59–66. [CrossRef]
30. Olsen, H.W.; Nichols, R.W.; Rice, T.L. Low gradient permeability measurements in a triaxial system. *Geotechnique* **1985**, *35*, 145–157. [CrossRef]
31. Wang, Q.; Cui, Y.J.; Tang, A.M.; Barnichon, J.D.; Saba, S.; Ye, W.M. Hydraulic conductivity and microstructure changes of compacted bentonite/sand mixture during hydration. *Eng. Geol.* **2013**, *164*, 67–76. [CrossRef]
32. Osipov, V.I.; Sokolov, V.N. Clays and Its Properties. In *Composition, Structure and Properties Formation*; GEOS: Moscow, Russia, 2013; p. 576. ISBN 978-5-89118-617-0.
33. Osipov, V.I.; Babak, V.G. The nature and mechanism of clay swelling. *Eng. Geol.* **1987**, *5*, 18–27.
34. Al-Taie, L.; Pusch, R.; Knutsson, S. Hydraulic properties of smectite rich clay controlled by hydraulic gradients and filter types. *Appl. Clay Sci.* **2014**, *87*, 73–80. [CrossRef]
35. Mondol, N.H.; Bjørlykke, K.; Jahren, J. Experimental compaction of clays: Relationship between permeability and petrophysical properties in mudstones. *Pet. Geosci.* **2008**, *14*, 319–337. [CrossRef]

*Article*

# In Situ Measurements of the Hydration Behavior of Compacted Milos (SD80) Bentonite by Wet-Cell X-ray Diffraction in an Opalinus Clay Pore Water and a Diluted Cap Rock Brine

Tobias Manzel [1,*], Carolin Podlech [1], Georg Grathoff [1], Stephan Kaufhold [2] and Laurence N. Warr [1]

[1]  Institute of Geography and Geology, University of Greifswald, 17489 Greifswald, Germany; carolin.podlech@uni-greifswald.de (C.P.); grathoff@uni-greifswald.de (G.G.); warr@uni-greifswald.de (L.N.W.)

[2]  Bundesanstalt für Geowissenschaften und Rohstoffe (BGR), 30655 Hannover, Germany; Stephan.kaufhold@bgr.de

*  Correspondence: tobias.manzel@stud.uni-greifswald.de

**Citation:** Manzel, T.; Podlech, C.; Grathoff, G.; Kaufhold, S.;Warr, L.N. In Situ Measurements of the Hydration Behavior of Compacted Milos (SD80) Bentonite by Wet-Cell X-ray Diffraction in an Opalinus Clay Pore Water and a Diluted Cap Rock Brine. *Minerals* **2021**, *11*, 1082. https://doi.org/10.3390/min11101082

Academic Editor: Keiko Sasaki

Received: 11 August 2021
Accepted: 26 September 2021
Published: 30 September 2021

**Publisher's Note:** MDPI stays neutral with regard to jurisdictional claims in published maps and institutional affiliations.



**Abstract:** Compacted bentonite is currently being considered as a suitable backfill material for sealing underground repositories for radioactive waste as part of a multi-barrier concept. Although showing favorable properties for this purpose (swelling capability, low permeability, and high adsorption capacity), the best choice of material remains unclear. The goal of this study was to examine and compare the hydration behavior of a Milos (Greek) Ca-bentonite sample (SD80) in two types of simulated ground water: (i) Opalinus clay pore water, and (ii) a diluted saline cap rock brine using a confined volume, flow-through reaction cell adapted for in situ monitoring by X-ray diffraction. Based on wet-cell X-ray diffractometry (XRD) and calculations with the software CALCMIX of the smectite d(001) reflection, it was possible to quantify the abundance of water layers (WL) in the interlayer spaces and the amount of non-interlayer water uptake during hydration using the two types of solutions. This was done by varying WL distributions to fit the CALCMIX-simulated XRD model to the observed data. Hydrating SD80 bentonite with Opalinus clay pore water resulted in the formation of a dominant mixture of 3- and 4-WLs. The preservation of ca. 10% 1-WLs and the apparent disappearance of 2-WLs in this hydrated sample are attributed to small quantities of interlayer K (ca. 8% of exchangeable cations). The SD80 bentonite of equivalent packing density that was hydrated in diluted cap rock brine also contained ca. 15% 1-WLs, associated with a slightly higher concentration of interlayer K. However, this sample showed notable suppression of WL thickness with 2- and 3-WLs dominating in the steady-state condition. This effect is to be expected for the higher salt content of the brine but the observed generation of $CO_2$ gas in this experiment, derived from enhanced dissolution of calcite, may have contributed to the suppression of WL thickness. Based on a comparison with all published wet-cell bentonite hydration experiments, the ratio of packing density to the total layer charge of smectite is suggested as a useful proxy for predicting the relative amounts of interlayer and non-interlayer water incorporated during hydration. Such information is important for assessing the subsequent rates of chemical transport through the bentonite barrier.

**Keywords:** bentonite; waste repositories; smectite; swelling; hydration; water content; Milos; interlayers

## 1. Introduction

Bentonites are currently of interest as backfill material for the underground sealing of nuclear waste repositories. They are particularly suitable for engineering a multi-barrier system in hard crystalline rocks such as granite, where a low-permeability buffer material is required to fill the gap between the host rock and the radioactive waste containers [1–3]. The primary function of the buffer is to prevent or significantly slow down the rate of radionuclide transport in the case of leakage from the high-level nuclear waste containers. Due to the extremely low permeability of bentonites and their high cation exchange capacity, these smectite-enriched materials are considered as suitable for retaining radioactive elements and thus preventing them from entering the host rock or biosphere [4,5].

The vital property of bentonite in forming a low permeability barrier is the ability of smectite to swell during hydration in aqueous solutions of varying electrolyte concentrations [6]. Hydrated bentonite can develop extremely low hydraulic conductivities in the order of $<10^{-11}$ m/s [7]. As bentonites are likely to be emplaced in the dry state either as compressed blocks or loser pellets (packing densities typically between 1.4 and 2.2 g/cm$^3$ [8]), it is important to study their hydration behavior once in contact with natural waters and to establish the mechanisms and rates of hydration during the early stages of saturation. Particularly relevant is to establish the mechanism by which water is stored in the bentonite and the way it affects the rate of subsequent chemical transport through the clay barrier.

There are three basic sites for non-crystalline water in hydrated bentonite [9]: (i) interlayer water adsorbed between two closely spaced negatively charged layers within smectite particles, (ii) adsorbed water on smectite particle surfaces and variably charged edge sites, and (iii) free pore water located in the spaces between grains. Bentonites dominated by interlayer water will be characterized by the lowest rates of chemical transport that reach the rates of diffusion [10,11] whereas bentonites with abundant free pore water will display higher rates of transport more characteristic of porous materials [12–14].

This study reports on the short-term hydration behavior of a Milos bentonite clay known as the SD80 sample, and evaluates its suitability when infiltrated by two types of solutions that simulate the natural groundwater of repository conditions in a mudrock formation and within a salt body. For this, the solutions used were an Opalinus clay pore water composition for the former, and a diluted cap rock brine for the latter. The hydration experiments were conducted as part of a long-term German research program known as the UMB (Umwandlungsmechanismen in Bentonitbarrieren) to aid the selection of the most suitable bentonite material in terms of their physical properties and mineral stability based on laboratory experimentation [12].

For the in situ experimental study of the hydration behavior of the SD80 bentonite, wet-cell diffractometry was used [9,15]. Compared to previously studied bentonites using this technique, the hydration patterns of the Milos bentonite shows some unusual features that may be related to some interlayer K as well as the generation of $CO_2$ gas from the dissolution of calcite. An assessment of related studies suggests the ratio of packing density to total layer charge provides a useful measure for predicting the hydration behavior of Ca- and Na-bentonites in advance of conducting a more detailed experimentation study.

## 2. Materials and Methods

For the experimental investigation, industrial bentonite from Milos, Greece (SD80) was used, for which the properties are well characterized (Table 1). Previous quantifications by X-ray diffraction (XRD) Rietveld analyses show the raw powder material contains 89% smectite and 11% accessory minerals such as feldspar and traces of quartz, calcite, pyrite, and baryte (Table 1). Assuming a pure dioctahedral nature, compositional analyses of the purified smectite fraction by energy dispersive X-ray analyses (EDX) produced a total interlayer charge of $-0.36$ e/phuc with $-0.06$ e/phuc distributed in the tetrahedral sheet and $-0.30$ e/phuc in the octahedral sheet [16]. The mineral formula calculation also indicates an interlayer content of 0.01 K, 0.03 Na, 0.09 Ca, and 0.06 Mg per half-unit cell (phuc), confirming Ca as the dominant exchangeable cation. The CEC of the bulk powder measured using the Cu-trien method [17,18] is 87 cmol/kg.

**Table 1.** Mineral assemblages and properties of SD80 bentonite compared to previously studied bentonite clays investigated by wet-cell X-ray diffractometry. Ant: anatase, Brt: baryte, Cal: calcite, Fsp: feldspar, Sme: smectite, Py: pyrite, Qz: quartz, n.d.: not determined, $\xi$: total layer charge, CEC: cation exchange capacity. IMA-CNMNC approved mineral symbols, after Warr [19].

| Sample | Sme | Fsp | Mca | Qz | Ant | Cal | Py | Brt | $\xi$ | CEC |
|---|---|---|---|---|---|---|---|---|---|---|
| | wt. % | wt. % | wt. % | wt. % | wt. % | wt. % | wt. % | wt. % | e/phuc | cmol/kg |
| SD80 (Ca-bentonite) [18] | 89 | 7 | n.d. | <1 | <1 | <1 | <1 | <1 | −0.36 | 87 |
| MX-80 (Na-bentonite) [20] | 76 | 5 | 3–4 | 5–6 | n.d. | <2 | <1 | n.d. | −0.28 | 70 |
| Ibeco-seal-80 (Na-bentonite) [21] | >80 | <3 | <3 | n.d. | n.d. | 8-12 | n.d. | n.d. | −0.33 | 82 |
| Tixoton-TE (Ca-bentonite) [21] | >80 | 5–6 | <2 | 8–9 | n.d. | <1 | n.d. | n.d. | −0.29 | 71 |

To evaluate the hydration behavior of the bentonite, two synthetic solutions were used as infiltrating fluids to simulate: (i) an Opalinus clay pore water (OPA), and (ii) a diluted cap rock brine (CAP) (Table 2). The Opalinus clay pore solution had a total salt concentration of 0.27 mol/L and a Na:Ca ratio of 8.1, whereas the diluted saline cap rock brine had a total salt concentration of 2.57 mol/L and a Na:Ca ratio of 78. The salinity of the two solutions was 19 g/L and 155 g/L, respectively, with starting pH values of 7.8 and 7.3 [12].

**Table 2.** Chemical composition of the Opalinus clay pore water (OPA) and the diluted cap rock brine (CAP) [12]. TDS = total dissolved solids.

| Dissolved Solids | Opalinus Clay Pore Water (OPA) [g/L] | Diluted Cap Rock Brine (CAP) [g/L] |
|---|---|---|
| NaCl | 12.39 | 145.87 |
| $CaCl_2$ | 2.89 | 3.55 |
| $Na_2SO_4$ | 1.99 | 5.40 |
| KCl | 1.27 | 0.37 |
| $MgCl_2$ | 1.62 | |
| $SrCl_2$ | 0.08 | |
| $NaHCO_3$ | 0.04 | |
| TDS | 20.23 | 155.19 |

Wet cells were used as confined volume reaction chambers to study the in situ hydration behavior of the SD80 bentonite powder [15]. A detailed description of the device and how it can be used to quantify the amount of interlayer and non-interlayer water uptake is given in previous publications [9,22]. An overview of the experimental set up is given in Figure 1. Before experimentation, the samples were equilibrated at laboratory conditions at 25 °C and ca. 50% relative humidity. In the SD80 experiments, the powder was introduced incrementally into the wet cell chamber and compacted using a metal cylinder of the same diameter. This way, relatively higher bulk densities of 1.48 and 1.50 g/cm$^3$ could be achieved. A thin X-ray transparent Kapton (polyimide) foil was used to cover the sample and reduce fluid evaporation (Figure 1a). This was fixed into place using a Teflon O-ring. Between XRD measurements, the wet cell was sealed with a Teflon lid that was held by a metal plate and screws. This ensured that a constant volume was maintained during hydration. For solution flow, two Teflon bottles were connected to each end of the wet cell using two-sided threads (Figure 1b). The upper bottle contained about 50 mL of the migrating solution and the bottom bottle was used to catch any percolating fluid. In the case of the SD80 bentonite experiments, no water accumulated in the lower bottle. The amount of inflow solution was determined by regular weighing of the well-cell holder.

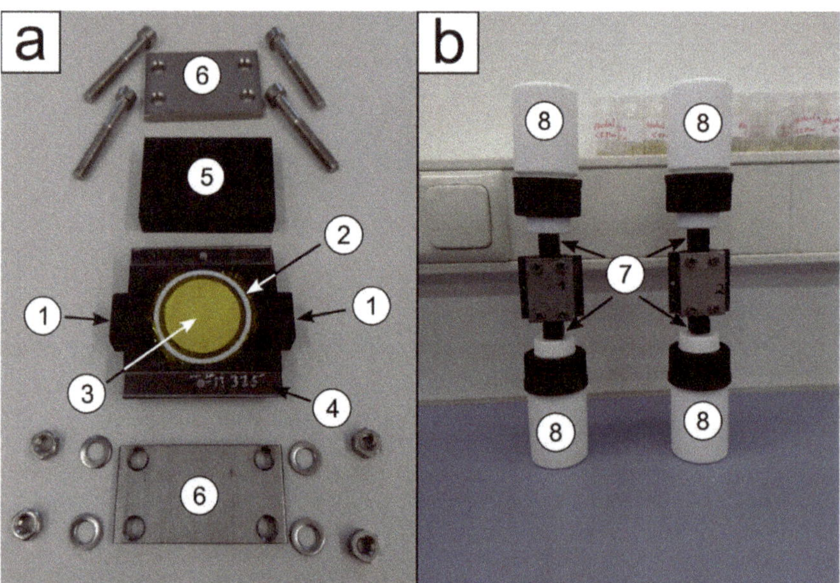

**Figure 1.** (**a**) Overview of the components of the wet cell. (**b**) Assembled wet cell assemblage. (1) Threaded inlets on each side of the cell. (2) Teflon O-ring seal. (3) Kapton foil (7.5 μm thick) used for sealing the sample chamber containing the compacted SD80 bentonite powder. (4) Modification bracket to mount the wet cell on the Bruker D8 Advance stage. (5) Teflon cover. (6) Metal plates and screws used to seal the cell with the Teflon cover to maintain a confined volume system. (7) Adaptors used to connect the wet cell and the Teflon bottles. (8) Teflon bottles to supply and capture the percolating solution.

For the XRD analysis, a Bruker D8 Advance diffractometer (D8 ADVANCE, Bruker, Billerica, MA, USA) with Fe-filtered CoKα radiation (40 kV, 30 mA) was used. The diffractometer was equipped with a 1D Lynx Eye Detector and a 0.5° divergence slit. The scans were collected from 3° to 50° 2θ with a scanning rate of 2° 2θ/min. The software CALCMIX (created by A. Plançon and V.A. Drits) [23], was applied to quantify the number of water layers (WLs) in the interlayer. For this, XRD patterns were matched using combinations of any three distinct WL configurations for smectite. The CALCMIX-calculated 001-reflection was adjusted to best fit the measured reflection by varying the percentage abundance of the specific WL configurations. It was assumed that the Reichweite $R$ is ordered and was set to $R = 1$ accordingly, and the particle thicknesses (the XRD-scattering domain sizes) had log-normal distributions with a mean around $N = 9$ as the average number of lattice layers. These assumptions represent a simplification of the variables used in previous studies [9,22], but comparisons of the sample XRD patterns used for the different studies produced consistent results within the errors of the method applied.

Based on the quantified percentages of the WLs and knowledge of the total amount of water inflow, it was possible to differentiate between interlayer water and non-interlayer water. Non-interlayer water is defined as the sum of adsorbed water and pore water not incorporated into interlayer sites. The interlayer water content was calculated as follows:

$$\sum_{n=1}^{4} n \cdot G_{wl} \cdot WL_n \tag{1}$$

where $n$ is equivalent to the number of water layers, $G_{wl} = 0.09$ cm$^3$/g is the water content per layer [22,24–26], and $WL_n$ is the number of water layers derived from CALCMIX. The difference between the total intake monitored by weight change and the interlayer water content describes the amount of non-interlayer water content. As no surface area data was available for the SD80 bentonite, no attempt was made in this study to differentiate between the amounts of surface and pore water.

## 3. Results

While no visual changes were observed in the bentonite powders during the first 48 h, the monitoring of the weight change indicated some water uptake of around 0.05 mL/g (Figure 2).

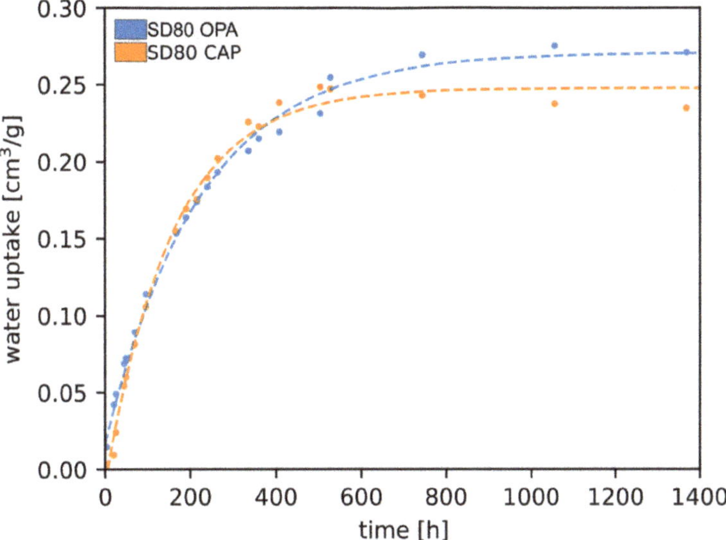


**Figure 2.** Comparison of the inflow of solution into the SD80 bentonite with Opalinus clay pore water (OPA), and diluted cap rock brine (CAP). Volumes were calculated from weight measurements after correction for salt content.

During the first 200 h of experimentation, both cells showed a similar rate of water intake with an average of $0.85 \times 10^{-3}$ mL/g·h. After 200 h, differences in the water uptake appeared, whereby SD80 with Opalinus clay pore water continued hydrating at a slow rate up to 1100 h before reaching its steady state. This was in contrast to the bentonite in diluted cap rock brine, which reached its maximum water uptake after 600 h. The slight decrease in water in the steady-state can be attributed to minor evaporation loss through the Kapton foil, which occurred at a rate that was faster than the inflow of water. At the end of the experimental run, when both SD80 samples of similar packing density reached a steady state of hydration, the sample reacted with Opalinus clay solution had taken in ca. 10% more water than the sample reacted with diluted cap rock brine (0.270 mL/g and 0.248 mL/g, respectively).

The first changes in sample appearance occurred after 167 h, when a radial darkening of samples occurred during water intake. This contrasts with the fast hydration front that crossed the samples observed by Warr and Berger [9] and indicates the wetting of the sample was slower and largely from below (Figure 3). That the Opalinus clay pore water sample showed clear darkening at the inlet indicated that a full wet state was not attained. In contrast, the diluted cap rock brine sample was fully darkened at the end of the experimental period, indicating a full wet state had been reached. Also, small gas bubbles could be observed in the diluted cap rock sample trapped between the clay and Kapton foil. This feature was not observed in the SD80 cell infiltrated by Opalinus clay pore water.

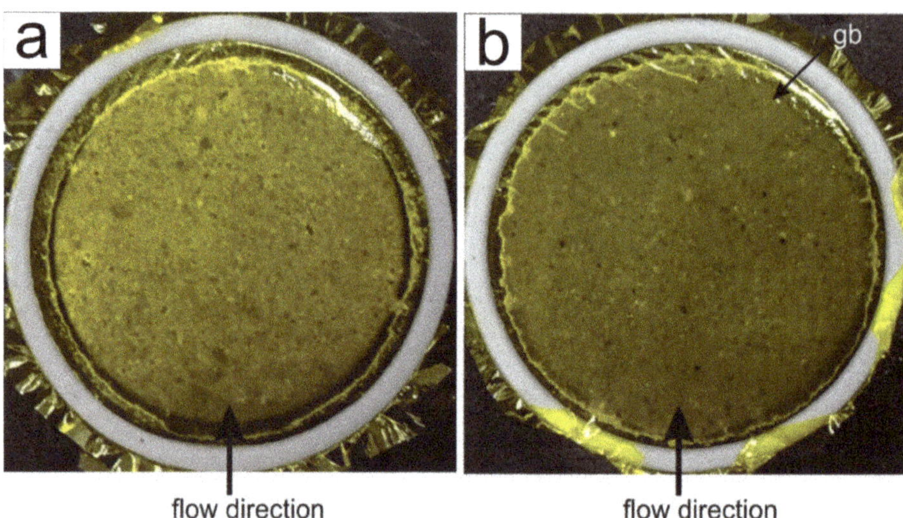


**Figure 3.** Photographs of the well cell experiments of hydrated SD80 bentonite close to the steady-state condition (**a**) infiltrated by Opalinus clay pore water and (**b**) infiltrated by diluted cap rock brine. gb = gas bubbles.

The measurement of the initial basal spacing of the dry SD80 powder determined by XRD before hydration was around 15 Å, as expected for a Ca-montmorillonite [27]. That indicates the dominance of the 2-WL structure in the interlayer under laboratory humidity conditions (ca. 50%). Modelling of the 001 basal planes using CALCMIX (Figure 4) indicated the 2-WL and 1-WL mixtures dominated the starting mixtures (Figure 5). The slight variation in abundance between the two samples probably reflected small differences in the air humidity at the time of measurement, whereby the wet cell that was infiltrated with diluted cap rock brine contained relatively less 2-WL and more 1-WL compared to the Opalinus clay solution-treated bentonite.

During the hydration of sample SD80 with the Opalinus clay solution, a rapid shift of the initial reflection towards higher basal spacing between 18 and 18.5 Å was observed after 69 h (Figure 4a). During the first 400 h of dehydration, this sample was characterized by the increase in 3-WLs at the expense of the less abundant 2-WL and 1-WL structures. After 450 h, when all 2-WLs had disappeared, some 3-WL structures grew to form 4-WL structures that varied in abundance between 18 and 25%. After 200 h, ca. 10% of 1-WLs remained intact during the remaining course of the experimentation and coexisted in a steady state with ca. 65% 3-WL and ca. 25% 2-WL.

A different pattern of interlayer hydration was observed in the diluted cap rock brine experiment (Figure 5b). In contrast to the first experiment, the basal spacing showed only a minor change towards large values with a shift of the initial reflection to a basal spacing of up to 15.2 Å (Figure 4b). CALCMIX calculations indicated the abundance of 2-WLs increased over the first 100 h and then stabilized at ca. 55%. This occurred at the expense of the 0 and 1 WLs, whereby the 0 WL structure only existed for the first hours of experimentation and the 1-WL structure decreased to a steady state of ca. 20%. The hydrated interlayer structure reached the steady-state condition after around 300 h with a mixture of ca. 55% 2-WL, ca. 25% 3-WL, and around 20% 1-WL. No WL structures >3 were detected.

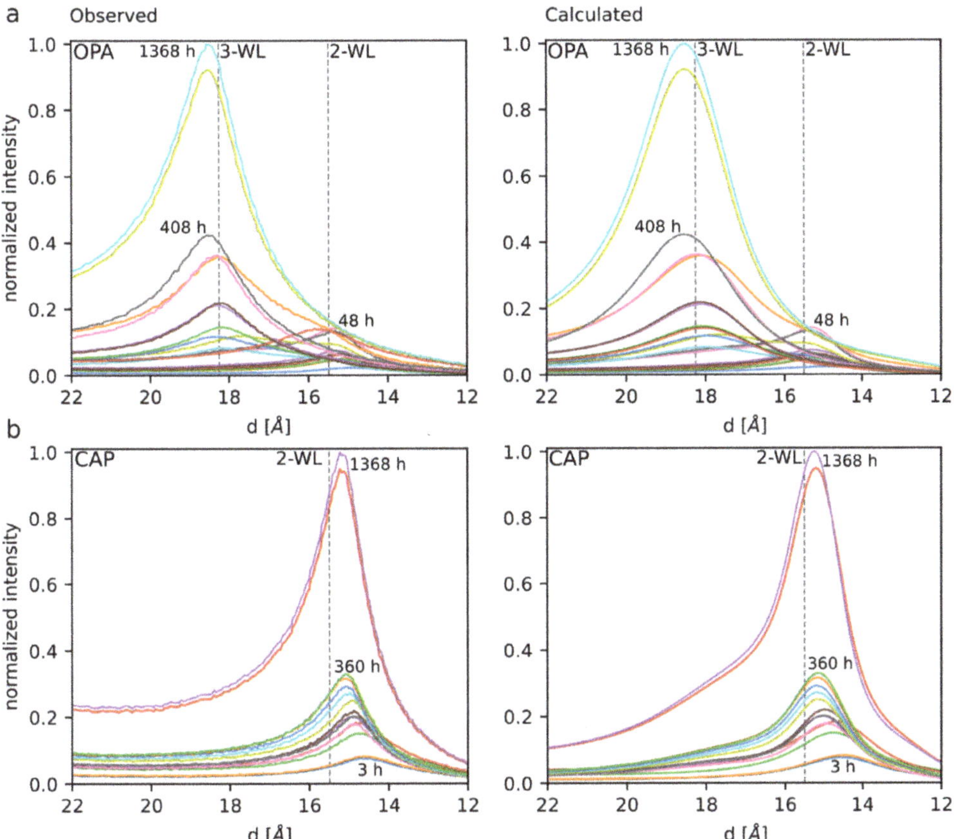



**Figure 4.** Comparison between observed and calculated XRD patterns for the varying hydration states of Milos bentonite. (**a**) SD80 infiltrated by Opalinus clay pore water, and (**b**) SD80 infiltrated by diluted cap rock brine.

Calculations of the amount of interlayer and non-interlayer water using Equation (1) showed major differences between the SD80 bentonite hydrated in the two types of solutions used. During the first 200 h when infiltrated by Opalinus clay pore water, the smectite incorporated a roughly equal amount of interlayer and non-interlayer water. After 200 h, more non-interlayer water instead of interlayer water entered the cell, resulting in a 57% and 43% mixture in a steady state. Overall, ca. 0.27 mL/g entered the SD80 bentonite by the end of the experiment.

The SD80 bentonite infiltrated with the diluted cap rock brine showed a similar rapid uptake of equal proportions of interlayer and non-interlayer water during the first 50 h of the experiment. However, after 50 h, the amount of interlayer water stopped at 0.05 mL/g and the uptake continued only as non-interlayer water. In the steady state, ca. 0.25 mL/g of total water content was comprised of ca. 80% non-interlayer water and just ca. 20% interlayer water. Compared with the Opalinus clay pore water hydrated sample, the diluted cap rock brine-infiltrated bentonite contained less than half the amount of interlayer water (0.05 mL/g compared to 0.12 mL/g).

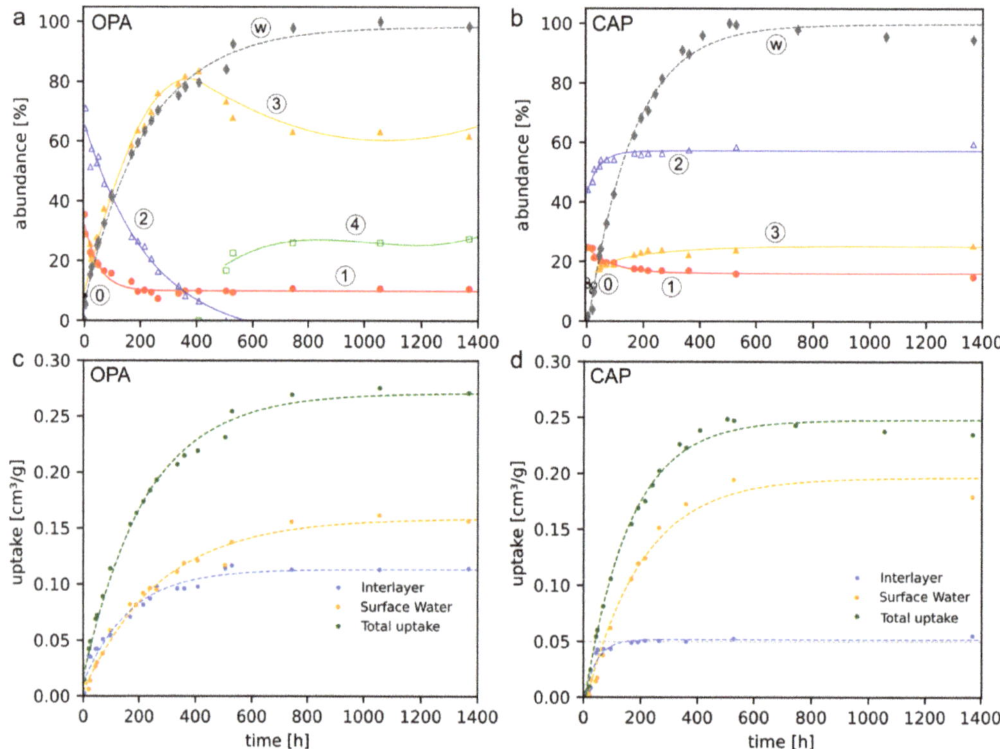

**Figure 5.** Relative abundance of water-layers developed (a,b) and the location of stored water (c,d) during the hydration of bentonite within the two wet cell experiments. The numbers of water layers (0, 1, 2, 3, 4) are shown in circles. (**a**) SD80 infiltrated by Opalinus clay pore water, (**b**) SD80 infiltrated by diluted cap rock brine, (**c**) partitioning of interlayer and non-interlayer (surface and pore) water in SD80 infiltrated by Opalinus clay pore water, (**d**) partitioning of interlayer and non-interlayer (surface and pore) water in SD80 hydrated by diluted cap rock brine. The total water uptake curve (w) is normalized to 100%.

## 4. Discussion

### 4.1. Hydration Mechanism of the Milos SD80 Bentonite

The overall hydration behavior of the Milos SD80 bentonite shows the expected pattern of interlayer expansion whereby the number of WLs increased until a steady state was reached. As the two hydration experiments were prepared with the same packing density and using the sample powdered material, the differences observed between the two wet cells can be solely attributed to the different chemistry of the infiltrating solutions used.

Considering the SD80 bentonite hydrated in Opalinus clay pore water (total dissolved solids (TDS) content = 19 g/L), the initial mixture of 2-, 1- and 0-WLs, expanded to 3- and 4-WLs. This is similar to the documented Ca-bentonite interlayer expansion in solutions of relatively high electrolyte concentration (13.9 g/L) where similar mixtures of 3-, 4-, and 2-WLs developed in a sample of low packing density [9]. However, the apparent disappearance of the 2-WL structure in the SD80 sample after 600 h while retaining ca. 10% 1-WLs is an unusual feature not seen in previously published experiments and requires explanation. As the CALCMIX program cannot model mixtures with more than three WL combinations, it may be partly an artifact, whereby the combination of 3-, 4-, and 1-WLs matched the patterns better than a mixture of 3-, 4-, and 2-WLs. Due to this limitation, some 2-WLs were likely retained in the steady state of hydration: albeit with a lower abundance than the remaining 1-WL structures. The precise reason why 1-WL structures remained more abundant than 2-WLs in the Opalinus clay pore water once hydrated indicates that some of the interlayers in the Milos smectite were inhibited from swelling. One clear possibility could be the presence of small amounts of K (+0.03 e/phuc) in the interlayer sites, which are less prone to expansion and are likely to retain the 1-WL structure, even in the water-saturated state [28]. As ca. 8% of the layer charge ($-0.36$ e/phuc) was occupied by $K^+$, this corresponds well with the ca. 10% of 1-WLs remaining in the steady-state condition. The Opalinus clay pore solution contains 0.002 mol/L KCl, and given the high preference of $K^+$ adsorption to interlayer sites, some cation exchange is a possibility. Based on exchangeable cations as measured by the atomic adsorption spectroscopy (AAS) of SD80 smectite treated in long-term batch experiments at various temperatures, the amount of $K^+$ was lower than the unaltered smectite material with a ca. 20–40% decrease in concentration [16]. In contrast, the interlayer K content of untreated and treated bentonite measured by SEM-EDX analyses showed no significant difference between the two samples. These differences may reflect variations in the amounts of exchangeable and fixed K that occurred following experimental treatment.

The idea that the remaining 1-WL structures are related to some K cations in the smectite interlayer of the SD80 sample is also supported by the bentonite infiltrated by diluted cap rock brine. In this case, the higher abundance of 1-WLs remaining after hydration was over 15%, and this result corresponds with the higher amount of KCl in the solution (0.005 mol/L), which was 2.5 times more than in the Opalinus clay pore water sample. Composition measurements of the interlayer cations in the SD80 bentonite analyses following batch reactor experiments at 25 °C also showed less (20–40%) interlayer K following treatment with diluted cap rock brine [13,18]. SEM-EDX analyses of similar materials measured slightly higher (+0.02 e/phuc) amounts of interlayer K following treatment with the same brine [18]. The higher concentration of K remaining after hydration explains the higher abundance of remaining 1-WLs in this sample.

In contrast to the SD80 bentonite treated with Opalinus clay pore water, the same bentonite hydrated in diluted cap rock brine with a very high electrolyte concentration (TDS content = 155 g/L) showed a notably low degree of interlayer expansion, with a mixture of 2-, 3- and 1-WLs. As this is the most saline solution yet studied by wet-cell diffractometry, the strong suppression of thicker water layers and the high amounts of non-interlayer water in this experiment were likely to result from the high salinity of this brine. Such saline solutions significantly reduce the thickness of the double diffuse layer, and inhibit interlayer expansion by minimizing the difference in the concentration of ions in the interlayer and the surrounding pore water [6].

Another feature of interest was the occurrence of gas bubbles in the SD80 bentonite experiment hydrated by the diluted cap rock brine, which were trapped in solution between the upper surface of the bentonite clay and the Kapton foil. This can be attributed to the release of $CO_2$ gas associated with the dissolution of the calcite present in the bentonite sample. This gas was detected in batch reaction experiments conducted using the same material, and was responsible for generating significant gas pressures [12]. Batch reactions after 1 year produced some of the highest swelling pressures in diluted cap solution, reaching values of 2.21 $\pm$ 0.01 MPa [6].

The reason why gas bubbles were only observed in the more saline solution and not in the Opalinus clay pore water remains unclear. The intensity of the XRD reflections for calcite was notably higher than the sample containing the $CO_2$ bubbles, so it may have been due to heterogeneity of the material and small-scale differences in calcite abundance (both <1%). Furthermore, the pH of the diluted cap rock brine was slightly lower (pH 7.3) than the more alkane (pH 7.8) Opalinus clay pore water, but these small differences are unlikely to give rise to significantly different calcite dissolution rates and the pH was strongly buffered by the smectite clay in these experiments. One possible explanation could relate to the significantly higher Na:Ca ratio of the CAP solution (78) compared to the

Opalinus clay pore water (8.1) and its higher molar concentration of $Ca^{2+}$, which is known to increase the rate of calcite dissolution in Na–Ca–Mg–Cl brines (Tables 2 and 3) [29].

**Table 3.** List of samples with corresponding parameters from this study [1]; Warr and Berger [2], 2007; Perdrial and Warr [3], 2011; Berger, 2008 [4] [9,22,30]. BD = bulk density, DD = dry density, ξ = total layer charge, SSA = specific surface area, TWU = total water uptake, TDS = total dissolved solids, NIW = non-interlayer water. * = Na-smectite, # = Ca-smectite, + = estimation for Millipore water. n.d. = not determined.

| Sample | BD g/cm$^3$ | DD g/cm$^3$ | ξ phuc$^{-1}$ | SSA m$^2$/g | TWU mL/g | TDS g/L | Na:Ca Ratio | 1-WL [%] | 2-WL [%] | 3-WL [%] | >3-WL [%] | NIW mL/g |
|---|---|---|---|---|---|---|---|---|---|---|---|---|
| #SD80 [1] | 1.502 | 1.47 | −0.36 | n.d. | 0.27 | 19 | 8.1 | 10 | 0 | 63 | 27 | 0.16 |
| #SD80 [1] | 1.476 | 1.44 | −0.36 | n.d. | 0.25 | 155 | 78 | 15 | 60 | 25 | 0 | 0.20 |
| #TTE [2] | 0.94 | 0.76 | −0.29 | 103.01 | 0.61 | 0.041 | 0.3 | 0 | 13 | 42 | 45 | 0.46 |
| #TTE [2] | 0.94 | 0.78 | −0.29 | 103.01 | 0.69 | 13.6 | 26.4 | 0 | 7 | 63 | 30 | 0.53 |
| * IS80 [2] | 1.15 | 1.08 | −0.33 | 55.33 | 0.35 | 0.041 | 0.3 | 29 | 11 | 60 | 0 | 0.18 |
| * IS80 [2] | 1.14 | 1.04 | −0.33 | 55.33 | 0.45 | 13.6 | 26.4 | 0 | 55 | 40 | 5 | 0.23 |
| * +MX80 [3] | 1.43 | 1.41 | −0.28 | 30.03 | 0.40 | 0.0011 | - | 0 | 15 | 77 | 8 | 0.05 |
| * +MX80 [3] | 1.35 | 1.23 | −0.28 | 30.03 | 0.55 | 0.0116 | - | 0 | 0 | 61 | 39 | 0.18 |
| * +MX80 [3] | 1.60 | 1.31 | −0.28 | 30.03 | 0.34 | 0.0116 | - | 0 | 10 | 79 | 11 | 0.06 |
| * Swy-2 [4] | 1.37 | X1.28 | −0.32 | 27.64 | 0.44 | 0.0011 | - | 13 | 63 | 24 | 0 | 0.18 |
| * Swy-2 [4] | 1.36 | X1.27 | −0.32 | 27.64 | 0.45 | 58.44 | Na only | 22 | 64 | 14 | 0 | 0.19 |

Although the controlling parameters of $CO_2$ release remain uncertain, the generation of gas pressure in confined volume experiments is likely to influence the swelling behavior of the bentonite and may affect the thickness of the WLs developed. Compared to other bentonite experiments, the combination of the high swelling pressures produced by this clay and the limited interlayer expansion whereby the 2-WL was retained as the most abundant form may have been influenced by the additional effects of the $CO_2$ gas pressure. The interaction between gas generation and the swelling behavior of smectite in confined volume systems is a topic that requires further experimental study.

*4.2. Predicting Bentonite Hydration in Confined Volume Systems Based on the Physical-Chemical Properties of the Bentonite*

The mechanisms of smectite hydration are generally well understood. In addition to the effects of packing density and the volume of space available, the process is influenced by (i) the interlayer charge and its distribution, (ii) the type of interlayer cations present, and (iii) the chemistry and concentration of dissolved ions in the solution. After compiling all the available results of experiments conducted by X-ray wet-cell diffractometry (Table 3), it is of interest to discuss the overall patterns of hydration for variably compacted bentonites and the main controlling parameters. Based on this assessment, we consider whether or not a diagrammatic plot can be developed as a predictive tool for assessing bentonite hydration without actually conducting time-consuming experimental hydration tests.

Plotting the packing density versus the total water uptake reveals the expected trend of increasing packing density and decreasing water uptake (Figure 6a). The lowest amount of solution inflow was achieved in the two SD80 bentonite experiments of this study due to the higher dry densities used. This overall tendency appears to be largely independent of whether Na or Ca dominate the interlayer of the smectite. The type of interlayer cation is, however, a primary factor in determining the amount of interlayer versus non-interlayer water incorporated during hydration. Considering the relationships between the dominant interlayer cations of smectites (Figure 6b) shows clearly that relatively more non-interlayer water is incorporated into Ca-smectites than into Na-smectites, as established in previous studies [9,22]. These relationships appear to be largely independent of solution Na:Ca ratio, and the total content of dissolved salts indicates that any exchange reactions that occur during the hydration and wetting of the bentonite are far from incomplete. This is not unexpected given the very low solution-to-bentonite ratios involved in these experiments.

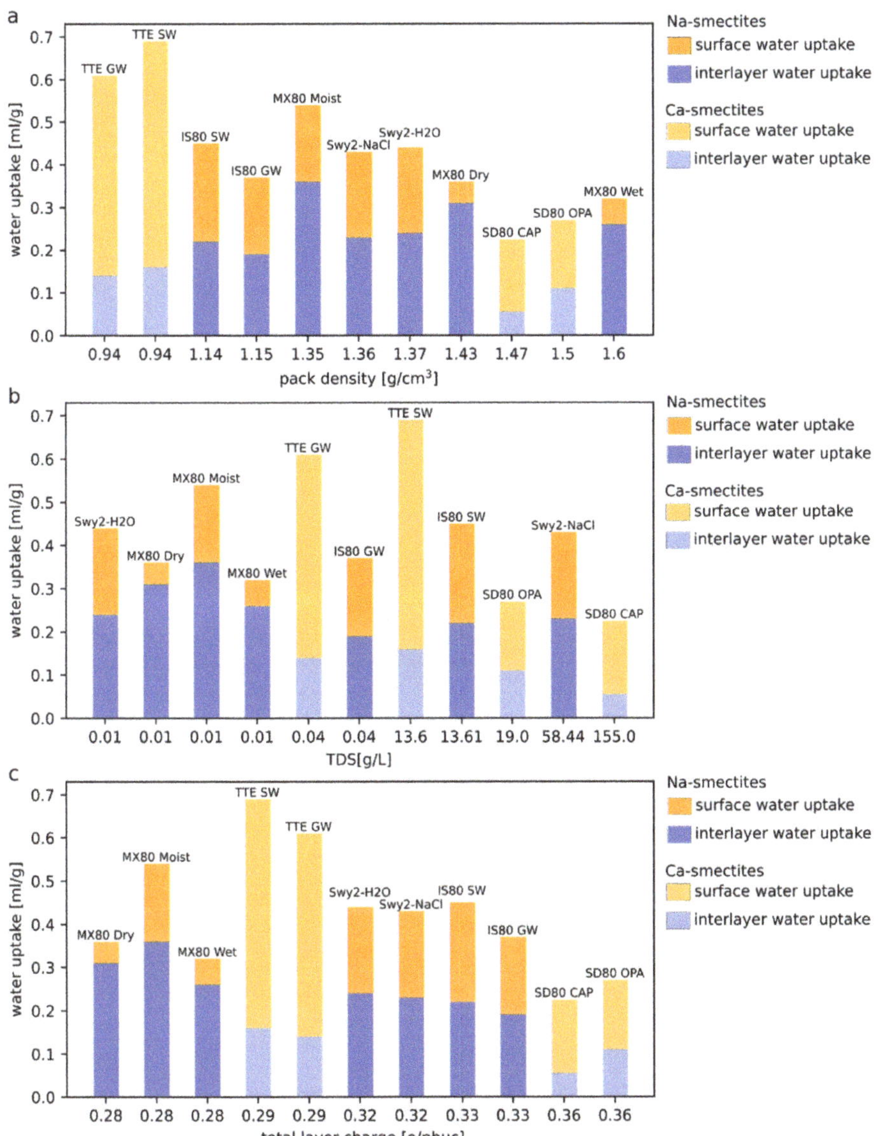

**Figure 6.** Summary of the amounts of interlayer and non-interlayer water in hydrated bentonite samples studied by wet-cell X-ray diffractometry [9,22,30]. (**a**) water uptake verses packing density, (**b**)water uptake verses total dissolved solids (TDS), (**c**) water uptake verse total layer charge.

The relationship between hydration and the total layer charge is less clear due to the small spread of charge values ranging between −0.28 to −0.36 e/phuc. The most water uptake occurs in Tixoton bentonite with a charge of −0.29 e/phuc and a decrease towards the higher layer charges of the SD80 bentonite. This pattern may, however, be partially explained by the differences in dry densities between these samples.

When considering the WL structures of all hydrated bentonite samples, some notable patterns can be recognized (Figure 7). For both Ca- and Na-smectites, higher dry packing densities lead to a lower number of WL structures due to the restriction of space, and, in the case of Na-smectites, due to the build of higher swelling pressures caused by the additional osmotic gradients characteristic of Na interlayers (Figure 7a). As revealed by this study of the SD80 bentonite sample, which is dominated by Ca in the interlayer, the reduction in the thickness of the double diffuse layer in the brine and the additional generation of $CO_2$ are predicted to have a similar effect in repressing the thickness of the WLs that may develop. These effects have so far received little attention in previous swelling experiments.

In all Ca-smectites studied, the dominant thicknesses were 2 or 3 WLs, and the 2-WL was more common when using solutions with a high TDS content (Table 2; Figure 7b). In contrast, Na-smecties developed abundant 3 or >3 WLs, except for the purified Wyoming montmorillonite sample [30], where the very high smectite abundance of this clay did not develop thicker WL structures due to the higher swelling pressures generated when using a purified smectite sample. In this case, the Na-smectite of the Swy-2 samples in 1M NaCl (58.44 g/L) developed a very similar hydration structure to the Ca-smectite of the SD80 sample, with the dominance of 2-WLs and the preservation of some 1-WL structures. The precise role of the total layer charge on the WL structure remains uncertain from the interlayer hydration patterns (Figure 7c), although there is a tendency for the higher layer-charged smectites to retain fewer WLs compared to the lower-layer charges that generally favor thicker structures.

Despite the multitude of parameters that can affect WL structure and the partitioning between interlayer and non-interlayer water, the four most important physical–chemical features that can be used to describe the hydration behavior of bentonites can be plotted (Figure 8). By calculating the ratios of the interlayer water/total water uptake versus the packing density/total layer charge, general linear relationships are observed for both Ca- and Na-smectites. Bentonites dominated by interlayer hydration, high packing densities, and low layer charge plot more in the upper right part of the curve, whereas bentonites with high amounts of more mobile non-interlayer water, low packing densities, and higher layer charge fall in the lower left part of the curve. The plot also reflects how Na-bentonites are characterized by systematically higher interlayer/total water uptake ratios for any given state of packing density/total layer charge. As there are more data points for the Na-bentonite correlation ($R^2 = 0.91$, n = 7), this line is considered to be the more accurate of the two curves. With only four data points, the Ca-bentonite line ($R^2 = 0.68$) represents only a rough approximation and may well run quasi-parallel to the Na-bentonite correlation. Such a plot may be used as a useful predictive tool for assessing the mobility of water in hydrated bentonites whereby the upper-right parts of the curve represent low diffusion-controlled transport rates and the lower-left parts of the curve present faster rates of chemical transport due to the abundance of more loosely held surface and pore water. In terms of hydration behavior, the SD80 Milos bentonite is viewed as one of the more suitable backfill materials containing Ca-smectite, whereas the MX80 bentonite appears to be the most favorable Na-smectite variety considered here.

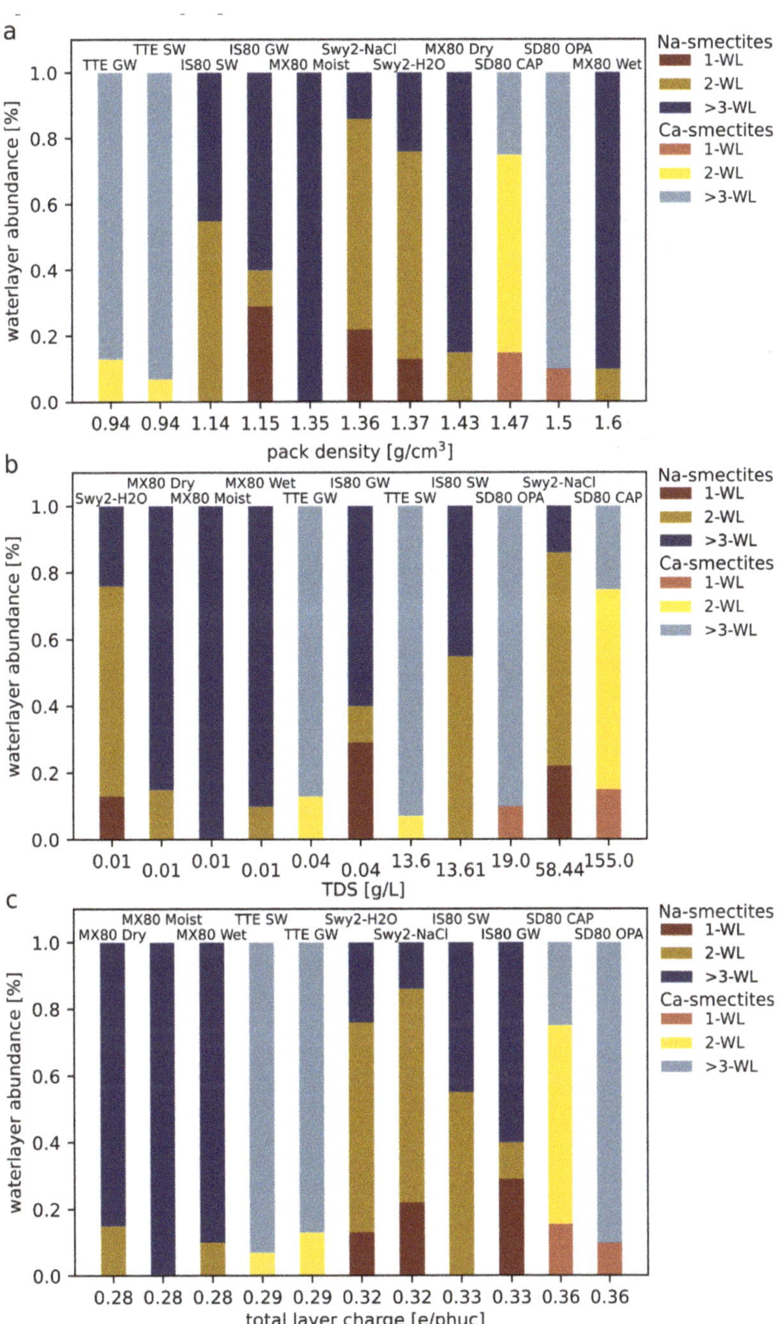

**Figure 7.** Summary of the number of water layers developed in hydrated bentonite samples studied by wet-cell X-ray diffractometry [9,22,30]. (**a**) water uptake verses packing density, (**b**) water uptake verses total dissolved solids (TDS), (**c**) water uptake verse total layer charge.

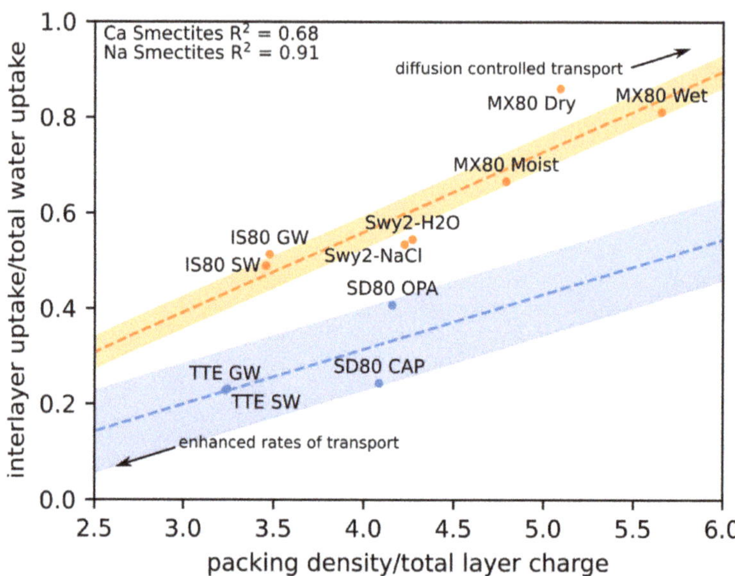


**Figure 8.** Plot showing the correlative relationships between the interlayer/total water uptake ratio versus the packing density/total layer charge ratios for Ca-(blue) and Na-(orange) bentonites (data source given in Table 3). The shaded areas correspond to the standard deviation of the data sets.

## 5. Conclusions

(1) The hydration of smectite studied by wet-cell X-ray diffractometry provides useful constraints for assessing the primary parameters of bentonite wetting that are controlled by the interacting parameters of the packing density, the total layer charge, and the type of interlayer cations.

(2) Hydration of the SD80 Milos bentonite shows some unusual features compared to published experiments. The retention of some 1-WLs in both the Opalinus clay pore water and the diluted cap rock brine is suggested to reflect K remaining in some interlayer sites. Also, the notably smaller thickness of WLs developed in the diluted cap rock brine probably resulted from the high total solid content of the solution, the generation of $CO_2$ bubbles, and the additional internal gas pressure within the bentonite clay, which is suggested to have additionally suppressed WL thicknesses.

(3) Whereas the chemistry and concentration of dissolved ions are also important, the overall patterns of bentonite hydration in many types of solutions indicate this factor to be more of a secondary effect during initial hydration and water saturation.

(4) For Ca- and Na-bentonite, determining the packing density/total layer charge ratio can be used as a predictive parameter for estimating the relative amount of interlayer-versus-non-interlayer water that will be stored in the system, which in turn will help assess the probable rates of chemical transport.

**Author Contributions:** Conceptualization, T.M. and L.N.W.; data curation, T.M. and C.P.; formal analysis, T.M. and C.P.; funding acquisition, G.G. and L.N.W.; investigation, T.M. and C.P.; methodology, T.M. and L.N.W.; project administration, G.G. and L.N.W.; resources, C.P. and S.K.; software, T.M.; supervision, G.G. and L.N.W.; visualization, T.M.; writing–original draft, T.M. and C.P.; writing–review and editing, T.M., C.P., G.G., S.K. and L.N.W. All authors have read and agreed to the published version of the manuscript.

**Funding:** This research is part of the joint project UMB, funded by the Federal Ministry of Economic Affairs and Energy (BMWi) under the grant number: 02 E 11344C. We also acknowledge support for the Article Processing Charge from the DFG (German Research Foundation, 393148499) and the Open Access Publication Fund of the University of Greifswald.

**Data Availability Statement:** The data presented in this study are partially available on request from the corresponding author.

**Acknowledgments:** Special thanks to Artur Meleshyn, who provided helpful feedback.

**Conflicts of Interest:** The authors declare no conflict of interest.

# References

1. Pusch, R. Use of Bentonite for Isolation of Radioactive Waste Products. *Clay Miner.* **1992**, *27*, 353–361. [CrossRef]
2. Villar, M.V.; García-Siñeriz, J.L.; Bárcena, I.; Lloret, A. State of the Bentonite Barrier after Five Years Operation of an in Situ Test Simulating a High Level Radioactive Waste Repository. *Eng. Geol.* **2005**, *80*, 175–198. [CrossRef]
3. Ye, W.-M.; Chen, Y.-G.; Chen, B.; Wang, Q.; Wang, J. Advances on the Knowledge of the Buffer/Backfill Properties of Heavily-Compacted GMZ Bentonite. *Eng. Geol.* **2010**, *116*, 12–20. [CrossRef]
4. Kaufhold, S.; Dohrmann, R. Stability of Bentonites in Salt Solutions | Sodium Chloride. *Appl. Clay Sci.* **2009**, *45*, 171–177. [CrossRef]
5. Kaufhold, S.; Dohrmann, R. Distinguishing between More and Less Suitable Bentonites for Storage of High-Level Radioactive Waste. *Clay Miner.* **2016**, *51*, 289–302. [CrossRef]
6. Bergaya, F.; Lagaly, G. General Introduction: Clays, Clay Minerals, and Clay Science. In *Developments in Clay Science*; Bergaya, F., Theng, B.K.G., Lagaly, G., Eds.; Handbook of Clay Science; Elsevier: Amsterdam, The Netherlands, 2006; Chapter 1; Volume 1, pp. 1–18.
7. Cho, W.J.; Lee, J.O.; Chun, K.S. The Temperature Effects on Hydraulic Conductivity of Compacted Bentonite. *Appl. Clay Sci.* **1999**, *14*, 47–58. [CrossRef]
8. Dixon, D.; Sanden, T.; Jonsson, E.; Hansen, J. *Backfilling of Deposition Tunnels: Use of Bentonite Pellets*; Swedish Nuclear Fuel and Waste Management Co.: Stockholm, Sweden, 2011.
9. Warr, L.N.; Berger, J. Hydration of Bentonite in Natural Waters: Application of "Confined Volume" Wet-Cell X-ray Diffractometry. *Phys. Chem. Earth Parts A/B/C* **2007**, *32*, 247–258. [CrossRef]
10. García-Gutiérrez, M.; Gormenzana, J.L.; Missana, T.; Mingarro, M.; Molinero, J. Overview of Laboratory Methods Employed for Obtaining Diffusion Coefficients in FEBEX Compacted Bentonite/Descripción de Los Métodos de Laboratorio Empleados Para Obtener Coeficientes de Difusión En Bentonita Compactada FEBEX. *J. Iber. Geol.* **2006**, *32*, 37–53.
11. Sample-Lord, K.M.; Shackelford, C.D. Solute Diffusion in Bentonite Pastes. *J. Geotech. Geoenviron. Eng.* **2016**, *142*, 04016033. [CrossRef]
12. Meleshyn, A. *Mechanisms of Transformation of Bentonite Barriers (UMB)*; Report GRS-390; GRS: Braunschweig, Germany, 2019.
13. Podlech, C.; Warr, L.N.; Grathoff, G.; Kasbohm, J. Umwandlungsmechanismsen in Bentonitbarrieren (UMB). Teilprojekt C: Tonmineralogische Arbeiten. 2019. Available online: https://edocs.tib.eu/files/e01fb20/1696279712.pdf (accessed on 17 June 2021).
14. Kremleva, A.; Krüger, S. Reaction Mechanisms in Bentonite Barriers: Part D: Quantum Mechanical Modelling of Fe-Containing Smectites. Teilprojekt D Tonmineralogische Arbeiten. 2019. Available online: https://edocs.tib.eu/files/e01fb20/1736573527.pdf (accessed on 20 September 2021).
15. Warr, L.N.; Hofmann, H. In Situ Monitoring of Powder Reactions in Percolating Solution by Wet-Cell X-ray Diffraction Techniques. *J. Appl. Crystallogr.* **2003**, *36*, 948–949. [CrossRef]
16. Podlech, C.; Matschiavelli, N.; Peltz, M.; Kluge, S.; Arnold, T.; Cherkouk, A.; Meleshyn, A.; Grathoff, G.; Warr, L.N. Ben tonite Alteration in Batch Reactor Experiments with and without Organic Supplements: Implications for the Disposal of Radioactive Waste. *Minerals* **2021**, *11*, 932. [CrossRef]
17. Lorenz, P.; Meier, L.; Kahr, G. Determination of the Cation Exchange Capacity (CEC) of Clay Minerals Using the Complexes of Copper (II) Ion with Triethylenetetramine and Tetraethylenepentamine. *Clays Clay Miner.* **1999**, *47*, 386–388.
18. Stanjek, H.; Künkel, D. CEC Determination with Cu-Triethylenetetramine: Recommendations for Improving Reproducibility and Accuracy. *Clay Miner.* **2016**, *51*, 1–17. [CrossRef]
19. Warr, L. IMA–CNMNC approved mineral symbols. *Mineral. Mag.* **2021**, *85*, 291–320. [CrossRef]
20. Eypert-Blaison, C.; Sauzéat, E.; Pelletier, M.; Michot, L.J.; Villiéras, F.; Humbert, B. Hydration Mechanisms and Swelling Behavior of Na-Magadiite. *Chem. Mater.* **2001**, *13*, 1480–1486. [CrossRef]
21. Hofmann, H.; Bauer, A.; Warr, L.N. Behavior of Smectite in Strong Salt Brines under Conditions Relevant to the Disposal of Low-to Medium-Grade Nuclear Waste. *Clays Clay Miner.* **2004**, *52*, 14–24. [CrossRef]
22. Perdrial, J.N.; Warr, L.N. Hydration Behavior of MX80 Bentonite in a Confined-Volume System: Implications for Backfill Design. *Clays Clay Miner.* **2011**, *59*, 640–653. [CrossRef]
23. Plançon, A.; Drits, V. Expert System for Structural Characterization of Phyllosilicates: I. Description of the Expert System. *Clay Miner.* **1994**, *29*, 33–38. [CrossRef]
24. Cases, J.; Bérend, I.; Besson, G.; Francois, M.; Uriot, J.; Thomas, F.; Poirier, J. Mechanism of Adsorption and Desorption of Water Vapor by Homoionic Montmorillonite. 1. The Sodium-Exchanged Form. *Langmuir* **1992**, *8*, 2730–2739. [CrossRef]

25. Cases, J.; Bérend, I.; François, M.; Uriot, L.; Michot, L.; Thomas, F. Mechanism of Adsorption and Desorption of Water Vapor by Homoionic Montmorillonite: 3. The $Mg^{2+}$, $Ca^{2+}$, $Sr^{2+}$ and $Ba^{2+}$ Exchanged Forms. *Clays Clay Miner.* **1997**, *45*, 8–22. [CrossRef]
26. Tambach, T.J.; Hensen, E.J.; Smit, B. Molecular Simulations of Swelling Clay Minerals. *J. Phys. Chem. B* **2004**, *108*, 7586–7596. [CrossRef]
27. Ferrage, E.; Lanson, B.; Sakharov, B.A.; Drits, V.A. Investigation of Smectite Hydration Properties by Modeling Experimental X-ray Diffraction Patterns: Part I. Montmorillonite Hydration Properties. *Am. Mineral.* **2005**, *90*, 1358–1374. [CrossRef]
28. Denis, J.H.; Keall, M.J.; Hall, P.L.; Meeten, G.H. Influence of Potassium Concentration on the Swelling and Compaction of Mixed (Na,K) Ion-Exchanged Montmorillonite. *Clay Miner.* **1991**, *26*, 255–268. [CrossRef]
29. Gledhill, D.K.; Morse, J.W. Calcite Dissolution Kinetics in Na–Ca–Mg–Cl Brines. *Geochim. Cosmochim. Acta* **2006**, *70*, 5802–5813. [CrossRef]
30. Berger, J. *Hydration of Swelling Clay and Bacteria Interaction. An Experimental in Situ Reaction Study*; Universite Louis Pasteur: Strasbourg, France, 2008.

*Article*

# Reaction and Alteration of Mudstone with Ordinary Portland Cement and Low Alkali Cement Pore Fluids

**Keith Bateman *, Yuki Amano, Mitsuru Kubota, Yuji Ohuchi and Yukio Tachi**

Radionuclide Migration Research Group, Japan Atomic Energy Agency, Ibaraki 319-1194, Japan; amano.yuki@jaea.go.jp (Y.A.); kubota.mitsuru@jaea.go.jp (M.K.); ohuchi.yuji@jaea.go.jp (Y.O.); tachi.yukio@jaea.go.jp (Y.T.)
* Correspondence: bateman.keith@jaea.go.jp

**Abstract:** The construction of a repository for the geological disposal of radioactive waste will utilize cement-based materials. Following closure, resaturation will result in the development of a highly alkaline porewater. The alkaline fluid will migrate and react with host rock, producing a chemically disturbed zone (CDZ) around the repository. To understand how these conditions may evolve, a series of batch and flow experiments were conducted with Horonobe mudstone and fluids representative of the alkaline leachates expected from a cementitious repository. Both ordinary Portland cement (OPC) and low alkali cement (LAC) leachates were examined. The impact of the LAC leachates was more limited than the OPC leachates, with experiments using the LAC leachate showing the least reaction and lowest long-term pH of the different leachate types. The reaction was dominated by primary mineral dissolution, and in the case of OPC leachates, precipitation of secondary calcium-silicate-hydrate (C-S-H) phases. Flow experiments revealed that precipitation of the secondary phases was restricted to close to the initial contact zone of the fluids and mudstone. The experimental results demonstrate that a combination of both batch and flow-through experiments can provide the insights required for the understanding of the key geochemical interactions and the impact of transport.

**Keywords:** radioactive waste; cement-clay interaction; OPC; LAC; alkaline leachate

**Citation:** Bateman, K.; Amano, Y.; Kubota, M.; Ohuchi, Y.; Tachi, Y. Reaction and Alteration of Mudstone with Ordinary Portland Cement and Low Alkali Cement Pore Fluids. *Minerals* **2021**, *11*, 588. https://doi.org/10.3390/min11060588

Academic Editors: Ana María Fernández, Stephan Kaufhold, Markus Olin, Lian-Ge Zheng, Paul Wersin and James Wilson

Received: 13 April 2021
Accepted: 26 May 2021
Published: 31 May 2021

**Publisher's Note:** MDPI stays neutral with regard to jurisdictional claims in published maps and institutional affiliations.

## 1. Introduction

The construction of a repository for geological disposal of radioactive waste will by necessity include the use of cementitious materials in a multiplicity of ways, such as fillers, liners, plugs, and seals [1–4]. Ordinary Portland cement (OPC)-based materials will be extensively used in the construction, and following closure, groundwater will saturate the repository and the use of OPC will result in the development of highly alkaline porewater (pH > 12.5), [5,6]. The alkaline fluid will migrate and react with the host rock to form a chemically disturbed zone (CDZ) around the repository [2]. A series of chemical gradients will develop over time and distance from the repository, disturbing the pH, redox, and fluid chemistry of the migrating fluid. It is important, particularly in the case of a radioactive waste repository, to understand the evolution of the CDZ in both time and space and subsequent impacts on the behavior and transport of any radionuclides in the CDZ.

The extent of the CDZ, beyond the host rock-cement interface, will depend upon several factors with both physical and chemical properties, i.e., host rock mineralogy, porosity/permeability, fracture density, groundwater composition, flow rates, and chemical buffering capacity. Previous studies have shown that in the host rock, silicate minerals dissolve in the highly alkaline pore fluid [2,7,8], followed by the precipitation of secondary minerals (calcium-silicate–hydrate, calcium-aluminum-silicate-hydrate (C-S-H, C-A-S-H), calcite) and aluminosilicates (zeolites, feldspars, feldspathoids) [9]. Other authors [10–12] have reported the interactions between cement pore fluids and argillaceous rocks, which demonstrated the alteration of argillaceous rocks by the high pH fluids and the buffering of the elevated pH by the clays and the reaction pathways. The degree of buffering and alteration of the primary minerals is dependent upon the starting mineralogy, i.e., clay mineral dissolution vs. carbonate reaction (including, when present, de-dolomitization [13]), cation exchange reactions, and the fluid chemistry, with $Ca(OH)_2$-dominated fluids tending to result in greater formation of C-S-H phases.

The general sequence of reaction of host rock with OPC leachates is generally well understood [2,7–9]. The high pH develops from the leaching of the concretes and cements used in the waste containment and repository construction, resulting in primary mineral dissolution, which partly mitigates the high pH in the CDZ. This is then followed by secondary phase formation with OPC, particularly C-(A-)S-H phases and zeolites. The extent of the zone of interaction is dependent not only upon the host rock but also upon the reactant fluid chemistry. To mitigate the degree of alteration of the host rock, 'low pH' or 'low alkali' cements (LACs) have been developed to provide a target porewater pH < 11. LACs are based on pozzolanic cements with 30–80 wt% of the OPC clinker replaced by siliceous materials, such as silica fume (SF), fly ash (FA), blast furnace slag, and/or metakaolin. The high siliceous content in the LAC lowers the pH of the porewater by several complementary mechanisms. Firstly, the OPC content is reduced, and the pozzolanic reaction of silica with portlandite and calcium increases the C-S-H gel content. However, the extent of the reaction of host rock with LACs is less well understood. Studies have often been limited to batch experiments [14,15] or modelling studies [16]. In batch experiments conducted with Freedland Ton clay and a 'low pH cement' pore fluid [14], the clay did not undergo significant degradation. In other work examining the reaction zones at interfaces of Opalinus claystone (OPA) with two different concrete types (OPC and LAC) [17], again there was little evidence for chemical alteration of the claystone. Other batch experimental studies [15] have concluded that zeolitic phases would form in the interaction zones between rock-forming minerals and low-pH cements. In addition, some studies have focused on the degradation of the low-pH cement itself, e.g., [18], with clay pore fluids, rather than the impact on the host rock.

Unlike OPC, there is no standard LAC composition; different compositions of LAC have been considered by different radioactive waste disposal implementers. For example, many European operators are considering the use of LACs based on cement, silica, and blast furnace slag [19], whilst the LAC formulation, being considered for use in Japan, uses fly ash in place of the blast furnace slag. In Japan, the Japan Atomic Energy Agency (JAEA) developed a low alkaline cement called HFSC424 (high content fly-ash silica fume cement), composed of fly-ash (FA) 40 wt%, silica fume (SF) 20 wt%, and OPC 40 wt% [20,21], which results in a target porewater pH < 11.

Many previous studies have used only batch experiments to study the interaction of alkaline fluids with host rocks, since they are particularly suited to long duration studies, being simpler and easier to set up and maintain. The effects of time can be studied by running multiple identical experiments for differing durations [22]. In addition, elevated temperatures can be used to enhance mineral dissolution kinetics, speeding up reaction progress so that they may be observed more easily in the laboratory. Previously, batch experiments have been used to study the interactions between cement pore fluids and argillaceous rocks [10–12,18]. These studies demonstrated that the high pH leachates reacted with the argillaceous rocks, leading to the buffering of the elevated pH of the cement pore fluids by the clays and other host rock minerals, e.g., dolomite [13]. However, batch experiments, whilst useful, do not precisely replicate the sequence of mineral changes observed in the CDZ of a radioactive waste repository, since with batch experiments, the effects of transport are not addressed. In addition, batch experiments can be affected by the presence of transient phases or constrained availability of key components.

Flow-through experiments adapted from chemical engineering studies [23] are a useful technique to study transport processes, allowing the investigation of the spatial as well as temporal changes. Flow experiments are by their nature more complex than batch experiments and are best suited to shorter timescales, usually only a few months duration.

Typically, a flow system set-up is comprised of a fluid reservoir, a pump to control flow rate, and the reactor itself with length scales of millimeters to meters. On completion of the experiment, it is possible to study the changes in mineralogy with distance by sectioning of the reacted solid. In addition, unlike batch experiments, it is possible to extract some physical properties from flow experiments, such as variations in porosity/permeability, using tracer tests. In addition, flow experiments can provide valuable well-constrained 'test cases' for the validation and calibration of reactive transport geochemical models. Such models may then be used, with increased confidence, to model the longer term evolution of the fluid rock interactions beyond that possible experimentally, which in the case of radioactive waste disposal extends to tens of thousands of years.

Flow-through column experiments have been used to investigate the reaction of cement pore fluids with single minerals and a variety of potential host rocks [24–26], but few previous studies have considered the impact on argillaceous rocks. This study, comparing the impact of leachates from traditional OPC-based materials with leachates from a LAC (specifically HFSC424) with a mudstone host rock used a combination of both batch experiments to provide an indication of the evolution of the alkaline leachate/mudstone system, together with flow-through experiments, which allowed the investigation of the spatial as well as temporal changes.

## 2. Materials and Methods

Samples of mudstone were collected from the Horonobe Underground Research Laboratory (URL) site, Hokkaido, Japan. The samples were taken from the gallery walls from the Koetoi formation, which is a massive and lithologically homogeneous, diatomaceous mudstone that contains amorphous silica (40–50 wt%), clay (17–25 wt%), quartz (7–10 wt%), and feldspar (5–10 wt%) [27,28]. The mudstone samples were crushed to <500 μm prior to being used in the experiments. Crushed materials were used, since the greater surface area allows greater (chemical) reaction and hence greater degree of reaction within the time constraints of a laboratory study. However, it is recognized that the crushing process could also result in the generation of 'experimental artefacts' by production of highly reactive 'fines'.

Cement pore fluids representative of the alkaline leachates expected from a cementitious repository [29,30] were used in the batch and flow experiments. The first fluid represented a 'young' OPC leachate pH ~13.4 with high [Na] and [K]. A second fluid, an 'evolved' OPC leachate, was saturated with respect to portlandite, pH ~12.5. The third fluid represented the leaching of HFSC424 concrete with Horonobe groundwater, pH ~11 [21]. Details of the concentration of ions in the fluids are given in Table 1. The OPC leachates were prepared from analytical grade reagents; Na and K were added as hydroxides, and Ca as CaO. The HFSC424 leachate was prepared, under a $N_2$ atmosphere, by equilibrating in equal masses, Horonobe groundwater, from the 07-V140-M03 borehole (located in the Koetoi Formation), with crushed HFSF424 concrete together, for 14 days before use.

**Table 1.** Initial concentrations of ions in the fluids used for the experiments.

| Leachate Type | Components (mg/L) | | | | | | |
|---|---|---|---|---|---|---|---|
| | pH at 24 °C | Na | K | Ca | $SiO_2$ | $Cl^-$ | $SO_4^{2-}$ |
| 'young' OPC leachate (Na-K-Ca-OH) | 13.4 | 1500 | 7300 | 60 | – | – | – |
| 'evolved' OPC leachate (saturated $Ca(OH)_2$) | 12.5 | | | 800 | – | – | – |
| HFSC424 leachate | 11 | 3300 | 1330 | 94 | 370 | 3300 | 760 |
| Horonobe ground water, borehole 07-V140-M03 | 7.9 | 2960 | 160 | 80 | – | 3400 | 585 |

## 2.1. Equipment

The batch experiments used in this study consisted of simple non-reactive equipment comprised of 50, 100, and 250 mL capacity polypropylene bottles. All experiments were conducted inside a glove box that was continuously flushed with $N_2$. The primary aim of the flushing with $N_2$ was to remove any carbon dioxide to prevent precipitation of $CaCO_3$ from the alkaline leachates; a consequence of this was that oxygen levels also remained low (<0.5%), but no attempt was made to impose redox control. The oxygen concentration inside the glove box was monitored using a JIKCO JKO-02LJD3 meter/sensor combination (ICHINEN JIKCO Co., Ltd., Shibaura, Japan). Fluid:solid (F:S) ratios, ranging from solid to fluid dominated systems, of 1:1, 10:1, and 100:1 (mL:g) were used, with typical durations for the experiments of 24 h, and 7, 28, and 56 days. All batch experiments were performed at lab temperature of ~24 °C. On termination of the batch experiments, the fluids were sampled within the $N_2$ glove box, and the solids were recovered by filtration and then vacuum dried before being prepared for subsequent mineralogical analysis.

As well as batch experiments, flow-through experiments with continuous collection of reacted fluids allowing the changes in fluid chemistry to be tracked over time were also conducted. As with the batch experiments, the reacted solid was only sampled on termination, but by sectioning of the reacted solids, it was possible to study the changes in mineralogy with distance. Two different approaches were used in this study for the flow experiments. The first was comprised of PEEK (polyetheretherketone) columns (300 mm long, and 7.5 mm i.d.) packed with crushed mudstone and connected to fluid reservoirs and sample collection bottles (Figure 1). Similar experimental equipment has been used in other studies [24–26] to examine the impact of alkaline fluids on potential host rock materials. The second approach consisted of a small flow cell (SFC) constructed from acrylic plastic and sealed by a combination of 'O-rings' and bolts (Figure 1). Filters and porous polypropylene disks (on both inlet and outlet sides) acted both to distribute the incoming fluid across the whole face of the mudstone sample and prevent blockage of the outlet tubing. An advantage of the SFC equipment was the larger diameter of the cell, which reduced the chance of blockages forming due to movement of 'fines', preventing flow through the cell. A slight disadvantage was the reduced sample length (only 10 mm compared to the 300 mm length of the columns). Table 2 gives a comparison between the physical parameters for the different flow equipment. For both flow setups, the volumetric flow rate was typically ~0.5 mL per hour (~12 mL per day). Flow for both types of equipment was controlled by a Cole-Parmer MASTERFLEX® peristaltic pump (Cole-Parmer, Vernon Hills, IL, USA). Although fast compared to the flow rates likely to be encountered in a radioactive waste repository, the flowrate chosen was a compromise between maintaining the lowest possible steady flow rate with the peristaltic pump and maximizing the number of pore volumes passing through the flow experiments in a laboratory time frame. Periodically, the reacted fluids from the flow experiments were sub-sampled and prepared for chemical analysis. As with the batch experiments, the flow experiments were again performed at lab temperature of ~24 °C.

**Table 2.** Comparison of the physical parameters for the different flow equipment.

|  | Column | Small Flow Cell (SFC) |
| --- | --- | --- |
| Total mudstone weight (g) | 10.3 | 3.14 |
| Length (mm) | 300 | 10 |
| Cross-sectional area (mm$^2$) | 44 | 314 |
| Volumetric flow rate (mL/h) | ~0.5 | ~0.5 |
| Residence time (h) | ~15 | ~3.5 |

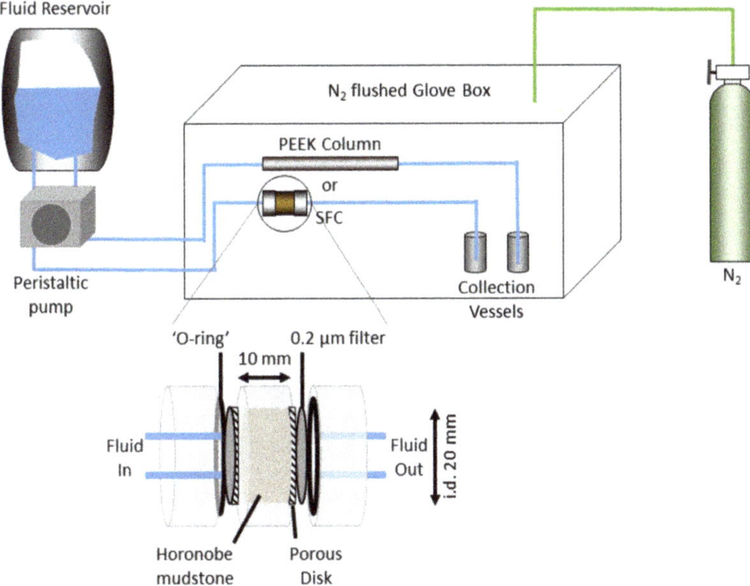



**Figure 1.** Schematic of flow set-up, with inset detail of small flow cell (SFC). Note: 'O-rings', porous disks, and filters were fitted to both sides of the SFC.

### 2.2. Analysis

All collected fluids were filtered using 0.2 μm syringe filters and then sub-sampled for determination of cations and anions. Typically, a 4 mL sample of the fluid was diluted two-fold with 18 MΩ demineralized water (Millipore Simplicity® ultrapure water system) and then acidified with concentrated $HNO_3$ (1% *v/v*) to preserve the sample. This sample was used for the analysis of major cations by a combination of ICP-OES (inductively coupled plasma-optical emission spectrometry) using a Shimadzu ICPE-9800 (Shimadzu Corporation, Kyoto, Japan), and ICP-MS (inductively coupled plasma-mass spectrometry), using a Perkin-Elmer NexION 300x (PerkinElmer, Inc., Waltham, MA, USA), both calibrated using matrix matched standards. A second subsample was taken for determination of major anions by IC (ion chromatography) using a Dionex ICS-5000 (Thermo Fisher Scientific, Waltham, MA, USA) ion chromatograph system calibrated using a mixed standard solution (Kanto Chemical Co., Inc, Tokyo, Japan). All fluid samples were stored at <5 °C until required for analysis.

The pH of the experimental fluids was determined immediately upon sampling using a DKK-TOA Corp. model HM-30P meter and combination electrode calibrated using DKK-TOA Corp. buffers at 4.01, 6.86, and 9.18 pH (Japanese standard), pH accurate to ±0.02 pH.

On completion of the flow experiments, columns were sectioned into ~15 mm long pieces using a small rotary cutting saw (Proxxon KG 50, Kiso Power Tool Co., Ltd, Osaka, Japan) and then vacuumed dried before being prepared for subsequent mineralogical analysis. The samples from the SFC were vacuumed dried, whilst still held in the central section, before being carefully extruded and sectioned into ~1.5 mm thick slices using a thin blade. Samples from the batch experiments were filtered to remove excess fluid and then vacuumed dried.

Once dried, the solid samples were prepared for petrographic analysis by a combination of scanning electron microscopy (SEM), using a JEOL JSM-6510 Series SEM, (JEOL Ltd., Tokyo, Japan) and X-ray diffraction (XRD) analysis (RigaKu SmartLab XRD, Rigaku Corporation, Tokyo, Japan), with a 9 kW X-ray source). Sub-samples for SEM analysis were prepared as either gold or carbon coated random mount stub samples. Techniques included SEM using both secondary electron imaging (SE) and backscattered electron (BSE) imaging, and element distribution analysis using energy-dispersive X-ray spectroscopy (EDS). Samples for XRD were prepared for analysis by taking a representative sub-sample and grinding it to a fine powder.

The saturation indices (SI = log (IAP/$K_s$), where IAP: ion activity product; $K_s$: solubility constant) of the primary and potential secondary minerals, in the reacted fluids, were calculated using the PHREEQC v3.6.3 geochemical code [31]. Calculations were performed using the JAEA thermodynamic database (TDB) [32]. JAEA-TDB version PHREEQC 19.dat (v1.2, 03.Mar.2020) was used for the calculations, being the latest version available at the time.

## 3. Results

### 3.1. Aqueous Chemistry

3.1.1. Changes in pH

The changes in pH with time for the batch and flow experiments are shown in Figure 2a. In the 1:1 F:S batch experiments, a large decrease in pH was observed, with the greatest change being seen in the experiments with the HFSC424 leachate (pH ~3.5) compared to the experiments with the OPC leachates. The pH reduction in the 10:1 F:S experiments was smaller, and again so in the 100:1 F:S experiments. In all cases, the pH drop was the least in the experiments with the 'young' OPC leachate and greatest with the HFSC424 leachate. In the 100:1 F:S experiments with the 'young' OPC leachate, the pH was only ~0.2 pH units below the initial leachate.

In all the flow experiments, there was an initial drop in pH before it later recovered to around ~0.2 pH units below that of the initial fluids. These initial changes in pH were similar to those seen in the equivalent 1:1 batch experiments, and the later 'steady-state' pH was similar to that observed in the 100:1 F:S batch experiments. Again, as with the batch experiments, the initial decrease was greatest in the experiments with the HFSC424 leachate.

Since the Horonobe mudstone contains a small amount of pyrite (<2 wt%) [27,28], and the samples used here had not been preserved in a reducing environment, it is highly likely that the low pH seen in some batches and all flow experiments were due to the dissolution of sulphate phases formed by the oxidation of pyrite during sample storage and preparation. Indeed, corresponding increases in [$SO_4{}^{2-}$] up to ~8000 mg/L in the 1:1 batch experiments, and in the first two days of the flow experiments, were seen in the fluids with low pH, and XRD analysis (Figures S1 and S2) of the unreacted mudstone indicated the presence of jarosite and/or gypsum. Previous analysis of cores from Horonobe [33] have also shown the oxidation of pyrite and the precipitation of calcium sulphates such as gypsum.

3.1.2. Changes in Na Concentration

The variation in Na concentration is shown in Figure 2b. With the 'young' OPC leachate, the concentrations of Na in the 100:1 and 10:1 F:S batch experiments (which were fluid dominated) were close to that of the initial leachate (~1500 mg/L), but the 1:1 F:S experiments showed increases in [Na] compared to the initial leachate, up to ~2600 mg/L. A similar increase in [Na] was seen in the 1:1 F:S experiments with the 'evolved' OPC leachate (from below detection to ~1500 mg/L), with a smaller ~200 mg/L increase seen in the 10:1 F:S but no change in the 100:1 F:S experiments.

With the HFSC424 leachate, again there was an increase in the [Na] in the reacted fluids in the 1:1 F:S experiments, to ~3600 mg/L, but in the 10:1 and 100:1 F:S experiments, there was a ~300 mg/L decrease. In the flow experiments (Figure 2b), there was an initial decrease in [Na] with both the 'young' OPC and HFSC424 leachates from ~1500 to ~1300 mg/L and from ~3300 to 1200 mg/L, respectively. The Na concentrations in both fluids then slowly increased over the next 16 d until the concentrations were close to but still below those of the original leachates.

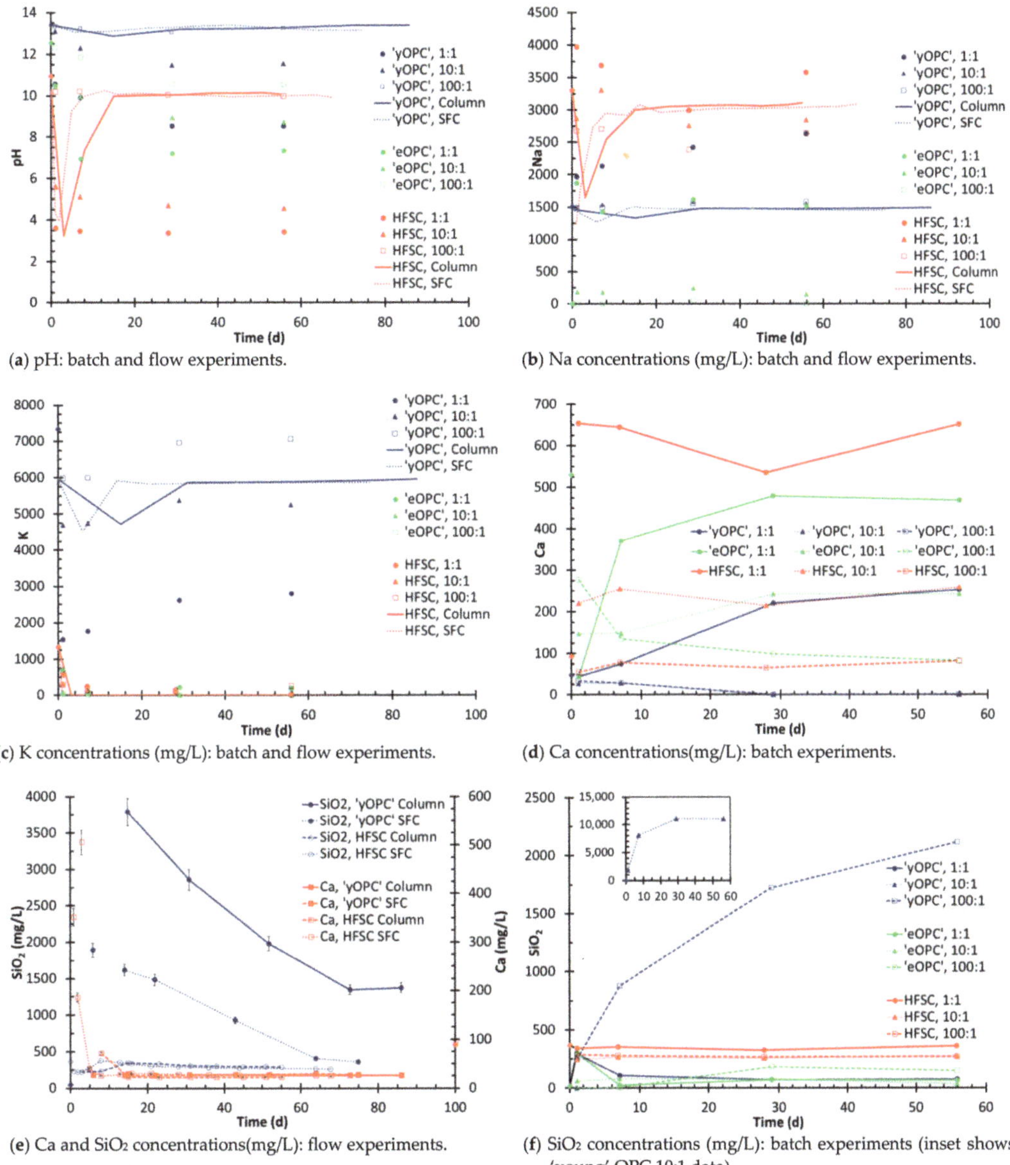

(**a**) pH: batch and flow experiments.

(**b**) Na concentrations (mg/L): batch and flow experiments.

(**c**) K concentrations (mg/L): batch and flow experiments.

(**d**) Ca concentrations (mg/L): batch experiments.

(**e**) Ca and SiO₂ concentrations (mg/L): flow experiments.

(**f**) SiO₂ concentrations (mg/L): batch experiments (inset shows 'young' OPC 10:1 data).

**Figure 2.** Changes in fluid chemistry with time. (**a**) pH batch and flow experiments; (**b**) [Na] batch and flow experiments; (**c**) [K] batch and flow experiments; (**d**) [Ca] batch experiments; (**e**) [Ca] and [SiO$_2$] flow experiments; (**f**) [SiO$_2$] batch experiments (inset shows 'young' OPC 10:1 data). Diamonds on the vertical axis indicate initial leachate concentrations. In the graphs where both batch and flow data are shown together (**a**–**c**), the single points indicate data from batch experiments, and the lines the flow experiments. Legend text: yOPC—'young' OPC leachate; eOPC—'evolved' OPC leachate; HFSC—HFSC424 leachate. For batch experiments, the F:S is indicated.

### 3.1.3. Changes in K Concentration

In the batch experiments (Figure 2c), with the 'young' OPC leachate, [K] decreased, most notably so in the 1:1 F:S experiments where concentrations dropped to ~2000 mg/L. Smaller decreases were seen in the higher F:S experiments, with [K] in the 100:1 F:S experiments being slightly below the initial concentrations. With the 'evolved' OPC leachate, [K] was typically <100 mg/L. In the case of the HFSC424 leachate, [K] decreased in all the batch experiments, from ~1330 to <150 mg/L.

In the flow experiments (Figure 2c) with the 'young' OPC leachate, [K] decreased initially by around 1300 mg/L, but in the later fluids, there was only a slight decrease compared to the initial leachate concentration. With the HFSC424 leachate, [K] in the reacted fluids flow experiments decreased from the initial concentration to <1 mg/L.

### 3.1.4. Changes in Ca Concentration

The behavior of Ca in the batch experiments was more complex than that of the other alkali metals (Figure 2d). With the 'young' OPC leachate, in the 1:1 F:S experiments, [Ca] increased to ~250 mg/L after 56 d. However, in the higher F:S experiments, [Ca] decreased to below that of the 'young' OPC leachate, i.e., <2 mg/L after 56 days. With the 'evolved' OPC leachate, [Ca] initially decreased significantly, the largest decreases to ~44 mg/L being seen after 24 h in the 1:1 experiments. In the longer duration 1:1 and 10:1 F:S experiments, [Ca] then increased to 470 and 240 mg/L, respectively, but remained below that of the original OPC leachate. In the 100:1 F:S experiments, [Ca] remained low (<100 mg/L) and did not recover with time.

In the flow experiments with the 'young' OPC leachate (Figure 2e), [Ca] decreased from the original leachate, stabilizing at ~25 mg/L after ~16 d. In the flow experiments with the HFSC424 leachate, [Ca] increased from ~100 mg/L to ~500 mg/L before decreasing to similar concentrations (~25 mg/L), as seen in the OPC leachate flow experiments.

### 3.1.5. Changes in Silica Concentration

In the 1:1 F:S batch experiments with the 'young' OPC leachate, silica concentrations (Figure 2f) initially increased to ~300 mg/L but then decreased over time to ~80 mg/L. However, with the 'young' OPC leachate, very high silica concentrations were observed in the 28 and 56 d, 10:1 F:S batch experiments, ~10,000 mg/L (inset in Figure 2f) and <2000 mg/L in the 100:1 experiments. Concentrations in the batch experiments with the 'evolved' OPC leachate were significantly lower, at <100 mg/L. In the HFSC424 leachate batch experiments, silica concentrations remained close to or lower than the original leachate.

In the flow experiments with the 'young' OPC leachate (Figure 2e), the dissolved silica concentration initially increased to ~3800 mg/L in the column experiment, and ~1900 mg/L in the SFC experiment; thereafter, the concentration of silica in the fluids decreased to ~1300 mg/L in the columns and <300 mg/L in the SFC. The higher silica concentrations observed in the columns, compared to the SFC experiments, reflects the greater mass of mudstone in the columns as well as the longer residence time (Table 2). In both HFSC424 leachate flow experiments (Figure 2e), silica concentrations were always lower than the original HFSC424 leachates, and significantly lower than with the 'young' OPC leachate.

### 3.1.6. Changes in Other Cations

Dissolved iron tracked the dissolution of sulphate (see Section 3.1.1) with increases in [total Fe] up to ~250 mg/L in the 1:1 batch experiments, and in the first two days of the flow experiments, coinciding with the changes in sulphate concentrations. In all the flow experiments following the peak in [total Fe], the concentration in the reacted fluids decreased to <0.01 mg/L by ~16 d of flow (equivalent to ~50 pore volumes).

With the exception of Rb and Sr, which tracked the concentrations of Na and Ca, respectively, the concentrations of other cations analyzed (e.g., Al, Ba, Cs, Cu, Li, Mg, Mn, Ni, Pb) were either below detection or showed no significant change from the original leachates.

### 3.1.7. Changes in Anions

As previously mentioned, increases in sulphate concentration, due to presence of pyrite, were seen in the fluids that had low pH values. However, chloride, which is a component of the Horonobe groundwater and therefore present in the HFSC424 leachate, behaved in a conservative manner, remaining close to the concentration of the initial HFSC424 leachate. The other anions analyzed (Br, F, $NO_3^-$, $NO_2^-$, and $PO_4^{3-}$) were either below detection or showed no significant change from the original leachates.

### 3.2. Solid Analysis

### 3.2.1. Batch Experiments

Samples from the 1:1 F:S batch experiments with the 'young' OPC leachate were analyzed by SEM, including EDS and XRD. There was evidence of primary material dissolution (removal of fine material present in the original mudstone), but SEM–EDS analysis found no evidence for the formation of secondary phases. XRD analysis (Figure S1) of the unreacted mudstone indicated the presence of jarosite and/or gypsum alongside the primary minerals. XRD analysis of the reacted solids (Figure S1) was less informative; the main peaks identified were those of the original primary minerals, though some peaks could be attributed to crystalline C-S-H phases, but the relative peak intensities were too low to make a definitive interpretation. In addition, since many C-S-H phases are amorphous gels, XRD is of limited value.

In the higher F:S 'young' OPC leachate experiments, again the absence of fines suggested primary mineral dissolution, but only in the experiments with the 'evolved' OPC leachates was evidence found by SEM EDS analysis for secondary C-(A-)S-H phases (Figure 3); semi-quantitative analysis gave Ca:Si ranging from 0.89 to 1.1. In the batch experiments with the HFSC424 leachates, no mineralogical evidence (SEM observations and analysis by SEM-EDS, and XRD) was found for the formation of any secondary C-S-H phases.

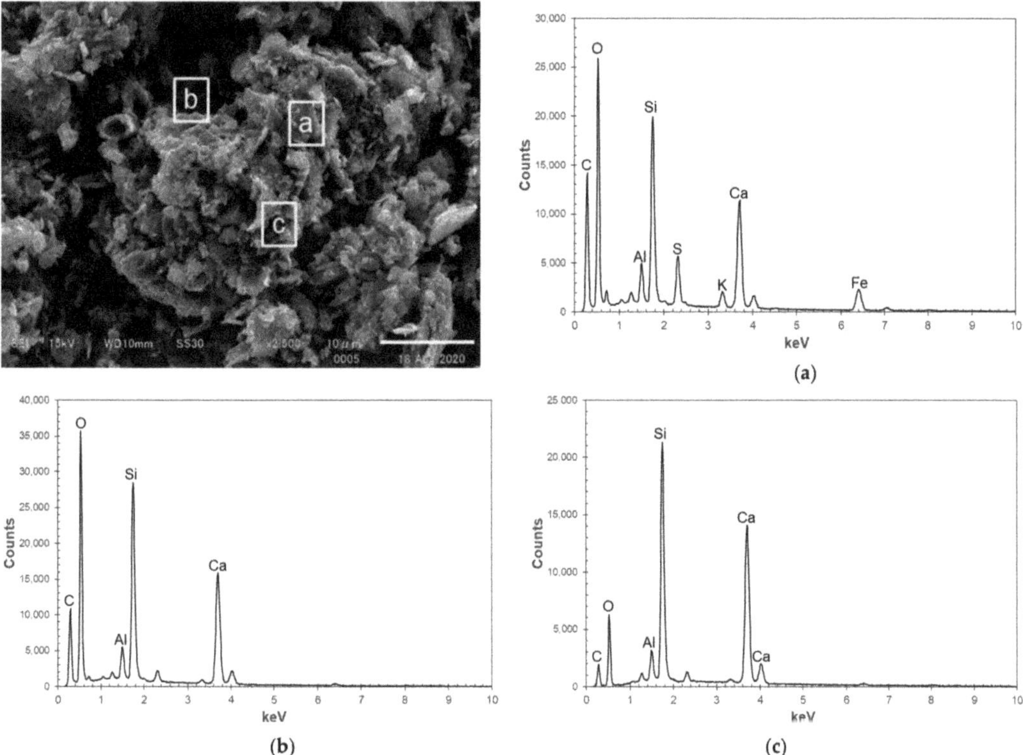

**Figure 3.** Secondary electron (SE) image and SEM-EDS analysis of reacted mudstone. Batch experiment 10:1 F:S with the 'evolved' OPC leachate. The white squares (**a–c**) in the SEM image indicate areas used for SEM-EDS analysis, and the corresponding figures (**a–c**) show the presence of primary minerals (**a**) and secondary C-S-H (**b,c**).

### 3.2.2. Flow Experiments

A summary of the evolution of the mineralogical analysis from the column experiment with the 'young' OPC leachate is shown in Figure 4. Section 1 of the column (~0–20 mm) under SEM analysis showed extensive precipitation of secondary C-S-H and C-A-S-H gel-like (hydrous) phases (Figure 4) with different Ca:Si ratios (from 0.5 to 1.4). However, SEM examination of the remaining sections of the column showed no evidence of further precipitation of any secondary phases, and only the presence of the primary mineral phases (Figure 4), some of which appeared to be free of fines, suggesting a degree of mineral dissolution.

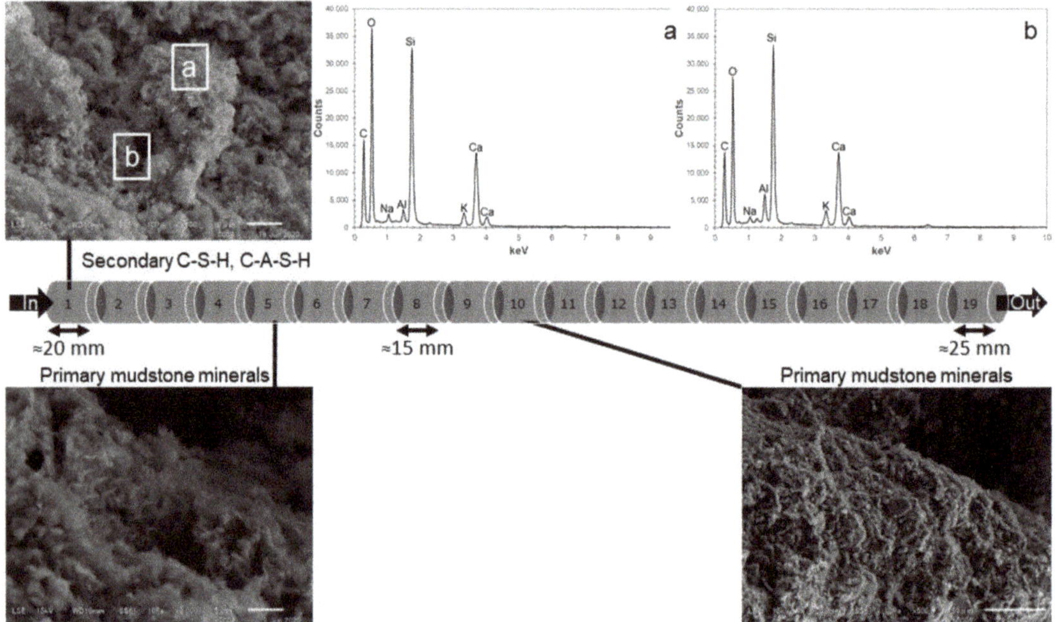

**Figure 4.** Summary of mineralogy, column experiment with the 'young' OPC leachate. Arrows indicate direction of flow; numerals refer to Section numbers; each Section (apart from Sections 1 and 19) ~15 mm long. All SEM photos are secondary electron (SE) images. White squares (**a**,**b**) in the SEM image from Section 1 indicate areas used for SEM-EDS analysis of C(A)-S-H phases.

The equivalent experiment was performed using the SFC with the 'young' OPC leachate (Figure 5). In Section 1 (~0–1.5 mm) under SEM observation, only primary minerals and diatom fragments present in the original mudstone [28] were visible (Figure 5); there was no evidence (SEM EDS element mapping) found for the formation of any secondary phases. However, in Section 2 (~1.5–3 mm), a variety of secondary C-S-H and C-A-S-H phases (Ca:Si 0.5–0.7) were observed (Figure 5). In examination by SEM of Sections 3–6 (3–10 mm), no secondary phase(s) were observed, and the primary mineral surfaces showed little sign of reaction.

In the SFC experiment using the HFSC424 leachate (Figure 6) in Section 1 (~0–2.5 mm) and Section 2 (~2.5–5 mm), unlike the equivalent experiment with the 'young' OPC leachate (Figure 5), only primary minerals, which were relatively free of fines, were present. Additionally, there was no mineralogical evidence (SEM observations and SEM-EDS element mapping) for the precipitation of any secondary C-S-H phases, though occasional very rare crystals with zeolitic-type compositions were detected by SEM-EDS analysis.

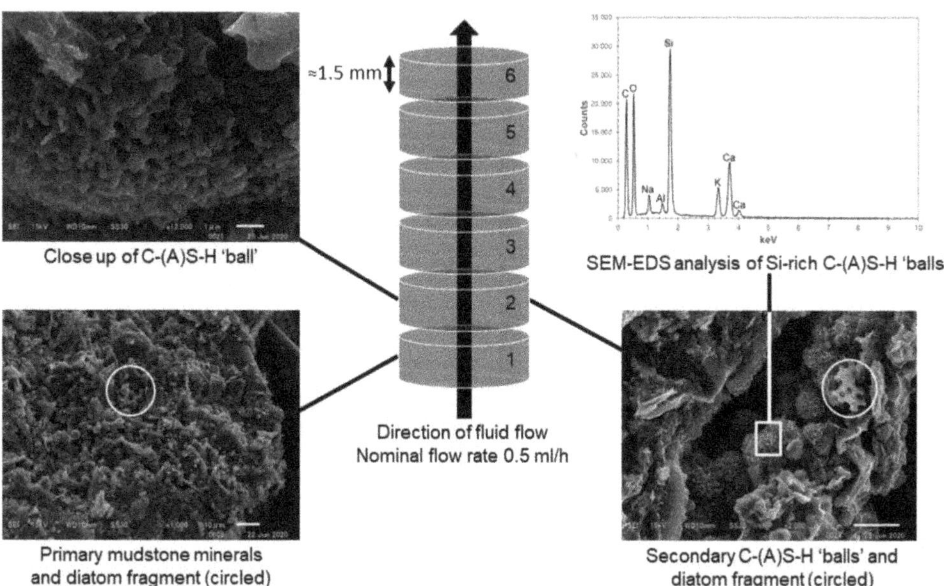



**Figure 5.** Summary of mineralogy, SFC experiment with the 'young' OPC leachate. Arrow indicates direction of flow; numerals refer to Section numbers. Each section was ~1.5 mm long. All SEM photos are secondary electron (SE) images. The white square in the SEM image from Section 2 indicates the area used for SEM-EDS analysis of C(A)-S-H phases.

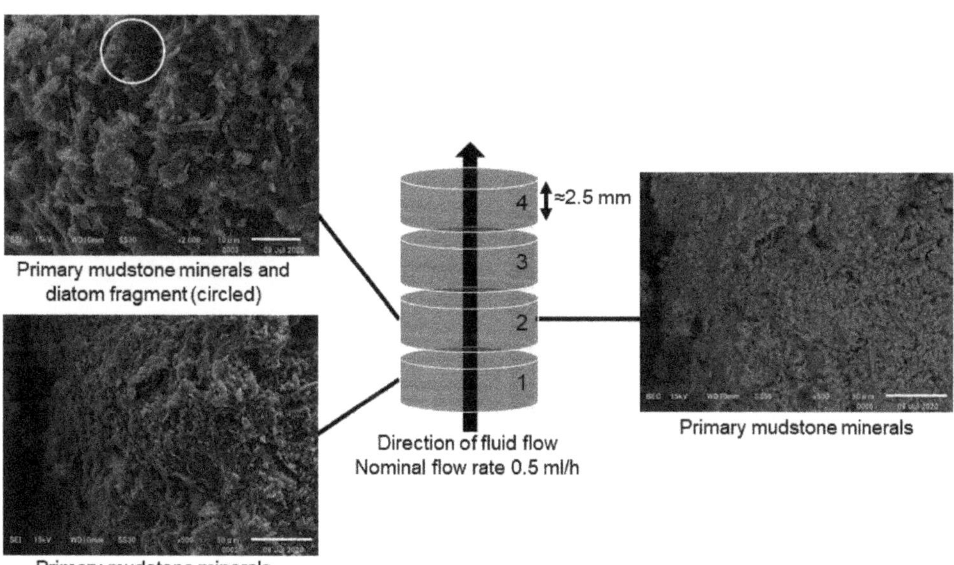



**Figure 6.** Summary of mineralogy, SFC experiment with the HFSC424 leachate. Arrow indicates direction of flow; numerals refer to Section numbers. Each section was ~2.5 mm long. All SEM photos are secondary electron (SE) images.

All of the solids from the flow experiments were also examined by XRD, but the results were inclusive with the main peaks present belonging to the primary minerals (e.g., SFC experiment with the 'young' OPC leachate, Figure S2), and as in the batch experiments, although some peaks could be attributed to C-S-H phases, the relative peak intensities were again too low to make a definite interpretation.

## 4. Discussion

### 4.1. Chemical Evolution and Buffering Behavior

In this study, using samples of Horonobe mudstone, both OPC and LAC leachate types showed a similar behavior with regard to pH trends (see Figure 2a and Section 3.1). The longer term reduction in pH was not as large as seen in some other studies of argillaceous clays, such as the Opalinus claystone (OPA) [13], but unlike the Horonobe mudstone, OPA also contained significant dolomite that reacted to form calcite and precipitate Mg-hydroxides, which aided the reduction in pH. However, the longer-term pH changes were similar to those seen in other studies using OPC type leachates with a variety of mineralogies [24–26].

With regard to the behavior of both Na and K (Figure 2b,c) in the absence of any identifiable K- or Na-bearing secondary precipitates, the initial decreases in their concentrations observed in both the batch experiments and the early stages of the flow experiments suggests possible ion exchange reactions with the clays in the mudstone. In the flow experiments, the latter recovery of [Na] and [K] (Figure 2b,c) suggests that all the accessible exchangeable sites may have been occupied. In the case of the Horonobe mudstone, Koetoi formation, studied here, smectite typically represents ~10 wt% of the primary minerals [27,28]. Ion exchange has been identified in other studies on argillaceous clays [13,34], in which it was demonstrated to have a significant impact on the fluid chemistry, but it was not possible to confirm the ion exchange by cation exchange capacity analysis in these experiments due to the small solid sample size (in particular for the flow experiments) available.

The trend in [Ca] (Figure 2d,e) was thus complicated by the apparent ion exchange reactions. The observed increases in [Ca] in the 1:1 F:S batch experiments (both 'young' OPC and HFSC424 leachates) and the initial stages of the flow experiments were likely due to ion exchange. However, with both 'young' OPC and HFSC424 leachates, the latter Ca concentrations in the flow experiments and those in the higher F:S batch experiments decreased, which would indicate the precipitation of Ca-bearing phase(s). In contrast, in all the batch experiments with the 'evolved' OPC leachate, [Ca] in the reacted fluids was always lower than the original leachate, again suggesting the formation of Ca-bearing secondary phase(s). Taken together with the changes in silica concentrations in the experiments, it is likely that the secondary phase(s) would be C-S-H phases.

The continued presence of dissolved silica in the flow experiments suggests that there was continued dissolution of the primary silicate phases present in the mudstone, and this was most significant with the highly alkaline 'young' OPC leachate.

### 4.2. Mineral Dissolution and Precipitation Processes

In terms of mineralogical observations, in all the experiments there was SEM evidence for primary mineral dissolution (Figures 3–6), but only in the flow experiments with the OPC leachate and in some of the higher F:S batch experiments was secondary phase formation observed. These were comprised of C-(A)S-H phases of differing morphologies (see Figures 3–5), from 'gel-like' to more 'fibrous' phases and C-S-H 'balls', of varying Ca:Si ratios. In the experiments with the HFSC424 leachates, there was little analytical mineralogical evidence to support secondary phase formation.

As noted above, the changes in [Ca] and the relationship with dissolved silica suggests the formation of Ca-silicate phases in most of the experiments, most likely secondary C-S-H and C-A-S-H phases, though they were only directly observed in the experiments with the OPC leachates. The elevated silica concentrations most likely reflect the dissolution of the primary amorphous silica and aluminosilicate minerals.

Aluminum concentrations would be expected to increase alongside silica due to dissolution of the primary aluminosilicates, so the absence of Al in the fluids suggests the preferential dissolution of the amorphous silica or possible incorporation in secondary minerals, such as C-A-S-H phases or zeolites.

To examine the relationship between the reacted fluid chemistry and mineralogy, the PHREEQCv3 geochemical code [31] was used to calculate mineral saturation indices (SI) in the reacted fluids. This showed that in the 1:1 OPC leachate batch experiments,

C-S-H phases were likely to be undersaturated, i.e., SI < 0 (Figure 7a), whereas in the higher F:S, OPC leachate batch experiments C–S–H phases were saturated (i.e., SI > 0). The SI calculations also suggested that zeolites (e.g., phillipsites and analcime) could be possible secondary phases (i.e., SI > 0) in these batch experiments (Figure 7a). The degree of saturation decreased with higher F:S ratios, and in some of the 100:1 F:S experiments, some zeolites, e.g., analcime, were undersaturated. However, no mineralogical evidence was found for the precipitation of zeolites. It is worth noting that a previous review of bentonite alteration by alkaline fluids [7] concluded that the available thermodynamic data for zeolites used in models tend to overestimate their stability.

In addition to C-S-H, M-S-H (magnesium silicate hydrate) have been identified in other studies [35–37] as potential products of cement-clay interactions (including LAC), though in some studies [15,18] they were only observed on the cement side of the interface and not within the mudstone. However, in the experiments conducted here, there was no mineralogical evidence found for their precipitation. The saturation state calculations suggested that only in the flow experiments with the 'young' OPC leachates were M-S-H phases saturated (Figure S3). In the batch experiments with all three fluids, only in the 100:1 F:S experiments (Figure S3) were some M-S-H phases found to be saturated.

In the flow experiments, the mineralogical observations showed that with the 'young' OPC leachates, there had been dissolution of the primary minerals and subsequent secondary C-(A-)S-H formation. The SI calculations for the reacted fluids (Figure 7d) indicated that, like the higher F:S OPC leachate batch experiments, C-S-H phases were likely to be saturated (i.e., SI > 0). However, unlike the batch experiments, most zeolites were undersaturated. In addition, the degree of undersaturation increased with time, i.e., number of pore volumes, which was a similar pattern to the batch experiments, where in the 100:1 F:S experiments, some zeolites were unsaturated (Figure 7a). The assessment of mineral saturation also showed that the primary mineral phases were slightly undersaturated, suggesting continued dissolution (i.e., SI < 0).

In the 'evolved' OPC batch experiments, the SI calculations suggested that as with the 'young' leachate, the C-S-H phases were undersaturated in the 1:1 experiments (i.e., SI < 0), but that as the F:S ratio (and pH) increased, they approached or exceeded saturation (i.e., SI > 0), and indeed C-S-H phases were observed in the higher F:S experiments (Figure 2b). Again, zeolites, although not observed, were predicted to be saturated in the 1:1 experiments, becoming less saturated with higher F:S ratios.

In the batch experiments with HFSC424 leachates, no mineralogical evidence was found for precipitation of any secondary phases. The saturation state calculations found that only in the reacted fluids from the 100:1 F:S experiments (Figure 7c) were some C-S-H phases close to saturation (i.e., SI > 0). In addition, zeolites were again identified as potential secondary phases in both the 10:1 and 100:1 experiments (Figure 7c), though unlike with the OPC leachate, zeolites were more saturated in the higher F:S experiments.

The saturation state calculations for the fluids from HFSC424 leachate flow experiments (Figure 7e), indicted that in the early fluids, like the 1:1 and 10:1 batch experiments, that C-S-H phases were undersaturated (i.e., SI < 0). In addition, as with the 100:1 batch experiments, in the later fluids some C-S-H phases were close to saturation (i.e., SI ~0) or even slightly saturated, which suggests that they could have precipitated, though no mineralogical evidence was found. Mass balance calculations for the precipitation of possible C-S-H phases, based on the difference in [Ca] between the incoming HFSC424 leachate and the reacted fluids, suggested that the maximum possible mass of C-S-H that could have formed would have been very low (i.e., Hillebrandite-type C-S-H < 0.01 g, Tobermorite 14Å-type < 0.02 g). The SI calculations also indicated that zeolites, as with the 1:1 batch experiments, were undersaturated (i.e., SI < 0). In the later fluids (Figure 7e), some zeolites, e.g., Ca-phillipsite, were slightly saturated, which was the same as the batch experiments (Figure 7c). Although no zeolites were seen in this study, other studies with LAC leachates [15] found that as well as C-S-H phases, zeolitic phases could also form in experiments with LAC leachates, but those studies [15] were conducted above ambient temperatures, which would have increased dissolution of the primary minerals, potentially leading to higher saturation states for both C-S-H and zeolitic phases. Primary mineral phases were undersaturated throughout, again suggesting continued dissolution.

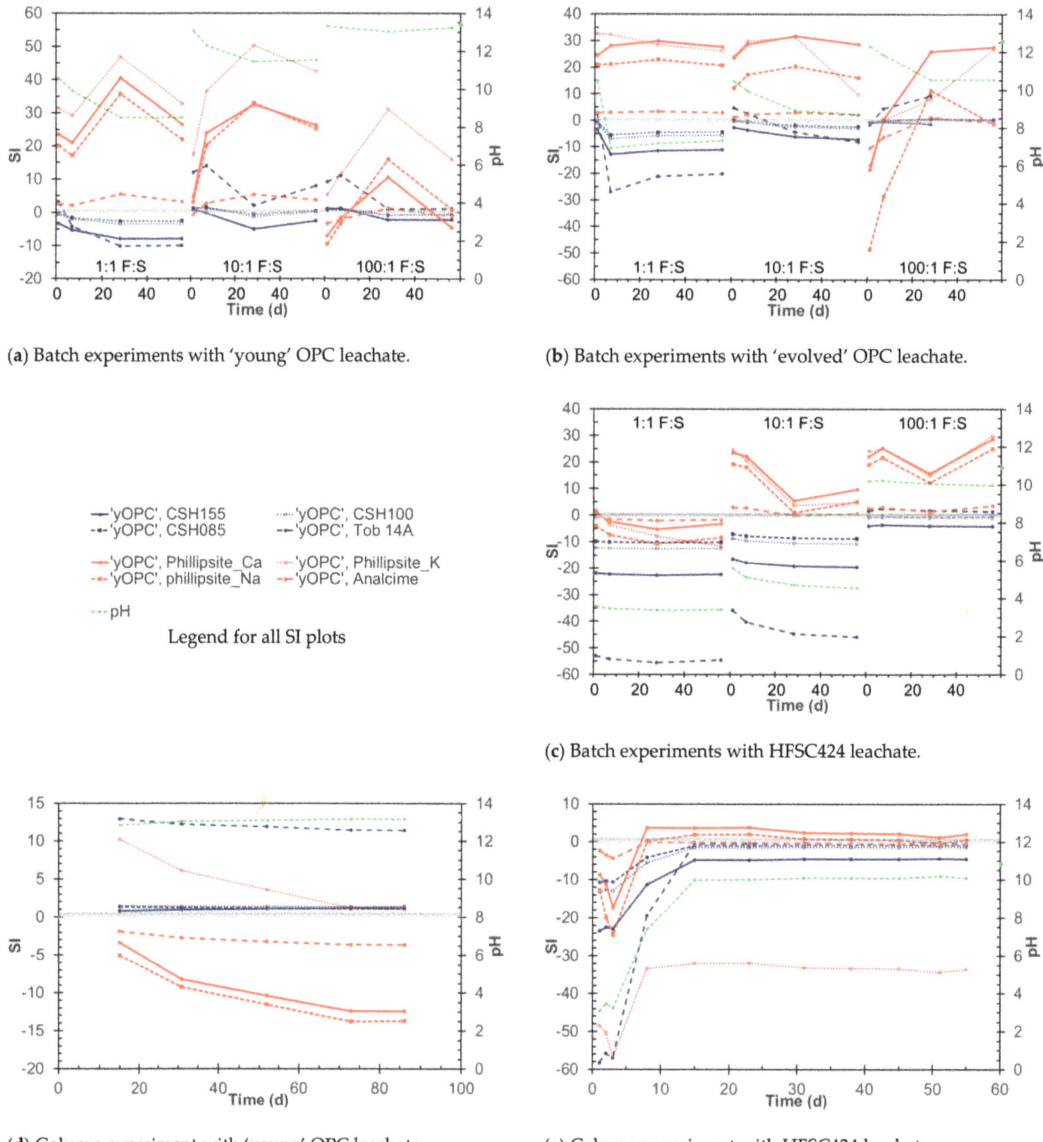



(**a**) Batch experiments with 'young' OPC leachate.

(**b**) Batch experiments with 'evolved' OPC leachate.

(**c**) Batch experiments with HFSC424 leachate.

(**d**) Column experiment with 'young' OPC leachate.

(**e**) Column experiment with HFSC424 leachate.

**Figure 7.** Indicative C-S-H and zeolite mineral saturation indices (SI) with time, together with pH. (**a**) Batch experiments with 'young' OPC leachate. (**b**) Batch experiments with 'evolved' OPC leachate. (**c**) Batch experiments with HFSC424 leachate. (**d**) Column experiment with 'young' OPC leachate. (**e**) Column experiment with HFSC424 leachate. Batch experiments are plotted from low to high F:S ratio for comparison with the flow experiments (C-S-H in blue, zeolites in red, pH in green). Shaded grey line indicates equilibrium state, i.e., SI = 0; below this line minerals will tend to dissolve, and above it they may precipitate.

### 4.3. Implications of Static vs. Dynamic Experiments

As can be seen by the comparison between the results of the batch and flow experiments, the 1:1 batch experiments are analogous to the initial conditions observed in flow experiments and thus representative of the early stage reactions. However, if the host rock contains a minor but highly reactive phase(s) that dominates early stage reaction but is not representative of the long-term evolution, as is the case with the Horonobe mudstone (i.e., dissolution of jarosite and/or gypsum), the low F:S batch experiments are liable to lead to unrealistic conclusions being drawn. In this case, the flow experiments give a better representation of the likely longer-term reactions, since transient reactions (i.e., initial changes in pH) are more easily identified. In addition, although the changes in [Na] and [K] suggested that ion exchange may have occurred in both batch and flow experiments, the higher concentrations of Na and K after ~20 d in the flow experiments suggest that it may only have a transient effect on the fluid chemistry.

The results from the fluid dominated, higher F:S ratio experiments were similar to the latter conditions seen in the flow experiments and give an indication of the longer-term evolution. However, as with the lower F:S batch experiments, there is the possibility of incorrect conclusions being drawn about the long term behavior if one of the species required for secondary phases formation is present only in one component (fluid or solid), and thus in the case of batch experiments it is finite. This is clearly illustrated in the higher F:S batch experiments with the 'young' OPC fluids, where extremely high concentrations of dissolved silica (Figure 3f) were present once the Ca from the original leachate was exhausted (Figure 3f).

The use of different configurations for flow experiments can also give better insight into the sequence of the reaction. In this study, the small flow cell (SFC) had advantages over the columns, for the fast reactive clay system evaluated here, in being able to better capture the spatial changes.

Batch reactors are well suited to long duration studies, being simpler and easier to maintain, and since they will tend towards equilibrium they are essential to the understanding of the long-term fluid–rock interactions. Flow experiments by introducing transport allows the investigation of the spatial and temporal extend of reactions, as well as being able to identify transient processes. In addition, they can be used to identify (key) components, which if constrained by availability, could influence the reaction sequence. Thus, the combination of static and dynamic approaches together with the linking of batch and flow data sets can provide an ideal mix of experimental methods with which to explore the long-term evolution of the alkaline fluid/mudstone system and identify key parameters and constraints. In addition, the configuration of the flow experiments, particularly in terms of length and sampling scale, needs to be appropriate to the (likely) reactivity of the experimental system under investigation. It is also recognized that laboratory experiments tend to yield faster reaction rates (due to use of crushed samples in the lab and the lower permeability, porosity, etc., in the field) compared to those observed in the field [38] and that allowance must be made for this when designing the experiments and extrapolating lab results to the field.

### 4.4. Implications for Spatial and Temporal Evolution

In terms of the spatial extent of the reactions, although there was evidence for continued dissolution of the primary minerals, i.e., trends in silica concentration, and the lack of fines in the reacted solids seen under SEM, the zone of precipitation in the flow experiments was limited to only the first few tens of millimeters in the case of the columns and ~3 mm for the SFC. Although the extent of the zone of precipitation was physically shorter in the SFC experiment compared to the column, comparison of the mass of Horonobe mudstone present in each set-up showed the actual extent of the reaction in each case to be similar. Section 1 of the column contained ~0.7 g of mudstone, whilst the first two sections of the SFC combined together contained ~0.6 g. One of the benefits of using dynamic flow experiments compared to the static batch experiments is in the determination of the extent of the zone of precipitation. In particular, the use of the SFC with its greater cross-sectional area and the ability to prepare thinner sliced samples is advantageous in this study, since it demonstrated that there was a slight delay before the precipitation of secondary phases, likely due to the need for silicate release from primary mineral dissolution and C-(A-)S-H nucleation.

Column experiments have been previously conducted [39], using samples of crushed smectite and OPA, to determine the extent of reaction using a similar simplified OPC leachate to this study, and although larger columns were used than those in this study, even after an 18-month duration the precipitation zone extended only ~20 mm into the OPA, where C-A-S-H phases were observed. The flow experiments presented here were of relatively short duration (<3 months) and were conducted at higher flow rates (potentially increasing the available Ca); nevertheless, the extent of the precipitation zone was similar, being observed only in the first sections (i.e., up to ~20 mm in the column, and ~3 mm with the SFC). The results of this study are also consistent with previous modelling and experimental studies on the extent of the reactions in argillaceous rocks [8,17,40], which all concluded that the zone of precipitation would be limited to only a few centimeters for in situ experiments and even more limited in the case of laboratory experiments [41]. Thus, the available data suggests that for mudstones, the extent of mineral precipitation is likely to be limited to the centimeter scale.

The experiments with the HFSC424 LAC leachates showed much less overall reaction compared to the OPC leachates, particularly in terms of the lack of observation of secondary phases.

There have been few other experimental studies [15,17,18,42] that have considered the impact of the use of LACs. The data available from this and these previous studies demonstrate that the primary objective of reducing pH and minimizing clay mineral dissolution can be achieved. However, some authors [17] have suggested that given the limited extent of reaction between OPC and argillaceous host rocks, the use of LACs is not necessary. Indeed, the use of OPC may favor secondary phase formation, especially that of C-S-H phases, which may then reduce transport of the leachates by reducing the permeability and porosity and so further limit the extent of CDZ. However, as well as changing the transport properties of the host rock, the secondary minerals formed by the reaction of OPC leachates with the host rock may also alter the sorption capacity with respect to radionuclides. Both C-S-H phases, and zeolites if present, may not exhibit an equally strong sorption capacity as the pristine host rock minerals, so reducing the potential for radionuclide retardation by sorption. In addition, sorption onto C-S-H can vary with Ca:Si ratio; for example, Cs sorbs well onto C-S-H phases with a low Ca:Si ratio [43]. This balance between reduced alteration and radionuclide migration may be an important consideration when deciding on the use or not of LACs. As well as influencing the radionuclide migration, the formation of secondary phases may limit access to the underlying reacting minerals by reducing the reactive surface area, since the secondary minerals will form a 'skin' on the primary mineral [17] but also by reducing the available porosity and permeability of the flowing system. The nature of this 'skin' of secondary minerals is not well characterized and will depend upon which secondary phases are precipitated, i.e., C-S-H and/or zeolites; the crystallinity of these secondary phases (for example C-S-H phases often initially form as gels becoming more crystalline with time [7]), which may affect fluid diffusion to the mineral surface; and also on the persistence of the skin, since the secondary phases formed may re-dissolve as the alkaline fluid chemistry evolves with time.

Flow experiments as used in this study can be useful here as the sequence of primary mineral dissolution, secondary mineral precipitation, and persistence can be directly observed. However, the experiments presented here employed crushed materials with high surface areas that will have enhanced the rate reaction and primary dissolution and are time limited.

## 5. Conclusions

In summary, both batch and flow-through experiments were conducted, reacting Horonobe mudstone with ordinary Portland cement (OPC) and low alkali cement (LAC) leachates to gain new insights into the reaction of host rock and such leachates. Typically, in the long term, pH decreased only slightly for all the leachates investigated. Major changes in the fluid chemistry were limited to a decrease in [Ca] due to precipitation of secondary C-(A-)S-H phases and the presence of dissolved silica from the continued dissolution of primary minerals.

The 'young' OPC leachate experiments showed the greatest reaction and alteration of the mudstone, but the zone of precipitation of C-(A-)S-H phases in the flow experiments was limited to only the first few tens of millimeters in the case of the columns, and ~3 mm for the small flow cell. Ion exchange reactions may have initially altered the fluid chemistry, but thereafter the reactions were dominated by primary mineral dissolution and the precipitation of secondary phases, particularly C-(A-)S-H.

It should be noted that unlike OPC there is no standard LAC composition, and that different compositions of LAC have been considered by different radioactive waste disposal implementers. It is also worth considering that disposal sites may have a combination of both OPC and LAC cements used for construction, so the effects of a mixed leachate or different spatially distributed leachates as well as different LAC compositions may need to be investigated and understood.

The results presented here demonstrate that a combination of both batch and flow-through experiments can provide the insights required for the understanding of the key geochemical interactions and the impact of transport, allowing the spatial as well as temporal evolution of the alkaline leachate/mudstone system to be successfully investigated. However, these experiments employed crushed materials, so the use of intact samples and in situ experiments with lower more realistic flow rates, together with data from monitoring boreholes in existing underground structures where cement and concretes have been deployed, is required to understand fully the spatial extent of the reactions to evaluate mineral evolution and the effect on radionuclide migration.

In addition, this research concentrated on the chemical and mineralogical effects within the mudstone and did not study the impact on the radionuclide migration, so further work with the altered mineral assemblage is required to determine the extent of the impact on the radionuclide behavior. The data from the experiments here can provide 'test cases' for the validation and calibration of reactive transport geochemical models, which can then be used to model the longer term evolution of the fluid rock interactions beyond what is possible experimentally. The experiments presented here suggest, together with other studies of argillaceous rocks, that in a mudstone, similar in composition to the Horonobe mudstone studied here, that the zone of greatest reaction in the CDZ is likely to be limited in extent, perhaps on the scale of centimeters.

**Supplementary Materials:** The following are available online at https://www.mdpi.com/article/10.3390/min11060588/s1, Figure S1: XRD analysis of mudstone samples from 10:1 F:S batch experiments with OPC leachates. Figure S2: XRD analysis of mudstone samples from SFC flow experiment with the 'young' OPC leachate. Figure S3 Indicative M-S-H saturation indices (SI) with time, together with pH.

**Author Contributions:** The individual contributions to this paper are as follows: Conceptualization and methodology, K.B., Y.A., and Y.T.; validation, K.B., M.K., and Y.O.; investigation and formal analysis, K.B., M.K., and Y.O.; resources, Y.T. and Y.A.; data curation, K.B. and Y.O.; writing—original draft preparation, K.B.; writing—review and editing, all authors. All authors have read and agreed to the published version of the manuscript.

**Funding:** This research received no external funding.

**Data Availability Statement:** The data used in this study is available from the authors upon request.

Acknowledgments: We thank Hikari Beppu and Takashi Endo for assistance with fluid chemical analysis. This study was supported by the JAEA Horonobe Underground Research Center, Hokkaido, Japan, providing the rock samples and background information.

Conflicts of Interest: The authors declare no conflict of interest.

## References

1. JNC. H12: Project to Establish the Scientific and Technical Basis for HLW Disposal in Japan, Project Overview Report and three Supporting Reports. JNC TN1410 2000–001~004; 2000. Available online: https://jopss.jaea.go.jp/pdfdata/JNC--TN1410--2000--003.pdf (accessed on 26 April 2021).

2. Baker, A.J.; Bateman, K.; Hyslop, E.K.; Ilett, D.J.; Linklater, C.M.; Milodowski, A.E.; Noy, D.J.; Rochelle, C.A.; Tweed, C.J. Research on the alkaline disturbed zone resulting from cement–water–rock reactions around a cementitious repository. In *NIREX REPORT N/054*; UK Nirex Ltd.: Harwell, UK, 2002.

3. JAEA. Second Progress Report on Research and Development for TRU Waste Disposal in Japan; Repository Design, Safety Assessment and Means of Implementation in the Generic Phase (TRU–2). JAEA–Review 2007–010, 2007/03; 2007. Available online: https://www.jaea.go.jp/04/be/documents/doc_02.html (accessed on 26 April 2021).

4. Falck, W.E.; Nilsson, K.-F. Geological Disposal of Radioactive Waste–Moving Towards Implementation. In *JRC Reference Reports, JRC45385 (EUR 23925 E)*; European Commission: Brussels, Belgium, 2009. Available online: http://ec.europa.eu/dgs/jrc/downloads/jrc_reference_report_2009_10_geol_disposal.pdf (accessed on 26 March 2021).

5. Atkinson, A. The time dependence of pH within a repository for radioactive waste disposal. In *UKAEA Report AERE–R 11777*; UKAEA: Harwell, UK, 1985.

6. Berner, U. Evolution of pore water chemistry during degradation of cement in a radioactive waste repository environment. *Waste Manag. Nucl. Fuel Cycle* **1992**, *12*, 201–219. [CrossRef]

7. Savage, D.; Walker, C.S.; Arthur, R.C.; Rochelle, C.A.; Oda, C.; Takase, H. Alteration of bentonite by hyperalkaline fluids: A review of the role of secondary minerals. *Phys. Chem. Earth* **2007**, *32*, 287–297. [CrossRef]

8. Savage, D. A review of analogues of alkaline alteration with regard to long–term barrier performance. *Mineral. Mag.* **2011**, *75*, 2401–2418. [CrossRef]

9. Moyce, E.B.A.; Rochelle, C.A.; Morris, K.; Milodowski, A.E.; Chen, X.; Thornton, S.; Small, J.S.; Shaw, S. Rock alteration in alkaline cement waters over 15 years and its relevance to the geological disposal of nuclear waste. *Appl. Geochem.* **2014**, *50*, 91–105. [CrossRef]

10. Bauer, A.; Berger, G. Kaolinite and smectite dissolution rate in high molar KOH solutions at 35 °C and 80 °C. *Appl. Geochem.* **1998**, *13*, 905–916. [CrossRef]

11. Claret, F.; Bauer, A.; Schäfer, T.; Griffault, L.; Lanson, B. Experimental Investigation of the Interaction of Clays with High–pH Solutions: A Case Study from the Callovo–Oxfordian Formation, Meuse–Haute Marne Underground Laboratory (France). *Clays Clay Miner.* **2002**, *50*, 633–646. [CrossRef]

12. Ramirez, S. Alteration of the Callovo–Oxfordian clay from Meuse–Haute Marne underground laboratory (France) by alkaline solution. I. A XRD and CEC study. *Appl. Geochem.* **2005**, *20*, 89–99. [CrossRef]

13. Adler, M.; Mäder, U.K.; Waber, H.N. Core infiltration experiment investigating high pH alteration of low–permeability argillaceous rock at 30 °C. In Proceedings of the 10th International Symposium Water Rock Interaction (WRI–10), Villasimius, Italy, 10–15 July 2001; pp. 1299–1302.

14. Pusch, R.; Zwahr, H.; Gerber, R.; Schomburg, J. Interaction of cement and smectitic clay–theory and practice. *Appl. Clay Sci.* **2003**, *23*, 203–210. [CrossRef]

15. Lothenbach, B.; Bernard, E.; Mäder, U. Zeolite formation in the presence of cement hydrates and albite. *Phys. Chem. Earth Parts A/B/C* **2017**, *99*, 77–94. [CrossRef]

16. Idiart, A.; Laviña, M.; Kosakowski, G.; Cochepin, B.; Meeussen, J.C.L.; Samper, J.; Mon, A.; Montoya, V.; Munier, I.; Poonoosamy, J.; et al. Reactive transport modelling of a low–pH concrete/clay interface. *Appl. Geochem.* **2020**, *115*, 104562. [CrossRef]

17. Mäder, U.; Jeni, A.; Lerouge, C.; Gaboreau, S.; Miyoshi, S.; Kimura, Y.; Cloet, V.; Fukaya, M.; Claret, F.; Otake, T.; et al. 5-year chemico-physical evolution of concrete-claystone interfaces, Mont Terri rock laboratory (Switzerland). *Swiss J. Geosci.* **2017**, *110*, 307–327. [CrossRef]

18. Dauzeres, A.; Le Bescop, P.; Sardini, P.; Cau Dit Coumes, C.; Brunet, F.; Bourbon, X.; Timonen, J.; Voutilainen, M.; Chomat, L.; Sardini, P. On the physico–chemical evolution of low–pH and CEM I cement pastes interacting with Callovo–Oxfordian pore water under its in situ $CO_2$ partial pressure. *Cem. Concr. Res.* **2014**, *58*, 76–88. [CrossRef]

19. Vehmas, T.; Maria Cruz Alonso, V.M.; Vašíček, R.; Rastrick, E.; Gaboreau, S.; Večerník, P.; Leivo, M.; Holt, E.; Fink, N.; Mouheb, N.A.; et al. Characterization of Cebama low–pH reference concrete and assessment of its alteration with representative waters in radioactive waste repositories. *Appl. Geochem.* **2020**, *121*, 104703, ISSN 0883–2927. [CrossRef]

20. Iriya, K.; Matsui, A.; Mihara, M. Study on applicability of HFSC for radioactive waste repositories. In Proceedings of the 7th International Conference on Radioactive Waste Management and Environmental Radiation, Nagoya, Japan, 26–30 September 1999.

21. Mihara, M.; Iriya, K.; Torii, K. Development of low–alkaline cement using pozzolans for geological disposal of long–lived radioactive waste. *Doboku Gakkai Ronbunshuu F* **2008**, *64*, 92–103. [CrossRef]

22. Rochelle, C.A.; Milodowski, A.E.; Bateman, K.; Moyce, E.B.A. A long–term experimental study of the reactivity of basement rock with highly alkaline cement waters: Reactions over the first 15 months. *Mineral. Mag.* **2016**, *80*, 1089–1113. [CrossRef]
23. Levenspiel, O. *Chemical Reaction Engineering*, 3rd ed.; Wiley: Hoboken, NJ, USA, 1998; ISBN 978-0-471-25424-9.
24. Bateman, K.; Coombs, P.; Pearce, J.M.; Wetton, P.D. Nagra/Nirex/SKB Column Experiments; Fluid Chemical and Mineralogical Studies. In *British Geological Survey Report WE/95/26*; British Geological Survey: Nottingham, UK, 2001.
25. Bateman, K.; Coombs, P.; Pearce, J.M.; Noy, D.J.; Wetton, P.D. Fluid Rock Interactions in the Disturbed Zone: Nagra/Nirex/SKB Column Experiments—Phase II. In *British Geological Survey Report WE/99/5*; British Geological Survey: Nottingham, UK, 2001.
26. Small, J.S.; Byran, N.; Lloyd, J.R.; Milodowski, A.E.; Shaw, S.; Morris, K. Summary of the BIGRAD Project and Its Implications for a Geological Disposal Facility. In *Report NNL*; 2016; Volume 16, p. 13817. Available online: https://rwm.nda.gov.uk/publication/summary--of--the--bigrad--project--and--its--implications--for--a--geological--disposal--facility/ (accessed on 26 March 2021).
27. Mazurek, M.; Eggenberger, U. Mineralogical analysis of core samples from the Horonobe area. In *RWI Technical Report 05–01*; Institute of Geological Sciences, University of Bern: Bern, Switzerland, 2005.
28. Hiraga, N.; Ishii, E. Mineral and Chemical Composition of Rock Core and Surface Gas Composition in Horonobe Underground Research Laboratory Project (Phase 1). In JAEA Technical Report JAEA–Data/Code 2007–022; Japan Atomic Energy Agency, Tokai–Mura, Japan. 2008. Available online: https://jopss.jaea.go.jp/pdfdata/JAEA-Data-Code-2007-022.pdf (accessed on 16 May 2021).
29. Savage, D.; Hughes, C.R.; Milodowski, A.E.; Bateman, K.; Pearce, J.; Rae, E.; Rochelle, C.A. The evaluation of chemical mass transfer in the disturbed zone of a deep geological disposal facility for radioactive wastes. I. Reaction of silicates with Calcium hydroxide fluids. In *Nirex Report NSS/R244*; UK Nirex Ltd: Harwell, UK, 1998.
30. Savage, D.; Bateman, K.; Hill, P.; Hughes, C.R.; Milodowski, A.E.; Pearce, J.; Rochelle, C.A. The evaluation of chemical mass transfer in the disturbed zone of a deep geological disposal facility for radioactive wastes. II. Reaction of silicates with Na–K–Ca–hydroxide fluids. In *Nirex Report NSS/R283*; UK Nirex Ltd: Harwell, UK, 1998.
31. Parkhurst, D.L.; Appelo, C.A.J. Description of Input and Examples for PHREEQC Version 3–A Computer Program for Speciation, Batch–Reaction, One–Dimensional Transport, and Inverse Geochemical Calculations. U.S. Geological Survey Techniques and Methods, Book 6, 2013, Chapter A43, p. 497. Available online: https://pubs.usgs.gov/tm/06/a43/ (accessed on 26 April 2021).
32. JAEA. *The Project for Validating Assessment Methodology in Geological Disposal System*; Annual Report for JFY2016, JAEA Technical Report; Japan Atomic Energy Agency (JAEA): Ibaraki, Japan, 2017. (In Japanese)
33. Mochizuki, A.; Ishii, E.; Miyakawa, K.; Sasamoto, H. Mudstone redox conditions at the Horonobe Underground Research Laboratory, Hokkaido, Japan: Effects of drift excavation. *Eng. Geol.* **2020**, *267*, 105496. [CrossRef]
34. Berner, U.; Kulik, D.A.; Kosakowski, G. Influence of a low–pH cement liner on the near field of a repository for spent fuel and high–level radioactive waste. *Phys. Chem. Earth* **2013**, *64*, 46–56. [CrossRef]
35. Bernard, E.; Lothenbach, L.; Cau–Dit–Coumes, C.; Chlique, C.; Dauzères, A.; Pochard, I. Magnesium and calcium silicate hydrates, part I: Investigation of the possible incorporation in calcium silicate hydrate (C–S–H) and of the calcium in magnesium silicate hydrate (M–S–H). *Appl. Geochem.* **2018**, *89*, 229–242. [CrossRef]
36. Bernard, E.; Lothenbach, B.; Chlique, C.; Wyrzykowski, M.; Dauzères, A.; Pochard, I. Cau–Dit–Coumes, C. Characterization of magnesium silicate hydrate (M–S–H). *Cem. Concr. Res.* **2019**, *116*, 309–330. [CrossRef]
37. Bernard, E.; Jenni, A.; Fisch, M.; Grolimund, D.; Mäder, U. Micro–X–ray diffraction and chemical mapping of aged interfaces between cement pastes and Opalinus Clay. *Appl. Geochem.* **2020**, *115*, 104538. [CrossRef]
38. White, A.F.; Brantley, S.L. The effect of time on the weathering of silicate minerals: Why do weathering rates differ in the laboratory and field? *Chem. Geol.* **2003**, *202*, 479–506. [CrossRef]
39. Taubald, H.; Bauer, A.; Schafer, T.; Geckeis, H.; Satir, M.; Kim, J.I. Experimental investigation of the effect of high–pH solutions on the Opalinus Shale and the Hammerschmiede Smectite. *Clay Miner.* **2000**, *35*, 515–524. [CrossRef]
40. Torres, E.; Turrero, M.J.; Escribano, A.; Martín, P.L. *Geochemical Interactions at the Concrete–Bentonite Interface of Column Experiments. Long–Term Performance of Engineered Barrier Systems PEBS: DELIVERABLE (D–N°: D2.3–6–1)*. 2013. Available online: https://www.pebs--eu.de/PEBS/EN/Downloads/D2_3_6_1.pdf?__blob=publicationFile&v=2 (accessed on 17 March 2021).
41. Shafizadeh, A.; Gimmi, T.; Van Loon, L.R.; Kaestner, A.P.; Mäder, U.K.; Churakov, S.V. Time–resolved porosity changes at cement–clay interfaces derived from neutron imaging. *Cem. Concr. Res.* **2020**, *127*, 105924. [CrossRef]
42. Huertas, F.; Farias, J.; Griffault, L.; Leguey, S.; Cuevas, J.; Ramirez, S.; Vigil de la Villa, R.; Cobeňa, J.; Andrade, C.; Alonso, M.C.; et al. *Effects of Cement of Clay Barrier Performance, ECOCLAY project: Final Report*; Report EU 19609; European Commission: Brussels, Belgium, 2000.
43. Missana, T.; García–Gutiérrez, M.; Mingarro, M.; Alonso, U. Comparison between caesium and sodium retention on calcium silicate hydrate (CSH) phases. *Appl. Geochem.* **2018**, *98*, 36–44, ISSN 0883-2927. [CrossRef]

*Article*

# Geochemical, Geotechnical, and Microbiological Changes in Mg/Ca Bentonite after Thermal Loading at 150 °C

Vlastislav Kašpar [1,2,*], Šárka Šachlová [1], Eva Hofmanová [1], Bára Komárková [3,4], Václava Havlová [1], Claudia Aparicio [5], Kateřina Černá [6], Deepa Bartak [6] and Veronika Hlaváčková [6]

[1] Fuel Cycle Chemistry Department, ÚJV Řež, a. s., Hlavní 130, 250 68 Husinec, Czech Republic; sarka.sachlova@ujv.cz (Š.Š.); eva.hofmanova@ujv.cz (E.H.); vaclava.havlova@ujv.cz (V.H.)

[2] Department of Water Landscape Conservation, Faculty of Civil Engineering, Czech Technical University in Prague, Thákurova 7, 166 29 Praha 6, Czech Republic

[3] Centre of Instrumental Techniques, Institute of Inorganic Chemistry of the Czech Academy of Sciences, Husinec-Řež 1001, 250 68 Husinec, Czech Republic; komarkova@iic.cas.cz

[4] Department of Chemistry, Faculty of Science, University of Ostrava, 30. dubna 22, 701 03 Ostrava, Czech Republic

[5] Department Material and Mechanical Properties, Centrum výzkumu Řež s.r.o., Hlavní 130, 250 68 Husinec, Czech Republic; claudia.aparicio@cvrez.cz

[6] Institute for Nanomaterials, Advanced Technologies and Innovation, Technical University of Liberec, Bendlova 7, 460 01 Liberec, Czech Republic; katerina.cerna1@tul.cz (K.Č.); deepa.bartak@tul.cz (D.B.); veronika.hlavackova@tul.cz (V.H.)

* Correspondence: vlastislav.kaspar@ujv.cz; Tel.: +420-266172087

**Citation:** Kašpar, V.; Šachlová, Š.; Hofmanová, E.; Komárková, B.; Havlová, V.; Aparicio, C.; Černá, K.; Bartak, D.; Hlaváčková, V. Geochemical, Geotechnical, and Microbiological Changes in Mg/Ca Bentonite after Thermal Loading at 150 °C. *Minerals* **2021**, *11*, 965. https://doi.org/10.3390/min11090965

Academic Editors: Ana María Fernández, Stephan Kaufhold, Markus Olin, Lian-Ge Zheng, Paul Wersin and James Wilson

Received: 30 July 2021
Accepted: 31 August 2021
Published: 3 September 2021

**Publisher's Note:** MDPI stays neutral with regard to jurisdictional claims in published maps and institutional affiliations.

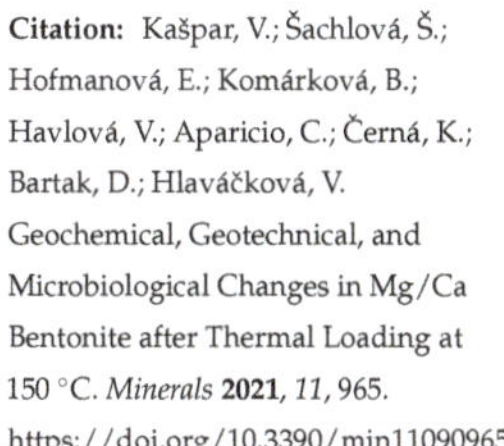

**Abstract:** Bentonite buffers at temperatures beyond 100 °C could reduce the amount of high-level radioactive waste in a deep geological repository. However, it is necessary to demonstrate that the buffer surrounding the canisters withstands such elevated temperatures, while maintaining its safety functions (regarding long-term performance). For this reason, an experiment with thermal loading of bentonite powder at 150 °C was arranged. The paper presents changes that the Czech Mg/Ca bentonite underwent during heating for one year. These changes were examined by X-ray diffraction (XRD), thermal analysis with evolved gas analysis (TA-EGA), aqueous leachates, Cs sorption, cation exchange capacity (CEC), specific surface area (SSA), free swelling, saturated hydraulic conductivity, water retention curves (WRC), quantitative polymerase chain reaction (qPCR), and next-generation sequencing (NGS). It was concluded that montmorillonite was partially altered, in terms of the magnitude of the surface charge density of montmorillonite particles, based on the measurement interpretations of CEC, SSA, and Cs sorption. Montmorillonite alteration towards low- or non-swelling clay structures corresponded well to significantly lower swelling ability and water uptake ability, and higher saturated hydraulic conductivity of thermally loaded samples. Microbial survivability decreased with the thermal loading time, but it was not completely diminished, even in samples heated for one year.

**Keywords:** magnesium bentonite; radioactive waste disposal; thermal loading; montmorillonite content; thermal analysis with evolved gas analysis; cation exchange capacity; specific surface area; saturated hydraulic conductivity; water retention curves; microbial survivability

## 1. Introduction

Spent nuclear fuel (SNF) generates significant heat via radioactive decay. After a period of aboveground cooling, a deep geological repository (DGR) is considered the only suitable way to dispose of such high-level radioactive waste. For many DGR concepts, the temperature on the interface between disposal canisters and bentonite buffer is typically designed to be lower than 100 °C. If a higher temperature is allowed (beyond 100 °C), the cooling and storage period could shorten, or more SNF assemblies could be loaded in the disposal canister. Both would positively affect the economy of radioactive waste disposal, especially when considering size optimisation [1]. However, possible changes in DGR designs must be justified, by proving that the performance targets for the engineered barrier safety functions are fulfilled, even at an elevated temperature, and confirming that the repository ensures post-closure radiation safety.

The thermal criterion of not exceeding a buffer peak temperature of 100 °C was chosen to avoid bentonite alteration [2]. Alteration here refers to mineral transformations of montmorillonite, leading to low- or non-swelling clay structures, which may not meet the buffer safety criteria for swelling pressure and hydraulic conductivity. Several possible processes, such as atomic substitutions (eventually, iron reduction/oxidation) in the mineral structure, layer charge redistribution, balanced by the compensating cations and congruent dissolution, are involved in the montmorillonite transformation [2]. The main alteration processes involve the transformation to illite/montmorillonite mixed layers, illite, beidellite, saponite, or chlorite [2,3].

Moreover, bentonite soluble components dissolve under hydrothermal conditions and the solutes may interact with silica released by the montmorillonite transformation, resulting in precipitation of new Si phases (so-called cementation), also contributing to a reduction of buffer plasticity and an increase of hydraulic conductivity [2,3]. Cementation can also be caused by non-Si compounds (e.g., sulphate and carbonate minerals), showing increased stability at higher temperatures [3].

Based on the critical review of the performance of the bentonite barrier, at temperatures beyond 100 °C, Wersin et al. [3] supported the idea of raising the thermal criterion to 120 °C; they identified higher thermal stability of bentonite under dry conditions (up to 300 °C) and noted the scarce data on hydraulic and mechanical properties of heat-exposed bentonites. The highest temperature yet applied in-situ was in the ABM5 experiment at Äspö Hard Rock Laboratory (Sweden). Temperatures up to 250 °C were achieved at the steel heater interface with compacted bentonites [1,4]. No significantly different chemical and mineralogical reactions were observed compared to the previous ABM tests or similar tests performed in the temperature range of 90–130 °C. In this test, the extent of magnesium enrichment, in the vicinity of the heater, observed in many large-scale tests, was small [4]. No data on the characterisation of bentonites from the ABM5 experiment from the geotechnical point of view is yet available. Complete ABM experiments provided geotechnical data for MX-80, Asha 505, Deponit CAN, and Friedland clay. Slightly lower swelling pressure was determined for heat-exposed materials Asha 505, Deponit CAN, and Friedland clay. No changes in hydraulic conductivities between the above reference and in-situ exposed materials were observed [5–7].

The influence of higher temperatures on bentonite performance and the lack of data from the thermo-hydro-mechanical (THM) behaviour of clay-based materials led to identifying this topic as a high priority subject within the framework of the EURAD project—WP7 HITEC. In general, two approaches are used to investigate thermally induced effects on clay-based materials.

The first approach examined HM properties at different temperatures. Geotechnical parameters of water-saturated compacted bentonites (FEBEX, MX-80, and Korean Ca-bentonite) were determined at temperatures ranging from 20 to 150 °C [8–12]. Decreasing swelling pressure and suction pressure and increasing hydraulic conductivity with temperature were assigned to the temperature-dependent water properties and physico-chemical interactions of water at the microscopic level [9]. However, once corrected for temperature-dependent water density and viscosity, the intrinsic permeability was not significantly temperature dependent [11,12]. This approach demonstrates buffer safety functions (namely swelling pressure and hydraulic conductivity) at all thermal stages within the DGR evolution.

In the second approach, the material property changes of preheated material were analysed. The material was mainly provided from in-situ experiments, which were designed to simulate the heat released from waste canisters containing radioactive waste. MX-80 bentonite, the "reference" one, and the one exposed to the temperature of 130 °C at the copper heater in the form of parcel samples trimmed to the experimental cells, or re-compacted parcel samples to the original bulk density up to 2000 kg/m$^3$, was subjected to the determination of hydraulic conductivity (among other determinations). Slightly lower hydraulic conductivity in the trimmed samples in comparison to the re-compacted ones was observed. However, no significant difference between samples from different positions (warm, cool) in the parcel was observed; therefore, the decrease could not be related to temperature gradients [13]. A mixture of FoCa clay with 35 wt.% sand and 5 wt.% graphite was used in OPHELIE mock-up experiment where a temperature up to 140 °C was maintained for 4.5 years [14]. Slightly higher hydraulic conductivity of exposed material was determined. The authors explain this fact with thermally induced changes of the sample microstructure. An increase in intermediate pore sizes within the samples with temperature, which caused an increase in water and the cross-sectional area available for water to flow, was indicated by mercury intrusion porosimetry. Moreover, water retention curves were determined. Thermally loaded samples showed lower water retention capacity than the initial ones for the lower suction levels. For suction pressure over 150 MPa, no difference was observed. Thermally loaded samples showed less pronounced hysteresis than the initial ones, which was also attributed to the microstructural changes mentioned above [14].

Natural clay materials (e.g., bentonites) also represent an important source of indigenous microorganisms [5,15]. The microbial activity might negatively affect the long-term safety of DGR as it can result in microbiologically influenced corrosion of the canister and possibly alter the bentonite mineralogy [16,17]. Microbial survival and growth are generally inhibited by thermal and radiation performance after deposition of the canister [17]. However, bacterial endospores that are microbial survival forms with reduced water content, and undetectable metabolic activities, can tolerate adverse environmental conditions, such as desiccation, wet and dry heat (above 100 °C), and UV and gamma irradiation for extended periods [18,19]. Several studies in the bentonite environment have also demonstrated microbial survivability. Masurat et al. [20,21] found a loss of sulphate-reducing bacteria (SRB) viability after loading MX-80 bentonite at 120 °C for 20 h. However, another experiment with even more prolonged heating for one week failed to eradicate SRB [22]. Similarly, the presence of indigenous SRB, which survived heat loading of Boom Clay at 120 °C for 48 h, was reported [23]. Eleven compacted bentonites were exposed to groundwater saturation and high temperatures (90–130 °C) in the in-situ experiment—Alternative Buffer Material at the Äspö Hard Rock Laboratory—for approximately one year. Microbial analyses of three bentonites within the first test package showed a detrimental effect of heat on bacterial activity. The bacterial growth was scarce; only mesophilic aerobic heterotrophs in the range of $10^2$–$10^3$ g$^{-1}$ wet weight were cultivated [5].

It can be presumed that, during the early stage of disposal, when the buffer is exposed to an elevated temperature induced by the radioactive waste and, concurrently, is not yet saturated by the groundwater, the microbial activity would be insignificant. However, the safety functions of bentonite may be affected at this stage. Therefore, it is crucial to identify potential changes in the safety functions of bentonite when exposed to elevated temperatures. This work explores Czech Mg/Ca bentonite before and after medium-term thermal loading at 150 °C in a powder state. Several techniques and methods were chosen for geochemical and geotechnical characterisation (e.g., XRD, TA-EGA, aqueous leachates, CEC, SSA, free swelling, saturated hydraulic conductivity, WRC). In addition, microbiological methods were used to investigate bacterial survivability in thermally loaded bentonite and the ability to regenerate from spores in favourable cultivation conditions.

## 2. Materials and Methods

### 2.1. Bentonite Powder

It is anticipated that Czech bentonites will be used in the construction of engineered barriers in the future Czech DGR project [24]. For this study, a commercial bentonite product originating from the Černý vrch deposit (Keramost PLC, Most, Czech Republic) was chosen. Bentonite Černý vrch (BCV) is dominated by montmorillonite, with prevailing divalent exchangeable cations (mainly magnesium). Basic characteristics of the BCV bentonite were summarised in an HITEC SotA report [1].

Since December 2019, approximately 10 kg of bentonite powder in open stainless-steel vessels have been exposed to a 150 °C temperature in a heating chamber (Binder, Tuttlingen, Germany). This set-up was designed to simulate an early post-closure state of the DGR; when radioactive decay generates a significant amount of heat, oxygen is still available, and the buffer is not saturated by the groundwater. Overall, thermal loading of bentonite powder was planned for two years. This paper deals with the characterisation of the initial BCV bentonite (denoted as BCV_IN) and two subsamples taken from the thermally loaded bentonite after a year and a half (denoted as BCV_0.5_y and BCV_1.0_y, respectively).

The first sampling revealed a colour change of bentonite on the surface, from light brown to dark brown (see Figure 1). The thin surface layer was sampled as a separate sample. The same procedure was then followed for the second sampling, after one year. This paper only focuses on the characterisation of the bentonite powder beneath the thin surface layer. This sample defines the barrier properties and allows evaluating the impact of the thermal loading. The effect of the potential thermally induced surface buffer layer near the containers on the DGR is expected to be negligible (regarding long-term safety) due to the minimal thickness. However, future research should focus on the properties of the thermally induced surface layer and the extent to which it will form in a compacted state.

**(a)**        **(b)**

**Figure 1.** (**a**) Thermal loading of bentonite powder in open vessels, and (**b**) visible colour change from light brown to dark brown at the powder surface observed during sampling.

### 2.2. Geochemical Characterisation

Bentonite powder samples were subjected to various determinations, providing insight into geochemical properties, such as the composition of major minerals, the identification of gaseous products evolved upon heating, or soluble phases released on contact with water. Cation exchange capacity (CEC), specific surface area (SSA), and caesium sorption determination were selected to characterise the clay component in bentonite.

Powder X-ray diffractometry (XRD) was conducted on an Empyrean third generation diffractometer (Malvern Panalytical B.V., Almelo, Netherlands) with the following measurement conditions: Co-$K_\alpha$ radiation, a PIXcel3D detector, a range 3.5–105° 2θ, a step of 0.026° 2θ, and a total counting time of 5.25 h. Randomly oriented powder mounts (with inner ZnO standard) and clay fraction (oriented specimens) were analysed. Minerals were identified using HighScore Plus version 4.8 (Malvern Panalytical B.V., Almelo, The Netherlands) software and PDF-4+ 2020 database. Quantitative analysis was performed using a graphical user interface Profex v.4.2.1 (Nicola Döbelin, Solothurn, Switzerland) [25].

Thermal analysis with evolved gas analysis (TA-EGA) was performed on SetSys Evolution's apparatus (Setaram, Calurie-et-Curie, France). The weight of the as-received samples was approximately 10 mg. Thermal degradations of bentonites were studied under an argon atmosphere (flow of 60 mL/min) in $\alpha$-$Al_2O_3$ crucible. The heating rate was set to 10 °C/min in the temperature range of 30–1000 °C. Gases that evolved during heating were analysed with a quadrupole mass spectrometer QMG 700 (Pfeiffer Vacuum, Asslar, Germany) connected to a thermal analyser through a Supersonic System (Setaram, Calurie-et-Curie, France). Specific fragments corresponding to water, carbon dioxide, and sulphur oxides, were recorded.

The soluble salts were identified based on aqueous leachates. The dried (105 °C) samples were contacted with deionised water (40 mL) at three solid-to-liquid ratios for seven days. The suspensions were centrifuged, and supernatants were filtered using a 0.2 µm syringe filter. The main cations in the filtrates were analysed using atomic absorption spectrometer SavantAA (GBC Scientific Equipment, Braeside, Australia). The total alkalinity of the filtrates was determined by potentiometric titration using Titralab TIM800 (Radiometer Analytical, Loveland, USA). The concentrations of sulphates, chlorides, and fluorides were determined by ion chromatography (ALS Global, Prague, Czech Republic).

The highest solid-to-liquid ratio was also tested for caesium sorption. After seven days of shaking, the bentonite suspension of 200 g/L, aliquots of CsCl spiked with $^{134}$Cs were added and interacted for another seven days. Two initial caesium concentrations of spiking aliquots (0.001 mol/L and 6.4 mol/L) were chosen. Aliquots from supernatants, separated by centrifugation, were determined for $^{134}$Cs activity on gamma counter 1480 Wizard 3 (Perkin Elmer, Waltham, USA). Blank samples (without solid phase) were prepared and processed in the same way to evaluate the distribution coefficient of Cs. Each sorption experiment under the given conditions was performed in duplicate.

The CEC and exchangeable cations were determined by the Cu(II)-triethylenetetramine (Cu-trien) method [26,27]. Cu-trien (0.01 mol/L) was mixed with dried (105 °C) samples in the solid-to-liquid ratio of 25 g/L. After interaction for 30 min, the suspensions were centrifuged, and supernatants were analysed. The $Cu^{2+}$ concentration was determined by UV/Vis spectrophotometer Specord 205 (Analytik Jena, Jena, Germany). The concentration of displaced cations ($Na^+$, $K^+$, $Ca^{2+}$, and $Mg^{2+}$) was determined by atomic absorption spectrometer SavantAA (GBC Scientific Equipment, Braeside, Australia). Two values of cation exchange capacity were derived: $CEC_{Vis}$ from copper depletion and $CEC_{SUM}$ by summing equivalent concentrations of displaced cations.

The specific surface area (SSA) was determined based on the sorption of ethylene glycol monoethyl ether (EGME) [28,29]. Moreover, 1 g of bentonite powder (dried at 105 °C) was mixed with 2 mL of EGME and equilibrated in a desiccator with $CaCl_2$-EGME solvate. The applied procedure [30] consisted of regular desiccator evacuation and monitoring of the samples until a constant weight was reached. The SSA was derived from the mass ratio of adsorbed EGME and dried bentonite sample.

### 2.3. Geotechnical Characterisation

Free swelling tests, determination of water retention curves (WRC), and saturated hydraulic conductivity were selected for geotechnical characterisation. Free swelling tests were performed with 10 mL deionised water and 0.6 g powders (previously dried at 105 °C). This solid-to-liquid ratio allows a better determination of the swell index in graduated cylinders for low swelling materials (such as Ca- and Mg-bentonites), highlighting the differences between the input and the thermally loaded bentonite samples. The standard ratio of 0.02 g/L [31] did not allow the determination of the swell index nor the comparison between the samples.

Bentonite dry densities of 1400, 1600, and 1800 kg/m$^3$ were chosen for determining saturated hydraulic conductivity and WRC. Bentonite powder was compacted in the cylindrical space of a cell body (diameter of 30 mm, thickness of 15 mm) using a hydraulic press MEGA 11-300 DM1S (FORM+TEST Seidner & Co., Riedlingen, Germany). In the WRC experiments, a hole was drilled for placing a wireless sensor (diameter of 17 mm, depth of 6 mm). The compacted sample was closed with end plates composed of stainless steel fabrics placed on the carbon composite percolator [32]. Three-quarter section views of the experimental cells are shown in Figure 2.

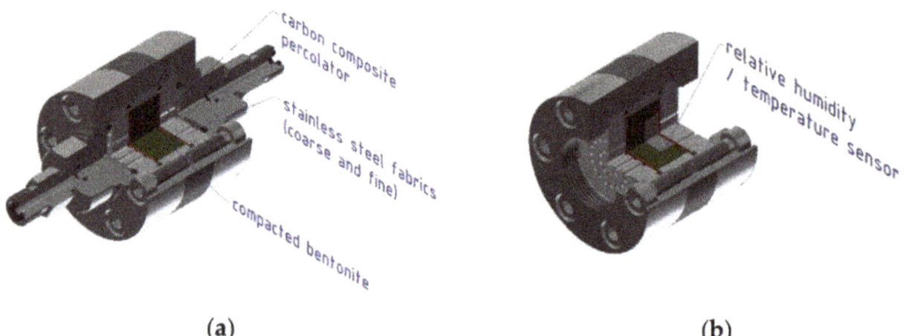


**Figure 2.** Experimental cell set-up for measuring (**a**) saturated hydraulic conductivity and (**b**) water retention curves.

Saturated hydraulic conductivity of compacted bentonite was determined according to ISO 17892-11:2019. The cells were placed into a desiccator filled with deionised water. The desiccator was regularly vacuumed for at least two weeks. After water saturation, the cell was connected via pressure hoses to the assembly for measuring saturated hydraulic conductivity consisted of an input and output pressure controller (HPDPC and ELDPC, GDS Instruments, Hook, UK). Hydraulic heads of 5, 8, and 10 MPa were applied for the dry density of 1400, 1600, and 1800 kg/m$^3$, respectively. The experiment was terminated when a steady-state water flow was reached, and the output water flow equalled the input water flow. The cell was dismantled, and the actual dry density was determined. The hydraulic conductivity was calculated according to Darcy's law and converted to a referential temperature of 10 °C according to the ISO standard.

The block/sensor method [33] was used to determine WRC. A wireless sensor with its internal memory, Hygrochron iButton (Maxim Integrated, San Jose, CA, USA), was applied for measuring relative humidity and temperature inside the compacted sample exposed to saturated air with water vapour inside a desiccator filled with water. The cell was regularly weighed until a constant mass was reached. When constant mass was achieved, the cells were dismantled and the sensors removed. The measured data of relative humidity and temperature were used to calculate the suction pressure according to the Kelvin equation for each time interval. WRC were then constructed as a dependence between suction pressure and gravimetric water content.

### 2.4. Microbiological Characterisation

Cultivation of bentonite suspensions in an anaerobic atmosphere rich in $H_2$, with a subsequent molecular genetic analysis, was used for studying microbial survivability in bentonite samples exposed to thermal loading. The bentonite suspensions were prepared from 2 g of naturally wet bentonite powders, mixed with 10 mL sterile Milli-Q water. Samples were incubated in an anaerobic workstation (Don Whitley Scientific, Bingley, United Kingdom) under an atmosphere containing 94% Ar and 6% $H_2$ at room temperature. Suspensions were sampled at three different cultivation times, immediately after phases mixing (0 d), after 14 and 28 days, and kept in a freezer until DNA isolation. Zero-point (0 d) samples and BCV_IN samples were prepared in duplicate, whereas the heat-loaded samples in triplicate.

The DNA from the bentonite samples was extracted by DNeasy® PowerMax® Soil Kit (QIAGEN, Germantown, TN, USA) according to the manufacturer's protocol. The extracted DNA was concentrated and purified by the DNA Clean & Concentrator Kit (Zymo Research, Irvine, CA, USA) and quantified by a Qubit 2.0 fluorometer (Life Technologies, Carlsbad, CA, USA). One background kit control without the sample was co-extracted together with samples in each DNA extraction batch to see the laboratory background and possible contaminations resulting from the kit or the laboratory. The extracted DNA was used to describe the microbial composition in the samples (16S rDNA amplicon sequencing, NGS) and for the relative quantification of microbial biomass in the samples (16S rDNA quantitative PCR). For these DNA-based analyses, protocols similar to [34] were followed.

## 3. Results

### 3.1. XRD

The bulk mineralogical composition of the bentonites is shown in Table 1. Bentonite samples are dominated by montmorillonite ($d_{060}$ ~ 1.50 Å), accompanied mainly by quartz and kaolinite. According to the position of the basal diffraction of the untreated sample ($d_{001}$ ~ 14.9 Å), Ca-montmorillonite was identified. The presence of saponite and nontronite was excluded in all samples based on an absence of $d_{060}$ spacing at 1.52 Å. Of all remaining crystalline phases, illite, goethite, anatase, calcite, aragonite, siderite, and sanidine were quantified in amount <5 wt.%. Analcime, ankerite, and augite were detected in very small amounts (bellow quantification limit). Measurements of as-received samples with ZnO inner standard indicated a reduction of amorphous phases in thermally loaded samples. Therefore, depending on the quantification method used, montmorillonite content decreased slightly or remained almost unchanged in thermally loaded samples. No significant bentonite alteration in terms of mineral transformation can be established considering the uncertainties of the XRD method, especially when dealing with clay minerals and trace minerals. However, basal diffraction of montmorillonite progressively disappeared, indicating increasing disorder of the montmorillonite layers and dehydration connected to the thermal loading.

**Table 1.** Semiquantitative XRD analysis of randomly oriented powder mounts (wt.%); bql—below quantification limit, nd—not detected.

| Phases | BCV_IN | BCV_0.5_y | BCV_1.0_y |
|---|---|---|---|
| Montmorillonite | 72 | 68 | 71 |
| Quartz | 10 | 10 | 10 |
| Kaolinite-1A | 6 | 8 | 6 |
| Illite-2M1 | 2 | 3 | 3 |
| Sanidine | 2 | 3 | 3 |
| Goethite [1] | 3 | 3 | 3 |
| Anatase | 2 | 2 | 2 |
| Calcite | 1 | 1 | 1 |
| Aragonite | 1 | 1 | 1 |
| Siderite | 1 | 1 | bql |
| Ankerite | bql | bql | bql |
| Augite | nd | bql | nd |
| Analcime | nd | bql | bql |
| Total | 100 | 100 | 100 |

[1] Amount of goethite may be underestimated due to the small size of crystallites.

### 3.2. TA-EGA

The results of simultaneously performed thermogravimetry (TG) and differential thermal analysis (DTA) supplemented with mass spectrometry identification of evolved gases can be seen in Figure 3. Variability of endothermic and exothermic processes in DTA curves suggests a complex composition of BVC bentonite. Four major mass losses were identified from the derivative thermogravimetric (dTG) curves: at approximately 30–200, 200–300, 300–600, and 600–800 °C. Water related mass losses were found for the first three processes. A shift of 240 °C peak of BCV_IN to higher temperatures was observed for thermally loaded samples. The same shift was also observed in the EGA curves of evolved water. Another difference in the EGA curve for $CO_2$ can be seen at 105 °C and in the region of 350–550 °C. A very low signal for $SO_2$ was registered around 800 °C, decreasing in order: BCV_IN > BCV_0.5_y > BCV_1.0_y.

### 3.3. Aqueous Leachates

Interaction of bentonite samples with deionised water at three solid-to-liquid ratios resulted in the leachate composition of studied ions are presented in Table 2. The dominant aqueous species were $Na^+$ and $HCO_3^-$. The $HCO_3^-$ source is the dissolution of carbonate phases, which are abundant in BCV bentonite. Dissolution of carbonate phases is followed by cation exchange reactions releasing cations, particularly $Na^+$, into the leachate [35,36]. The concentrations of all ions increase with increasing bentonite content. All ion concentrations, except for chloride and sulphate, also increase with the heating period of bentonite powder. It follows that the reactivity of soluble phases has changed. The thermal loading somehow affected the bentonite, possibly in terms of mineral transformation or recrystallisation. More considerable changes in bentonite/water reactivity were detected in the first heating period.

Differences in leachable cation concentrations between the heated and input samples were found mostly for calcium, then potassium and magnesium. The released $Na^+$ concentration is comparable for all materials corresponding to the lowest capability of sodium to compete at montmorillonite exchange sites. Powder heating at 150 °C resulted in the most pronounced increase in leachable fluoride concentrations, up to 2.2 times. The bicarbonate concentration in leachates from the heated samples was up to 1.6 times higher than from the BCV_IN sample. The concentration of leachable $SO_4^{2-}$ and $Cl^-$, on the other hand, decreased with the thermal loading of bentonite powders, indicating a decrease in the solubility of the chloride and sulphate phases.

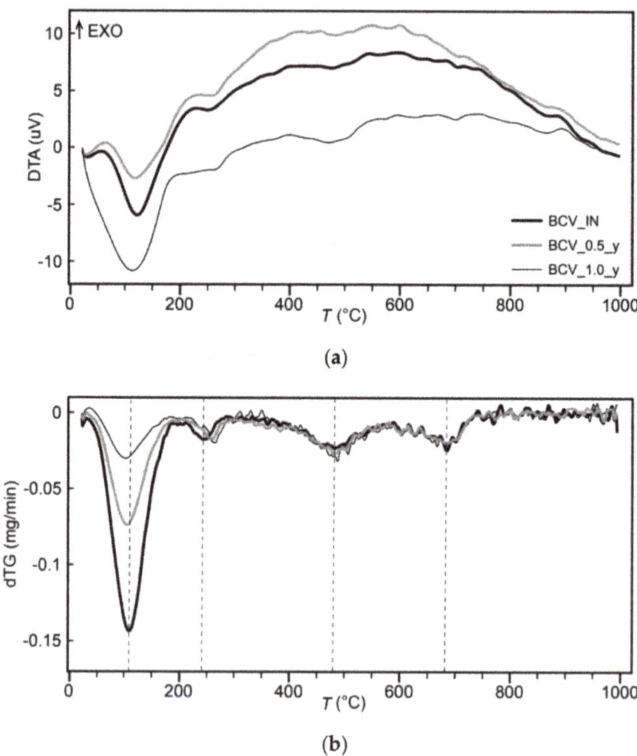


**Figure 3.** *Cont.*

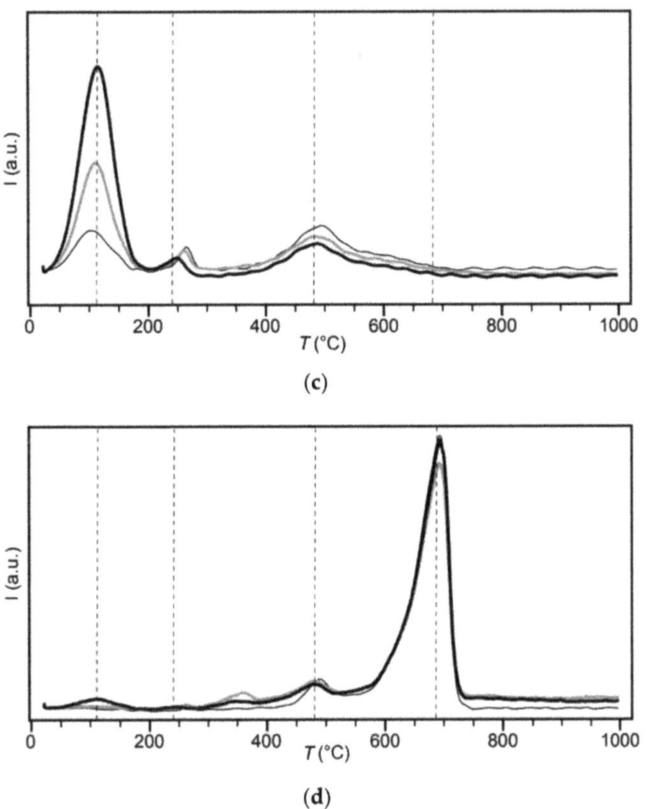

**Figure 3.** TA-EGA. Differential thermal analysis (**a**), derivative thermogravimetry curves (**b**), EGA curves for water (**c**), and $CO_2$ (**d**) of the powder samples.

**Table 2.** Aqueous leachates data for the studied BCV samples at three different solid-to-liquid ratios.

| mmol/L | IN 25 g/L | 0.5_y 25 g/L | 1.0_y 25 g/L | IN 112.5 g/L | 0.5_y 112.5 g/L | 1.0_y 112.5 g/L | IN 200 g/L | 0.5_y 200 g/L | 1.0_y 200 g/L |
|---|---|---|---|---|---|---|---|---|---|
| $Na^+$ | 1.4 | 1.5 | 1.5 | 3.8 | 4.1 | 4.1 | 4.9 | 5.5 | 5.9 |
| $K^+$ | 0.13 | 0.26 | 0.32 | 0.26 | 0.51 | 0.60 | 0.33 | 0.62 | 0.81 |
| $Ca^{2+}$ | 0.061 | 0.19 | 0.24 | 0.10 | 0.22 | 0.26 | 0.13 | 0.27 | 0.33 |
| $Mg^{2+}$ | 0.11 | 0.23 | 0.22 | 0.23 | 0.36 | 0.33 | 0.29 | 0.49 | 0.49 |
| $HCO_3^-$ | 1.7 | 2.4 | 2.6 | 3.8 | 5.1 | 5.2 | 4.8 | 6.6 | 7.3 |
| $SO_4^{2-}$ | 0.065 | 0.062 | 0.060 | 0.25 | 0.23 | 0.21 | 0.44 | 0.36 | 0.35 |
| $Cl^-$ | 0.011 | 0.009 | 0.009 | 0.032 | 0.024 | 0.021 | 0.053 | 0.040 | 0.040 |
| $F^-$ | 0.028 | 0.061 | 0.059 | 0.089 | 0.14 | 0.12 | 0.097 | 0.16 | 0.14 |

### 3.4. Caesium Sorption

The used methodology of sorption experiments covered two extremes of the Cs concentration range: 0.0008 and 62 mmol/L of the initial carrier concentration, resulting in the distribution coefficients presented in Table 3. In both cases, the lowest decrease in Cs concentration was observed in the supernatant of the BCV_1.0_y sample. This sample contained the largest amount of leachable ions that may compete in the sorption process. The lowest $K_d$ for BCV_1.0_y could be caused by cation competition or reflects an alteration of the sorption sites. Cs sorption experiments at a wide concentration range of carrier and competing cations on bentonites, smectite-rich clays, and illite-rich clay identified $K^+$ as the major competitive cation, followed by $Ca^{2+}$ and finally by $Na^+$ at low Cs concentration [37]. In our experiments, the main competing cation was $Na^+$; the concentrations of other competing cations (see Table 2) were much lower than those affecting the Cs sorption process [37]. Competition of $Cs^+$ with leachable cations for sorption sites can also be ruled out in experiments with the applied high concentration, where caesium represents the dominant cation. Decreasing $K_d$ values thus indicate a reduction in the sorption sites for Cs when bentonite powder was thermally loaded.

**Table 3.** Caesium distribution coefficients $K_d$ at two different carrier concentrations (low~0.0008 mmol/L Cs, high~62 mmol/L Cs).

| Bentonite Sample | $K_d$_low | $K_d$_high |
|---|---|---|
| | (mL/g) | |
| BCV_IN | $5322 \pm 533$ | $29.28 \pm 0.66$ |
| BCV_0.5_y | $4478 \pm 379$ | $28.66 \pm 0.65$ |
| BCV_1.0_y | $2621 \pm 164$ | $27.73 \pm 0.62$ |

*3.5. Cation Exchange Capacity, Specific Surface Area*

The results of the CEC and SSA measurements are shown in Table 4. Both determined parameters decrease upon heating bentonite at 150 °C. While the SSA changed substantially after the first heating phase and then remained unaffected, the CEC steadily declined. The difference between the $CEC_{SUM}$ and $CEC_{Vis}$, stating the proportion of minerals dissolution on the CEC value, increased. The rising amount of dissolved cations with the heating period of bentonite powder is consistent with the results of the aqueous leachates (see Table 2). Thus, the cation exchange experiment provides summary information about the cation replacements in the montmorillonite interlayers and the interaction of Cu-trien with the sample. Since these two phenomena cannot be discriminated, Figure 4 displays two variants of data processing. The first one does not allow the dissolution of any minerals or impurities. The concentration of displaced cations was related to the $CEC_{SUM}$ value. The latter allows the same interaction of Cu-trien with the sample as water at the same solid-to-liquid ratio. The concentration of displaced cations was reduced by subtracting aqueous leachable cations at 25 g/L and was related to the reduced $CEC_{SUM}$ value. No measurable amount of displaced iron in supernatants (in the CEC determination or in aqueous leachates) was detected.

**Table 4.** Cation exchange capacity and specific surface area data; $n$ corresponds to the number of replicates.

| Bentonite Sample | $CEC_{Vis}$ | $CEC_{SUM}$ | $CEC_{SUM}$-$CEC_{Vis}$ | $n$ | SSA | $n$ |
|---|---|---|---|---|---|---|
| | | (meq/100 g) | | | (m$^2$/g) | |
| BCV_IN | $59.2 \pm 1.4$ | $62.6 \pm 1.0$ | $3.5 \pm 1.7$ | 13 | $468 \pm 18$ | 14 |
| BCV_0.5_y | $55.0 \pm 1.5$ | $59.2 \pm 1.4$ | $4.3 \pm 2.1$ | 12 | $436 \pm 4$ | 6 |
| BCV_1.0_y | $50.9 \pm 1.0$ | $58.3 \pm 2.1$ | $7.4 \pm 2.3$ | 9 | $435 \pm 7$ | 6 |

Depending on the CEC data processing and, thus, considering the dissolution reactions followed by the cation exchange reactions, the sodium population has significantly changed, because $Na^+$ was identified as the major water-leachable cation. Nevertheless, the trends in the exchangeable cations are identical regardless of the CEC data processing. Most of the exchangeable sites in BCV bentonite were occupied by magnesium and calcium. The exchangeable cations in thermally loaded samples varied with heating time. Magnesium proportion decreased at the expense of calcium and potassium; sodium remained unaffected upon heating.

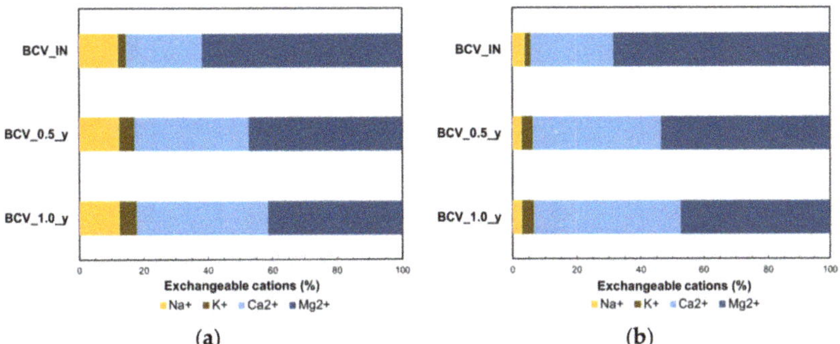

**Figure 4.** Exchangeable cations displaced by Cu-trien (**a**) and corrected by aqueous leachable cations (**b**).

### 3.6. Geotechnical Parameters

Free swelling of dried powders in water resulted in different volumes of swollen material. Steady-state volumes of BCV_IN, BCV_0.5_y, and BCV_1.0_y were 4.8, 2.1, and 1.8 mL, respectively.

Table 5 presents the results of saturated hydraulic conductivity measurements. Thermally loaded samples showed higher saturated hydraulic conductivities; the longer the loading period, the higher the saturated hydraulic conductivity. For all dry densities, saturated hydraulic conductivity increased ca 1.3 times when thermally loaded for half a year and ca 1.6 times when thermally loaded for one year.

**Table 5.** Saturated hydraulic conductivity $k_{10}$ data for studied samples at actual bentonite dry density $\rho_d$.

| Bentonite Sample | $\rho_d$ (kg/m$^3$) | $k_{10}$ (m/s) |
| --- | --- | --- |
| | 1420 | $2.8 \cdot 10^{-13}$ |
| BCV_IN | 1581 | $1.0 \cdot 10^{-13}$ |
| | 1784 | $4.1 \cdot 10^{-14}$ |
| | 1427 | $3.7 \cdot 10^{-13}$ |
| BCV_0.5_y | 1605 | $1.3 \cdot 10^{-13}$ |
| | 1730 | $6.3 \cdot 10^{-14}$ |
| | 1434 | $4.1 \cdot 10^{-13}$ |
| BCV_1.0_y | 1587 | $1.6 \cdot 10^{-13}$ |
| | 1773 | $6.1 \cdot 10^{-14}$ |

Wetting paths of water retention curves are displayed in Figure 5. It can be seen that the water retention ability of BCV bentonite decreases when thermally loaded. The biggest difference was found between the unloaded sample and the loaded ones; thermal loading time seemed to have a lesser impact. This trend was more significant for lower water content.

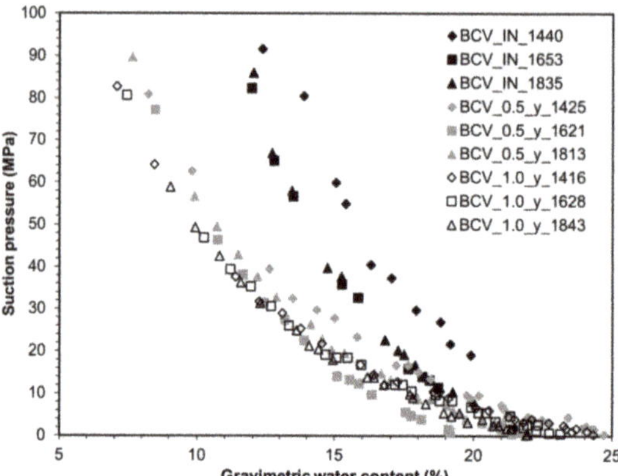

**Figure 5.** Water retention curves of compacted bentonite samples at given dry density (kg/m$^3$).

### 3.7. Microbiological Characterisation

Generally, the DNA extraction resulted in very low DNA yields, as revealed by Qubit measurement. Therefore, the qPCR method by 16S rDNA marker was used to better distinguish between the samples and estimate the changes in gene copy number in the studied samples. The quantification cycle (Cq) values were related to the Cq values of freshly mixed suspensions (0 d) for each bentonite powder to quantify the magnitude of relative change in gene copy numbers. The PCR efficiency for the 16S rDNA marker was estimated beforehand by measuring the slope of curves constructed from a serial dilution of template DNA from several environmental samples. Relative quantification, displayed in Figure 6, showed an increase in the relative abundance of 16S rDNA copies in all samples after 28 days of cultivation in anaerobic conditions. The relative increase varied among samples. The highest relative increase was detected in BCV_0.5_y samples after 28 days of cultivation. A slightly lower relative increase was found for BCV_IN. The lowest relative increase in the gene copies was determined for BCV_1.0_y. Our data also show that 14 days of recovery time for restoring microbial activity from dormant stages is insufficient for most samples.

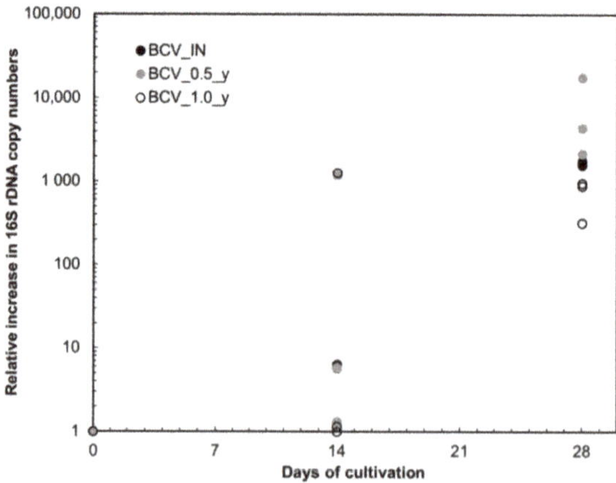

**Figure 6.** qPCR analysis of the 16S rDNA gene in anaerobically cultivated bentonite suspensions.

NGS sequencing technique enabled describing the microbial composition in the suspensions. The results of NGS sequencing (see Figure 7) were consistent with the qPCR analyses. The low NGS signal was found for the studied samples with low 16S rDNA copy numbers and microbial abundance.

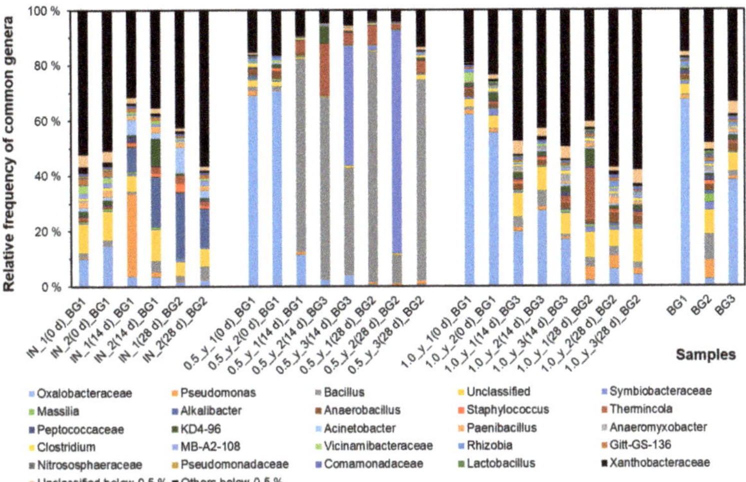



**Figure 7.** Relative abundance of detected microbial genera by NGS sequencing. The sample notation reflects the number from replicates, the cultivation time, and background (BG) controls co-extracted with each batch.

Various facultative anaerobic nitrate-reducing (NRB) genera such as *Alkalibacter, Acinetobacter, Bacillus, Pseudomonas, Paenibacillus, Symbiobacterium*, or *Massilia*, or iron-reducing (IRB) genus *Thermincola* were detected in BCV_IN samples. All of these genera are typical for Czech Mg/Ca bentonite [38]. Much lower genera diversity was found in the BCV_0.5_y samples. After 14 or 28 days of cultivation, the suspensions contained mostly the NRB genus *Bacillus* or the IRB genus *Thermincola*. In the samples 0.5_y_3(14 d)_BG3 and 0.5_y_2(28 d)_BG2, the presence of an unspecified bacterial clone from the family *Symbiobacteraceae* was detected. Sequencing of DNA from BCV_1.0_y samples resulted in poor NGS results. Only sample 1.0_y_1(28 d)_BG2 showed an increased frequency of the thermophilic genus *Thermincola* compared with the background controls (BG) co-extracted with the samples. This result indicates that, although the relative increase in 16S rDNA copy numbers in the BCV_1.0_y samples cultivated for 28 days was detected, this increase did not result in sufficiently high absolute gene copy numbers for successful NGS analysis. Microbial composition of all zero-point samples was very similar to the pattern detected in co-extracted background controls, which reflects their very low DNA content and lack of DNA signal similarly to the situation observed in 1.0_y samples. The only difference was observed in IN_1(0 d) and IN_2(0 d) samples, with lower proportion of *Oxalobacteraceae* family members compared to BG1 control. However, the relative frequencies of particular taxa are rather sensitive to several biases introduced during DNA extraction and PCR in low-DNA samples, and we consider the observed difference insignificant.

## 4. Discussion

The applied geochemical characterisation, targeting both the overall composition (in terms of mineralogy, thermolabile, and soluble phases) and the clay component, allowed investigating the effect of heating on individual components. The XRD analysis of BCV bentonite provided a substantial number of various phases, which may be affected by the exposition to the high temperatures, i.e., montmorillonite component, other clay minerals and non-clay minerals. Based on the XRD analysis, considering its uncertainties in detecting variations in the structural chemical composition of clay minerals, it is not possible to deduce any significant mineral transformation upon heating.

The processes that the bentonite underwent up to 150 °C are indicated by the results of thermal analysis. Besides the apparent loss of water content in as-received samples, TA-EGA identified a release of $CO_2$ from the BCV_IN sample. Several $CO_2$-related mass losses were recorded when heating the bentonite samples to higher temperatures, up to 1000 °C. The $CO_2$ release can be attributed mainly to the carbonate phases, which XRD and aqueous leachates detected. Assigning peaks to given minerals is difficult due to superimpositions of double carbonates with calcite [39,40]. The presence of various carbonates in BCV might be reflected in the broad $CO_2$ peak at 600–800 °C. The main difference in thermal behaviour between BCV_IN and thermally loaded samples was found in the shift of $H_2O$ and $CO_2$ peaks. These shifts, and release of $CO_2$ from the BCV_IN sample up to 150 °C, indicate some alteration process during bentonite heating. Altered reactivity of carbonates was also detected by aqueous leachates. A significantly higher concentration of leachable $HCO_3^-$ from thermally loaded samples indicates an intense alteration of carbonate minerals, supported by the increased leaching of cations.

The impact of bentonite heating to 150 °C can also be inferred based on the leaching of fluorides, sulphates, and chlorides. A change, though to a minor extent, in sulphate content was also registered by TA-EGA. Nevertheless, no mineral phases containing fluorides, sulphates, and chlorides were detected by XRD. However, they may be present as impurities in minerals and/or in accessory amounts. For example, fluoride is commonly found in montmorillonites [5,41,42]. It can substitute the hydroxyl groups in the octahedral sheets [41]. The most significant increase in leachable concentration with exposure to elevated temperature has been determined for fluoride. An increased release of $F^-$ into the solution might indicate montmorillonite lattice modification.

The effect of thermal loading on clay component was examined in detail using the determination of SSA and CEC and sorption experiments with caesium. The thermal loading caused a decrease in the SSA. A decrease in SSA upon heating up to 200 °C was also observed for Barmer bentonite [43]. A lower specific surface area may indicate recrystallisation of minerals or montmorillonite collapse. Lower amorphous phases in thermally loaded samples may indicate recrystallisation of minerals, often leading to textural coarsening. Coarsening then may facilitate the release of species when contacted with water. Higher leachable concentrations were found for thermally loaded samples. However, since EGME penetrates in the montmorillonite interlayers due to its polarity, the SSA decrease can be more attributed to the montmorillonite alteration than to recrystallisation of minerals.

Montmorillonite lattice modification in thermally loaded samples can be deduced from the experiments with selectively interacting cations (such as Cu-trien cation or $Cs^+$). The $^{134}Cs$ distribution coefficient at both studied carrier concentration decreased in the order: IN > 0.5_y > 1.0_y. The same trend was determined for the CEC determination with Cu-trien. Decreasing CEC due to extensive drying was observed in various bentonites and one illite/montmorillonite clay and was attributed to the type of exchangeable cations and their fixation [44]. Generally, the origin of the CEC is a permanent negative charge of montmorillonite clay layers due to isomorphous substitutions [45]. Altered reactivity of the applied selective cations, interacting by the cation exchange, clearly indicates a change in surface charge density of clay particles. Such a partial alteration can result in the formation of new phases (low-swelling or even non-swelling) at the expense of montmorillonite content.

To support the finding of reduced montmorillonite content, which could not be confirmed using XRD, data analysis of CEC, SSA was performed. The interlocking of SSA, CEC and smectite content is illustrated in Figure 8. Data selection (see Table 6) was mainly restricted by existing SSA values measured according to the EGME method. Previously published data were compared with the results from the current study. Five groups of bentonites can be distinguished corresponding to their smectite content. The lowest SSA and CEC values are typical for bentonites containing up to 50 wt.% of smectite. The highest

SSA and CEC values are typical for bentonites containing almost 100 wt.% of smectite. The $CEC_{Vis}$ and SSA values of the BCV_IN sample correspond well with the correlation matching the group of bentonites containing 50–80 wt.% of smectite.

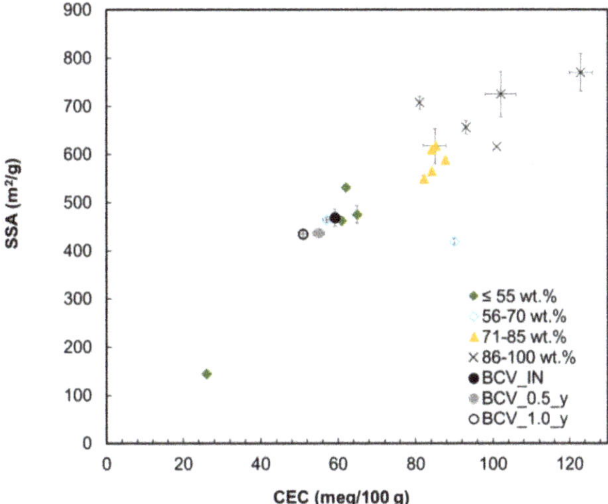


**Figure 8.** Correlation between CEC and SSA for various bentonites depending on their smectite content (wt.%).

**Table 6.** Data on smectite content, SSA and CEC of selected bentonites.

| Bentonite Sample | Location | Smectite Content (wt.%) | SSA (m²/g) | CEC (meq/100 g) |
|---|---|---|---|---|
| SAz-1 | Cheto (USA) | 98 [46] | 770 ± 39 [30] | 123 ± 3 [47] |
| SWy-2 | Wyoming (USA) | 75 [46] | 618 ± 37 [30] | 85 ± 3 [47] |
| MX-80 | Wyoming (USA) | 81.3 [6] | 610 ± 3 [6] | 84 [6] |
| FEBEX | Cortijo de Archidona (Spain) | 92 [48] | 725 ± 47 [48] | 102 ± 4 [48] |
| SPA 050 | Cortijo de Archidona (Spain) | 88 [37] | 656 ± 14 [30] | 93.1 [37] |
| SPA 04 | Cortijo de Archidona (Spain) | 75 [37] | 588 ± 4 [30] | 87.5 [37] |
| SPA 051 | Cortijo de Archidona (Spain) | 54 [37] | 475 ± 19 [30] | 64.9 [37] |
| Volclay KWK 20-80 | Südchemie (Germany) | 90 [49] | 708 ± 14 [30] | 81 [49] |
| Friedland clay | Friedland (Germany) | 31.8 [6] | 144 ± 2 [6] | 26 [6] |
| Deponit CAN | Milos (Greece) | 72.1 [6] | 550 ± 5 [6] | 82 [6] |
| Asha | Kutch (India) | 67.1 [6] | 419 ± 6 [6] | 90 [6] |
| VSK 11 10.5 | Jastrabá F. (Slovakia) | 89 [50] | 616 [50] | 101 [50] |
| VSK 11 30.5 | Jastrabá F. (Slovakia) | 52 [50] | 462 [50] | 61 [50] |
| VSK 11 31.5 | Jastrabá F. (Slovakia) | 78 [50] | 565 [50] | 84 [50] |
| RO-M | Rokle (Czech Republic) | 59 [49] | 464 ± 5 [30] | 57 ± 1 [49] |
| BM | Rokle (Czech Republic) | 52 [51] | 531 ± 6 [30] | 62 [51] |

The most interesting finding of this study is in the composition of exchangeable cations. Magnesium dominated bentonite lost a significant amount of exchangeable $Mg^{2+}$ when heated. The proportion of $Ca^{2+}$ and $K^+$ increased with heating, while $Na^+$ remained unaffected (see Figure 4). These changes in exchangeable cations took place in the solid state, without the influence of external fluids. In comparison, in the in situ experiments with access to groundwater, a gradually increasing proportion of the divalent cations towards the heater was observed [13]. Only water originally present in BCV_IN and escaping due to thermal loading could act as a transport medium. From this point of view, it is necessary to analyse the top layers of bentonite, which were excluded from current research due to the small amount (see Section 2.1). The loss of exchangeable $Mg^{2+}$ can be explained by increased fixation between charged surfaces [44] or crystal lattice transformation. Magnesium in the montmorillonite structure is located in the octahedral sheet together with Al and Fe [13]. It can be expected that when the montmorillonite lattice is damaged, $Mg^{2+}$ is released from the structure, and the surface charge is redistributed. A very slight decrease in the mean layer charge density for montmorillonite was detected in compacted bentonite (Almeria, Spain) exposed in situ to 100 °C [52]. The change in surface charge density was deduced from our results of CEC, SSA and Cs sorption. Moreover, 1.5–2 times higher Mg concentrations were determined in the aqueous leachates from the thermally loaded samples compared to the input sample. Therefore, we are inclined to explain the changes in exchangeable cations to the partial lattice modifications. Plötze et al. [52], however, observed Mg enrichment in samples from the heater region, as in many large-scale tests [4,13]. The mechanism behind the Mg enrichment is not yet clear [4]. Both mechanisms, magnesium enrichment and depletion (as observed in this study), would be worth investigating thoroughly in future research.

The studied geotechnical parameters (free swelling, saturated hydraulic conductivity, and water retention curves) showed worsening with the thermal loading time. All these parameters are related to the behaviour of water in bentonite (maximal expansion of expandable structures in free swelling test, water uptake and transport through confined clay particles). As the water uptake ability of montmorillonite depends critically on the magnitude of the surface charge density of the particles [45], the observed geotechnical behaviour indicates changes in surface charge distribution. The resulting reduction of expandable clay structures towards low- or non-swelling clay structures upon heating was therefore subjected to data analysis of saturated hydraulic conductivity. Figure 9 displays hydraulic conductivities converted to a referential temperature of 10 °C, according to ISO 17892-11, for BCV bentonite and selected bentonites (from Table 6) tested by Karnland et al. [53]. A correlation between saturated hydraulic conductivity and montmorillonite content can be seen. The higher the $k_{10}$, the lower the montmorillonite content. The $k_{10}$ values for BCV can be categorised between Rokle bentonite and Friedland clay, which corresponds well to their low montmorillonite content. The saturated hydraulic conductivity of Deponit CAN, MX-80 and Asha is one order of magnitude smaller due to higher montmorillonite content.

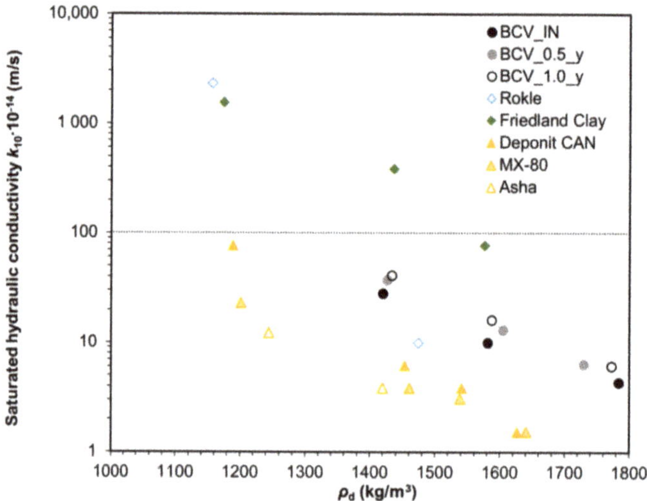


**Figure 9.** Saturated hydraulic conductivity of BCV compared to data for Friedland clay, Rokle, Deponit CAN, MX-80, and Asha [53].

A shift of measured saturated hydraulic conductivity for BCV towards clay material of lower montmorillonite content with thermal loading period can be clearly seen. Nevertheless, BCV meets an acceptance safety limit for saturated hydraulic conductivity of buffer lower than $10^{-12}$ m/s [17], even after the thermal loading. It can also be noticed in Figure 9 that the differences between input and thermally loaded BCV samples were more dramatic for lower dry densities, whereas for dry densities over 1700 kg/m$^3$, the differences induced by the thermal loading seem to be reduced. Based on this observation, it can be hypothesised that no significant effect of thermal loading on saturated hydraulic conductivity [13] might be due to the high density of bentonite samples. On the other hand, the most notable difference between BCV samples was found in free swelling tests, where the dry density was very low. Therefore, it seems that the thermally induced changes manifest more in less dense systems. For further investigation, higher dry densities or, and vice-versa, lower ones than those used (e.g., 2000 kg/m$^3$ and 1000 kg/m$^3$) would be promising to determine saturated hydraulic conductivity and WRC to highlight the impact of the thermal loading on buffer mass.

Water retention ability for all dry densities decreased with the thermal loading period. This behaviour was more relevant for lower water content; for water content above 20%, the water retention ability of both thermally unloaded and thermally loaded samples became closer to each other. This trend was also observed by [9]. The most significant difference is between the BCV_IN samples and the BCV_0.5_y ones; the difference between BCV_0.5_y and BCV_1.0_y is minor. It indicates that most of the material changes took place during the first half-year of thermal loading. This fact corresponds well with SSA values, where between BCV_IN and BCV_0.5_y, a crucial decrease was observed, whereas between BCV_0.5_y and BCV_1.0_y no visible change was detected.

Microbiological experiments under very hospitable conditions in a suspended state (i.e., without space and water availability limitation) with continuous electron donor supply in the form of H$_2$ present in an anaerobic atmosphere showed that heating of bentonite powder at 150 °C for up to one year did not result in bentonite sterilisation. However, a reduction of microbial load with increasing thermal loading time was observed. This reduction was especially true for the BCV_1.0_y samples, where microbial recovery was lower than BCV_IN and BCV_0.5_y. Our experiment further showed that microbial recovery from the dormant stages is a lengthy and probably partially stochastic procedure. After 14 days of cultivation, the signal in relative quantification seems to be rather random in the case of thermally loaded samples, while after 28 days of cultivation, the pattern was much more consistent, and this cultivation time proved to be more suitable for the detection of microbial activity restoration in bentonite samples at given environmental conditions.

Although we showed that bacteria could at least partially resume their activity after medium-term heat loading when cultivated at suitable conditions, the results imply that a substantial reduction in microbial load or even complete sterilisation of bentonite powder could be possible when the heating time exceeds one year. An equilibrium between an increase of the temperature limit at the canister surface resulting in bentonite sterilisation and the potential risk of losing favourable geochemical and geotechnical properties should therefore be addressed in the DGR performance assessment.

## 5. Conclusions

A medium-term thermal loading of bentonite powder resulted in detectable changes of properties, important directly or indirectly for the safety performance as a barrier material in a deep geological repository. No significant bentonite alteration, in terms of mineralogical composition, was observed by XRD. The thermal loading resulted in progressive disappearance of basal diffraction of montmorillonite, indicating increasing disorder of the montmorillonite layers. The impact of thermal loading on the overall bentonite composition was registered by aqueous leachates and TA-EGA mainly for carbonate phases, then for sulphates and fluorides.

Based on the determination of cation exchange capacity, total surface area and caesium distribution coefficient and interpretation of these parameters, it was concluded that montmorillonite had been partially altered in terms of the magnitude of the surface charge density of montmorillonite particles. Montmorillonite alteration towards low- or non-swelling clay structures was also supported by the determined geotechnical behaviour. Swelling ability and water uptake ability of thermally loaded samples were lower, and saturated hydraulic conductivity higher than for the input bentonite material.

This study has shown that XRD is a suitable method for mineralogical characterisation, but it was found to be insensitive for detecting slight variations of minerals in bentonite. In order to identify the altered montmorillonite fraction in more detail, a future investigation will focus on the clay fraction separation and its detailed analysis. We also plan to sample the heated bentonite powder after two years of thermal loading. Therefore, the current methods applied in this study to characterise bentonite properties will be amended to capture an alteration of clay minerals, especially by monitoring silica, aluminium, iron, and magnesium, which could confirm the hypothesis of thermal stress damage to the montmorillonite lattice.

The thermal loading also resulted in a reduction of microbial survivability and, thus, affected possible future microbial activity in bentonite. Thermal loading could play an important role in a permanent reduction of microbial activity in the bentonite buffer. However, more extended experiments focused on the stress-induced hindrances to microbial growth are needed.

**Author Contributions:** Conceptualisation and methodology, V.K., E.H., and V.H. (Václava Havlová); validation, V.K. and E.H.; formal analysis and investigation, V.K., Š.Š., E.H., B.K., C.A., D.B., V.H. (Veronika Hlaváčková), and K.Č.; data curation, V.K.; writing—original draft preparation, V.K., Š.Š., B.K., and K.Č.; writing—review and editing, all authors; visualisation, V.K.; supervision, E.H. and V.H. (Václava Havlová); project administration V.K.; funding acquisition, V.K. and V.H (Václava Havlová). All authors have read and agreed to the published version of the manuscript.

**Funding:** This research was funded by the European Commission European Joint Programme on Radioactive Waste Management, European Union's Horizon 2020 research and innovation programme 2014–2018, under grant agreement number 847593. The APC was co-funded by Správa úložišť radioaktivních odpadů (SÚRAO). Výstup byl vytvořen za finanční účasti SÚRAO (SO2020-017). The microbiological analyses were funded by the Technology Agency of the Czech Republic project under grant agreement number TK02010169.

**Data Availability Statement:** The data used in this study are available from the authors upon request.

**Acknowledgments:** The authors warmly thank the Centre of Experimental Geotechnics, Faculty of Civil Engineering CTU in Prague, namely Kateřina Černochová, Jiří Svoboda, and Radek Vašíček, for providing thermally loaded bentonite.

**Conflicts of Interest:** The authors declare no conflict of interest. The funders had no role in the study's design, in the collection, analyses, or interpretation of data, in the writing of the manuscript, or in the decision to publish the results.

## References

1. Villar, M.V.; Armand, G.; Conil, N.; de Lesquen, C.; Herold, P.; Simo, E.; Mayor, J.C.; Dizier, A.; Li, X.; Chen, G.; et al. D7.1 HITEC. In *Initial State-of-the-Art on THM Behaviour of i) Buffer Clay Materials and of Ii) Host Clay Materials*; EURAD Project, Horizon: Brussels, Belgium, 2020; p. 222.
2. Leupin, O.X.; Birgersson, M.; Karnland, O.; Korkeakoski, P.; Sellin, P.; Mäder, U.; Wersin, P. *NAGRA Technical Report 14–12: Montmorillonite Stability under Near-Field Conditions*; National Cooperative for the Disposal of Radioactive Waste: Wettingen, Switzerland, 2014; p. 104.
3. Wersin, P.; Johnson, L.H.; McKinley, I.G. Performance of the bentonite barrier at temperatures beyond 100 °C: A critical review. *Phys. Chem. Earth Parts A/B/C* **2007**, *32*, 780–788. [CrossRef]
4. Kaufhold, S.; Dohrmann, R.; Ufer, K.; Svensson, D.; Sellin, P. Mineralogical analysis of bentonite from the ABM5 heater experiment at Äspö Hard Rock Laboratory, Sweden. *Minerals* **2021**, *11*, 669. [CrossRef]

5.  Svensson, D.; Dueck, A.; Nilsson, U.; Olsson, S.; Sandén, T.; Lydmark, S.; Jägerwall, S.; Pedersen, K.; Hansen, S. SKB technical report TR-11-06: Alternative buffer material. In *Status of the Ongoing Laboratory Investigation of Reference Materials and Test Package 1.*; Swedish Nuclear Fuel and Waste Management Co: Stockholm, Sweden, 2011; p. 146.

6.  Kumpulainen, S.; Kiviranta, L. *POSIVA Working Report 2011-41: Mineralogical and Chemical and Physical Study of Potential Buffer and Backfill Materials from ABM Test Package 1*; POSIVA Oy: Olkiluoto, Finland, 2011; p. 50.

7.  Kumpulainen, S.; Kiviranta, L.; Korkeakoski, P. Long-term effects of an iron heater and äspö groundwater on smectite clays: Chemical and hydromechanical results from the in situ alternative buffer material (ABM) test package 2. *Clay Miner.* **2016**, *51*, 129–144. [CrossRef]

8.  Villar, M.V.; Gutiérrez-Álvarez, C.; Martín, P.L. Low-suction water retention capacity of bentonite at high temperature. In *E3S Web of Conferences*; EDP Sciences: Les Ulis, France, 2020; Volume 195, p. 04010. [CrossRef]

9.  Villar, M.V.; Gómez-Espina, R.; Lloret, A. Experimental investigation into temperature effect on hydro-mechanical behaviours of bentonite. *J. Rock Mech. Geotech. Eng.* **2010**, *2*, 71–78. [CrossRef]

10. Jacinto, A.C.; Villar, M.V.; Gómez-Espina, R.; Ledesma, A. Adaptation of the van Genuchten expression to the effects of temperature and density for compacted bentonites. *Appl. Clay Sci.* **2009**, *42*, 575–582. [CrossRef]

11. Cho, W.J.; Lee, J.O.; Chun, K.S. The temperature effects on hydraulic conductivity of compacted bentonite. *Appl. Clay Sci.* **1999**, *14*, 47–58. [CrossRef]

12. Daniels, K.A.; Harrington, J.F.; Zihms, S.G.; Wiseall, A.C. Bentonite permeability at elevated temperature. *Geosciences* **2017**, *7*, 3. [CrossRef]

13. Karnland, O.; Olsson, S.; Dueck, A.; Birgersson, M.; Nilsson, U.; Hernan-Håkansson, T.; Pedersen, K.; Nilsson, S.; Eriksen, T.E.; Rosborg, B. SKB technical report TR-09-29: Long term test of Buffer Material at the Äspö Hard Rock Laboratory, LOT Project. In *Final Report on the A2 Test. Parcel*; Svensk Kärnbränslehantering AB: Stockholm, Sweden, 2009; p. 279.

14. Van Humbeeck, H.; Verstricht, J.; Li, X.L.; De Cannière, P.; Bernier, F.; Kursten, B. *EURIDICE Report 09-134: The OPHELIE Mock-up Final Report*; EIG EURIDICE: Mol, Belgium, 2009; p. 197.

15. Fru, E.C.; Athar, R. In situ bacterial colonization of compacted bentonite under deep geological high-level radioactive waste repository conditions. *Appl. Microbiol. Biotechnol.* **2008**, *79*, 499–510. [CrossRef]

16. Enning, D.; Garrelfs, J. Corrosion of iron by sulfate-reducing bacteria: New views of an old problem. *Appl. Environ. Microbiol.* **2014**, *80*, 1226–1236. [CrossRef] [PubMed]

17. Andersson, J.; Vira, J.; Salo, J.-P.; Morén, L.; Pers, K.; Pastina, B.; Hellä, P. *Posiva SKB Report 01: Safety Functions, Performance Targets and Technical Design Requirements for a KBS-3V Repository*; Svensk Kärnbränslehantering AB: Forsmark, Sweden; Posiva Oy: Eurajoki, Finland, 2017; p. 120.

18. Nicholson, W.L.; Munakata, N.; Horneck, G.; Melosh, H.J.; Setlow, P. Resistance of bacillus endospores to extreme terrestrial and extraterrestrial environments. *Microbiol. Mol. Biol. Rev.* **2000**, *64*, 548–572. [CrossRef]

19. Setlow, P. Spores of bacillus subtilis: Their resistance to and killing by radiation, heat and chemicals. *J. Appl. Microbiol.* **2006**, *101*, 514–525. [CrossRef]

20. Masurat, P.; Eriksson, S.; Pedersen, K. Microbial sulphide production in compacted wyoming bentonite MX-80 under in situ conditions relevant to a repository for high-level radioactive waste. *Appl. Clay Sci.* **2010**, *47*, 58–64. [CrossRef]

21. Masurat, P.; Eriksson, S.; Pedersen, K. Evidence of Indigenous sulphate-reducing bacteria in commercial wyoming bentonite MX-80. *Appl. Clay Sci.* **2010**, *47*, 51–57. [CrossRef]

22. Bengtsson, A.; Pedersen, K. Microbial sulphide-producing activity in water saturated wyoming MX-80, asha and calcigel bentonites at wet densities from 1500 to 2000 Kg M−3. *Appl. Clay Sci.* **2017**, *137*, 203–212. [CrossRef]

23. Bengtsson, A.; Pedersen, K. Microbial sulphate-reducing activity over load pressure and density in water saturated boom clay. *Appl. Clay Sci.* **2016**, *132–133*, 542–551. [CrossRef]

24. Hausmannová, L.; Hanusová, I.; Dohnálková, M. *SÚRAO Technical Report 309/2018: Summary of the Research of Czech Bentonites for Use in the Deep Geological Repository-Up To 2018*; Správa úložišť radioaktivních odpadů: Praha, Czech Republic, 2018; p. 63.

25. Doebelin, N.; Kleeberg, R. Profex: A graphical user interface for the rietveld refinement program BGMN. *J. Appl. Cryst.* **2015**, *48*, 1573–1580. [CrossRef]

26. Meier, L.P.; Kahr, G. Determination of the cation exchange capacity (CEC) of clay minerals using the complexes of copper (II) ion with triethylenetetramine and tetraethylenepentamine. *Clays Clay Miner.* **1999**, *47*, 386–388. [CrossRef]

27. Ammann, L.; Bergaya, F.; Lagaly, G. Determination of the cation exchange capacity of clays with copper complexes revisited. *Clay Miner.* **2005**, *40*, 441–453. [CrossRef]

28. Carter, D.L.; Heilman, M.D.; Gonzales, C.L. Ethylene glycol monoethyl ether for determining surface area of silicate minerals. *Soil Sci.* **1965**, *100*, 356–360. [CrossRef]

29. Carter, D.L.; Mortland, M.M.; Kemper, W.D. Specific Surface. In *Methods of Soil Analysis*; John Wiley & Sons: Hoboken, NJ, USA, 1986; pp. 413–423. ISBN 978-0-89118-864-3.

30. Brázda, L.; Červinka, R. Determination of specific surface area of clay minerals by EGME method. In *Proceedings of the 4th BELBaR Annual Workshop, Berlin, Germany, 15 September 2016.

31. ASTM. *D5890-19 Test Method for Swell Index of Clay Mineral Component of Geosynthetic Clay Liners*; ASTM International: West Conshohocken, PA, USA, 2019.

32. Gondolli, J.; Večerník, P. The uncertainties associated with the application of through-diffusion, the steady-state method: A case study of strontium diffusion. *Geol. Soc. Spec. Publ.* **2014**, *400*, 603–612. [CrossRef]

33. Villar, M.V. Water retention of two natural compacted bentonites. *Clays Clay Miner.* **2007**, *55*, 311–322. [CrossRef]

34. Černoušek, T.; Shrestha, R.; Kovářová, H.; Špánek, R.; Ševců, A.; Sihelská, K.; Kokinda, J.; Stoulil, J.; Steinová, J. Microbially influenced corrosion of carbon steel in the presence of anaerobic sulphate-reducing bacteria. *Corros. Eng. Sci. Technol.* **2019**, *55*, 127–137. [CrossRef]

35. Filipská, P.; Zeman, J.; Všianský, D.; Honty, M.; Škoda, R. Key Processes of long-term bentonite-water interaction at 90 °C: Mineralogical and chemical transformations. *Appl. Clay Sci.* **2017**, *150*, 234–243. [CrossRef]

36. Sasaki, Y.; Shibata, M.; Yui, M.; Ishikawa, H. Experimental Studies on the Interaction of Groundwater with Bentonite. *MRS Online Proc. Libr.* **1995**, *353*, 337–344. [CrossRef]

37. Vejsada, J.; Hradil, D.; Řanda, Z.; Jelínek, E.; Štulík, K. Adsorption of cesium on czech smectite-rich clays—A comparative study. *Appl. Clay Sci.* **2005**, *30*, 53–66. [CrossRef]

38. Černá, K.; Černoušek, T.; Polívka, P.; Ševců, A. Survival of Microorganisms in Bentonite Subjected to Different Levels of Irradiation and Pressure. In Proceedings of the Microbiology in Nuclear Waste Disposal, Stockholm, Sweden, 7–9 May 2019; p. 57, No 661880.

39. Steudel, A.; Mehl, D.; Emmerich, K. Simultaneous thermal analysis of different bentonite–sodium carbonate systems: An attempt to distinguish alkali-activated bentonites from raw materials. *Clay Miner.* **2013**, *48*, 117–128. [CrossRef]

40. Kaufhold, S.; Emmerich, K.; Dohrmann, R.; Steudel, A.; Ufer, K. Comparison of Methods for Distinguishing Sodium Carbonate Activated from Natural Sodium Bentonites. *Appl. Clay Sci.* **2013**, *86*, 23–37. [CrossRef]

41. García-Romero, E.; Lorenzo, A.; García-Vicente, A.; Morales, J.; García-Rivas, J.; Suárez, M. On the structural formula of smectites: A review and new data on the influence of exchangeable cations. *J. Appl Cryst.* **2021**, *54*, 251–262. [CrossRef] [PubMed]

42. Thomas, J.; Glass, H.D.; White, W.A.; Trandell, R.M. Flouride content of clay minerals and argillaceous earth materials. *Clays Clay Miner.* **1977**, *25*, 278–284. [CrossRef]

43. Kale, R.C.; Ravi, K. Influence of thermal loading on index and physicochemical properties of barmer bentonite. *Appl. Clay Sci.* **2018**, *165*, 22–39. [CrossRef]

44. Kaufhold, S.; Dohrmann, R. Effect of extensive drying on the cation exchange capacity of bentonites. *Clay Miner.* **2010**, *45*, 441–448. [CrossRef]

45. Birgersson, M.; Hedström, M.; Karnland, O.; Sjöland, A. Chapter 12-bentonite buffer: Macroscopic performance from nanoscale properties. In *Geological Repository Systems for Safe Disposal of Spent Nuclear Fuels and Radioactive Waste*; Woodhead Publishing: Duxford, UK, 2017; ISBN 978-0-08-100652-8.

46. Chipera, S.J.; Bish, D.L. Baseline studies of the clay minerals society source clays: Powder X-ray diffraction. *Clays Clay Miner.* **2001**, *49*, 398–409. [CrossRef]

47. Borden, D.; Giese, R.F. Baseline studies of the clay minerals society source clays: Cation Exchange capacity measurements by the ammonia-electrode method. *Clays Clay Miner.* **2001**, *49*, 444–445. [CrossRef]

48. Fernandez, A.M.; Cuevas, J.; Rivas, P. Pore water chemistry of the febex bentonite. In Proceedings of the Scientific Basis for Nuclear Waste Management XXIV: Materials Research Society Symposium Proceedings, Sydney, Australia, 27–31 August 2000; Volume 663. [CrossRef]

49. Vejsada, J. The uncertainties associated with the application of batch technique for distribution coefficients determination—A case study of cesium adsorption on four different bentonites. *Appl. Radiat. Isot.* **2006**, *64*, 1538–1548. [CrossRef] [PubMed]

50. Osacký, M.; Binčík, T.; Paľo, T.; Uhlík, P.; Madejová, J.; Czímerová, A. Mineralogical and Physico–Chemical Properties of Bentonites from the Jastrabá Formation (Kremnické Vrchy Mts., Western Carpathians). *Geol. Carpathica* **2019**, *70*, 433–445. [CrossRef]

51. Vejsada, J.; Jelínek, E.; Řanda, Z.; Hradil, D.; Přikryl, R. Sorption of cesium on smectite-rich clays from the bohemian massif (Czech Republic) and their mixtures with sand. *Appl. Radiat. Isot.* **2005**, *62*, 91–96. [CrossRef]

52. Plötze, M.; Kahr, G.; Dohrmann, R.; Weber, H. Hydro-mechanical, geochemical and mineralogical characteristics of the bentonite buffer in a heater experiment: The HE-B project at the Mont Terri Rock laboratory. *Phys. Chem. Earth Parts A/B/C* **2007**, *32*, 730–740. [CrossRef]

53. Karnland, O.; Olsson, S.; Nilsson, U. *SKB Technical Report TR-06-30: Mineralogy and Sealing Properties of Various Bentonites and Smectite-Rich Clay Materials*; Svensk Kärnbränslehantering AB: Stockholm, Sweden, 2006; p. 112.

*Article*

# Mineralogical, Geochemical and Geotechnical Study of BCV 2017 Bentonite—The Initial State and the State following Thermal Treatment at 200 °C

František Laufek [1,*], Irena Hanusová [2], Jiří Svoboda [3], Radek Vašíček [3], Jan Najser [4] and Magdaléna Koubová [1], Michal Čurda [1], František Pticen [1], Lenka Vaculíková [5], Haiquan Sun [4,6,7] and David Mašín [4]

[1] Czech Geological Survey, Geologická 6, 15200 Prague, Czech Republic; magdalena.koubova@geology.cz (M.K.); michalcurda@centrum.cz (M.Č.); frantisek.pticen@geology.cz (F.P.)
[2] Czech Radioactive Waste Repository Authority (SÚRAO), Dlážděná 6, 11000 Prague 1, Czech Republic; hanusova@surao.cz
[3] Faculty of Civil Engineering, Czech Technical University, Thakurova 7, 16629 Prague 6, Czech Republic; svobodaj@fsv.cvut.cz (J.S.); radek.vasicek@cvut.cz (R.V.)
[4] Faculty of Science, Charles University, Albertov 6, 12843 Prague 2, Czech Republic; jan.najser@natur.cuni.cz (J.N.); haiquansun@zju.edu.cn (H.S.); david.masin@natur.cuni.cz (D.M.)
[5] Institute of Geonics, Academy of Sciences of the Czech Republic, Studentská 1768, 70800 Ostrava-Poruba, Czech Republic; lenka.vaculikova@ugn.cas.cz
[6] Hainan Institute of Zhejiang University, Sanya 572025, China
[7] Ocean College, Zhejiang University, Zhoushan 316021, China
* Correspondence: frantisek.laufek@geology.cz; Tel.: +420-251085210

**Citation:** Laufek, F.; Hanusová, I.; Svoboda, J.; Vašíček, R.; Najser, J.; Koubová, M.; Čurda, M.; Pticen, F.; Vaculíková, L.; Sun, H.; et al. Mineralogical, Geochemical and Geotechnical Study of BCV 2017 Bentonite—The Initial State and the State following Thermal Treatment at 200 °C. *Minerals* **2021**, *11*, 871. https://doi.org/10.3390/min11080871

Academic Editors: Ana María Fernández, Stephan Kaufhold, Markus Olin, Lian-Ge Zheng, Paul Wersin and James Wilson

Received: 30 June 2021
Accepted: 7 August 2021
Published: 12 August 2021

**Publisher's Note:** MDPI stays neutral with regard to jurisdictional claims in published maps and institutional affiliations.

**Abstract:** Bentonites are considered to be the most suitable materials for the multibarrier system of high-level radioactive waste repositories. Since BCV bentonite has been proved to be an ideal representative of Czech Ca-Mg bentonites in this respect, it has been included in the Czech Radioactive Waste Repository Authority (SÚRAO) buffer and backfill R&D programme. Detailed knowledge of processes in the material induced by thermal loading provides invaluable assistance regarding the evolution of the material under repository conditions. Samples of both original BCV 2017 bentonite and the same material thermally treated at 200 °C were characterised by means of chemical analysis, powder X-ray diffraction, infrared spectroscopy, thermal analysis, cation exchange capacity, specific surface area (BET) measurements, the determination of the swell index, the liquid limit, the swelling pressure and water retention curves. The smectite in BCV 2017 bentonite comprises Ca-Mg montmorillonite with a significant degree of $Fe^{3+}$ substitution in the octahedral sheet. Two main transformation processes were observed following heating at 200 °C over 27 months, the first of which comprised the dehydration of the montmorillonite and the subsequent reduction of the 001 basal distance from 14.5 Å (the original BCV 2017) to 9.8 Å, thus indicating the absence of water molecules in the interlayer space. The second concerned the dehydration and partial dehydroxylation of goethite. With the exception of the dehydration of the interlayer space, the PXRD and FTIR study revealed the crystallochemical stability of the montmorillonite in BCV 2017 bentonite under the selected experimental conditions. The geotechnical tests indicated no major changes in the mechanical properties of the thermally treated BCV 2017 bentonite, as demonstrated by the similar swelling pressure values. However, the variation in the swell index and the gradual increase in the liquid limit with the wetting time indicated a lower hydration rate. The retention curves consistently showed the lower retention capacity of the thermally treated samples, thus indicating the incomplete re-hydration of the thermally treated BCV 2017 exposed to air humidity and the difference in its behaviour compared to the material exposed to liquid water.

**Keywords:** bentonite; smectite; crystal structure; water in the smectite interlayer; XRD; mineralogical changes; thermal treatment; BET; swell index; liquid limit; swelling pressure; water retention curves

---

## 1. Introduction

Bentonites are considered to be the most suitable materials for the multibarrier system of high-level radioactive waste (HLRW) repositories [1–4]. The favourable behaviour of bentonites is mainly influenced by the presence of smectites having unique physical and physico-chemical properties [3,5]. The stability of smectites is a key factor for all concepts that consider the use of bentonites in the engineered barriers of nuclear waste repositories [3]. The physico-chemical properties of smectites can be deteriorated by interaction with groundwater coming from the surrounding rock mass and by the heat generated from the radioactive decay of the waste [5–9].

The Czech deep geological repository concept is based on the use of bentonites and montmorillonite-rich clays of Czech origin as the buffer and backfill materials. Research work has been conducted to date primarily with respect to four Czech bentonites, each of which is of the calcium/magnesium (Ca/Mg) type. The first bentonite to be tested was extracted from the Rokle deposit, followed by industrially processed B75 bentonite, BAM—a mixture of Rokle, Stránce and Černý vrch bentonites and BCV—bentonite from the Černý vrch deposit.

Following the pilot characterisation of the various types of bentonites and montmorillonite-rich clays, experiments focused on the construction and operation of physical models of the buffer (Mock-Up-CZ and Mock-up Josef). The Mock-Up-CZ experiment comprised a physical model that followed the KBS-3V arrangement using Czech Rokle bentonite in the form of compressed blocks. A heater enabled the heating of the bentonite layer to a maximum of 95 °C. This physical model conducted in a laboratory environment was artificially saturated with synthetic granitic water. A detailed description of the experiment and the results can be found in [10].

The Mock-up Josef experiment [11] was installed in the real rock environment of the Josef Underground Research Laboratory. The project continues to provide valuable information on the behaviour of B75 bentonite that has been subjected to continuous loading in a similar way to that anticipated under deep geological repository conditions. The dismantling of this project is planned for 2022.

The BCV 2017 bentonite considered in this study comprises an industrially processed (dried and milled) bentonite extracted from the Černý vrch deposit and processed at the Keramost Ltd. Obrnice plant, Most, Czech Republic.

BCV bentonite was first subjected to testing in 2017 [12]. The basic characteristics of this material have been summarised in [13]. BCV bentonite is currently being used in the HotBENT full-scale experiment [14], the Interaction physical in situ models experiment underway at the Bukov underground research facility [15], the BEACON European project [16] and the Engineering barrier 200C project [17].

The Czech Radioactive Waste Repository Authority (SÚRAO) anticipates that this material will be subjected to extensive testing in a number of upcoming research projects and demonstration and full-scale experiments aimed at determining the bentonite to be used in the future Czech deep geological repository for used nuclear fuel.

Most concepts currently consider temperatures up to 100 °C to prevent water boiling and to prevent possible changes in the bentonite which could lead to decreased performance. The aim of presented research is to investigate the potential for increasing the temperature limit of the bentonite buffer. This could lead to significant cost savings due to increased disposal capacity. Moreover, it has the potential to enhance the safety margin in case of emergency events.

This paper presents a laboratory experiment based on the long-term heating of BCV 2017 bentonite at 200 °C. Special attention was drawn to investigate what changes in terms of impact on the barrier performance will happen at temperatures up to 200 °C. Even if there are mineralogical changes happening, it does not mean that material is losing its required properties (such as sealing or retention action) to perform its function fully (or in reduce manner). It is important to note that an increase in temperature up to 200 °C affects only a very small part of the barrier. A detailed description of the various properties affected by thermal treatment will provide invaluable assistance regarding the evolution of the material under repository conditions.

## 2. Materials and Methods

### 2.1. Material

The subject of this study comprised BCV 2017 bentonite (denoted "original BCV 2017") and the same material subjected to heating at 200 °C for 12, 15 and 27 months ("thermally-treated BCV 2017").

### 2.2. Analytical Techniques

The chemical and mineralogical composition of the experimental material (before and after heating at 200 °C) was studied via the powder X-ray diffraction (PXRD), infrared spectroscopy (FTIR), cation exchange capacity (CEC), thermal analysis and specific surface area measurement (BET). Selected geotechnical characteristics including the swell index, liquid limit, swelling pressure and water retention curves were subsequently determined.

### 2.2.1. Chemical Analysis and Powder X-ray Diffraction (PXRD)

The chemical composition of the original and thermally treated BCV 2017 bentonite was determined via standard wet chemistry in the laboratories of the Czech Geological Survey, Prague. The material was not subjected to any form of treatment prior to analysis except for drying at room temperature and grinding by McCrone mill. The chemical composition of solid samples was determined by standard techniques for silicate analysis summarized by Dempírová et al. [18]. $SiO_2$, $Al_2O_3$ and FeO were determined by titration. $TiO_2$, $Fe_2O_3$, MgO, MnO, CaO, $Na_2O$ and $K_2O$ were analysed by atomic absorption spectrometry (AAS) after decomposition of silicate matrix in a mixture of HF, $HNO_3$ and $H_2SO_4$. $P_2O_5$ was measured photometrically. Carbonate carbon was measured by infrared spectroscopy as $CO_2$ evolved from the sample by its reaction with orthophosphoric acid. C and S were determined by infrared spectroscopy after heating of the sample to 1350–1450 °C. The humidity ($H_2O^-$) was quantified as a loss in weight by drying at 110 °C until a constant mass of the sample was reached. For determination of the structural water ($H_2O^+$), the sample is heated at 1050 °C until a constant mass was reached. The $H_2O^+$ content was calculated by subtracting of volatile components ($CO_2$, C, S) from the weight loss obtained by heating at 1050 °C. The relative $2\sigma$ uncertainties did not exceed 1% ($SiO_2$), 2% (FeO), 5% ($Al_2O_3$, $K_2O$, $Na_2O$), 7% ($TiO_2$, MnO, CaO), 6% (MgO) and 10% ($Fe_2O_3$, $P_2O_5$) [18].

The mineralogical composition of the original and the thermally treated BCV 2017 bentonite was studied via powder X-ray diffraction (PXRD) analysis. The diffraction data were collected in a Bragg–Brentano reflection geometry on a Bruker D8 diffractometer equipped with a LynxEye XE detector. CoKα radiation and a 10 mm variable divergence slit were applied. The data were collected in the angular range 4–80° of 2Θ, with a 0.015° step size and a time of 5 s per step. The semiquantitative phase analysis was conducted by means of the Rietveld method applying the BGMN program with Profex 4.0 [19] as the graphical user interface. The models of the crystal structures used in the refinement were taken from the Profex 4.0 program internal database. The amorphous phase content was not quantified.

### 2.2.2. Infrared Spectroscopy

The infrared (IR) spectra of both the original and the thermally treated BCV 2017 material were recorded on a Nicolet 6700 FTIR spectrometer (Thermo Scientific, Waltham, MA, USA) equipped with a DTGS/KBr detector. Two IR spectroscopy techniques were used for the evaluation of the structural changes in the BCV 2017 due to long-term heating at 200 °C, i.e., the pressed bromide (KBr) pellet and the attenuated total reflection (ATR) methods. Samples of 2 and 0.5 mg were homogenised in an agate mortar with 200 mg of optically pure KBr for the recording of the optimal spectra in the 4000–3000 $cm^{-1}$ and 4000–400 $cm^{-1}$ regions, respectively. A total of 64 scans were taken with a resolution of

$4 \text{ cm}^{-1}$ for each spectrum. Prior to measurement, the KBr disks intended for the spectral region of $4000–3000 \text{ cm}^{-1}$ were heated at 150 °C for 24 h so as to eliminate the effect of surface-bound water. The reflective ATR technique was included in the experimental study since it allows for the direct measurement of powdered samples. The ATR spectra were collected using a single-reflective ATR accessory/Smart Orbit Diamond Crystal/ in the absorbance mode. A total of 32 scans were taken in the range $4000–400 \text{ cm}^{-1}$ with a spectral resolution of $4 \text{ cm}^{-1}$ for each spectrum. The final IR spectra were adjusted by means of ATR correction.

### 2.2.3. Cation Exchange Capacity (CEC)

The cation exchange capacity was determined using the Cu(II)-triethylenetetramine method according to [20]. Thus, 250 mg of the sample was mixed with 10 mL of 0.01 M of Cu(II)-triethylenetetramine. The samples were shaken for at least 30 min and then centrifuged at 3000 rpm for 10 min. A volume of 3 mL of the supernatant was then transferred into cuvettes and the concentration of the Cu(II) complex was measured via spectrophotometry. The solution absorbance was measured at 577 nm. The amount of the copper complex adsorbed was calculated from the concentration differences. The proportion of the main exchangeable cations ($Na^+$, $K^+$, $Ca^{2+}$, $Mg^{2+}$) was analysed in the supernatant solution by means of atomic absorption (Cu, Mg) and atomic emission (Ca, Na, K) spectroscopy.

### 2.2.4. Thermal Analysis Measurement

Simultaneous thermal analysis (TG-DTA) combined with evolved gas analysis (EGA) was employed for the study of the bentonite samples. The TG-DTA analysis was conducted using a Setsys Evolution device (Setaram, Caluire, France) coupled with an Omnistar Mass Spectrometer (Pfeiffer, Asslar, Germany). The experiments were performed within a temperature range of 20 °C to 1000 °C in an air atmosphere with a flow rate of $20 \text{ mL·min}^{-1}$. The heating rate was $10 \text{ K·min}^{-1}$. The evolved gases $CO_2$ and $H_2O$ were monitored by means of the Mass Spectrometer.

### 2.2.5. Specific Surface Area (BET)

The specific surface area and the distributions of the volume mesopores were measured using a 3Flex analyser (Micromeritics, Norcross, GA, USA) employing the gas sorption technique (the adsorption of nitrogen). The adsorption isotherms were fitted applying the Brunauer–Emmett–Teller (BET) method for the specific surface area.

### 2.2.6. Saturation of Thermally Treated BCV 2017 from the Aqueous Phase

The ability of certain 2:1 phyllosilicates, including smectites, to incorporate interlayer water molecules and the subsequent change in basal spacing has been studied extensively in the past (e.g., [21–24] and references therein). In order to demonstrate the ability of montmorillonite to incorporate water molecules following long-term heating at 200 °C, experiments were performed involving direct saturation from the aqueous phase. One gram of thermally treated BCV 2017 was added to 25 mL of demineralised water and mixed during the experiment. The bentonite reacted with the water in a beaker covered with foil for 44 days at a temperature of 21 °C. As the sample incorporated water, it was necessary to add demineralised water regularly up to the initial level. The bentonite material was analysed after 7, 14, 22, 30, 44 and 54 days of saturation by means of PXRD. Qualitative descriptions were compiled, i.e., on the position ($d_{001}$) and full width at half maximum (FWHM) of the 001 montmorillonite reflection for the characterisation of the smectite hydration.

### 2.2.7. Swell Index

The determination of the swell index (SI, mL/2 g) was performed according to the ASTM D5890—11 Standard Test Method for the Swell Index of the Clay Mineral Component of Geosynthetic Clay Liners [25]. The resulting value indicated a volume of 2 g of dried material following one day of free swelling (24 h).

The tests involved the dispersion of approx. 2 g samples of dry bentonite in 100 mL graduated cylinders in increments of 0.1 mL. The material was added at intervals of at least 10 min so as to allow for full hydration and the settlement of the clay at the bottom of the cylinder. The process continued until the entire 2 g samples had been added to the cylinders. The samples were then covered and protected from disturbance for a period of 16–24 h, following which the levels of the settled and swollen clay were recorded to the nearest 0.5 mL. The final volumes relating to 2 g of the dry material were then calculated. Five samples/cylinders were processed at a time so as to allow for the calculation of the average values. Some of the samples have been left additional time (up to 10 days) in dispersion in order to evaluate influence of longer saturation time.

### 2.2.8. Liquid Limit

The liquid limit ($w_L$; %) comprises the water content at which a soil changes from the liquid to the plastic state. The determination of the liquid limit was performed via the fall cone method according to standard ČSN EN ISO 17892-12 [26]. A cone with a tip angle of 30° and a mass of 80 g was used for testing purposes. In this case, the liquid limit equalled the water content at a penetration (depth) of 20 mm.

Distilled water was used for the wetting of the samples before and during the tests. The bentonite samples were carefully mixed and patiently kneaded with water so as to attain an initial penetration of 15 mm (according to the standard) and maintained in this state for 24 h. More water was then added, and the sample kneaded so as to once more attain 15mm penetration prior to the commencement of the testing process.

### 2.2.9. Swelling Pressure

The swelling pressure was measured in a constant volume cell apparatus that allows for the combination of the investigation of both the water permeability and the total pressure. The cells were designed for bottom-up saturation and the continuous monitoring of the evolution of total pressure. The tops and bottoms of the samples were fitted with sintered steel permeable plates so as to prevent the leaching ("mobilisation") of the material. The piston and force sensor for the measurement of the total (or swelling) pressure of the bentonite are positioned between the upper flange of the chamber and the upper surface of the sample. Force sensors were connected to a central data logger.

The bentonite BCV 2017 was uniaxially compacted into the hollow steel cylinders that formed the central part of the cell. Distilled water was used as the saturation medium. The test continued until the flow and the total pressure stabilised. The water pressure source was subsequently disconnected so as to allow for the determination of the swelling pressure.

### 2.2.10. Water Retention Curves

The water retention curves were determined by means of the commonly used vapour equilibrium method [27]. The relative humidity in the closed containers was investigated via the application of eight different saturated salt solutions [28] and was found to vary in the range 12.0–97.6%. The unique relationship between the imposed relative humidity and total suction is given by Kelvin's law. Samples with three differing initial dry densities (1.27, 1.60 and 1.90 g/cm$^3$) were prepared from bentonite powder by means of static uniaxial compaction under controlled loading. The compacted samples were carefully cut into pieces with irregular dimensions and typical volumes of 2–5 cm$^3$ and placed in vessels containing the various salt solutions. Three sets of samples were tested. The first set was prepared from original BCV 2017 bentonite with its natural water content of 11%. The second set was prepared in the same way; however, the pieces of compacted bentonite had been subjected to one year of thermal loading at 200 °C. The third set of samples was compacted from thermally treated powder (1 year, 200 °C). The heated powder used for the latter set of samples was equilibrated at a relative humidity of 43% (i.e., the standard laboratory environment) prior to compaction so as to replicate the compaction conditions of the first two sets of samples.

## 3. Results and Discussion

### 3.1. Chemical and Mineralogical Composition of BCV 2017

A comparison of the chemical compositions of the original and thermally treated bentonite BCV 2017 (27 months at 200 °C) is provided in Table 1. The results indicate a relatively high content of $Fe_2O_3$, e.g., in comparison with that of Bentonite 75 (5.57 wt.% [29]). The chemical composition of the original and thermally treated materials were found to be similar, with the exception of the differing hydration water contents ($H_2O^-$), which was a direct consequence of the heating of the material at 200 °C. As indicated by the PXRD study (see below), the heating of the material at 200 °C resulted in the complete dehydration of the interlayer space of the montmorillonite and its structural collapse to the 9.8 Å structure. Hence, the $H_2O^-$ content of thermally treated material is significantly lower than that for the original one.

**Table 1.** Chemical composition of the original and the thermally treated bentonite BCV 2017.

| Wt.% | Original BCV 2017 | Thermally Treated BCV 2017 |
|---|---|---|
| $SiO_2$ | 51.86 | 52.39 |
| $TiO_2$ | 2.34 | 2.60 |
| $Al_2O_3$ | 15.56 | 15.13 |
| $Fe_2O_3$ | 11.41 | 13.69 |
| FeO | 0.14 | 0.10 |
| MgO | 2.82 | 3.01 |
| MnO | 0.20 | 0.24 |
| CaO | 2.83 | 2.99 |
| $Na_2O$ | 0.37 | 0.41 |
| $K_2O$ | 1.02 | 0.95 |
| $P_2O_5$ | 0.51 | 0.53 |
| F | 0.12 | 0.10 |
| $CO_2$ | 1.68 | 1.60 |
| C | 0.17 | 0.11 |
| S | <0.010 | 0.02 |
| $H_2O^+$ | 9.06 | 6.34 |
| Total | 100.09 | 100.22 |
| $H_2O^-$ | 9.23 | 1.48 |

The lower structural water content ($H_2O^+$) in the thermally treated material was related to the transformation of goethite to hematite which occurred during the long-term heating of the material at 200 °C.

The mineralogical composition of the original BCV 2017 consists of clay minerals—montmorillonite, kaolinite and illite. Among non-clay minerals, quartz, goethite, Mg-bearing calcite, siderite, anatase and ankerite were detected. Since the smectite hydration state is characterised by the evolution of (001) basal-spacing ($d_{001}$) under variable relative humidity [23,24,30–32], a variation was observed in the $d$-value of the (001) basal spacing between 14.1 and 14.7 Å depending on the relative humidity (RH) in the laboratory. For an RH of 40%, the 001 reflection appeared at 14.5 Å. The usual hydration state of smectite has been described in terms of three-layer types that evince differing layer thicknesses that correspond to the common hydration states for smectites: dehydrated layers (0W, $d_{001}$ = 9.7–10.2 Å), monohydrated layers (1W, $d_{001}$ = 11.6–12.9 Å) and bihydrated layers (2W, $d_{001}$ = 14.9–15.7 Å) [32]. The $d_{001}$ value of 14.5 Å observed for montmorillonite in the BCV 2017 is an intermediate value between those basal spacing corresponding to the discrete 1W and 2W hydration states. Consequently, it indicates presence of high amount of bihydrated layers (2W) and a small amount of monohydrated layers (1W) in the montmorillonite structure at an RH of 40%.

The mineralogical compositions of the original and the thermally treated BCV are shown in Table 2, while Figure 1 provides a comparison of the powder X-ray diffraction patterns (PXRD). The PXRD pattern reflects two main changes in the samples of the thermally treated BCV 2017 compared to the original material. It is clear from Figure 1 that the most remarkable change concerned the shift in the (001) montmorillonite reflection from 14.5 to 9.8 Å for the material subjected to heating accompanied by an intensity decrease. This basal spacing corresponds to dehydrated montmorillonite, i.e., montmorillonite without the presence of water molecules in the interlayer. A second feature observed in the PXRD patterns is disappearance of goethite in the material subjected to heating and its dehydration to hematite according to the equation:

$$2\ \alpha\text{-FeOOH} \rightarrow \alpha\text{-Fe}_2\text{O}_3 + \text{H}_2\text{O} \tag{1}$$

**Table 2.** Semiquantitative PXRD phase analysis of the original and the thermally treated BCV 2017 bentonite (in wt.%).

| Wt.% | Original BCV 2017 | Thermally Treated BCV 2017 (200 °C) |
|---|---|---|
| Anatase | 2.3 | 2.6 |
| Quartz | 11.4 | 11.8 |
| Montmorillonite | 69.7 | 70.7 * |
| Mg-calcite | 3.7 | 3.3 |
| Goethite | 3.1 | n.d. |
| Hematite | n.d. | 1.7 |
| Kaolinite | 5.0 | 5.5 |
| Ankerite | 0.6 | 0.6 |
| Siderite | 0.5 | 0.4 |
| Illite | 3.7 | 3.4 |

* Corresponds to dehydrated montmorillonite (0W). n.d. not detected.

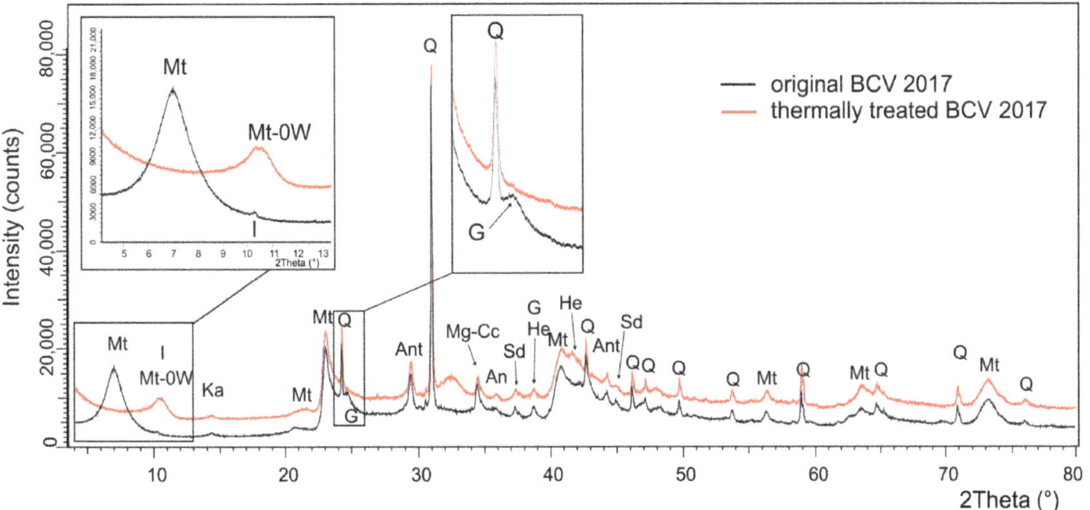


**Figure 1.** Comparison of the PXRD patterns of the original and thermally treated (200 °C) BCV 2017 (CoKα radiation). The left inset illustrates the change in the 001 reflection of montmorillonite following thermal treatment; the right inset illustrates the disappearance of the goethite reflection following heating. (Mt = montmorillonite, Mt-0W = dehydrated montmorillonite, Ka = kaolinite, Q = quartz, I = illite, Ant = anatase, G = goethite, He = hematite, Mg-Cc = Mg-calcite, Sd = siderite, An = ankerite).

The newly formed hematite exhibits broad peaks in its PXRD pattern which indicate that this phase is present as micro to nano-domains. Ruan et al. [33] described the formation of hematite from goethite at 240 °C and the further development of the hematite structure from 250° to 300 °C. De Faria and Lopes [34] and Gonzáles et al. [35] indicated a temperature interval of 260–280 °C for this dehydration process. Conversely, Walter et al. [36] observed the dependence of the goethite-hematite transformation temperature on the particle size and determined a transformation temperature of 191.7 °C for small goethite particles (with a BET specific surface of 149 m$^2 \cdot$g$^{-1}$) according to high-temperature XRD measurements. It is worth noting that a temperature of 250 °C was observed for the goethite-hematite transformation by the thermal analysis (see below) of an original BCV 2017 sample. However, the long-term heating of BCV 2017 at 200 °C over 27 months was sufficient to lead to the dehydration of goethite.

An apparent consequence of this reaction comprises a change in the colour of the sample following long-term heating. Original BCV 2017 is yellow-coloured while heated BCV 2017 exhibits a pale brown colour with red shading (Figure 2).

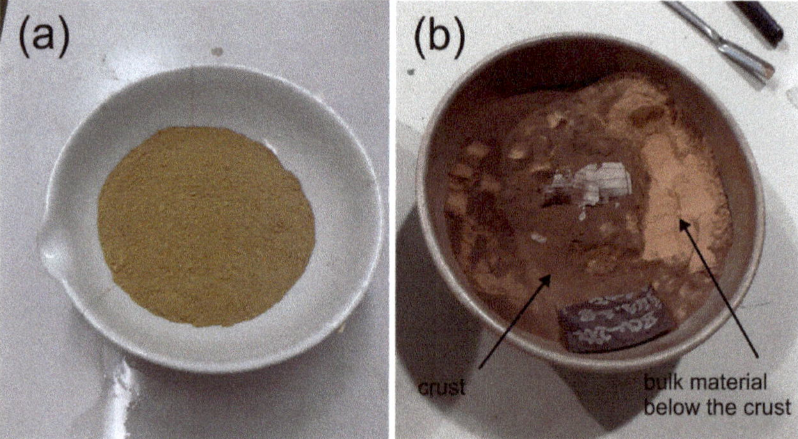


**Figure 2.** Comparison of (**a**) original and (**b**) thermally treated BCV 2017 at 200 °C. Note the presence of a thin dark crust on the surface of the thermally treated material.

It is interesting to note that a thin dark crust formed over the bulk of the BCV 2017 during heating at 200 °C (Figure 2). The crust exhibited a dark brown colour and was apparently coarser than the bulk material (Figure 3). From the mineralogical point of view, the crust appeared to be identical to the bulk of the thermally treated material; the PXRD study revealed no notable differences between the crust and the bulk material. Its formation was most likely related to the sinte ring process that acted upon the surface layer of the BCV 2017 sample during heating. Similar observations were made concerning a raw bentonite material extracted from the Rokle deposit (Kadaň, Chomutov District Czech Rep.) following heating at elevated temperatures (up to 500 °C) [37]. Data from silicate analysis and XRD data are available in Supplementary Material (Silicate analysis, XRD data).

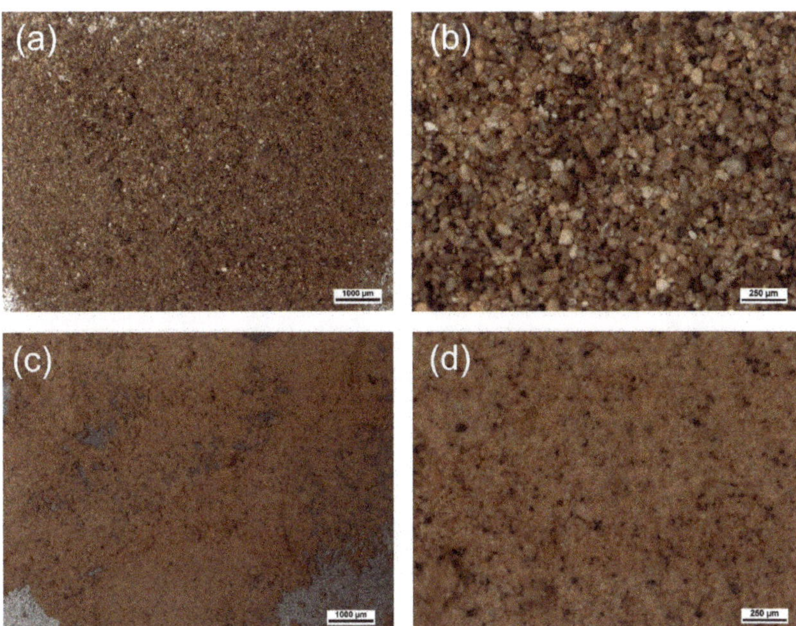

**Figure 3.** Microscopic images of thermally treated BCV 2017: (**a,b**) a crust that formed on the surface of the sample; (**c,d**) bulk material. Note the difference in grain size between the crust and bulk material.

### 3.2. Infrared Spectroscopy (IR)

The IR spectra of the original and the thermally treated BCV 2017 (200 °C) bentonite samples measured via the KBr pellet and ATR techniques are shown in Figure 4a,b. The IR spectroscopy confirmed the presence of montmorillonite as the dominant mineral phase in both bentonite samples. In addition to montmorillonite, kaolinite, a quartz admixture and traces of carbonates were also detected in the samples. A detailed list of the spectral bands that correspond to the montmorillonite and other minerals present in the original and thermally treated BCV 2017 are listed in Table 3. The assignment of the bands was performed according to [38–41]. The OH deformation band at around 874 cm$^{-1}$ (AlFeOH) in all the types of IR spectra of both samples indicates the significant substitution of octahedral Al$^{+3}$ by Fe$^{+3}$ in the montmorillonite structure. Other features in the spectra that are common to dioctahedral montmorillonites include a Si–O stretching band at around 1037 cm$^{-1}$ and Al–O–Si and Si–O–Si deformations at around 524 and 469 cm$^{-1}$, respectively. It is clear from Figure 4 that the IR spectra of the original and thermally treated materials are very similar; the intensities of the absorption bands and their positions remained virtually unchanged. Subsequently, the IR spectroscopic study also proved the crystallochemical stability of the montmorillonite in BCV 2017 bentonite under selected experimental conditions. As was indicated by the PXRD study, the main mineralogical change in the thermally treated compared to the original material comprised the dehydration of the interlayer water in the montmorillonite and the subsequent reduction of the $d_{001}$ spacing to 9.8 Å. Unlike the PXRD and TG/DTA thermal analysis, conventional IR spectroscopy did not directly reveal the dehydration of the interlayer. Montmorillonites could contain water in several forms: interlayer water that form part of hydration envelopes of interlayer cations, adsorbed water surrounding the outside surface of montmorillonite, free water in voids between soil particles and hydroxyl water [38,42,43]. The overlap of the bands originating from the structural OH groups and the OH groups from the adsorbed and interlayer water in the stretching vibrations region (3700–3000 cm$^{-1}$) complicated the detection of the dehydration of the interlayer by conventional IR approach.

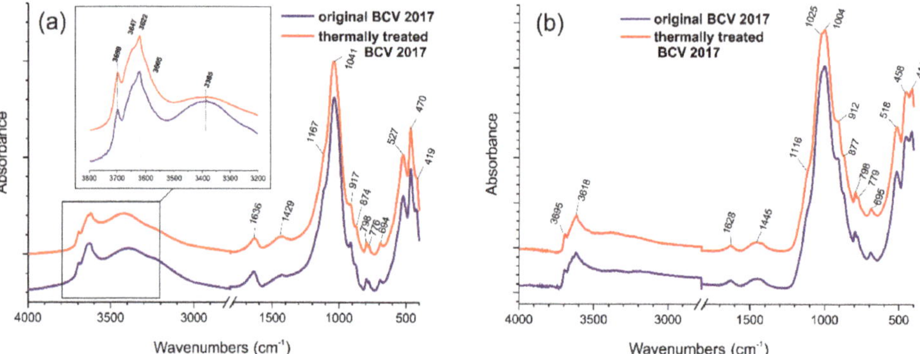

**Figure 4.** Infrared spectra of the original and the thermally treated BCV 2017 obtained via the (**a**) KBr pellet and (**b**) the ATR techniques.

**Table 3.** Positions of the absorption bands (cm$^{-1}$) in the IR spectra of the original and the thermally treated (200 °C) samples of BCV 2017 acquired via the by KBr and ATR measuring techniques, and their assignments.

| Assignment of the Spectral Bands | Original BCV 2017 | | Thermally Treated BCV 2017 | |
|---|---|---|---|---|
| | KBr | ATR | KBr | ATR |
| Kaolinite (minor phase) OH stretching of the inner-surface hydroxyl groups | 3698 * | 3695 | 3699 * | 3695 |
| Kaolinite (minor phase) OH stretching of the inner-surface hydroxyl groups | 3647 * | 3643 | 3647 * | - |
| OH stretching of the structural hydroxyl groups | 3623 * | 3618 | 3622 * | 3618 |
| Chlorite (admixture) | 3565 * | - | - | - |
| OH stretching of the adsorbed water molecules | 3385 * | - | 3385 * | - |
| OH deformation of the adsorbed water molecules | 1646 | 1626 | 1636 | 1628 |
| Carbonates (minor phases) | 1424 | 1454 | 1429 | 1445 |
| Quartz (admixture) | 1170 | - | 1167 | - |
| Si−O stretching (longitudinal mode) | 1118 | 1116 | 1116 | 1116 |
| In plane Si−O stretching | 1037 | 1022 | 1041 | 1025 |
| In plane Si−O stretching | - | 1004 | - | 1004 |
| Al−Al−OH deformation | 912 | 916 | 917 | 912 |
| Al−Fe−OH deformation | 877 | 875 | 874 | 877 |
| Quartz (minor phase) | 798 | 798 | 798 | 798 |
| Quartz (minor phase) | 776 | 779 | 776 | 779 |
| Quartz (minor phase) | 695 | 692 | 694 | 695 |
| Al−O−Si deformation | 524 | 518 | 527 | 518 |
| Si−O−Si deformation | 469 | 453 | 470 | 458 |
| Si−O deformation | 422 | 416 | 420 | 414 |

* KBr pellet prepared from 2.0 mg of the sample with a 200 mg/KBr pellet heated overnight at 150 °C. - not observed.

### 3.3. Thermal Analysis

The differential thermal analysis (DTA) and thermogravimetric analysis (TG) curves together with the mass spectroscopy signals for $CO_2$ and water molecules for the original and the thermally treated (200 °C, 27 months) BCV 2017 bentonite are shown in Figure 5. It is clear from the Figure 5 that five and six main reactions were observed for the thermally treated and the original BCV 2017, respectively. The first endothermal reaction occurred in the temperature range 30–200 °C, with an endothermic peak centred at 100 °C and resulting mass losses of 10.4 and 3.4 wt.% for the original and the thermally treated BCV

2017 samples, respectively. Concerning the original material, this reaction corresponded to the dehydration of the external surface, the pores and the interlayer space in the montmorillonite [44,45]. According to our PXRD study, the montmorillonite in the thermally treated BCV 2017 shows $d_{001}$ position 9.8 Å, which indicates a state in which no water molecules are present in the interlayer ([32] and references therein), while the original material contained monohydrated and bihydrated water layers in the interlayer space. Consequently, the reaction in the thermally treated material includes dehydration of the external surface and pores, which resulted in a significant weight loss during the reaction. A second endothermic reaction, which was observed only for the original material, occurred at around 250 °C and represented the dehydroxylation of the goethite to hematite according to Equation (1). This thermal effect was absent in the thermally treated BCV 2017 since the goethite had already been transformed into hematite during the 27 months of heating. The long-term heating of the material at 200 °C resulted in the transformation of the goethite to hematite at a temperature lower than the 250 °C indicated by the thermal analysis. This can be explained by the fact that the thermal analysis experiment is highly dynamic, and the detected temperature of a process taking place in the sample is also dependent on the heating rate during the experiment. The long-time experiment is close to the equilibrium conditions. The equilibrium temperature could be found by thermal analysis performed at a (virtual) heating rate zero K·min$^{-1}$, which is impossible.

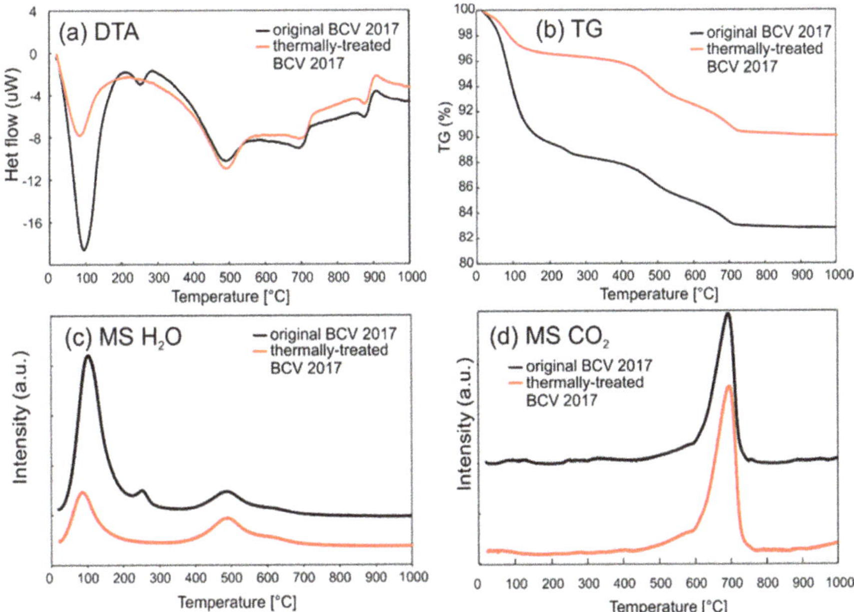

**Figure 5.** Comparison of (**a**) DTA, (**b**) TG curves and mass spectroscopy signals for (**c**) water and (**d**) carbon dioxide for the original and the thermally treated BCV 2017.

The third endothermic reaction, observed for both the original and the thermally treated BCV between 450 and 550 °C, corresponded to the dehydroxylation of the montmorillonite structure. According to Drits et al. [46,47], trans-vacant smectites have dehydroxylation temperatures of between 500 and 550 °C, while cis-vacant variants have a dehydroxylation temperature of around 700 °C. This suggests that the trans-vacant montmorillonite configuration prevails in BCV 2017 bentonite. This is in agreement with the observed $Fe^{3+}$ substitution in the octahedral sheet in montmorillonite as confirmed by the IR spectroscopy, since this substitution on the octahedral sheets of the dioctahedral phyllosilicates prefers to form trans-vacant forms [48–51]. The fourth endothermic reaction, with a peak at around 700 °C and related significant release of $CO_2$ visible for both samples, indicated the decomposition of the Mg-Ca carbonate [52]. The fifth endothermic reaction, with a peak at approximately 880 °C, was related to the disappearance of the layered structure of the montmorillonite in both types of BCV 2017 samples [53]. This reaction was accompanied by a negligible loss of weight. The sixth exothermic effect that occurred in the range 890–920 °C was associated with the development of high-temperature phases including spinel and affected both the original and the thermally treated BCV 2017 samples. Thermal analysis data are provided in Supplementary Materials (thermal analysis data).

### 3.4. Cation Exchange Capacity (CEC)

The interlayer cations in the montmorillonite in the original BCV 2017 consist mainly of $Mg^{2+}$ (38.7 meg/100 g) and $Ca^{2+}$ (19.9 meg/100 g), with some minor amounts of $Na^+$ (7.2 meg/100 g) and $K^+$ (2.1 meg/100 g). The CEC values determined for the original and the thermally treated materials are shown in Figure 6. Accordingly, the smectite type in BCV 2017 bentonite can be characterized as Ca-Mg montmorillonite. The CEC values decreased from an initial value of 60.9 meg/100 g to 54.5 meg/100 g following thermal treatment. It is assumed that the change in the CEC values before and after thermal treatment was related to the collapse of the interlayer structure avoiding further cation replacements.

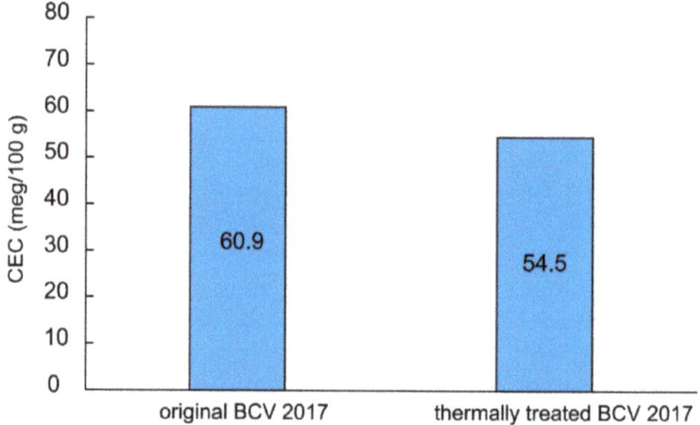

**Figure 6.** Cation exchange capacity (CEC) of the original and the thermally treated BCV 2017.

### 3.5. Saturation of Thermally Treated BCV 2017 from the Aqueous Phase

Figure 7a,b shows the evolution of the PXRD 001 profiles of the montmorillonite in the thermally treated BCV 2017 during saturation from the aqueous phase of the sample a function of the saturation time. The values of basal spacing ($d_{001}$) together with the full width at half maximum (FWHM) of the 001 reflection are presented in Figure 7. The montmorillonite in the thermally treated BCV 2017 showed the $d_{001}$ value of 9.8 Å, which corresponds to the hydration state without the presence of water molecules in the interlayer (0W) [32]. After 7 days of saturation in water, the montmorillonite evinced the $d_{001}$ value of 13.12 Å, which indicated the intermediate hydration state, i.e., between the 1W and 2W discrete hydration states. In accordance with a study by Ferrage et al. [31], the heterogeneity of the hydration state led to the interstratification of the various different layer types, which produced increased FWHM value of 2.22° 2θ compared to an FWHM value of 1.11° 2θ for the thermally treated material. After 15 days of saturation, the 001 reflection appeared at 14.45 Å, thus indicating the increasing proportion of the 2W layers in the montmorillonite structure; the FWHM value decreased to 1.96° 2θ. After 30 days of saturation, a slight shift in the position of the 001 reflection to 14.66 Å was observed followed by a slight decrease in the FWHM value to 1.89° 2θ. After 44 days, the 001 position was observed to occur at 15.1 Å and the FWHM was seen to have further decreased to 1.57° 2θ, thus indicating bihydrated layers (2W) within the montmorillonite structure. No notable further changes in the 001 profile were observed by PXRD after 54 days of saturation.

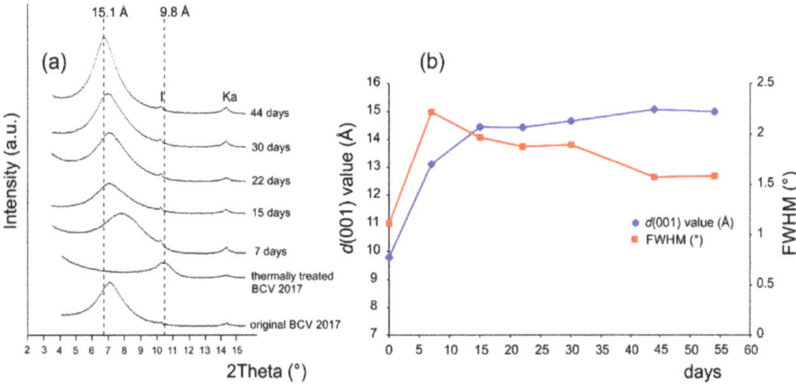

**Figure 7.** Evolution of (**a**) the PXRD profile, (**b**) the *d* and FWHM of the 001 montmorillonite reflection for the thermally treated BCV 2017 sample during saturation from the aqueous phase (CoKα radiation). The days indicate the duration of saturation. I: illite, Ka: kaolinite.

As is evident from Figure 7, relatively broad and often asymmetric peaks occurred in the PXRD during saturation with no apparent indication of homogenous stepwise hydration. In accordance with Holmboe et al. [54] and Ferrage et al. [32], the evolution of hydration exhibits a somewhat heterogeneous character that indicates the coexistence of different layer types (1W and 2W) in the interlayer space. It is interesting to note that the saturation of the thermally treated BCV 2017 for 44 days resulted in montmorillonite with bihydrated layers (2W) in the crystal structure. Conversely, the same thermally treated material placed at under standard room conditions (~40% RH, 21 °C) evinced different saturation behaviour. The PXRD patterns exhibited broad and diffuse 001 profiles that indicated significantly slower hydration than that induced via saturation from the liquid water. The behaviour also depended on the relative humidity, a topic that will form the subject of a separate study.

### 3.6. Specific Surface Area (BET)

The original BCV 2017 material is characterized by a BET-$N_2$ specific area of 91 $m^2/g$, which falls within the broad distribution of BET-$N_2$ specific area values for Ca-Mg-dominated bentonites, which range from 50 to 130 $m^2/g$ [55]. A slightly decreased value of 89 $m^2/g$ was determined for the thermally treated material. BET-$N_2$ values are known to depend on the exchangeable cation population [56], the microporosity and the accessible areas of the interlayer [55]. Lang et al. [57] noted that the removal of absorbed water during heating provides new adsorptive sites for $N_2$ molecules. Conversely, the dehydration of the interlayer space induces a decrease in the interlayer space. Consequently, changes in the specific surface area values depend on the competing effects of the dehydration of the external surface and the interlayer space. With respect to the thermally treated BCV 2017, the reduction of the interlayer space is most likely compensated for by the creation of new sites for $N_2$ molecules as a consequence of the removal of the absorbed water. Thus, the BET-$N_2$ specific surface area values were observed to be very similar for the original and the thermally treated materials.

The isotherms determined for the original and the thermally treated BCV 2017 are displayed in Figure 8. Both samples exhibit the type IV isotherm according to the IUAPC classification with noticeable H4-type hysteresis loops that are characteristic of slit-shaped pores. These findings suggest that both BCV 2017 samples comprise mesoporous materials with the limited contribution of microporosity. The micropore volumes, estimated from the t-plot method, are 0.0143 and 0.0093 cm$^3$/g for the original and the thermally treated material. These values represent 13.4% and 8.6% of the total pore volume (up to 40.31 nm). The total pore volume values (up to 40.31 nm), obtained from the N$_2$ adsorption isotherm, are 0.1065 and 0.1085 cm$^3$/g, respectively.

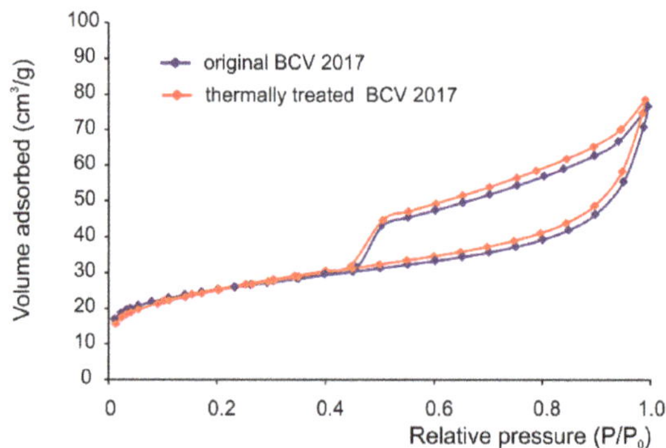


**Figure 8.** Absorption and desorption isotherms for original and thermally treated bentonite BCV 2017.

### 3.7. Swell Index and Liquid Limit

Swell index tests were performed on the original and the thermally treated materials (200 °C; 15 and 27 months). Two types of samples were investigated in the case of 27 months, i.e., a "bulk" sample that represented most of the volume of the treated material and a top surface layer that comprised the thin dark crust that formed on the surface of the bentonite BCV 2017 during thermal treatment (see Figures 2 and 3). The results are presented in Table 4; although a certain decrease (approx. 25%) is evident due to the thermal treatment, no differences were detected between the bulk material and the crust thermally treated for 27 months. The results indicate a reduction in the free swelling ability following treatment; however, it is important to bear in mind the short duration (24 h) of the testing process. To address that issue, some of the samples were left in cylinder up to 10 days in total in order to observe if an additional swelling occurs over time (see values with @ in Table 4 denoting time from test start). The treated samples exhibit delayed start of additional swelling. This indicates that some of the swelling capacity may be recuperated over time.

**Table 4.** Swell index and liquid limit of the BCV 2017.

| Treatment | Swell Index SI (mL/2g) | Liquid Limit $w_L$ (%) |
|---|---|---|
| original | 7.5@1d (7.12.2020)<br>8.3@1d, 8.8@2d, 9.3@7d, 9.4@10d (14.6.2021) | 130%;<br>139% [3,4] |
| 200 °C; 15 months | 5.5@1d, 5.5@2d, 5.6@7d, 6.5@10d (bulk, 14.6.2021) | 111% (wetted 24 h);<br>118% (wetted 10 days) |
| 200 °C; 27 months | 5.7@1d (bulk)<br>5.8@1d (top layer) | x |

x not measured, @ time from test start.

The liquid limit was determined for the original and the thermally treated materials (200 °C; 15 months; Table 4). In the case of the treated sample, an additional 10-day wetting test was also performed. Bentonite samples are usually wetted for 24 h prior to the determination of the liquid limit. As the result showed lower value for thermally treated material, and indication of slower resaturation was discussed (see Section 3.5), total time of 10 days was used for wetting before liquid limit test. Water was added each day and the samples kneaded so as to maintain the water content at 15 mm penetration. The visual observation revealed that a significantly higher amount of water was required than usual. The results indicate a lower (immediate) water adsorption ability following thermal treatment, with a partial resaturation ability when a time of 10 days was allowed.

*3.8. Swelling Pressure*

The swelling pressure results for the original and the thermally treated BCV 2017 (200 °C; 15 month) are presented in Figure 9. The data concerning the original BCV 2017 material were taken from [13,16], H2020 Beacon project [16], IE Bukov [15,58]. Even though the results were determined by two independent laboratories, i.e., at the Charles University (CU) and the Czech Technical University (CTU), they are consistent with each other, which suggests that the measurements are not biased due to some hidden experimental problems. Figure 9 presents three results for the thermally treated material (red squares).

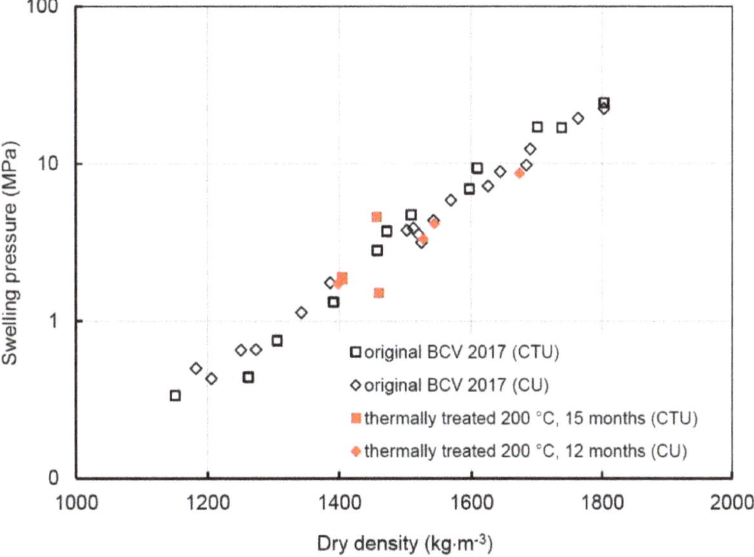


**Figure 9.** Dependency of swelling pressure on dry density—the original BCV 2017 material in black and the thermally treated material in red. Results were obtained by laboratories at Charles University (CU) and Czech Technical University (CTU).

The results for the thermally treated material do not indicate a significant shift from the original material; however, the scatter of the results appears to be more extensive.

A further consideration concerns the evolution of (total) pressure following the commencement of saturation. Samples with dry densities of approx. 1400–1500 kg·m$^{-3}$ were selected for comparison purposes. Figure 10 clearly indicates the significantly more rapid process with respect to the treated material, i.e., approx. two days was sufficient to attain at least 90% of the final total pressure. The original material required approx. 3 times longer (6 days) to attain the same pressure level. The temporary "sagging" of the pressure was absent during the early phase of the saturation of the thermally treated samples.

The potential slower water adsorption due to thermal treatment may have been the reason of the higher level of permeability of the sample at the beginning of the test; the probable consequence comprises the more uniform and more rapid saturation throughout the whole of the volume of the thermally treated sample.

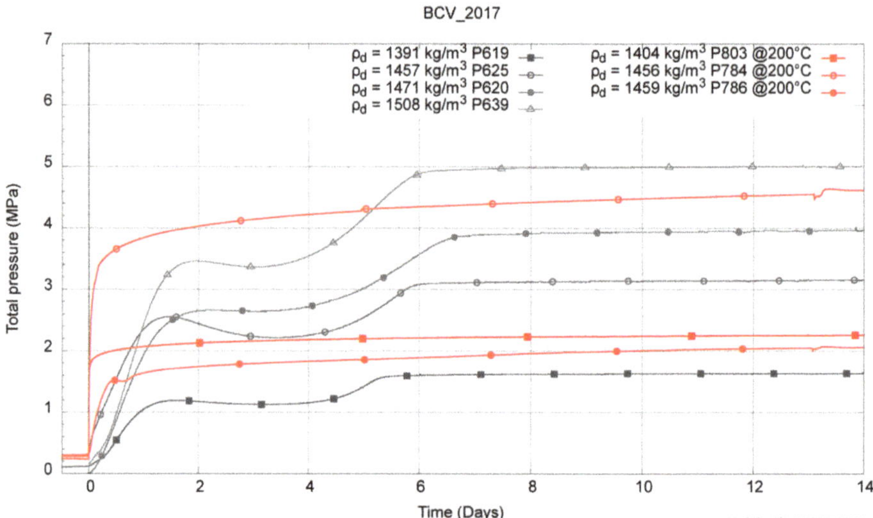

**Figure 10.** Evolution of the swelling pressure—the original material (black) and the thermally treated (red) materials; dry density of approx. 1400–1500 kg·m$^{-3}$.

### 3.9. Water Retention Curves

Figure 11 provides a summary of three sets of water retention curves. It reveals the negligible effect of the initial density on the water content, which can be explained by the double porosity structure typical for bentonites [59,60]. At high suctions, water is concentrated predominantly in micropores inside the aggregates, which are not significantly influenced by compaction [61].

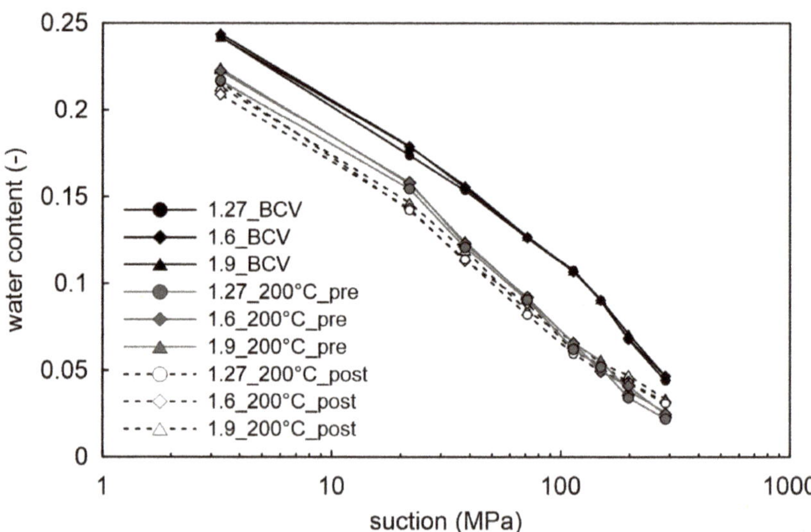

**Figure 11.** Water retention curves determined for three sets of samples of BCV 2017. The samples compacted before thermal treatment are marked "pre", while the samples compacted after thermal loading are marked "post".

The results clearly demonstrate the lower retention capacity of the thermally treated samples than that of the original BCV 2017 bentonite. The reduction was approximately constant over the whole of the suction range (3–287 MPa), which corresponds to relative humidities of 12–98%. Only a small difference was observed between the samples compacted before and after thermal treatment. Results are compiled in Supplementary Materials (data water retention curves).

The results indicate that thermally treated bentonite BCV 2017 fully rehydrates when it is exposed to free water (Figure 7) and its mechanical behaviour is reaching similar values compared to original BCV 2017 bentonite as demonstrated by swelling pressures (Figure 9) or overall trend in liquid limit determination (Table 4). On the other hand, when exposed to air humidity, re-hydration process is only partial and takes significantly longer time (Section 3.5, Figure 10). This behaviour is very likely related to the irreversible collapse of the interlayer structure of some smectite crystallites after heating.

## 4. Conclusions

Both the original BCV 2017 bentonite and the material thermally treated at 200 °C were characterised via chemical, mineralogical and thermal analysis, infrared spectroscopy, the specific area (BET), the swell index, the liquid limit, the swelling pressure and water retention curves. The results of the various methods were subsequently quantitatively analysed and compared, which led to the following main conclusions:

(i)    The original BCV 2017 bentonite contains around 69.7 wt.% of montmorillonite, which is characterised by $Fe^{3+}$ substitution in the octahedral sheet and Mg and Ca prevailing cations in the interlayer space.

(ii)    The BCV 2017 bentonite underwent two main transformation processes following thermal treatment at 200 °C over 27 months. The first transformation process comprised the dehydration of the smectite (i.e., the removal of the interlayer water) and a reduction in the 001 basal spacing from 14.5 Å (the original BCV 2017) to 9.8 Å following thermal treatment, which indicated the absence of water molecules in the interlayer space. The second reaction concerned the -dehydroxylation of goethite to hematite, which was related to the apparent change in the colour of the samples from yellow to pale brown with red shading.

(iii)    The PXRD and IR spectroscopic study revealed the crystallochemical stability of the montmorillonite in the BCV 2017 bentonite under selected experimental conditions. With the exception of the dehydration of the interlayer space, no deterioration occurred in the montmorillonite crystal structure during the experiment.

(iv)    Heating at 200 °C was found to have little effect on the BET-N2 specific area. The isotherms determined for the original and the thermally treated BCV 2017 were of the IV type with H4 hysteresis loops that indicated that both types of BCV 2017 samples comprise mesoporous materials with the limited contribution of microporosity.

(v)    The saturation of the thermally treated BCV 2017 (i.e., with a $d_{001}$ of 9.8 Å, 0W) from the aqueous phase resulted in montmorillonite with bihydrated layers (2W) in the interlayer space with a $d_{001}$ spacing of 15.1 Å.

(vi)    The geotechnical tests indicated no major changes in the mechanical properties of the thermally treated BCV 2017 bentonite as demonstrated by their similar swelling pressure values. However, the variation in the swell index and the gradual increase in the liquid limit over the wetting time indicated a lower hydration rate.

(vii)    The retention curves consistently pointed to the lower retention capacity of the thermally treated samples, which indicated the incomplete re-hydration of the thermally treated bentonite exposed to air humidity and a difference in the behaviour of the material compared to the bentonite exposed to liquid water

Despite indications of slower resaturation of BCV 2017 bentonite after thermal treatment followed by higher level of permeability during saturation phase, swelling ability remained unchanged which forms important result for the overall buffer function. The heating at 200 °C causes dehydration of the interlayer water from the montmorillonite structure; however, the thermally treated BCV 2017 bentonite fully rehydrates when it is exposed to free water. The complex multidisciplinary dataset describing behaviour of BCV, a suitable representative of Czech Ca-Mg bentonites, is an important milestone towards potential usage of higher temperatures in the deep repository. The results indicate that temperature limit of 200 °C could be feasible.

**Supplementary Materials:** The following are available online at https://www.mdpi.com/article/10.3390/min11080871/s1, PXRD data of original and thermally treated BCV 2017, thermal analysis data, silicate analysis data, retention curves data, specific surface measurements data (BET), protocols from liquid limits and swell index determination.

**Author Contributions:** Conceptualisation F.L., I.H., J.S., R.V., J.N. and D.M.; investigation F.L., J.S., R.V., J.N., M.Č., M.K., F.P., L.V. and H.S.; writing—original draft preparation F.L., I.H., J.S., R.V. and J.N. All authors have read and agreed to the published version of the manuscript.

**Funding:** This study was supported by the Engineered barrier 200C project (no. TK01030031) awarded by the Technology Agency of the Czech Republic. The fifth and eleventh authors were supported by project no. TK01010063 awarded by the Technology Agency of the Czech Republic.

**Acknowledgments:** The authors would like to thank M. Lhotka (UCT Prague) for the specific surface measurements, J. Havlín (UCT Prague) for the thermal analysis measurements, K. Černochová (CTU Prague) for swell index and swelling pressure measurements and D. Nádherná (CTU Prague) for liquid limit measurements.

**Conflicts of Interest:** The authors declare no conflict of interest.

## References

1. NAGRA. *Project "Gewähr 1985", Feasibility and Safety Studies for Final Disposal of Radioactive Wastes in Switzerland*; Nagra: Wettingen, Switzerland, 1985.
2. SKB Swedish Nuclear Fuel and Waste Management Company. *Treatment and Final Disposal of Nuclear Waste: Programme for Encapsulation Deep Geologic Disposal, and Research, Development and Demonstration*; RD&D Programme 95; SKB Swedish Nuclear Fuel and Waste Management Company: Stockholm, Sweden, 1995.
3. Pusch, R. Transport of radionuclides in smectite clay. In *Environmental Interactions of Clays*; Parker, A., Rae, J.E., Eds.; Springer: Berlin/Heidelberg, Germany, 1998; pp. 7–35.
4. Savage, D.; Lind, A.; Arthur, R.C. *Review of the Properties and Uses of Bentonite as a Buffer and Backfill Material*; SKI Report 99:9; Swedish Nuclear Fuel and Waste Management Company: Stockholm, Sweden, 1999.
5. Borrelli, R.A.; Thivent, O.; Ahn, J. Parametric studies on confinement of radionuclides in the excavated damaged zone due to bentonite type and temperature change. *Phys. Chem. Earth* **2013**, *65*, 32–41. [CrossRef]
6. Pusch, R.; Kasbohm, J.; Pacovsky, J.; Cechova, Z. Are all smectite clays suitable as "buffers"? *Phys. Chem. Earth* **2007**, *32*, 116–122. [CrossRef]
7. Villar, M.V.; Lloret, A. Influence of temperature on the hydro-mechanical behaviourof a compacted bentonite. *Appl. Clay Sci.* **2004**, *26*, 337–350. [CrossRef]
8. Ye, W.M.; Zhang, Y.W.; Chen, B.; Zheng, Z.J.; Chen, Y.G.; Cui, Y.J. Investigation on compression behaviour of highly compacted GMZ01 bentonite with suction and temperature control. *Nucl. Eng. Des.* **2012**, *252*, 11–18. [CrossRef]
9. Ye, W.M.; Zheng, Z.J.; Chen, B.; Chen, Y.G.; Cui, Y.J.; Wang, J. Effects of pH and temperature on the swelling pressure and hydraulic conductivity of compacted GMZ01 bentonite. *Appl. Clay Sci.* **2014**, *101*, 192–198. [CrossRef]
10. Svoboda, J.; Vašíček, R. Preliminary geotechnical results from the Mock-Up-CZ experiment. *Appl. Clay Sci.* **2010**, *47*, 139–146. [CrossRef]
11. Šťástka, J. Mock-up Josef Demonstration Experiment. *Tunel* **2014**, *23*, 65–73.
12. Červinka, R.; Vašíček, R.; Večerník, P.; Kašpar, V. *Complete Characterization of Bentonite BCV Technical Report SÚRAO TZ419/2019*; SÚRAO: Prague, Czech Republic, 2018.
13. Hausmannová, L.; Hanusová, I.; Dohnálková, M. *Summary of the Research of Czech Bentonites for Use in the Deep Geological Repository—Up to 2018*; Technical Report SÚRAO 309/2018/ENG; SÚRAO: Prague, Czech Republic, 2018.
14. Kober, F.; Vomvoris, S.; Finsterle, S. *HotBENT—High Temperature Buffer Experiment: Preliminary Design Study—Overview of Results*; Arbeitsbericht NAB17-29; Nagra: Wettingen, Switzerland, 2018.
15. Svoboda, J.; Vašíček, R.; Smutek, J.; Hausmannová, L.; Franěk, J.; Rukavičková, L.; Večerník, P. Interaction Experiment at the Bukov URF. In Proceedings of the 4th International Conference entitled "Underground Construction Prague 2019", Prague, Czech Republic, 3–5 June 2019.

16. Baryla, P.; Bernachy-Barbe, F.; Bosch, J.; Campos, G.; Carbonell, B.; Daniels, K.; Ferrari, A.; Guillot, W.; Gutiérrez-Álvarez, C.; Harrington, J.; et al. *Bentonite Mechanical Evolution-Experimental Work for the Support of Model Development and Validation*; Technical Report D4.1/2, Euratom Research and Training Programme, No. 745942; European Union: Maastricht, The Netherlands, 2019.

17. Svoboda, J.; Vašíček, R.; Šťástka, J.; Kruis, J.; Krejčí, T.; Mašín, D.; Najser, J.; Franěk, J.; Rukavičková, L.; Laufek, F.; et al. *TK01030031-Engineered Barrier 200C-Annual Report*; CTU: Prague, Czech Republic, 2020.

18. Dempírová, L.; Šikl, J.; Kašičková, R.; Zoulková, V.; Kříbek, B. The evaluation of precision and relative error of the main components of silicate analyses in Central Laboratory of the Czech Geological Survey. *Zpr. Geol. Výzk. V Roce* **2009**, *2010*, 326–330. (In Czech)

19. Döbelin, N.; Kleeberg, R. Profex: A graphical user interface for the Rietveld refinement program BGMN. *J. Appl. Crystallogr.* **2015**, *48*, 1573–1580. [CrossRef]

20. Ammann, L.; Bergaya, F.; Lagaly, G. Determination of the cation Exchange capacity of clays with copper complexes revisited. *Clay Miner.* **2005**, *40*, 441–453. [CrossRef]

21. Nagelschmidt, G. On the lattice shrinkage and structure of montmorillonite. *Z. Krist.* **1936**, *93*, 481–487. [CrossRef]

22. Bradley, W.F.; Grim, R.E.; Clark, G.F. A study of the behavior of montmorillonite upon wetting. *Z. Krist.* **1937**, *97*, 216–222.

23. Méring, J.; Glaeser, R. Sur le rôle de la valence des cations échangeables dans la montmorillonite. *Bull. Soc. Fr. Minéral. Cristallogr.* **1954**, *77*, 519–530. [CrossRef]

24. Ferrage, E.; Lanson, B.; Malikova, N.; Plançon, A.; Sakharov, B.A.; Drits, V.A. New insights on the distribution of interlayer water in bi-hydrated smectite from X-ray diffraction profile modeling of 00l reflections. *Chem. Mater.* **2005**, *17*, 3499–3512. [CrossRef]

25. ASTM International. *ASTM D5890-11th Standard Test Method for Swell Index of Clay Mineral Component of Geosynthetic Clay Liners*; ASTM International: West Conshohocken, PA, USA, 2011; 7p.

26. *ČSN EN ISO 17892-12: Determination of Consistency Limits*; ISO: Geneva, Switzerland, 2018.

27. Delage, P.; Howat, M.D.; Cui, Y.J. The relationship between suction and swelling properties in a heavily compacted unsaturated clay. *Eng. Geol.* **1998**, *50*, 31–48. [CrossRef]

28. OIML. *The Scale of Relative Humidity (RH) of Air Certified Against Saturated Salt Solutions*; The International Organization of Legal Metrology: Paris, France, 1996.

29. Filipská, P.; Zeman, J.; Všijanský, D.; Honty, M.; Škoda, R. Key processes of long-term bentonite-water interaction at 90 °C: Mineralogical and chemical transformations. *Appl. Clay Sci.* **2017**, *150*, 234–243. [CrossRef]

30. Harward, M.E.; Brindley, G.W. Swelling properties of synthetic smectites in relation to lattice substitutions. *Clays Clay Miner.* **1965**, *13*, 209–222. [CrossRef]

31. Ferrage, E.; Lanson, B.; Sakharov, B.A.; Drits, V.A. Investigation of smectite hydration properties by modelling of X-ray diffraction profiles. Part Montmorillonite hydration properties. *Am. Mineral.* **2005**, *90*, 1358–1374.

32. Ferrage, E.; Kirk, C.A.; Cressey, G.; Cuadross, J. Dehydration of Ca-montmorillonite at the crystal scale. Part 2 Mechanism and kinetics. *Am. Mineral.* **2007**, *92*, 1007–1017. [CrossRef]

33. Ruan, H.D.; Frost, R.L.; Kloprogge, J.T. The behavior of hydroxyl units of synthetic goethite and its dehydroxylated producthematite. *Spectrochim. Acta A* **2001**, *57*, 2575–2586. [CrossRef]

34. De Faria, D.L.A.; Lopes, F.N. Heated goethite and natural hematite: Can Raman spectroscopy be used to differentiate them? *Vib. Spectrosc.* **2007**, *45*, 117–121. [CrossRef]

35. González, G.; Sagarzazu, A.; Villalba, R. Study of the mechano-chemical transformation of goethite to hematite by TEM and XRD. *Mater. Res. Bull.* **2000**, *35*, 2295–2308. [CrossRef]

36. Walter, D.; Buxbaum, G.; Laqua, W. The mechanism of the thermal transformation from goethite to hematite. *J. Therm. Anal. Calorim.* **2001**, *63*, 733–748. [CrossRef]

37. Přikryl, R.; Hanus, R.; Kolaříková, I.; Vejsada, J.A. *Verification of Substitution of Bentonites by Montmorillonitic Clays*; Technical Report SÚRAO 10/2002/Wol.; Posiva Oy: Eurajoki, Finland, 2004.

38. Russel, J.D.; Fraser, A.R. Infrared methods. In *Clay Mineralogy Spectroscopic and Chemical Determinative Methods*; Wilson, M.J., Ed.; Chapman & Hall: London, UK, 1994; pp. 11–67.

39. Madejová, J.; Komadel, P. Baseline studies of the clay minerals society source clays: Infrared spectroscopy. *Clays Clay Min.* **2001**, *49*, 410–432. [CrossRef]

40. Vaculíková, L.; Plevová, E.; Ritz, M. Characterization of montmorillonites by infrared and Raman spectroscopy for preparation of polymer-clay nanocomposites. *J. Nanosci. Nanotechnol.* **2019**, *19*, 2775–2781. [CrossRef]

41. Stríček, I.; Šucha, V.; Uhlík, P.; Madejová, J.; Galko, I. Mineral stability of Fe-rich bentonite in the Mock-UP-CZ experiment. *Geol. Carpathica* **2009**, *60*, 431–436. [CrossRef]

42. Velde, B. *Introduction to Clay Minerals: Chemistry, Origins, Uses and Environmental Significance*; Chapman & Hall: London, UK, 1992.

43. Wang, H.; Shirakawabe, T.; Komine, H.; Ito, D.; Gotoh, T.; Ichikawa, Y.; Chen, Q. Movement of water in compacted bentonite and its relation with swelling pressure. *Can. Geotech. J.* **2020**, *57*, 921–932. [CrossRef]

44. Guggenheim, S.; Van Groos, A.K. Baseline studies of the clay minerals society source clays: Thermal analysis. *Clay Clay Miner.* **2001**, *49*, 433–443. [CrossRef]

45. Tajeddine, L.; Gailhanou, H.; Blanc, P.; Lassin, A.; Gaboreau, S.; Vieillard, P. Hydration–dehydration behavior and thermodynamics of MX-80 montmorillonite studied using thermal analysis. *Thermochim. Acta* **2015**, *604*, 83–93. [CrossRef]

46. Drits, V.A.; Besson, G.; Muller, F. An improved model for structural transformations of heat-treated aluminous dioctahedral 2:1 layer silicates. *Clays Clay Miner.* **1995**, *43*, 718–731. [CrossRef]

47. Drits, V.A.; Lindgreen, H.; Salyn, A.L.; Ylagan, R.; McCarty, D.K. Semiquantitative determination of trans-vacant and cis-vacant 2:1 layers in illites and illite-smectites by thermal analysis and X-ray diffraction. *Am. Mineral.* **1998**, *83*, 1188–1198. [CrossRef]

48. Drits, V.A.; McCarty, D.K.; Zviagina, B.B. Crystalchemical factors responsible for the distribution of octahedral cations over trans- and cis-sites in dioctahedral 2:1 layer silicates. *Clays Clay Miner.* **2006**, *54*, 131–152. [CrossRef]

49. Wolters, F.; Lagaly, G.; Kahr, G.; Nüesch, R.; Emmerich, K. A comprehensive characterization of dioctahedral smectites. *Clays Clay Miner.* **2009**, *57*, 115–133. [CrossRef]

50. Kaufhold, S.; Kremleva, A.; Krüger, S.; Roesch, N.; Emmerich, K.; Dohrmann, R. Crystal-chemical composition of dioctahedral smectites: An energy-based assessment of empirical relations. *ACS Earth Space Chem.* **2017**, *1*, 629–636. [CrossRef]

51. Wang, X.; Li, Y.; Wang, H. Structural Characterization of Octahedral Sheet in Dioctahedral Smectites by Thermal Analysis. *Minerals* **2020**, *10*, 347. [CrossRef]

52. Karathanasis, A.D. Thermal Analysis of Soil Minerals. In *Methods of Soil Analysis Part 5*; Ulery, A.L., Dree, L.R., Eds.; Soil Science Society of America, Inc.: Medison, WI, USA, 2008; pp. 117–160.

53. Číčel, B.; Krantz, G. Mechanism of montmorillonite structure degradation by percussive grinding. *Clay Miner.* **1981**, *16*, 151–162. [CrossRef]

54. Holmboe, M.; Wold, S.; Jonsson, M. Porosity investigation of compacted bentonite using XRD profile modeling. *J. Contam. Hydrol.* **2012**, *128*, 19–32. [CrossRef] [PubMed]

55. Kaufhold, S.; Dohrmann, R.; Klinkenberg, M.; Siegesmund, S.; Ufer, K. N2-BET specific surface area of bentonites. *J. Colloid Interface Sci.* **2010**, *349*, 275–282. [CrossRef]

56. Rutherford, D.W.; Chiou, C.T.; Eberl, D.D. Effects of exchanged cation on the microporosity of montmorillonite. *Clay Clay Miner.* **1997**, *45*, 534–543. [CrossRef]

57. Lang, L.Z.; Xiang, W.; Huang, W.; Cui, D.S.; Schanz, T. An experimental study on oven-drying methods for laboratory determination of water content of a calcium-rich bentonite. *Appl. Clay Sci.* **2017**, *150*, 153–162. [CrossRef]

58. Svoboda, J.; Vašíček, R.; Pacovská, D.; Šťástka, J.; Franěk, J.; Rukavičková, L.; Večerník, P.; Červika, R.; Nahodilová, R.; Laufek, F.; et al. *Interakční Experiment-Přípravné a Podpůrné Práce [Research Report]*; SÚRAO: Prague, Czech Republic, 2019.

59. Romero, E.; Della Vecchia, G.; Jommi, C. An insight into the water retention properties of compacted clayey soils. *Géotechnique* **2011**, *61*, 313–328. [CrossRef]

60. Mašín, D.; Khalili, N. Swelling phenomena and effective stress in compacted expansive clays. *Can. Geotech. J.* **2016**, *53*, 134–147. [CrossRef]

61. Sun, H.; Mašín, D.; Najser, J.; Neděla, V.; Navrátilová, E. Bentonite microstructure and saturation evolution in wetting–drying cycles evaluated using ESEM, MIP and WRC measurements. *Géotechnique* **2018**, *69*, 713–726. [CrossRef]

*Article*

# Bentonite Alteration in Batch Reactor Experiments with and without Organic Supplements: Implications for the Disposal of Radioactive Waste

Carolin Podlech [1,*], Nicole Matschiavelli [2], Markus Peltz [1], Sindy Kluge [2], Thuro Arnold [2], Andrea Cherkouk [2], Artur Meleshyn [3], Georg Grathoff [1] and Laurence N. Warr [1]

1. Institute of Geography and Geology, University of Greifswald, 17489 Greifswald, Germany; markus.peltz@uni-greifswald.de (M.P.); grathoff@uni-greifswald.de (G.G.); warr@uni-greifswald.de (L.N.W.)
2. Institute of Resource Ecology, Helmholtz-Zentrum Dresden-Rossendorf, 01328 Dresden, Germany; n.matschiavelli@hzdr.de (N.M.); s.kluge@hzdr.de (S.K.); t.arnold@hzdr.de (T.A.); a.cherkouk@hzdr.de (A.C.)
3. Gesellschaft für Anlagen- und Reaktorsicherheit (GRS) gGmbH, 38122 Braunschweig, Germany; artur.meleshyn@grs.de
* Correspondence: carolin.podlech@uni-greifswald.de

**Citation:** Podlech, C.; Matschiavelli, N.; Peltz, M.; Kluge, S.; Arnold, T.; Cherkouk, A.; Meleshyn, A.; Grathoff, G.; Warr, L.N. Bentonite Alteration in Batch Reactor Experiments with and without Organic Supplements: Implications for the Disposal of Radioactive Waste. *Minerals* **2021**, *11*, 932. https://doi.org/10.3390/min11090932

Academic Editors: Ana María Fernández, Stephan Kaufhold, Markus Olin, Lian-Ge Zheng, Paul Wersin and James Wilson

Received: 24 June 2021
Accepted: 20 August 2021
Published: 27 August 2021

**Publisher's Note:** MDPI stays neutral with regard to jurisdictional claims in published maps and institutional affiliations.

**Abstract:** Bentonite is currently proposed as a potential backfill material for sealing high-level radioactive waste in underground repositories due to its low hydraulic conductivity, self-sealing ability and high adsorption capability. However, saline pore waters, high temperatures and the influence of microbes may cause mineralogical changes and affect the long-term performance of the bentonite barrier system. In this study, long-term static batch experiments were carried out at 25 °C and 90 °C for one and two years using two different industrial bentonites (SD80 from Greece, B36 from Slovakia) and two types of aqueous solutions, which simulated (a) Opalinus clay pore water with a salinity of 19 g·L$^{-1}$, and (b) diluted cap rock solution with a salinity of 155 g·L$^{-1}$. The bentonites were prepared with and without organic substrates to study the microbial community and their potential influence on bentonite mineralogy. Smectite alteration was dominated by metal ion substitutions, changes in layer charge and delamination during water–clay interaction. The degree of smectite alteration and changes in the microbial diversity depended largely on the respective bentonite and the experimental conditions. Thus, the low charged SD80 with 17% tetrahedral charge showed nearly no structural change in either of the aqueous solutions, whereas B36 as a medium charged smectite with 56% tetrahedral charge became more beidellitic with increasing temperature when reacted in the diluted cap rock solution. Based on these experiments, the alteration of the smectite is mainly attributed to the nature of the bentonite, pore water chemistry and temperature. A significant microbial influence on the here analyzed parameters was not observed within the two years of experimentation. However, as the detected genera are known to potentially influence geochemical processes, microbial-driven alteration occurring over longer time periods cannot be ruled out if organic nutrients are available at appropriate concentrations.

**Keywords:** bentonite; smectite; layer charge; metal substitution; SEM–EDS; microbial diversity; organic supplements

## 1. Introduction

Bentonite clay is a natural material consisting predominantly of dioctahedral smectites, mainly montmorillonite of the montmorillonite-beidellite series [1,2]. Due to its very low hydraulic conductivity, high swelling capability and strong buffering capacity when hydrated, the material is considered to be ideal for sealing the underground repository space between canisters containing high-level radioactive waste (HLW) and the host rock formation [3,4]. Despite the versatility of this material, questions remain on the long-term stability of the bentonite in a repository setting, in particular when subjected to conditions of chemical disequilibrium caused by the circulation of formational ground waters and/or when in contact with corrodible metal canisters [5,6] or supporting cementitious materials [7,8]. Only if repositories remain intact for up to one million years, as foreseen in Germany [4], will any leakage of residual radioactive substances into biosphere be prevented.

The chemical stability of smectite is important when it comes to the longevity of the bentonite barrier. A key process for mineral transformation is the smectite-to-illite conversion, whereby the needed $K^+$ cations may be supplied by dissolved accessory minerals or formation groundwater [9]. Thus, the stability of smectite in contact with salt solutions has been the topic of numerous studies that considered variations in pH, salinity and/or temperature conditions [10–12].

An additional aspect of the bentonites long-term stability is the potential influence of microbial activity in environments within the repository setting. Whereas it is generally accepted that such activity will not be of importance in highly compacted bentonite materials (dry density $\geq 1600$ kg·m$^{-3}$) [13], where the tight pore space and very low hydraulic conductivities will limit transport and mobility, there will be niches in the repository site where erosion and particle transport could lead to less compacted material and the development of bentonite clay slurries [14].

Microbial activity has been shown to affect the properties of smectitic clay minerals as well as the adsorption of metals and actinides by a number of processes, including the mobilization and immobilization of toxic elements and radionuclides [15,16]. It is also known that the presence of sulfate-reducing bacteria (SRB) can lead to the formation of hydrogen sulfides ($H_2S$) and promote corrosion of metals [17,18]. As a result, $Fe^{3+}$ is reduced to $Fe^{2+}$ [19], whereby the metal Fe can destabilize dioctahedral smectites by $Fe^{2+}$ migration into the interlayer [20] and thus influence the smectites adsorption behavior, relevant to metal and radionuclides retardation [21].

The main objective in this study was to determine the alteration mechanisms of bentonites influenced by temperature, solution chemistry and microbial diversity using energy dispersive X-ray (EDX) spectroscopy analyses of extracted and purified smectite fractions as the primary method. For this purpose, we analyzed two sets of bentonite samples (SD80, B36) that differ in their mineralogical and chemical properties (i.e., accessory content, Fe content in smectite and layer charge distribution).

The series of slurry bentonites prepared with and without microbial substrates that were experimentally investigated reveal that minor but permanent changes in the smectite composition did occur that are attributable to the tetrahedral substitution of $Si^{4+}$ by $Al^{3+}$ and more variable octahedral substitution between $Fe^{3+}$, $Mg^{2+}$ and $Al^{3+}$. Our findings also emphasize that the microbial diversity depended on the bentonite itself as well as the applied conditions. However, there is no evidence that the observed smectite alterations were induced by microbial activity. Such alterations are more likely influenced by the addition of organic substrates, which are known to act as electron-donors and promote protonation/deprotonation surface reactions.

## 2. Materials and Methods

### 2.1. Experimental Set-Up

For the batch-experiments, two uncompacted European Ca-bentonites (SD80 from Milos, Greece and B36 from Lieskovec, Slovakia) were used. In order to obtain representative samples, both bentonites were mechanically homogenized prior to the experiments. Both bentonite samples were kindly provided by Stephan Kaufhold (BGR Hannover), whereby the B36 bentonite has been extensively characterized in previous studies [22,23].

Long-term batch experiments were performed at the "Gesellschaft für Anlagen-und Reaktorsicherheit" (GRS, Braunschweig) using two temperatures and two types of solution, which were prepared under ambient atmosphere (Figure 1, Table S1). Bentonites were reacted in sealed glass vessels (flushed with argon 4.6, 99.996% purity, 2–3 min) with either a synthetic Opalinus clay pore water (OPA) [24] with a salinity of 19 g·L$^{-1}$ or a diluted cap rock solution (CAP) [25] with a salinity of 155 g·L$^{-1}$. The solid to solution ratio by weight was set at 1:2. The bentonite slurries were exposed to two different temperatures (25 °C and 90 °C) for periods of one and two years without compaction or shaking. To study microbial diversity, batches supplemented with organic substrate (+S) were compared to control batches without organic substrate (−S). The substrate-mix was comprised of 0.05 mol·L$^{-1}$ sodium lactate, 0.05 mol·L$^{-1}$ sodium acetate, 0.003 mol·L$^{-1}$ methanol and 0.0001 mol·L$^{-1}$ anthraquione-2,6-disulfonate (AQDS) to stimulate the indigenous, viable microbial community. The complete experimental setup, including gases, substrates, pore waters and bentonites, was not assembled under sterile conditions. However, the amount of microbial contamination in previous studies using the same setup prepared under sterile conditions was shown to be not significant [26].

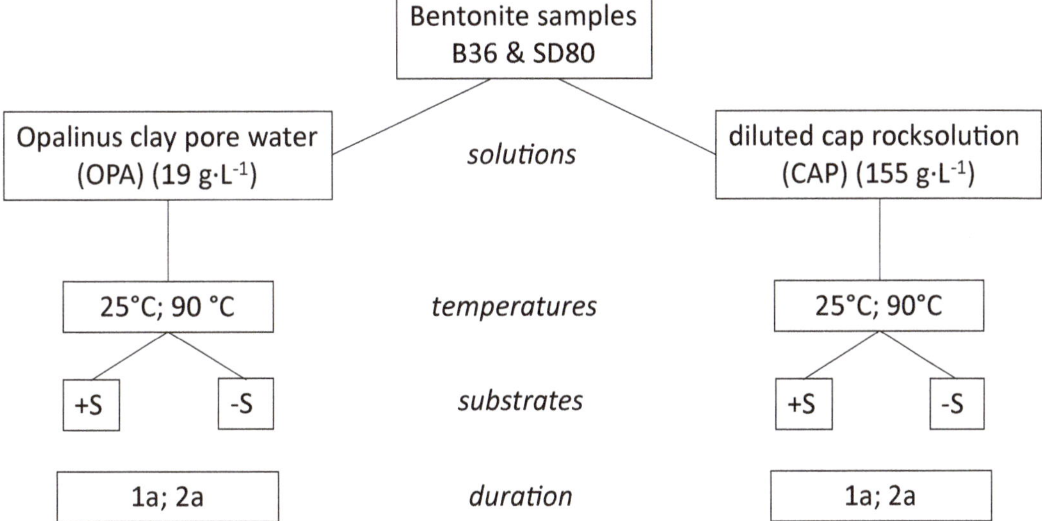


**Figure 1.** Flowchart illustrating the general setup of the bentonite batch-experiments. OPA: synthetic Opalinus clay pore water, CAP: synthetic diluted cap rock solution, +S: with substrate (Na-lactate, Na-acetate, methanol, AQDS), −S: without substrate, 1a: incubation for one year, 2a: incubation for two years.

The sampling of batches was carried out at the GRS laboratories. Bentonite slurries were extracted under sterile conditions for analyzing the microbial diversity. Therefore, up to 50 mL of bentonite suspension was transferred with a sterile spatula into sterile tubes. The respective samples were stored at −70 °C prior to extraction of the DNA. Afterwards, some supernatant solution was filtered and collected for ICP-OES analyses and 5–10 g of bentonite slurry sampled for mineralogical analysis. Subsequently, these samples as well as the raw materials were washed with distilled water using dialysis tubes (Spectra/Por® 1 dialysis membrane with a molecular weight cut-off of 6–8 kD) to remove excess salts such as those introduced by the OPA or CAP solutions. Afterwards, the samples were dried in an oven at 60 °C.

## 2.2. Bentonite Characterization (XRF, CEC, XRD)

The dialyzed bentonite powder was investigated using the following techniques. The major element oxide composition was analyzed by X-ray fluorescence spectroscopy (XRF) using a Philips PW2404 X-ray spectrometer (Tables S2 and S3). The water content was determined by oven drying an 800 mg bentonite sample at 105 °C until constant weight was achieved after about 7 days followed by heating the sample to 1050 °C for 1 h. The total change expressed as the loss-on-ignition (LOI) was determined based on the differences in mass. Afterwards, the samples were melted with a flux ($LiBO_2$ + $Li_2B_4O_7$) at 1100 °C to prepare fused tablets. Analytical measurements were calibrated against USGS standards (BHVO-2, AGV-2 and RGM-1).

The cation exchange capacity (CEC) was determined using the Cu(II)-trien method [27,28]. CEC measurements were repeated three times for each sample. The results were used to calculate mean values and standard deviations. Afterwards, the Cu(II)-trien exchanged cations (EC) $Na^+$, $K^+$, $Ca^{2+}$ and $Mg^{2+}$ were measured by atomic absorption spectroscopy (AAS).

The mineralogical composition of the bentonites was determined by X-ray diffraction analysis (XRD) using a Bruker D8 Advance θ–θ diffractometer with Fe-filtered CoKα radiation (40 kV, 30 mA). The diffractometer was equipped with a 0.5° divergence slit and a 1D LynxEye detector. Scans were collected between 4° and 80° 2θ for randomly oriented powders (scanning rate 1° 2θ/min) and between 4° and 50° 2θ for preferred oriented samples (scanning rate 3° 2θ/min) with a step size of 0.02° 2θ. The oriented samples were measured in the air-dried (AD) state and after ethylene glycol (EG) saturation. The XRD quantification were carried out with the Rietveld program BGMN and the user-interface Profex 4.0 [29].

### 2.3. Smectite Characterization (SEM-EDX)

The chemical composition of smectites was studied by scanning electron microscopy (SEM) in combination with EDX spectroscopy. The elemental abundance of purified smectite fractions were quantified as oxides and used to calculate the structural formulae and consequently the layer charge.

For sample purification, 200 mg of bulk material was weighed and placed in centrifuge tubes in 40 mL of deionized water. For initial removal of any remaining salts, the samples were sedimented by high speed centrifugation and the double distilled water replaced. Afterwards, the sample was redispersed using an ultrasonic bath. The smectite separation was performed by repeated centrifugation using an Eppendorf 5810 R centrifuge (800 rpm, 18 min) to separate the <1 µm size fraction. After each centrifugation, the supernatant was transferred to a clean centrifuged tube and topped up with deionized water prior to ultrasonic treatment, and then the centrifugation repeated. This cycle of treatment was repeated 11 times until no impurities were detected by XRD. One drop of the purified smectite suspension was then mounted on a polished pure graphite rod to obtain a homogenous flat layer of smectite particles. Before measurement, the samples were sputtered with palladium (Pd) to minimize charging effects.

SEM–EDX analyses were performed using a Zeiss Auriga scanning electron microscope equipped with an Oxford instruments X-MAX 80 mm$^2$ EDX silicon drift detector and a field emission cathode. Electron images were taken using a secondary electron (SE2) detector. Smectite elemental chemistry was determined on the basis of EDX mappings at 15 kV and a working distance of 5 mm. Elemental maps of homogenous, flat sample areas were obtained at 6000-fold magnification, a dwell time of 100 µs and by using an automatic drift correction. Thirty frames were collected for each map and three maps were obtained from independent areas of each sample, which were averaged to obtain the smectite composition (Tables S4 and S5). By using this routine, any areas of obvious contamination could be detected and excluded from the quantification. Data acquisition was performed with the software INCA (Oxford instruments) including pulse pile-up, escape peak and ZAF corrections. Oxygen measurements were not used for quantifications, and the results were instead normalized to 100% assuming that all cations were present as oxides.

The elemental mapping approach was used to overcome common problems of EDX analyses as described in detail by Christidis [30]. For quantifications, high precision, accuracy and sample homogeneity are mandatory. To reach appropriate precisions, Williams and Carter [31] suggested using total counts of at least $10^4$ above the background for each characteristic peak. This was achieved by using the aforementioned settings, which were further optimized to minimize potential loss of mobile cations such as alkali elements (e.g., $K^+$ or $Na^+$) due to beam damage [32,33], which is particularly evident during point analyses and worsens with measuring time. Although some local migration of cations may still occur during mapping even with short dwell times, it is assumed that the cations do not leave the mapped area of the specimen and that the averaged values remain constant throughout the measurement period.

As no commercially available EDX standards exist that are suitable for smectite calibration, standardless quantifications based on the vendor supplied pure element calibration of the detector were carried out [34]. The elemental-wise accuracy of the measurements was determined by applying the same measurement routine to ASTIMEX silicate standards with similar elemental compositions. The mean absolute deviations from the expected values in wt.% for analyses were 0.04 wt.% between 0 and 1 wt.%, 0.51 wt.% between 1 and 10 wt.% and 0.73 wt.% between 10 and 100 wt.%. Based on all standard measurements, the element specific mean deviations in wt.% were 0.31 for Na (0–5 wt.%), 0.09 for K (0–5 wt.%), 0.33 for Ca (0–5 wt.%), 0.07 for Mg (0–5 wt.%), 0.28 for Al (0–16 wt.%), 0.61 for Si (0–35 wt.%) and 0.46 for Fe (0–10 wt.%).

The chemical composition of the purified smectite samples was used to calculate the structural formula (SF) based on the method of Stevens [35]. The SF calculation was carried out on the basis of 11 oxygen atoms per half unit cell (phuc), where $H_2O$ and F are not considered. For the calculation it was assumed that the tetrahedral sheet consists of 4 atoms (Al + Si), whereas the occupancy of the octahedral sheet is assumed to be fixed with 2 atoms (Al + Fe + Mg). The exchangeable cations Na, K, Ca and excess Mg that cannot be incorporated into an octahedral site were assigned to the interlayer. AAS measurements of exchangeable cations after Cu(II)-trien exchange also confirmed the presence of Mg in the interlayer of both samples (Section 3.1). As the true oxidation state of Fe is unknown, $Fe^{3+}$ was assumed, as it is the dominant form in dioctahedral smectites [36]. Trace amounts of $TiO_2$ (<1 wt.% XRF) were related to anatase and/or rutile and thus excluded from the structural formula calculations. Based on the accuracy, precision and sample homogeneity of this method, changes of $\geq 0.02$ equivalents per half unit cell ($e \cdot phuc^{-1}$) are considered detectable.

### 2.4. Solution Chemistry (pH, ICP-OES)

The pH measurements of the supernatants were performed at 25 °C after short term experimentation (up to 30 days) in a glove box with argon atmosphere and after long term batch experimentation (after one and two years) in glass vessels. The pH values of both solutions (CAP, OPA) were taken as reference values.

The cation concentrations (Si, Mg, Ca, Na, K) of the solutions before and after experimentation were analyzed using an iCAP 7400 Duo inductively coupled plasma optical emission spectrometer (ICP-OES) (Thermo Fisher Scientific). Fe- and Al contents could not be determined as they were below detection limit. The chloride concentration was determined potentiometrically using the titration system Titrando 857 (Metrohm).

### 2.5. DNA Extraction, Amplification of 16S rRNA Gene and Sequencing

For analyzing the microbial diversity within the respective samples, DNA extraction and amplification of the V4-region was conducted according to the protocol of Matschiavelli et al. [37] without modifications.

Retrieved 16S rRNA gene sequences are available from the NCBI database with the bioproject accession number PRJNA758205 (https://www.ncbi.nlm.nih.gov/; accessed on 27 August 2021).

### 3. Results

### 3.1. Characterization of Starting Material

Both bentonites differed in their chemical and mineralogical composition (Tables 1 and 2). The main components of SD80 are smectite and feldspar, whereby quartz, anatase, calcite, pyrite and baryte occur as traces with <1 wt.% (Figure 2a). Sample B36 mainly consists of smectite, feldspar and quartz with minor amounts of cristobalite, kaolinite and anatase (Figure 2b). In comparison to SD80, B36 showed higher concentrations of $SiO_2$ and $K_2O$, which are associated with larger amounts of accessory minerals (~35 wt.%), i.e., feldspar, quartz and cristobalite. The 060-reflection of the randomly oriented powders were between 0.149 and 0.152 nm (Table 3) and correspond to dioctahedral smectite [38]. The 001-reflection of the oriented slides in air-dried condition (~60% RH) was located around 1.5 nm, which is characteristic of smectite with bivalent interlayer cations ($Ca^{2+}$ and/or $Mg^{2+}$) hydrated by two water layers [39]. The higher $Fe_2O_3$ content is related to Fe-rich smectite in B36 (Table 4). These results are similar to those obtained by Ufer et al. [22], who characterized mineralogically (XRD) and geochemically (XRF) various Greek Milos bentonites (cf. SD80) as well as bentonites from Slovakia (cf. B36).

**Table 1.** XRF main elemental composition of the bulk material (data in oxide wt.%). LOI: loss-on-ignition.

| Sample | LOI | $SiO_2$ | $Al_2O_3$ | $Fe_2O_3$ | MgO | CaO | $K_2O$ | $TiO_2$ | $Na_2O$ | MnO | $P_2O_5$ | Sum |
|---|---|---|---|---|---|---|---|---|---|---|---|---|
| SD80 | 16.5 | 51.7 | 17.9 | 4.8 | 2.9 | 2.5 | 0.9 | 0.7 | 0.7 | 0.1 | 0.2 | 99.4 |
| B36 | 13.9 | 58.6 | 16.3 | 7.4 | 1.5 | 1.2 | 1.5 | 0.8 | 0.4 | 0.1 | 0.1 | 101.8 |

**Table 2.** XRD quantification results based on Rietveld-Refinement. Ant: anatase, Brt: baryte, Cal: calcite, Crs: cristobalite, Fsp: feldspar, Kln: kaolinite, Sme: smectite, Py: pyrite, Qz: quartz, n.d.: not determined (data in wt.%). IMA-CNMNC approved mineral symbols after Warr [38].

| Sample | Sme | Fsp | Qz | Kln | Crs | Ant | Cal | Py | Brt |
|---|---|---|---|---|---|---|---|---|---|
| SD80 | 89 | 7 | <1 | n.d. | n.d. | <1 | <1 | <1 | 1 |
| B36 | 65 | 15 | 12 | 4 | 4 | <1 | n.d. | n.d. | n.d. |

**Table 3.** XRD parameter of both bentonites. RP: random powder, AD: air dried, EG: ethylene glycol-saturated, FWHM: full-width-at-half-maximum.

| Sample | $d_{001}$ [nm] | $d_{060}$ [nm] | $d_{001}$ [nm] | | $d_{001}$ FWHM [Δ° 2θ] | |
|---|---|---|---|---|---|---|
| | RP | RP | AD | EG | AD | EG |
| SD80 | 1.52 | 0.150 | 1.45 | 1.68 | 1.8 | 0.6 |
| B36 | 1.50 | 0.150 | 1.48 | 1.68 | 1.4 | 0.8 |

**Table 4.** Structural formula of purified smectites (<1 μm) based on elemental composition determined by EDX (values are rounded and given in e·phuc$^{-1}$).

| Sample | Tetrahedral Cations | | $\xi_{tet}$ | Octahedral Cations | | | $\xi_{oct}$ | Interlayer Cations | | | | $\xi$ | $\xi_{tet}$ |
|---|---|---|---|---|---|---|---|---|---|---|---|---|---|
| | $Si^{4+}$ | $Al^{3+}$ | | $Al^{3+}$ | $Fe^{3+}$ | $Mg^{2+}$ | | $Ca^{2+}$ | $Na^+$ | $K^+$ | $Mg^{2+}$ | | % |
| SD80 | 3.94 | 0.06 | −0.06 | 1.44 | 0.26 | 0.30 | −0.30 | 0.09 | 0.01 | 0.03 | 0.06 | −0.36 | 17 |
| B36 | 3.79 | 0.21 | −0.21 | 1.37 | 0.46 | 0.16 | −0.16 | 0.11 | 0.01 | 0.03 | 0.06 | −0.38 | 56 |

Based on EDX-derived compositions of purified mineral fractions and by applying the smectite classification scheme of Emmerich et al. [40], the SD80 smectite was classified as a low charged beidellitic montmorillonite with a tetrahedral charge of 17% and a total layer charge of $\xi = -0.36$ e·phuc$^{-1}$ and the smectite of B36 as a medium charged Fe-rich montmorillonitic beidellite with 56% tetrahedral charge and a total layer charge of $\xi = -0.38$ e·phuc$^{-1}$ (Table 4).

The SD80 bentonite, with almost 90 wt.% smectite content, had a higher CEC value of $87 \pm 2.2$ cmol·kg$^{-1}$ compared to the $54 \pm 2.0$ cmol·kg$^{-1}$ of the B36 bentonite with 65 wt.% smectite (Table 5). Both bentonite interlayers were dominated by bivalent cations with $\geq 50\%$ $Ca^{2+}$ and about 30% $Mg^{2+}$. For SD80, the sum of exchangeable cations was equal to the determined CEC but differed slightly for sample B36, whereby the measured CEC was higher than the sum of EC. The latter could be an indication of excess adsorption of the Cu-trien index cation on clay mineral edges.

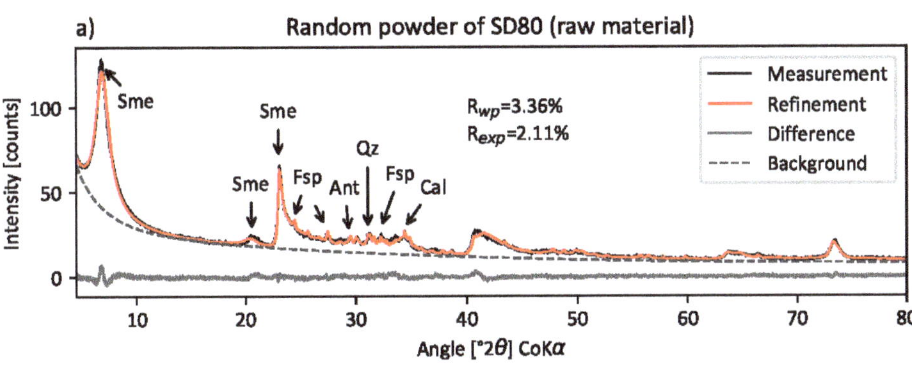


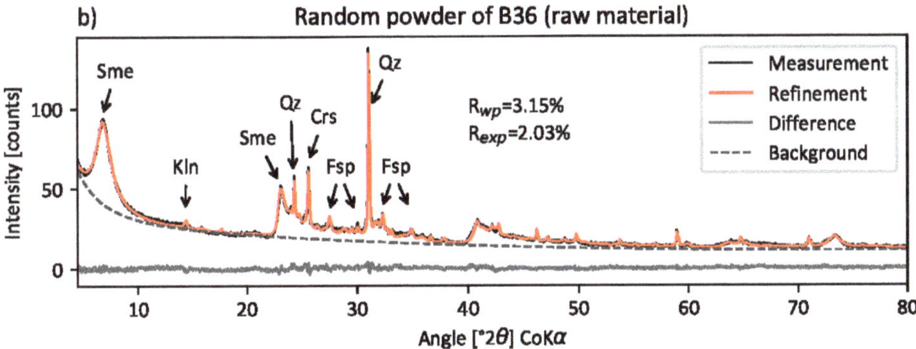


**Figure 2.** X-ray diffraction pattern of random powders from (**a**) SD80 and (**b**) B36 with Ant: anatase, Cal: calcite, Crs: cristobalite, Fsp: feldspars, Kln: kaolinite, Sme: smectite, Qz: quartz, $R_{wp}$: weighted residual square sum, $R_{exp}$: theoretical minimum value for $R_{wp}$.

**Table 5.** CEC values and exchangeable cations (EC) of both bentonites after Cu-trien saturation (sample weight 150 mg/60 mL).

| Sample | CEC | $Na^+$ | $K^+$ | $Ca^{2+}$ | $Mg^{2+}$ | $\Sigma_{cations}$ |
|---|---|---|---|---|---|---|
| | $cmol \cdot kg^{-1}$ | $cmol \cdot kg^{-1}$ | $cmol \cdot kg^{-1}$ | $cmol \cdot kg^{-1}$ | $cmol \cdot kg^{-1}$ | $cmol \cdot kg^{-1}$ |
| SD80 | $87 \pm 2.2$ | $17 \pm 0.4$ | $2 \pm 0.06$ | $43 \pm 2.4$ | $23 \pm 2.3$ | $86 \pm 2.6$ |
| B36 | $54 \pm 2.0$ | $1 \pm 0.03$ | $1 \pm 0.04$ | $30 \pm 1.4$ | $14 \pm 0.4$ | $46 \pm 1.8$ |

*3.2. Characterization of the Reacted Bentonites*

3.2.1. Visual Changes of the Bentonite Batches

After long-term experimentation, the bentonite batches showed distinct color changes. The one-year samples reacted in OPA solution had a blueish coloration (Figure 3a,b), whereas the corresponding two-year batches appeared more beige or greyish (Figure 3c,d). In addition, the B36 samples appeared brownish with a characteristic green coloration at the bentonite–solution interface, which indicates potential Fe-reduction [41] (Figure 3d). Within the substrate-bearing SD80 bentonite, the occurrence of black spots as well as the formation of gas bubbles (Figure 3a) and horizontal fissures due to gas release (Figure 3c) were observed in some of the reacted samples. The development of layers due to sedimentation is highlighted by a dotted line in the respective samples in Figure 3b–d.

bentonite SD80          bentonite B36

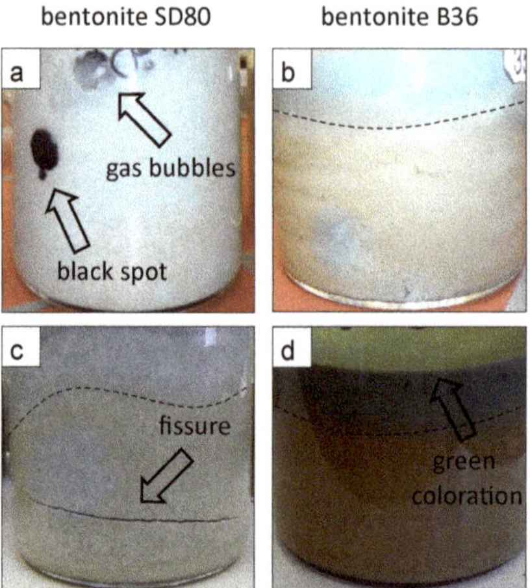


**Figure 3.** Color changes of both batches (SD80, B36) subjected to Opalinus clay pore water at 25 °C supplemented with substrate. Bentonite SD80 showed (**a**) the occurrence of black spots and gas bubbles after one year and (**c**) a horizontal fissure after two years. Bentonite B36 showed (**b**) a blueish coloration after one year and (**d**) a characteristic green coloration at the bentonite–solution interface after two years.

3.2.2. Microbial Diversity Analysis

In order to investigate a potential microbial effect on the here tested mineralogical parameters of B36 and SD80, the microbial diversity of the respective slurry experiments was analyzed and compared to the respective bentonite raw materials. In general, the concentration of extracted DNA was very low and for most samples almost below the detection limit (Table S6). Successful extraction and sequencing of DNA was achieved for seven samples, including the raw material of SD80 and B36 (Figure 4).

The sequencing results showed that the microbial diversity of the two initial bentonites were different from each other, and that the diversity changed in response to the applied conditions. Supplemented batches that contained SD80 bentonite and that were reacted in OPA solution at 25 °C for one and two years showed the formation of black precipitates and fissures (Figure 3a) as well as the dominance of spore-forming and sulfate-reducing bacteria from the genera *Desulfosporosinus*, *Desulfotomaculum* and *Desulfitobacterium* (Figure 4). The detected genera could reduce the sulfate present in order to form hydrogen sulfide [17,42]. In the equivalent batches with B36 bentonite, no "typical" sulfate reducers were detected. In the respective batches, microorganisms from the genus *Bacillus* were identified, which are known to form spores and to be resistant against harsh conditions [43]. Furthermore, microorganisms from the genera *Marinobacter* were identified in the one-year incubated batches that contained B36 bentonite, CAP solution and substrates (Figure 4). Some species from that genus are known to be resistant against the saline conditions that characterized the respective batches [44].

The results indicate that the microbial "inventory" of the two tested bentonites differed from each other and, thus, also the metabolic potential of the respective microbial species. Further, it is noteworthy that both bentonites contained a high abundance of unknown genera, whose metabolic potential and influence on the mineral assemblages were uncertain. As experiments were conducted under nonsterile conditions, the input of additional microbes could not be excluded. Nonetheless, if active, the detected genera may have influenced the solution chemistry, i.e., by the reduction of sulfate [45]. To assess if the detected genera led to any changes in smectite structure and composition, purified smectites of supplemented and control samples were analyzed and compared by XRD and SEM–EDX.

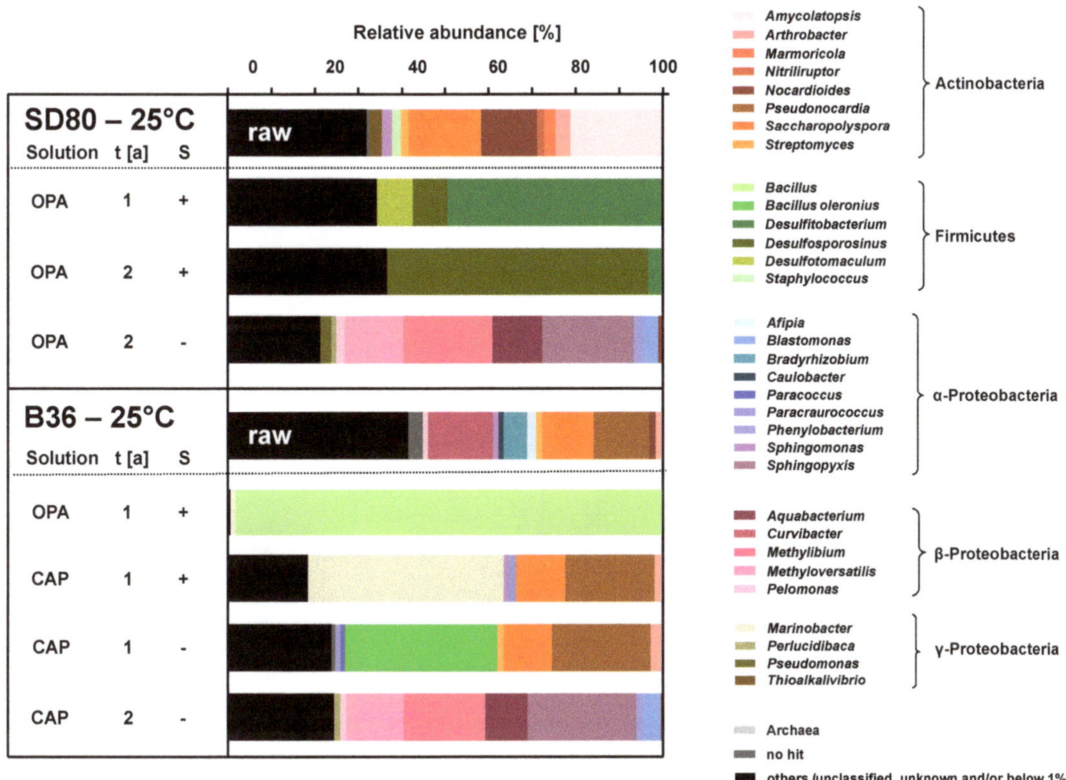



**Figure 4.** Microbial diversity in B36 and SD80 raw materials and samples incubated at 25 °C. Shown is the relative abundance of detected genera and their dependence on the added substrates (+S) and incubation time (t in years). The microbial community was analyzed by amplifying and sequencing the V4-region of the 16S rRNA gene via MiSeq Illumina (modified from Matschiavelli et al. [26]).

### 3.2.3. X-ray Diffraction

The random powder XRD patterns of both reacted bentonites showed no neoformed minerals. Minor quantitative differences in the mineralogical composition between the initial and the reacted samples were attributed to heterogeneous sampling. XRD patterns of EG-saturated (001)-reflections revealed some structural differences between samples that were largely independent of the interlayer cation configuration and water content (Tables S7 and S8).

The SD80 samples subjected to OPA solution did not show significant changes in FWHM or $d$ values over time or with increasing temperature (Figure 5a,b). The $d_{001}$ values were at 1.69 nm, and the FWHM ranged between 0.5 and 0.6 Δ° 2θ. However, the XRD patterns of SD80 treated with CAP solution (Figure 5c,d) did show minor differences with $d_{001}$ values between 1.67 and 1.69 nm, whereby the FWHM was constant at 0.6 Δ° 2θ. Only the $d_{001}$ of SD80 reacted for one year at 90 °C without substrate (–S) showed a more significant shift towards smaller $d$ values with $d_{001}$ = 1.65 nm. Overall, the $d_{001}$ reflection intensity of the reacted SD80 smectites notably decreased with increasing temperature conditions.

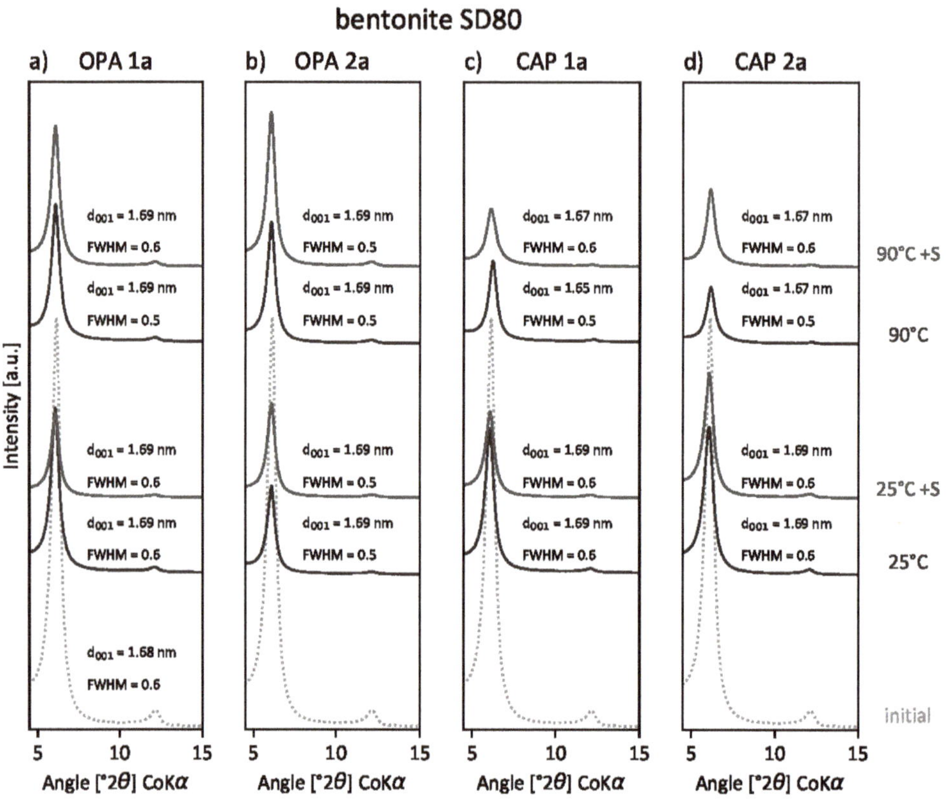



**Figure 5.** XRD patterns (oriented slides of purified smectite; ethylene glycol-saturated) of the initial SD80 bentonite and reacted samples at two different temperatures (25 and 90 °C) with substrate (+S) and controls. (**a,b**) OPA solution for 1 and 2 years and (**c,d**) CAP solution for 1 and 2 years. Besides for an overall decrease of reflection intensity, no changes in d-spacing and full-width-at-half-maximum (FWHM) were observed. FWHM in Δ° 2θ. OPA: Opalinus clay pore water, CAP: diluted cap rock solution.

The B36 samples treated with OPA solution did not show significant changes (Figure 6a,b). The (001)-reflection of smectite varied between 1.69 and 1.71 nm, and the FWHM remained constant at 0.8 Δ° 2θ. B36 samples in contact with the CAP solution (Figure 6c,d) showed larger variations in *d* values (1.64–1.72 nm) and FWHM values (0.9–1.1 Δ° 2θ). The highest increase in the FWHM up to 1.1 Δ° 2θ was accompanied by a shift of the $d_{001}$ towards smaller values (1.64 nm and 1.66 nm). This was observed in the two-year sample set reacted at 90 °C both with and without substrate. The increase of FWHM correlating with decreasing *d* values and lower reflection intensities indicates a decrease in crystallite thickness [39], which likely reflects the interlayer exchange of $Ca^{2+}$ by $Na^+$ and/or $K^+$.

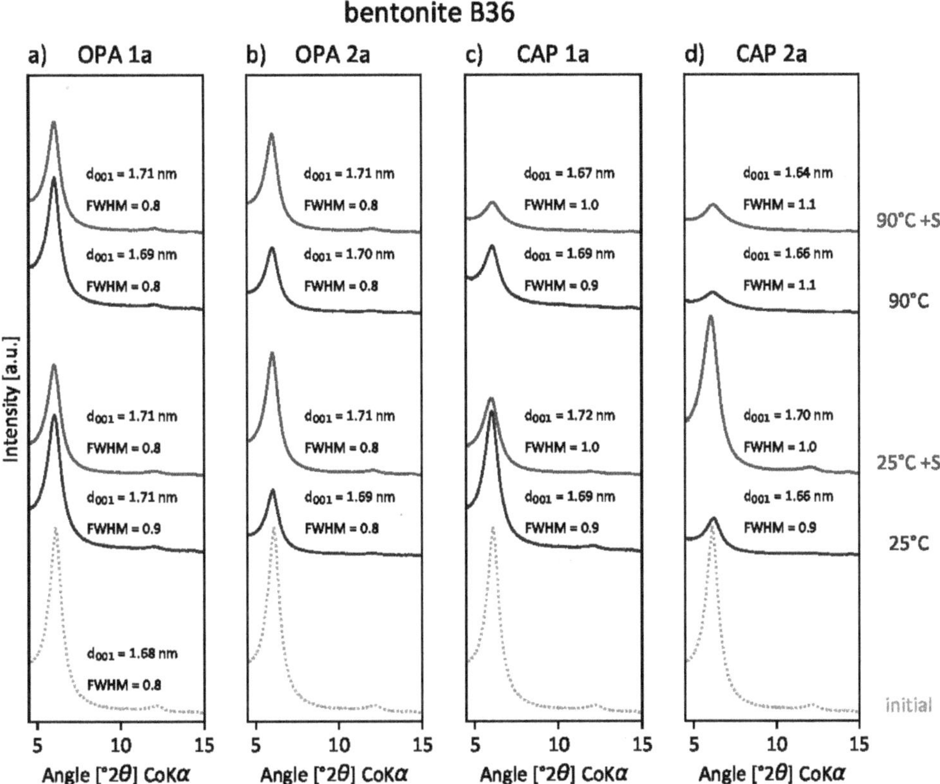



**Figure 6.** XRD patterns (oriented slides of purified smectite; ethylene glycol-saturated) of the initial B36 bentonite and reacted samples at two different temperatures (25 and 90 °C) with substrate (+S) and controls. (**a,b**) OPA solution for 1 and 2 years and (**c,d**) CAP solution for 1 and 2 years. Samples reacted in CAP solution show a continuous decrease of $d_{(001)}$ in combination with an increase of FWHM (in $\Delta°\ 2\theta$) with higher temperature and duration. OPA: Opalinus clay pore water, CAP: diluted cap rock solution.

### 3.2.4. Smectite Layer Charge Distribution

The initial sample of SD80 is classified as a low charged smectite with a total layer charge of $-0.36$ equivalents per half unit cell (e·phuc$^{-1}$) and 17% tetrahedral charge (Table 4). The SD80 smectite subjected to OPA solution showed minor changes in the total layer charge with a trend towards elevated layer charges (Figure 7a). The increase in total LC was mainly due to the tetrahedral substitution of $Si^{4+}$ by $Al^{3+}$ along with an overall decrease in octahedral $Fe^{3+}$ and an increase in octahedral $Mg^{2+}$.

Smectite samples of SD80 treated with CAP solution showed commonly increased layer charges by up to 0.05 e·phuc$^{-1}$ (Figure 7b). Overall, only minor changes in octahedral $Mg^{2+}$ ($-0.01$ to $+0.02$ e·phuc$^{-1}$) and octahedral $Fe^{3+}$ ($-0.05$ to 0.00 e·phuc$^{-1}$) were observed for SD80 smectites in CAP solution. Changes in the total layer charge were mainly attributed to the tetrahedral substitution of $Si^{4+}$ by $Al^{3+}$, which was observed for four samples with a layer charge difference higher than 0.04 e·phuc$^{-1}$. This includes three substrate-bearing samples (one-year samples at 25 °C; two year samples at 25 °C and 90 °C) and one sample reacted for two years at 90 °C without substrate that did not show an additional increase in octahedral $Al^{3+}$.

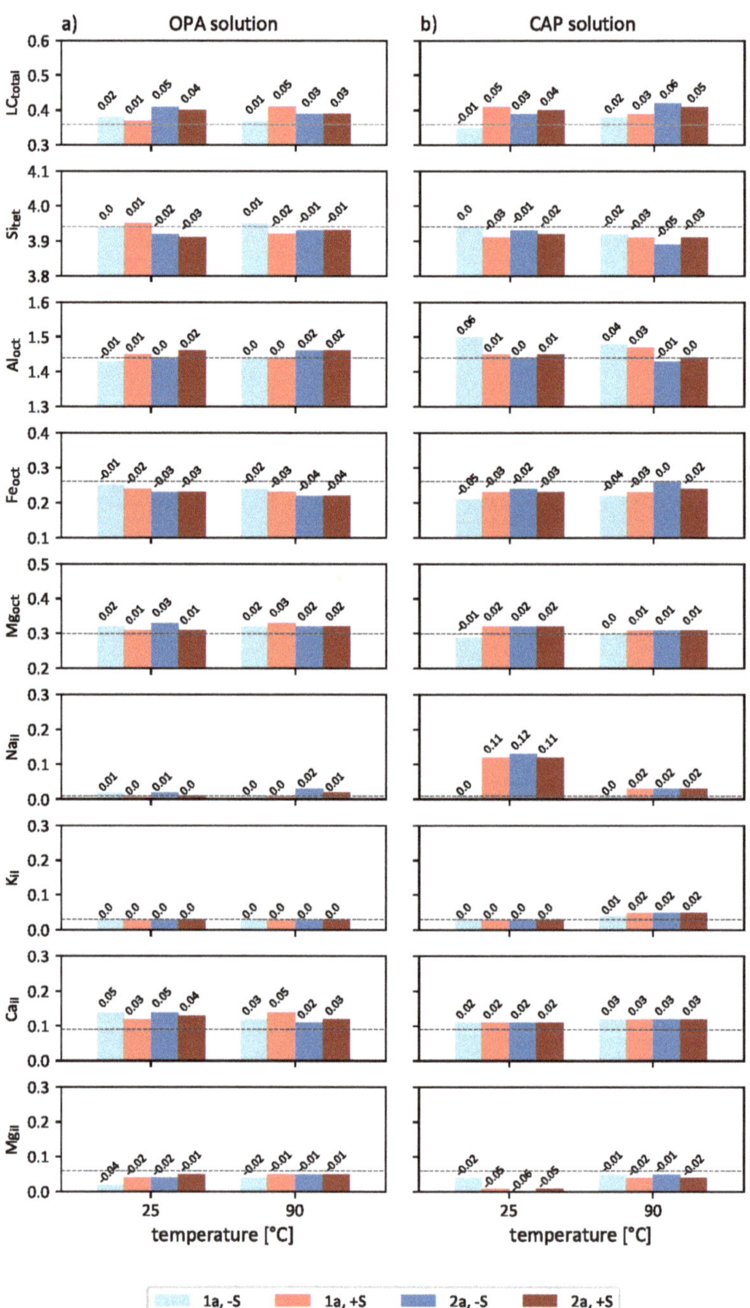

**Figure 7.** Layer charge and cation distribution of SD80 batch experiments with (**a**) OPA solution and with (**b**) CAP solution; tet: tetrahedral sheet, oct: octahedral sheet, il: interlayer. Values above the bars are displaying the rounded deviation (in e·phuc$^{-1}$) from the initial sample (dotted line).

The EDX-based structural formula was also used to determine the interlayer cation distribution. The interlayer of the initial SD80 smectite was dominated by $Ca^{2+}$ (0.09 e·phuc$^{-1}$) and $Mg^{2+}$ (0.06 e·phuc$^{-1}$), followed by small amounts of $K^+$ (0.03 e·phuc$^{-1}$) and $Na^+$ (0.01 e·phuc$^{-1}$) (Table S9). In contact with the OPA solution, only minor changes in $Na^+$, $K^+$ and $Mg^{2+}$ content were observed, with differences of up to $\pm0.02$ e·phuc$^{-1}$ (Figure 7a). An exception was the one-year control sample reacted at 25 °C, which showed a decrease of 0.04 e·phuc$^{-1}$ for $Mg^{2+}$. Overall, the $Ca^{2+}$ content increased between 0.02 e·phuc$^{-1}$ and 0.05 e·phuc$^{-1}$.

In contrast, the SD80 samples subjected to CAP solution showed an uptake of $Na^+$ into the interlayer with a decreasing amount of substitution with increased temperature (Figure 7b). Compared to the initial sample, the $Na^+$ in bentonites reacted at 25 °C increased by 0.11–0.12 e·phuc$^{-1}$. The only exceptions were the one-year control samples reacted at 25 °C and 90 °C, which did not show any differences in $Na^+$ content. The $K^+$ content of the 25 °C samples also showed no change, and only minor changes were observed for 90 °C samples with an increase by up to +0.02 e·phuc$^{-1}$. Overall, enhanced amounts of interlayer $Ca^{2+}$ (0.02–0.03 e·phuc$^{-1}$) and depleted amounts of $Mg^{2+}$ (0.01–0.06 e·phuc$^{-1}$) were observed under elevated temperatures.

In comparison, the increased LC in OPA-reacted SD80 samples correlated with the uptake of interlayer $Ca^{2+}$ at 25 °C and 90 °C, whereas in CAP-reacted samples, a constant increase in $Ca^{2+}$ was observed at 25 °C and 90 °C. Additionally, CAP samples at 25 °C showed a noticeable increase in $Na^+$, which correlated with LC, and a decrease in interlayer $Mg^{2+}$. At 90 °C, this trend was diminished but accompanied by an increase in $K^+$. These trends indicate solution-dependent changes whose implications will be discussed later.

The initial sample of B36 is classified as a medium charged smectite with a total layer charge of $-0.38$ e·phuc$^{-1}$ and 56% tetrahedral charge (Table 4). After treatment with OPA solution, no or only minor changes (0.00 to $-0.04$ e·phuc$^{-1}$) in the total layer charge were observed for the two-year samples reacted at 90 °C and the one-year sample sets. Only the two-year samples reacted at 25 °C showed increased layer charges by up to 0.05 e·phuc$^{-1}$ (Figure 8a). The latter are related to tetrahedral substitution of $Al^{3+}$ by $Si^{4+}$ along with an octahedral substitution of $Al^{3+}$ by $Fe^{3+}$, which results in an overall increase of tetrahedral charge (up to 0.06 e·phuc$^{-1}$).

The total LC of B36 subjected to CAP solution showed only minor changes at 25 °C, whereas samples reacted at 90 °C had increased layer charges up to 0.06 e·phuc$^{-1}$ in control samples and up to 0.14 e·phuc$^{-1}$ in substrate-bearing samples (Figure 8b). The only exception was the two-year control sample at 25 °C, which showed similar changes compared to the control samples reacted at 90 °C. The 90 °C samples displayed an overall decrease of tetrahedral $Si^{4+}$ between $-0.11$ e·phuc$^{-1}$ and $-0.14$ e·phuc$^{-1}$. Whereas no changes in octahedral charge were observed for the substrate-bearing samples ($Fe^{3+}$ substitutes for $Al^{3+}$), the octahedral charge decreased up to 0.08 e·phuc$^{-1}$ in the control samples. This was caused by a significant increase of $Al^{3+}$ (up to 0.16 e·phuc$^{-1}$) along with a decrease of $Fe^{3+}$ ($-0.08$ e·phuc$^{-1}$) and $Mg^{2+}$ (up to $-0.08$ e·phuc$^{-1}$) in the octahedral sheet.

The interlayer distribution of the initial B36 smectite was dominated by $Ca^{2+}$ (0.11 e·phuc$^{-1}$) and $Mg^{2+}$ (0.06 e·phuc$^{-1}$) followed by $K^+$ (0.03 e·phuc$^{-1}$) and $Na^+$ (0.01 e·phuc$^{-1}$) (Table S10). In OPA solution experiments, the B36 samples showed slight changes in interlayer cation content ($-0.03$ to 0.02 e·phuc$^{-1}$) (Figure 8a). The two-year sample set at 90 °C showed enhanced $Na^+$ contents greater than 0.03 e·phuc$^{-1}$.

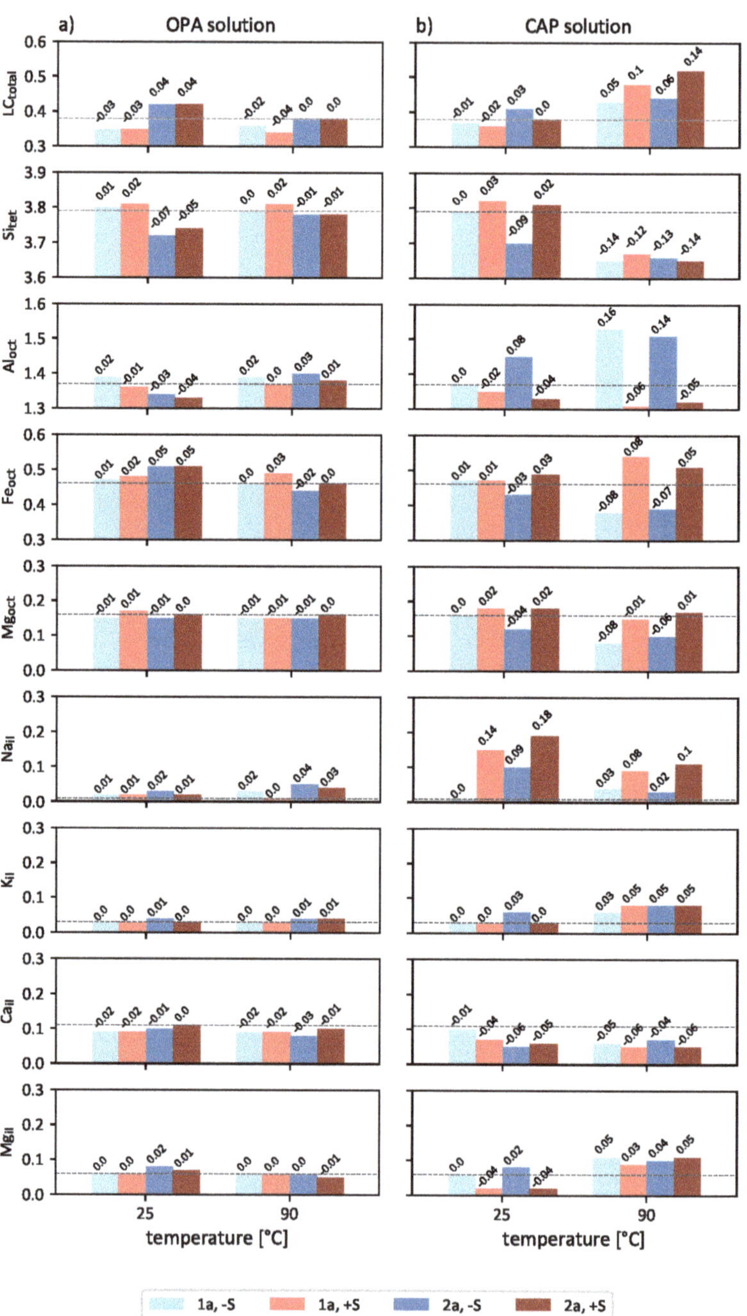



**Figure 8.** Layer charge and cation distribution of B36 batch experiments with (**a**) OPA solution and with (**b**) CAP solution; tet: tetrahedral sheet, oct: octahedral sheet, il: interlayer. Values above the bars are displaying the rounded deviation (in e·phuc$^{-1}$) from the initial sample (dotted line).

For B36 samples subjected to CAP solution, an overall uptake of $Na^+$ cations between 0.08 e·phuc$^{-1}$ to 0.18 e·phuc$^{-1}$ was observed (Figure 8b). The only exceptions were control samples, which reacted for one year at 25 °C, and for the one- and two-year samples reacted at 90 °C, which showed no or only minor increases of up to 0.03 e·phuc$^{-1}$. A more significant increase of interlayer $K^+$ between 0.03 e·phuc$^{-1}$ and 0.05 e·phuc$^{-1}$ was observed for the high temperature samples (90 °C) as well as for a control batch reacted for two years at 25 °C. For all samples, the $Ca^{2+}$ content decreased by up to $-0.06$ e·phuc$^{-1}$. The amount of interlayer $Mg^{2+}$ increased for the 90 °C samples up to $+0.05$ e·phuc$^{-1}$, whereas the substrate-bearing samples at 25 °C showed a decrease in $Mg^{2+}$ of $-0.04$ e·phuc$^{-1}$. The control batches reacted at 25 °C showed no or only slight changes ($+0.02$ e·phuc$^{-1}$) in interlayer $Mg^{2+}$.

The most noticeable changes in the B36 bentonite were found in CAP-reacted samples and involved an increased uptake of $Na^+$ into the interlayer at 25 °C and substrate-influenced tetrahedral and octahedral changes at 90 °C. These changes point towards higher reactivity of B36 in comparison to SD80 and will be discussed in detail later.

As solutions were not renewed during the batch experimentation, a complete exchange of interlayer cations, e.g., $Na^+$ for $Ca^{2+}$ in contact with the CAP solution was not expected.

### 3.2.5. Solution Chemistry of the Supernatant

The solution chemistry of the supernatant before and after experimentation was analyzed to estimate the dissolution and cation exchange behavior of both bentonites as well as the change of pH values. During the short-term experiments, pH values for SD80 in OPA solution varied between 7.2 and 7.4 (Table 6). The pH values of SD80 in CAP solution showed an initial decrease from pH = 7.3 to 7.1, which further decreased to pH = 6.9 after 1 day. In contrast, the pH values of B36 were characterized by a strong decrease in both solutions. In contact with OPA solution, the pH immediately decreased from 7.8 to 6.3 within 1 day. Subjected to CAP solution, the pH decreased from 7.3 to 5.7 and remained at 5.0 between 1 and 30 days.

**Table 6.** Results of pH short-term measurements of both bentonites (SD80, B36) in a glove box at 25 °C with a solid to solution ratio of 1:2. The pH values of both solutions (OPA, CAP) without bentonite are also used as a reference for long-term measurements.

| Sample | Solution | 0 d | 1 d | 8 d | 30 d |
|---|---|---|---|---|---|
| - | OPA | 7.8 | - | - | - |
| SD80 | OPA | 7.4 | 7.2 | 7.3 | - |
| B36 | OPA | 6.3 | 5.4 | 5.6 | 5.7 |
| - | CAP | 7.3 | - | - | - |
| SD80 | CAP | 7.1 | 6.9 | 6.9 | 6.9 |
| B36 | CAP | 5.7 | 5.0 | 5.0 | 5.0 |

For long-term experimentation (one and two years), an overall increase of Mg concentrations and an increase of Si concentrations with increasing temperature were observed for all solutions, which indicates temperature enhanced silicate dissolution. Higher Na concentrations in substrate-bearing samples were attributed to the additional input of sodium by Na-lactate and Na-acetate (Tables 7 and 8).

**Table 7.** Solution chemistry of SD80 bentonite batches and measured pH values (n.d. = not determined due to the complete adsorption of the solution by smectite). DUR: duration, SOL: solution, Su: substrates.

| Sample | DUR | T | SOL | Su | pH | Si | Mg | Ca | Na | K | S | Cl |
|---|---|---|---|---|---|---|---|---|---|---|---|---|
| | (a) | (°C) | | | | $(mmol \cdot L^{-1})$ | $(mmol \cdot L^{-1})$ | $(mmol \cdot L^{-1})$ | $(mmol \cdot L^{-1})$ | $(mmol \cdot L^{-1})$ | $(mmol \cdot L^{-1})$ | $(mmol \cdot L^{-1})$ |
| | 0 | 25 | OPA | − | 7.8 | $0.1 \pm 0.1$ | $14.5 \pm 0.6$ | $25.9 \pm 1.4$ | $226 \pm 11$ | $1.7 \pm 0.1$ | $14.7 \pm 1.1$ | $308 \pm 13$ |
| SD80 | 1 | 25 | OPA | − | n.d. | n.d. | 35.0 | 46.4 | 274 | n.d. | 22.1 | 372 |
| SD80 | 1 | 25 | OPA | + | n.d. | 1.8 | 41.7 | 39.7 | 379 | 3.1 | 0.3 | n.d. |
| SD80 | 1 | 90 | OPA | − | n.d. | n.d. | n.d. | n.d. | n.d. | n.d. | n.d. | n.d. |
| SD80 | 1 | 90 | OPA | + | n.d. | n.d. | n.d. | n.d. | n.d. | n.d. | n.d. | n.d. |
| SD80 | 2 | 25 | OPA | − | n.d. | 0.9 | 36.3 | 42.5 | 254 | 3.4 | 21.0 | n.d. |
| SD80 | 2 | 25 | OPA | + | n.d. | n.d. | n.d. | n.d. | n.d. | n.d. | n.d. | n.d. |
| SD80 | 2 | 90 | OPA | − | n.d. | 2.5 | 31.4 | 56.7 | 313 | 6.9 | 26.2 | 447 |
| SD80 | 2 | 90 | OPA | + | n.d. | n.d. | n.d. | n.d. | n.d. | n.d. | n.d. | n.d. |
| | 0 | 25 | CAP | − | 7.3 | $0.2 \pm 0.2$ | n.d. | $11.6 \pm 3.0$ | $2389 \pm 70$ | $6.0 \pm 0.6$ | $16.4 \pm 0.4$ | $2562 \pm 5$ |
| SD80 | 1 | 25 | CAP | − | 7.4 | 0.3 | 63.2 | 93.9 | 2352 | 7.3 | 22.1 | 2642 |
| SD80 | 1 | 25 | CAP | + | 7.4 | 0.6 | 82.0 | 97.7 | 2773 | 7.4 | 23.8 | 2652 |
| SD80 | 1 | 90 | CAP | − | 7.4 | 1.3 | 64.6 | 113.5 | 2729 | 9.5 | 18.9 | 2877 |
| SD80 | 1 | 90 | CAP | + | 7.2 | 1.7 | 78.0 | 117.2 | 2642 | 10.1 | 23.6 | 2750 |
| SD80 | 2 | 25 | CAP | − | 7.4 | 0.5 | 69.1 | 99.3 | 2481 | 7.3 | 23.5 | 2636 |
| SD80 | 2 | 25 | CAP | + | 7.6 | 0.7 | 79.8 | 96.3 | 2503 | 7.7 | 22.5 | 2667 |
| SD80 | 2 | 90 | CAP | − | 7.5 | 1.5 | 64.9 | 98.1 | 2446 | 10.0 | 14.6 | 2907 |
| SD80 | 2 | 90 | CAP | + | n.d. | 1.0 | 79.9 | 136.0 | 2698 | 9.8 | 20.5 | n.d. |

**Table 8.** Solution chemistry of B36 bentonite batches and measured pH values (n.d. = not determined due to the complete adsorption of the solution by smectite), DUR: duration, SOL: solution, Su: substrates.

| Sample | DUR | T | SOL | Su | pH | Si | Mg | Ca | Na | K | S | Cl |
|---|---|---|---|---|---|---|---|---|---|---|---|---|
| | (a) | (°C) | | | | (mmol·L$^{-1}$) | (mmol·L$^{-1}$) | (mmol·L$^{-1}$) | (mmol·L$^{-1}$) | (mmol·L$^{-1}$) | (mmol·L$^{-1}$) | (mmol·L$^{-1}$) |
| | 0 | 25 | OPA | − | 7.8 | 0.1 ± 0.1 | 14.5 ± 0.6 | 25.9 ± 1.4 | 226 ± 11 | 1.7 ± 0.1 | 14.7 ± 1.1 | 308 ± 13 |
| B36 | 1 | 25 | OPA | − | 7.3 | 1.1 | 28.7 | 50.9 | 232 | 2.5 | 15.6 | 353 |
| B36 | 1 | 25 | OPA | + | 6.9 | 1.4 | 30.4 | 60.7 | 283 | 2.4 | 15.9 | 345 |
| B36 | 1 | 90 | OPA | − | 5.1 | 2.7 | 22.5 | 58.6 | 231 | 3.5 | 14.7 | 358 |
| B36 | 1 | 90 | OPA | + | 5.2 | 3.1 | 25.9 | 59.3 | 300 | 3.1 | 14.5 | 347 |
| B36 | 2 | 25 | OPA | − | 7.1 | 1.1 | 28.4 | 58.3 | 232 | 2.8 | 15.1 | 355 |
| B36 | 2 | 25 | OPA | + | 7.5 | 1.1 | 27.7 | 60.2 | 282 | 2.1 | 11.8 | 333 |
| B36 | 2 | 90 | OPA | − | 5.0 | 3.0 | 27.3 | 62.0 | 250 | 4.5 | 16.5 | 385 |
| B36 | 2 | 90 | OPA | + | 5.2 | 3.2 | 24.1 | 55.5 | 288 | 3.8 | 14.6 | 342 |
| | 0 | 25 | CAP | − | 7.3 | 0.2 ± 0.2 | n.d. | 11.6 ± 3.0 | 2389 ± 70 | 6.0 ± 0.6 | 16.4 ± 0.4 | 2562 ± 5 |
| B36 | 1 | 25 | CAP | − | 6.7 | 0.5 | 35.8 | 96.1 | 2351 | 7.5 | 16.4 | 2590 |
| B36 | 1 | 25 | CAP | + | 7.5 | 0.7 | 40.9 | 107.8 | 2617 | 7.1 | 17.6 | 2621 |
| B36 | 1 | 90 | CAP | − | 4.5 | 2.2 | 38.6 | 107.3 | 2672 | 8.4 | 18.1 | 2787 |
| B36 | 1 | 90 | CAP | + | 4.7 | 2.0 | 36.7 | 107.6 | 2637 | 7.5 | 17.6 | 2659 |
| B36 | 2 | 25 | CAP | − | 7.0 | 0.6 | 34.5 | 92.7 | 2514 | 6.8 | 17.5 | 2576 |
| B36 | 2 | 25 | CAP | + | 7.1 | 0.7 | 38.1 | 93.1 | 2336 | 6.9 | 15.6 | 2597 |
| B36 | 2 | 90 | CAP | − | 4.6 | 2.4 | 37.8 | 101.1 | 2575 | 8.8 | 15.2 | 2777 |
| B36 | 2 | 90 | CAP | + | 4.8 | 2.5 | 36.2 | 97.7 | 2429 | 7.6 | 16.3 | 2615 |

The pH and most ICP-OES measurements for SD80 in contact with OPA solutions were not feasible due to the complete adsorption of water molecules by smectite particles. SD80 samples reacted in CAP solution showed pH values that varied between 7.2 and 7.6. Solutions of supplemented SD80 samples showed slightly higher Mg concentrations than the control samples. A significant decrease in the S concentration from about 15 mmol·L$^{-1}$ to almost 0 mmol·L$^{-1}$ was observed for the one-year sample reacted at 25 °C in OPA solution (+S) in which sulfate-reducing bacteria were detected as aforementioned (Table 7). Solutions of sample B36 showed decreasing pH values with increasing temperature (Table 8).

## 4. Discussion

### 4.1. Smectite Alteration Mechanisms

The structural alteration of the smectite, which was observed in both sets of bentonite experiments, can be attributed to three main mechanisms: (i) interlayer cation exchange, (ii) tetrahedral substitution of $Si^{4+}$ by $Al^{3+}$ and (iii) octahedral substitution of $Fe^{3+}$ and $Mg^{2+}$ by $Al^{3+}$ and to a lesser extent of $Al^{3+}$ by $Fe^{3+}$. To varying degrees, these mechanisms are controlled by experimental conditions, which are temperature, duration of the experiments, solution chemistry, the mineralogical composition of the bentonites and whether or not the experiments are supplemented with organic supplements. Due to the complexity of the bentonite assemblages and their microbial compositions, it was not possible to differentiate any single parameter as being responsible for the smectite alteration, so probable causes for the observed types of alteration are discussed below.

### 4.1.1. Interlayer Cation Exchange

The reactivity of bentonites in saline solutions has been studied extensively in the past [10,12,46]. As reported by Kaufhold and Dohrmann [11], the degree of interlayer cation exchange is determined by the ionic strength of the solution and the solid to solution ratio. Although both solutions used in this study are Na-dominated (Table S1), significant $Na^+$ uptake into the interlayer was only observed for CAP-reacted samples at 25 °C and partly at 90 °C, whereas samples reacted in OPA solution showed only minor interlayer exchange. In comparison to other studies (e.g., [47]), these slight changes are likely to reflect the low solid to solution ratio (1:2) in combination with non-stirred bentonite slurries.

At 90 °C, both smectites reacted in CAP solution showed an increase of interlayer $K^+$, which indicates an increase of collapsed layers. For B36, this is supported by the trends observed for EG-saturated 001-reflections (Figure 9), which show a continuous decrease of $d_{(001)}$ in combination with an increase of the FWHM at higher temperature and longer duration. This points towards a decrease in X-ray scattering domain sizes, which likely resulted from particle delamination [39]. Delamination is known to occur during the interlayer exchange of $Ca^{2+}$ by $Na^+$ when treated with NaCl-dominated saline solutions [48]. This is supported by the continuous decrease of basal spacing in combination with an increase of FHWM with both temperature and duration for sample B36. This effect is intensified in the presence of the organic substrate, which also contained higher amounts of $Na^+$ in the interlayer compared to the control samples (Table S10), which was possibly derived from the added Na-lactate and Na-acetate.

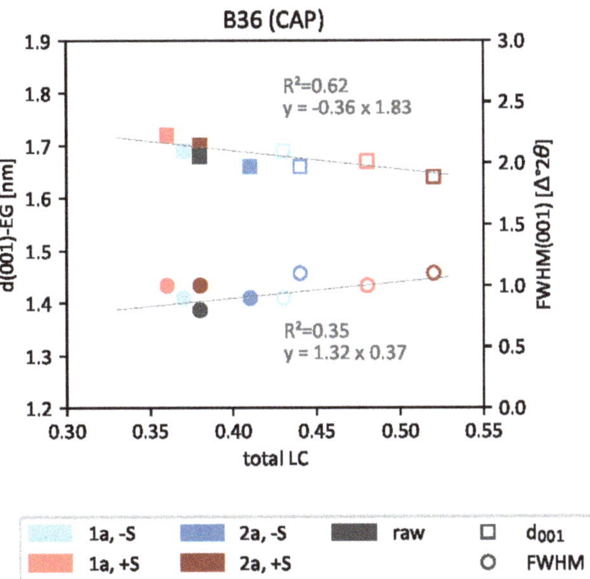



**Figure 9.** Plot of 001-reflection parameters against layer charge of ethylene glycol-saturated B36 samples reacted in CAP solution at 25 °C (circle filled) and 90 °C (circle unfilled). FWHM: full-width-at-half-maximum, LC: layer charge, –S: without substrate, +S: with substrate.

As layer charge is known to influence smectite colloidal properties, an increase in the total layer charge can lead to a decrease in crystalline swelling [49,50], whereby the swelling also decreases with increasing salinity of the respective pore water [46]. However, the changes in the 001-reflection (intensity, position, FWHM) of EG-saturated smectites are largely attributable to the reversible effects of interlayer cation exchange and particle delamination and therefore are not considered to represent a permanent structural change.

In comparison to B36 samples reacted at 25 °C, the 90 °C samples showed a significant decrease of the pH to values between 5.0–5.2 (OPA) and 4.5–4.8 (CAP), which was independent of the addition of substrate (Table 8). Kaufhold et al. [51] argued that the hydrolysis of exchanged interlayer $Ca^{2+}$ could lead to a minor decrease in pH values. However, an increase of $Ca^{2+}$ in both solutions was observed for all samples. Although no carbonate phases were detected by XRD in the bulk materials, minor amounts were found in other studies (SD80: 2.8 wt.%, B36: 0.35 wt.%) [25] and, if present in small amounts in these samples, could further explain the increase in $Ca^{2+}$. Therefore, interlayer $Ca^{2+}$ or any additional $Ca^{2+}$ in the solution is not considered to be the likely source of acidification. Beaufort et al. [52] observed decreasing pH values in hydrothermal experiments at 200 °C with purified smectites, which indicates that the alteration of smectite itself could influence the solution pH. However, the exact source of acidification remains unclear but is possibly related to deprotonation reactions that may occur during heating [53] and/or to the potential reduction of octahedral $Fe^{3+}$. The latter can lead to the adsorption of dissociated $H_2O$, whose $H^+$ can be subsequently substituted by $Na^+$ [54]. The decrease in pH likely did not result in enhanced montmorillonite dissolution, as dissolution rates within the observed pH range are generally of the same order of magnitude [55].

### 4.1.2. Tetrahedral and Octahedral Charge Distribution

The determination of the layer charge distribution based on EDX analyses can provide information about the early stage alteration of smectites, which cannot be detected or differentiated in the analyses of the bulk material (XRD, XRF). The interpretation of structural changes solely based on the total layer charge must be taken with caution, as elemental changes <0.02 e·phuc$^{-1}$ lie within the margin of error, and these errors will accumulate when summing up the total charge. As a result, more reliable interpretations are made by considering the charge distribution (Figures 10 and 11). During smectite alteration in the bentonite-solution experiments, a layer charge increase mainly occurred as a result of tetrahedral substitution of $Si^{4+}$ by $Al^{3+}$, as observed in both samples. A correlation between the increasing degree of smectite alteration and reaction time was observed in both SD80 and B36 bentonites reacted in OPA solutions at 25 °C. While both samples showed no alteration of the tetrahedral sheet after one year, a slight but significant decrease in tetrahedral $Si^{4+}$ was detected after two years both with and without substrate. The tetrahedral (TET) and octahedral (OCT) charge of samples reacted in OPA solution at 90 °C differed only slightly from the initial sample.

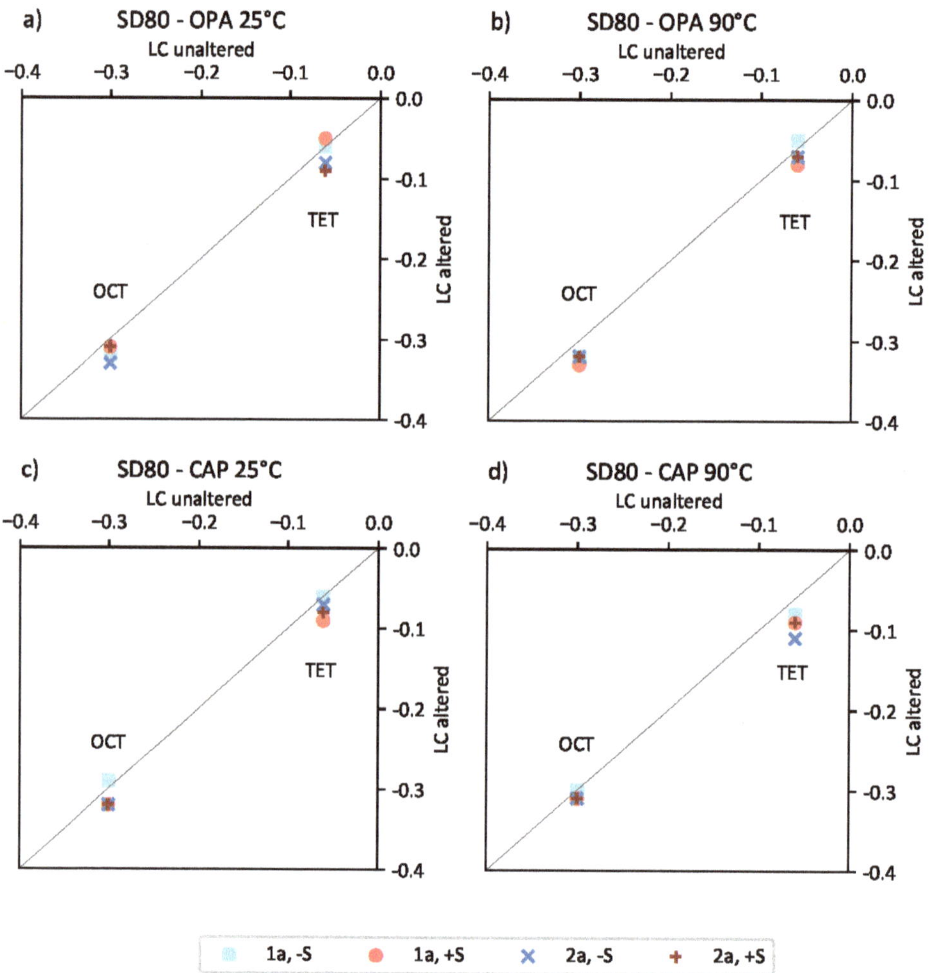


**Figure 10.** Scatter plots for SD80 smectites that compare the altered tetrahedral (TET) and octahedral (OCT) charges to the initial charge distribution for samples reacted in (**a,b**) Opalinus clay pore water (OPA) and (**c,d**) diluted cap rock solution (CAP). Overall, only minor changes were observed after experimentation. LC: layer charge, −S: control samples without substrate, +S: samples with substrate.

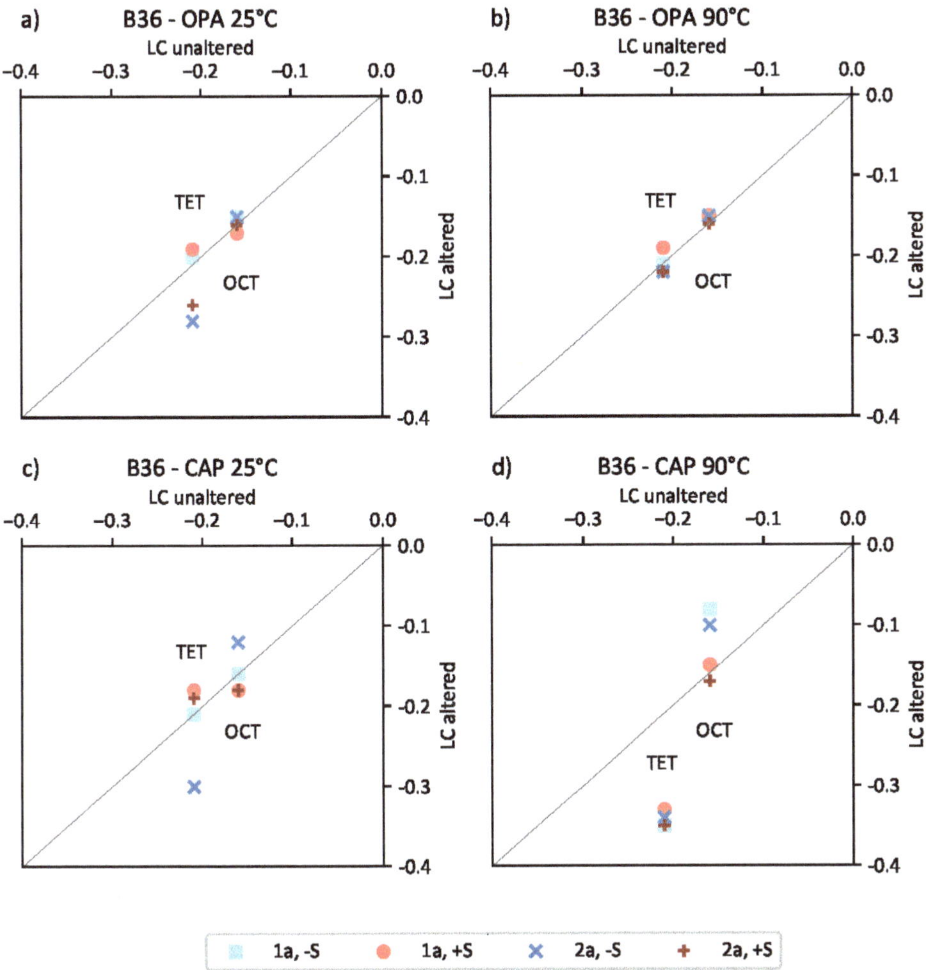



**Figure 11.** Scatter plots for B36 smectites that compare the altered tetrahedral (TET) and octahedral (OCT) charges to the initial charge distribution for samples reacted in (**a,b**) Opalinus clay pore water (OPA) and (**c,d**) diluted cap rock solution (CAP). An increase of the tetrahedral charge was observed for samples reacted in OPA solution at 25 °C after two years (**a**) and in the corresponding control sample reacted in CAP solution (**c**). B36 samples reacted in CAP solution at 90 °C show the highest changes, with an overall increase in tetrahedral charge along with a decrease in octahedral charge for the control samples. LC: layer charge, –S: control samples without substrate, +S: samples with substrate.

Where sample SD80 generally showed only minor variations (Figure 10), more significant changes were seen in the B36 bentonite, especially when subjected to CAP solution at 90 °C (Figure 11d). It should be noted that the control sample reacted at 25 °C after two years showed similar changes as the equivalent sample at 90 °C (Figure 11c). In contrast to MX80 studies by Herbert and co-workers [10], no surplus of Si was observed, indicating montmorillonite alteration toward kaolinite or pyrophyllite compositions as favored in closed systems. In this sample set, a significant increase in the tetrahedral layer charge due to substitution of $Si^{4+}$ by $Al^{3+}$ was detected. This resulted in an overall increase of layer charge and an alteration toward a beidellite reaction product. The additional Al needed for this "beidellitization" process may be provided by the localized dissolution and precipitation of montmorillonite [56] or by the dissolution of accessory minerals, such as unstable feldspar grains [57].

Considering the octahedral charge distribution, differences between control and supplemented samples were observed. Thus, the control samples displayed a significant increase in tetrahedral charge, along with a decrease of octahedral charge, due to the $Fe^{3+}$ and $Mg^{2+}$ substitution by $Al^{3+}$. This indicates the formation of high charged beidellitic layers ($\xi \geq 0.43$ e·phuc$^{-1}$, TET: 77–81%). Whereas, the substrate-bearing samples, with the overall highest layer charges $\geq 0.48$ e·phuc$^{-1}$, showed a similar increase of tetrahedral charge (TET: 67–69%), there was little change in the octahedral charge. Here, the octahedral $Al^{3+}$ was substituted by $Fe^{3+}$, and the interlayer $Na^+$ was slightly increased.

Beaufort and co-workers [52] found that the formation of beidellitic smectite is accompanied with the release of excess Mg in solution, which subsequently led to the formation of trioctahedral smectite. However, no evidence for trioctahedral precipitates was found in the XRD patterns of this study (e.g., $d_{060}$ reflection remained at 0.150 nm).

Although higher resolution SEM images (30,000-fold magnification) and XRD studies of the purified samples showed no indication of nanoparticle contamination, traces of accessory silicate phases (e.g., amorphous silica, cristobalite) cannot be fully ruled out. If present, they would result in Si-contamination of the compositional analyses. As the amount of Si detected in the purified smectite fraction was in most cases not sufficient to fill all tetrahedral metal sites (4 atoms per unit formula), it is assumed that the effect of any impurities was negligible and had no or little influence on the smectite compositions presented. Furthermore, random powder XRD measurements showed no formation of new minerals occurred during experimentation, which may interfere with analyses of the purified fractions.

Due to the different natures of the two bentonites studied, including their different smectite compositions, it is considered likely that the smectites, which were both subjected to the same experimental conditions, reacted differently in each material. In this context, the low charged SD80 as a beidellitic montmorillonite showed nearly no structural change, whereas B36, as a medium charged montmorillonitic beidellite, became more beidellitic with increasing temperature and reacted in the highly saline CAP solution. This corresponds to the findings of Nguyen-Thanh and co-workers [58], who suggested that reactivity of smectite increases with a higher degree of octahedral substitution. Furthermore, reactions with accessory minerals (e.g., K-feldspar, pyrite, calcite), the abundance of which varied between the bentonites, are also likely to have affected the smectite alterations, although such effects could not be determined in this study.

### 4.2. Microbial Diversity and Its Potential Influence on the Mineralogy

Microbial activity in a future HLW repository may influence the bentonite barrier performance by various processes, such as leading to potential gas and sulfide production or changes in redox potential [45]. Considering the results of EDX analyses, no microbial induced smectite alteration was observed in batches treated with substrate. Although the experimental setup was prepared under nonsterile conditions and with a diversity of substrates, the results of extracted DNA indicate that the batches did not develop diverse assemblages of microorganisms. Thus, the applied conditions favored the formation of specialized microorganisms that could adapt to these conditions. The detected genera, which dominated especially the SD80 slurry experiments with OPA solution, have been found in former bentonite slurry experiments [59,60], indicating that these microorganisms originated from the SD80 bentonite.

In general, bentonites are known for their low biomass [23], so that the concentration of extracted DNA was very low (Table S6). The DNA extraction was also hindered by the adsorption of negatively charged DNA to the cationic surfaces of the smectite particles, a mechanism enhanced with decreasing pH due to the increase in positively charged edge sites [61,62].

However, SD80 samples treated with substrate and reacted at 25 °C in OPA solution showed the highest amount of extracted DNA. Within these samples, spore-forming and sulfate-reducing bacteria are dominant. The latter are able to reduce sulfate in order to form $H_2S$ [17,42], which is known for promoting metal corrosion [17,18]. The formation of black spots in the respective sample could be an indication of FeS precipitation (Figure 3a), as similarly observed by Pedersen and co-workers who found a blackening of suspensions with a pH between 7 to 5 [19]. However, these black spots are local precipitates that were not detectable in the sampled material by XRD or SEM–EDX techniques.

The extreme conditions found in saline pore waters (e.g., CAP solution) promote the dominance of specialized microorganisms, i.e., the dominance of *Marinobacter* and *Bacillus* species in the respective B36 samples. Although no smectite alteration in correlation with microbes occurred, color changes in the substrate-bearing B36 batches could be an indication of $Fe^{3+}$ reduction (Figure 3d) [41]. In short-term experiments (98 days, 30 °C) with B36 reacted in OPA solution, the green coloration of the bentonite and supernatant pore water was also observed by Matschiavelli et al. [26]. Moreover, the measurements of the iron concentrations showed an increase of $Fe^{2+}$ and a decrease of $Fe^{3+}$ [26].

Although no typical iron-reducing bacteria were detected within the sample sets studied, other non-identified species as well as some of the detected genera (e.g., *Bacillus* [63]) may change the Fe oxidation state due to changing redox conditions or by direct or indirect iron reduction via AQDS [64,65]. This may lead to layer charge increases [56] and the consequences described in Section 4.1.1. Such changes, however, will not be detected by the EDX method applied in this study, which assumes all iron to be in the $Fe^{3+}$ state.

In general, the utilized organic substrates lactate, acetate and methanol are used for various anaerobic metabolisms as the energy sources [66–68], whereas the humic analogue AQDS serves as an electron shuttle [64]. It should be noted that the concentrations of the substrates added to the batches are higher compared to those expected in a natural bentonite environment or in the deep geological repository itself. To accelerate the microbial processes and to shorten the incubation period, the concentration was increased without knowing whether the tolerable concentration would be exceeded. Matschiavelli et al. [37] pointed out that a lactate or acetate concentration of 50 mM is possibly too high for soil organisms, which is also likely for the organic mixture concentrations used in this study. Inherent microorganisms may be affected, since they are not used to coping with these high concentrations in their natural environment. However, a general mix of substrates at lower concentrations is more relevant to a future deep geological repository setting, since it has been shown that organics such as lactate, acetate, formate and malate are present in pore waters that can potentially enter the repository [69]. Considering the respective environmental conditions, the addition of the correct concentrations of natural organic and humic acids to bentonite needs to be optimized in future experiments as potential substrates and electron shuttles for stimulating microbial metabolisms.

### 4.3. Implications for a Real Repository Scenario

The static batch-experiments with bentonite slurries performed in this study are not representative for the initial conditions of a repository site in terms of the degree of compaction, where tight pore space and low hydraulic conductivities will limit transport processes [13]. However, the importance of these experiments to the long-term performance of the bentonite barrier is considered to be relevant. Over time, niches in the repository are likely to form, where erosion and particle transport will lead to less compacted material [14]. This may result from the infiltration of groundwater through fissures or cavities and along contact zones, leading to the development of localized bentonite clay slurries and changing environmental conditions (e.g., salinity, pH), which may affect the bentonites stability.

Within the experimental period, no clear evidence for mineral transformations (e.g., smectite-to-illite conversion) of the smectite samples was observed. However, for the B36 smectite reacted at 90 °C in higher saline CAP solutions, an uptake of interlayer $K^+$ is recognizable. This is accompanied by an increase in layer charge, which is an initial step that could lead to the fixation of potassium. Additionally, XRD results point towards a decrease in swelling capability, which is indicated by lower *d* values and broader basal reflections (FWHM) in the respective samples (Figure 9). These findings indicate that in long-term experiments, illitization may possibly take place, which would lead to a reduction or even a loss in the swelling and sealing capacity of the bentonite buffer.

Kaufhold and Dohrmann [4] highlighted several key factors that make a bentonite less suitable for HLW disposal. These included mineral alteration, loss of swelling capability, availability of soluble or reactive phases (such as pyrite or gypsum), a high structural Fe content in the smectite as well as a high layer charge density. Considering these parameters, our mineralogical results indicate that the Fe-rich montmorillonitic–beidellite B36 bentonite with its higher abundance of accessory minerals (~35 wt.%) is likely to be less suitable as barrier material than the bentonite SD80 with beidellitic–montmorillonite as the main component (~90 wt.%). In this context, the B36 material showed clear signs of becoming more beidellitic when reacted at higher temperatures in the diluted cap rock solution compared to the SD80 bentonite that showed close to no structural change in any of the experiments conducted. As the highest increase in layer charge was observed for the substrate-bearing B36 samples at 90 °C in CAP solution, the addition of organic and humic acids that can act as electron-donors and promote protonation/deprotonation surface reactions [70] may well have contributed to this advanced state of alteration.

The relevance of microbial activity to the repository scenario, based on similar 1-year bentonite slurry experiments, has been previously discussed [37]. Indigenous microorganisms were shown to evolve under different temperature and substrate conditions but had no significant effect on the analyzed biogeochemical parameters [37,71]. In this follow-up study on two different bentonites, similar slurry experiments with an additional focus on mineralogical changes were compared with and without the addition of organic supplements. Similar microbial results were obtained when adding substrate, but the detected genera did not appear to significantly affect the mineralogical properties of the SD80 and B36 bentonites within the two years of experimentation at low temperatures. A notable result was also the very low amount of DNA in samples reacted in CAP solution or in OPA solution without organic supplements. This emphasizes the point that the availability of nutrients and the salinity are likely to be limiting factors for growth in a slurry environment. It is therefore important in future experimental studies to investigate more nutrient poor systems in waters of variable salinity over time periods of several years in order to gain a more accurate understanding of the potential influence of microbes in the long-term repository setting.

## 5. Conclusions

Static batch experiments were carried out for one and two years at 25 °C and 90 °C using two different bentonites (SD80 from Greece, B36 from Slovakia) and two types of saline solutions, which simulated (a) Opalinus clay pore water with a salinity of 19 g·L$^{-1}$ and (b) diluted cap rock solution with a salinity of 155 g·L$^{-1}$. The bentonites were supplemented with and without organic substrates to study the microbial diversity and their potential influence on the bentonite mineralogy.

1.  After experimentation, no neoformation of minerals was observed. Mineralogical and chemical changes can be attributed to interlayer cation exchange reactions, particle delamination and tetrahedral as well as octahedral metal ion substitutions. These changes are more pronounced at higher salinity and elevated temperatures.
2.  The initial charge distribution determines the reactivity of the smectite, with octahedral charge dominated smectites (e.g., SD80) being less susceptible to these alterations. However, the influence of accessory minerals (e.g., feldspar, calcite, pyrite) on the environment and smectite alteration should not be neglected with regard to the long-term stability of the bentonite barrier.
3.  Considering the microbial influence on a potential HLW repository, the detected genera in SD80 appear to be more important than the specialized microorganisms

detected in bentonite B36 due to their potential to reduce sulfate in order to form H₂S, and thus, promoting the corrosion of metal canisters. Further, it should be noted that the microbial diversity changed with respect to the bentonite and to the applied conditions used in this study. As a result, bentonite-inherent microorganisms may have a potential negative long-term effect on the barrier system. This should be considered when selecting bentonites as buffer material.

4. The reaction kinetics of smectite alteration as well as the precise role of microbes could not be determined due to the complexity of bentonite mineral assemblages and the large number of influencing factors. Further experimentation using simpler mineral mixtures and the addition of single substrates (hydrogen gas, lactate or acetate) at lower concentrations are required. The measurement and quantification of metabolites, e.g., the formation and consumption of organic acids and gases, is necessary to understand further the microbial metabolic potential within the bentonites and its impact on the barrier system.

**Supplementary Materials:** The following are available online at https://www.mdpi.com/article/10.3390/min11090932/s1, Table S1: Recipes for synthetic OPA and CAP solution, Table S2: XRF results of SD80, Table S3: XRF results of B36, Table S4: EDX raw data SD80, Table S5: EDX raw data B36, Table S6: Concentrations of extracted DNA (ng/µL), Table S7: XRD results of SD80, Table S8: XRD results of B36, Table S9: EDX-based SF SD80, Table S10: EDX-based SF B36

**Author Contributions:** Conceptualization, C.P., L.N.W., G.G., A.C., T.A., A.M.; methodology, C.P., L.N.W., N.M., A.C.; validation, C.P., N.M., M.P.; formal analysis, C.P.; investigation, C.P., N.M., S.K.; resources, A.M., L.N.W., G.G., T.A., A.C., N.M.; writing—original draft preparation, C.P., L.N.W., N.M.; writing—review and editing, all authors; visualization, C.P., N.M., M.P.; supervision, L.N.W., G.G., A.C.; project administration, L.N.W., T.A., A.M.; funding acquisition, L.N.W., T.A., A.M. All authors have read and agreed to the published version of the manuscript.

**Funding:** This research is part of the joint project "UMB" funded by the Federal Ministry of Economic Affairs and Energy (BMWi) under the grant number 02 E 11344C. We also acknowledge the support for the Article Processing Charge from the DFG (German Research Foundation, 393148499) and the Open Access Publication Fund of the University of Greifswald.

**Institutional Review Board Statement:** Not applicable.

**Informed Consent Statement:** Not applicable.

**Data Availability Statement:** The data presented in this study are partially available upon request from the corresponding author.

**Acknowledgments:** We would like to thank Manfred Zander (UG) for the introduction to electron microscopy and its sample preparation, Veronika Prause (GRS) for the batch preparation and pH measurements, Karin Grupe (GRS) for ICP-OES measurements, Jennifer Drozdowski (HZDR) for her help with sampling and Stephan Kaufhold (BGR) for providing the bentonites.

**Conflicts of Interest:** The authors declare no conflict of interest.

## References

1. Gilg, H.A.; Kaufhold, S.; Ufer, K. Smectite and bentonite terminology, classification, and genesis. In *Bentonites: Characterization, Geology, Mineralogy, Analysis, Mining, Processing and Uses*; Kaufhold, S., Ed.; Schweizbart Science Publishers: Stuttgart, Germany, 2021; ISBN 978-3-510-96859-6. in press.
2. Christidis, G.E.; Huff, W.D. Geological aspects and genesis of bentonites. *Elements* **2009**, *5*, 93–98. [CrossRef]
3. Sellin, P.; Leupin, O.X. The use of clay as an engineered barrier in radioactive-waste management—A review. *Clays Clay Miner.* **2013**, *61*, 477–498. [CrossRef]
4. Kaufhold, S.; Dohrmann, R. Distinguishing between more and less suitable bentonites for storage of high-level radioactive waste. *Clay Miner.* **2016**, *51*, 289–302. [CrossRef]
5. Savage, D. An Assessment of the Impact of the Degradation of Engineered Structures on the Safety-Relevant Functions of the Bentonite Buffer in a HLW Repository. NTB-13-02, Wettingen, Switzerland. 2014. Available online: https://inis.iaea.org/search/search.aspx?orig_q=RN:48088313 (accessed on 21 June 2021).

6.  Kaufhold, S.; Klimke, S.; Schloemer, S.; Alpermann, T.; Renz, F.; Dohrmann, R. About the corrosion mechanism of metal iron in contact with bentonite. *ACS Earth Space Chem.* **2020**, *4*, 711–721. [CrossRef]
7.  Gaucher, E.C.; Blanc, P. Cement/clay interactions—A review: Experiments, natural analogues, and modeling. *Waste Manag.* **2006**, *26*, 776–788. [CrossRef] [PubMed]
8.  Yokoyama, S.; Shimbashi, M.; Minato, D.; Watanabe, Y.; Jenni, A.; Mäder, U. Alteration of bentonite reacted with cementitious materials for 5 and 10 years in the Mont Terri Rock Laboratory (CI Experiment). *Minerals* **2021**, *11*, 251. [CrossRef]
9.  Pusch, R. Transport of radionuclides in smectite clay. In *Environmental Interactions of Clays: Clays and the Environment*; Parker, A., Rae, J.E., Eds.; Springer: Berlin/Heidelberg, Germany, 1998; pp. 7–35, ISBN 978-3-662-03651-8.
10. Herbert, H.-J.; Kasbohm, J.; Sprenger, H.; Fernández, A.M.; Reichelt, C. Swelling pressures of MX-80 bentonite in solutions of different ionic strength. *Phys. Chem. Earth Parts A/B/C* **2008**, *33*, S327–S342. [CrossRef]
11. Kaufhold, S.; Dohrmann, R. Stability of bentonites in salt solutions | sodium chloride. *Appl. Clay Sci.* **2009**, *45*, 171–177. [CrossRef]
12. Hofmann, H.; Bauer, A.; Warr, L.N. Behavior of smectite in strong salt brines under conditions relevant to the disposal of low- to medium-grade nuclear waste. *Clays Clay Miner.* **2004**, *52*, 14–24. [CrossRef]
13. Stroes-Gascoyne, S.; Hamon, C.J.; Maak, P. Limits to the use of highly compacted bentonite as a deterrent for microbiologically influenced corrosion in a nuclear fuel waste repository. *Phys. Chem. Earth Parts A/B/C* **2011**, *36*, 1630–1638. [CrossRef]
14. Missana, T.; Alonso, U.; Fernández, A.M.; García-Gutiérrez, M. Colloidal properties of different smectite clays: Significance for the bentonite barrier erosion and radionuclide transport in radioactive waste repositories. *Appl. Geochem.* **2018**, *97*, 157–166. [CrossRef]
15. Meleshyn, A. *Microbial Processes Relevant for Long-Term Performance of Radioactive Waste Repositories in Clays*; GRS-291; GRS: Köln, Germany, 2011.
16. Dong, H.; Jaisi, D.P.; Kim, J.; Zhang, G. Microbe-clay mineral interactions. *Am. Mineral.* **2009**, *94*, 1505–1519. [CrossRef]
17. Muyzer, G.; Stams, A.J.M. The ecology and biotechnology of sulphate-reducing bacteria. *Nat. Rev. Microbiol.* **2008**, *6*, 441–454. [CrossRef]
18. El Mendili, Y.; Abdelouas, A.; Bardeau, J.-F. Insight into the mechanism of carbon steel corrosion under aerobic and anaerobic conditions. *Phys. Chem. Chem. Phys.* **2013**, *15*, 9197–9204. [CrossRef] [PubMed]
19. Pedersen, K.; Bengtsson, A.; Blom, A.; Johansson, L.; Taborowski, T. Mobility and reactivity of sulphide in bentonite clays–Implications for engineered bentonite barriers in geological repositories for radioactive wastes. *Appl. Clay Sci.* **2017**, *146*, 495–502. [CrossRef]
20. Lantenois, S.; Lanson, B.; Muller, F.; Bauer, A.; Jullien, M.; Plançon, A. Experimental study of smectite interaction with metal Fe at low temperature: 1. Smectite destabilization. *Clays Clay Miner.* **2005**, *53*, 597–612. [CrossRef]
21. Soltermann, D. Ferrous Iron Uptake Mechanisms at the Montmorillonite-Water Interface under Anoxic and Electrochemically Reduced Conditions. Ph.D. Thesis, ETH, Zurich, Switzerland, 2014.
22. Ufer, K.; Stanjek, H.; Roth, G.; Dohrmann, R.; Kleeberg, R.; Kaufhold, S. Quantitative phase analysis of bentonites by the Rietveld method. *Clays Clay Miner.* **2008**, *56*, 272–282. [CrossRef]
23. Kaufhold, S.; Stucki, J.W.; Finck, N.; Steininger, R.; Zimina, A.; Dohrmann, R.; Ufer, K.; Pentrák, M.; Pentráková, L. Tetrahedral charge and Fe content in dioctahedral smectites. *Clay Miner.* **2017**, *52*, 51–65. [CrossRef]
24. Wersin, P.; Leupin, O.X.; Mettler, S.; Gaucher, E.C.; Mäder, U.; de Cannière, P.; Vinsot, A.; Gäbler, H.E.; Kunimaro, T.; Kiho, K.; et al. Biogeochemical processes in a clay formation in situ experiment: Part A–Overview, experimental design and water data of an experiment in the Opalinus Clay at the Mont Terri Underground Research Laboratory, Switzerland. *Appl. Geochem.* **2011**, *26*, 931–953. [CrossRef]
25. Meleshyn, A. *Mechanisms of Transformation of Bentonite Barriers (UMB)*; GRS-930; GRS: Köln, Germany, 2019.
26. Matschiavelli, N.; Drozdowski, J.; Kluge, S.; Arnold, T.; Cherkouk, A. Joint project: Umwandlungsmechanismen in Bentonitbarrieren—Subproject B: Einfluss von mikrobiellen Prozessen auf die Bentonitumwandlung, HZDR-103, Dresden, Germany. 2019. Available online: https://www.hzdr.de/db/!Publications?pNid=head&pSelMenu=0&pSelTitle=29398 (accessed on 21 June 2021).
27. Meier, L.P.; Kahr, G. Determination of the Cation Exchange Capacity (CEC) of clay minerals using the complexes of copper (II) Ion with Triethylenetetramine and Tetraethylenepentamine. *Clays Clay Miner.* **1999**, *47*, 386–388. [CrossRef]
28. Stanjek, H.; Künkel, D. CEC determination with Cu-triethylenetetramine: Recommendations for improving reproducibility and accuracy. *Clay Miner.* **2016**, *51*, 1–17. [CrossRef]
29. Doebelin, N.; Kleeberg, R. Profex: A graphical user interface for the Rietveld refinement program BGMN. *J. Appl. Crystallogr.* **2015**, *48*, 1573–1580. [CrossRef]
30. Christidis, G.E. Validity of the structural formula method for layer charge determination of smectites: A re-evaluation of published data. *Appl. Clay Sci.* **2008**, *42*, 1–7. [CrossRef]
31. Williams, D.B.; Carter, C.B. *Transmission Electron Microscopy: A Textbook for Materials Science*, 2nd ed.; Springer: New York, NY, USA; London, UK, 2009; ISBN 978-0-387-76501-3.
32. Velde, B. Electron microprobe analysis of clay minerals. *Clay Miner.* **1984**, *19*, 243–247. [CrossRef]
33. Van der Pluijm, B.A. Analytical electron microscopy and the problem of potassium diffusion1. *Clays Clay Miner.* **1988**, *36*, 498–504. [CrossRef]

34. Newbury, D.E.; Ritchie, N.W.M. Electron-excited X-ray microanalysis by energy dispersive spectrometry at 50: Analytical accuracy, precision, trace sensitivity, and quantitative compositional mapping. *Microsc. Microanal.* **2019**, *25*, 1075–1105. [CrossRef]
35. Stevens, R.E. A system for calculating analyses of micas and related minerals to end members. *US Geol. Surv. Bull.* **1946**, *950*, 101–119.
36. Wolters, F. Classification of Montmorillonites. Ph.D. Thesis, Universität Karlsruhe, Karlsruhe, Germany, 2005.
37. Matschiavelli, N.; Kluge, S.; Podlech, C.; Standhaft, D.; Grathoff, G.; Ikeda-Ohno, A.; Warr, L.N.; Chukharkina, A.; Arnold, T.; Cherkouk, A. The year-long development of microorganisms in uncompacted bavarian bentonite slurries at 30 and 60 °C. *Environ. Sci. Technol.* **2019**, *53*, 10514–10524. [CrossRef]
38. Warr, L.N. IMA-CNMNC approved mineral symbols. *Mineral. Mag.* **2021**, *85*, 291–320. [CrossRef]
39. Moore, D.M.; Reynolds, R.C. *X-ray Diffraction and the Identification and Analysis of Clay Minerals*, 2nd ed.; Oxford University Press: Oxford, UK, 1997; ISBN 978-0-19-508713-0.
40. Emmerich, K.; Wolters, F.; Kahr, G.; Lagaly, G. Clay Profiling: The classification of montmorillonites. *Clays Clay Miner.* **2009**, *57*, 104–114. [CrossRef]
41. Stucki, J.W. A review of the effects of iron redox cycles on smectite properties. *Comptes Rendus Geosci.* **2011**, *343*, 199–209. [CrossRef]
42. Klemps, R.; Cypionka, H.; Widdel, F.; Pfennig, N. Growth with hydrogen, and further physiological characteristics of *Desulfotomaculum* species. *Arch. Microbiol.* **1985**, *143*, 203–208. [CrossRef]
43. Nicholson, W.L.; Munakata, N.; Horneck, G.; Melosh, H.J.; Setlow, P. Resistance of Bacillus endospores to extreme terrestrial and extraterrestrial environments. *Microbiol. Mol. Biol. Rev.* **2000**, *64*, 548–572. [CrossRef] [PubMed]
44. Gauthier, M.J.; Lafay, B.; Christen, R.; Fernandez-Linares, L.C.; Acquaviva, M.; Bonin, P.; Bertrand, J.-C. Marinobacter hydrocarbonoclasticus gen. nov., sp. nov., a new, extremely halotolerant, hydrocarbon-degrading marine bacterium. *Int. J. Syst. Bacteriol.* **1992**, *42*, 568–576. [CrossRef] [PubMed]
45. Pedersen, K. Microbial Processes in Radioactive Waste Disposal, SKB TR-00-04, Stockholm, Sweden. 2000. Available online: https://www.osti.gov/etdeweb/servlets/purl/21330059 (accessed on 14 June 2021).
46. Savage, D. *The Effects of High Salinity Groundwater on the Performance of Clay Barriers*; SKI-R-05-54; Swedish Nuclear Power Inspectorate: Stockholm, Sweden, 2005.
47. Tertre, E.; Prêt, D.; Ferrage, E. Influence of the ionic strength and solid/solution ratio on Ca (II)-for-Na+ exchange on montmorillonite. Part 1: Chemical measurements, thermodynamic modeling and potential implications for trace elements geochemistry. *J. Colloid Interface Sci.* **2011**, *353*, 248–256. [CrossRef] [PubMed]
48. Warr, L.; Berger, J. Hydration of bentonite in natural waters: Application of "confined volume" wet-cell X-ray diffractometry. *Phys. Chem. Earth Parts A/B/C* **2007**, *32*, 247–258. [CrossRef]
49. Christidis, G.E.; Blum, A.E.; Eberl, D.D. Influence of layer charge and charge distribution of smectites on the flow behaviour and swelling of bentonites. *Appl. Clay Sci.* **2006**, *34*, 125–138. [CrossRef]
50. Laird, D.A. Influence of layer charge on swelling of smectites. *Appl. Clay Sci.* **2006**, *34*, 74–87. [CrossRef]
51. Kaufhold, S.; Dohrmann, R.; Koch, D.; Houben, G. The pH of aqueous bentonite suspensions. *Clays Clay Miner.* **2008**, *56*, 338–343. [CrossRef]
52. Beaufort, D.; Berger, G.; Lacharpagne, J.C.; Meunier, A. An experimental alteration of montmorillonite to a di + trioctahedral smectite assemblage at 100 °C and 200 °C. *Clay Miner.* **2001**, *36*, 211–225. [CrossRef]
53. Heller-Kallai, L. Protonation–deprotonation of dioctahedral smectites. *Appl. Clay Sci.* **2001**, *20*, 27–38. [CrossRef]
54. Drits, V.A. A Model for the mechanism of $Fe^{3+}$ to $Fe^{2+}$ reduction in dioctahedral smectites. *Clays Clay Miner.* **2000**, *48*, 185–195. [CrossRef]
55. Rozalen, M.; Huertas, F.J.; Brady, P.V. Experimental study of the effect of pH and temperature on the kinetics of montmorillonite dissolution. *Geochim. Cosmochim. Acta* **2009**, *73*, 3752–3766. [CrossRef]
56. Birgersson, M.; Karnland, O.; Korkeakoski, P.; Sellin, P.; Mäder, U.; Wersin, P. *Montmorillonite Stability Under Near-Field Conditions*; NTB-14-12; National Cooperative for the Disposal of Radioactive Waste (NAGRA): Wettingen, Switzerland, 2014; pp. 1015–2636.
57. Ye, W.-M.; He, Y.; Chen, Y.G.; Chen, B.; Cui, Y.-J. Thermochemical effects on the smectite alteration of GMZ bentonite for deep geological repository. *Environ. Earth Sci.* **2016**, *75*, 906. [CrossRef]
58. Nguyen-Thanh, L.; Herbert, H.-J.; Kasbohm, J.; Hoang-Minh, T.; Mählmann, R.F. Effects of chemical structure on the stability of smectites in short-term alteration experiments. *Clays Clay Miner.* **2014**, *62*, 425–446. [CrossRef]
59. Stroes-Gascoyne, S.; Sergeant, C.; Schippers, A.; Hamon, C.J.; Nèble, S.; Vesvres, M.-H.; Barsotti, V.; Poulain, S.; Le Marrec, C. Biogeochemical processes in a clay formation in situ experiment: Part D–Microbial analyses–Synthesis of results. *Appl. Geochem.* **2011**, *26*, 980–989. [CrossRef]
60. Fru, E.C.; Athar, R. In situ bacterial colonization of compacted bentonite under deep geological high-level radioactive waste repository conditions. *Appl. Microbiol. Biotechnol.* **2008**, *79*, 499–510. [CrossRef] [PubMed]
61. Cai, P.; Huang, Q.; Zhang, X.; Chen, H. Adsorption of DNA on clay minerals and various colloidal particles from an Alfisol. *Soil Biol. Biochem.* **2006**, *38*, 471–476. [CrossRef]
62. Greaves, M.P.; Wilson, M.J. The adsorption of nucleic acids by montmorillonite. *Soil Biol. Biochem.* **1969**, *1*, 317–323. [CrossRef]
63. Hobbie, S.N.; Li, X.; Basen, M.; Stingl, U.; Brune, A. Humic substance-mediated Fe (III) reduction by a fermenting Bacillus strain from the alkaline gut of a humus-feeding scarab beetle larva. *Syst. Appl. Microbiol.* **2012**, *35*, 226–232. [CrossRef]

64. Bai, Y.; Mellage, A.; Cirpka, O.A.; Sun, T.; Angenent, L.T.; Haderlein, S.B.; Kappler, A. AQDS and Redox-Active NOM enables microbial Fe (III)-mineral reduction at cm-scales. *Environ. Sci. Technol.* **2020**, *54*, 4131–4139. [CrossRef] [PubMed]
65. Jaisi, D.P.; Kukkadapu, R.K.; Eberl, D.D.; Dong, H. Control of Fe (III) site occupancy on the rate and extent of microbial reduction of Fe (III) in nontronite. *Geochim. Cosmochim. Acta* **2005**, *69*, 5429–5440. [CrossRef]
66. Pankhania, I.P.; Spormann, A.M.; Hamilton, W.A.; Thauer, R.K. Lactate conversion to acetate, $CO_2$ and $H_2$ in cell suspensions of Desulfovibrio vulgaris (Marburg): Indications for the involvement of an energy driven reaction. *Arch. Microbiol.* **1988**, *150*, 26–31. [CrossRef]
67. Thauer, R.K.; Möller-Zinkhan, D.; Spormann, A.M. Biochemistry of acetate catabolism in anaerobic chemotrophic bacteria. *Annu. Rev. Microbiol.* **1989**, *43*, 43–67. [CrossRef]
68. Daniels, L.; Sparling, R.; Sprott, G. The bioenergetics of methanogenesis. *Biochim. Biophys. Acta (BBA)—Rev. Bioenerg.* **1984**, *768*, 113–163. [CrossRef]
69. Courdouan, A.; Christl, I.; Meylan, S.; Wersin, P.; Kretzschmar, R. Isolation and characterization of dissolved organic matter from the Callovo–Oxfordian formation. *Appl. Geochem.* **2007**, *22*, 1537–1548. [CrossRef]
70. Ramos, M.E.; Huertas, F.J. Adsorption of lactate and citrate on montmorillonite in aqueous solutions. *Appl. Clay Sci.* **2014**, *90*, 27–34. [CrossRef]
71. Grigoryan, A.A.; Jalique, D.R.; Medihala, P.; Stroes-Gascoyne, S.; Wolfaardt, G.M.; McKelvie, J.; Korber, D.R. Bacterial diversity and production of sulfide in microcosms containing uncompacted bentonites. *Heliyon* **2018**, *4*, e00722. [CrossRef]

*Article*

# Ex and In Situ Reactivity and Sorption of Selenium in Opalinus Clay in the Presence of a Selenium Reducing Microbial Community


Nele Bleyen [1,*], Joe S. Small [2], Kristel Mijnendonckx [1], Katrien Hendrix [1], Achim Albrecht [3], Pierre De Cannière [4], Maryna Surkova [4], CharlesWittebroodt [5] and Elie Valcke [1]

[1] SCK CEN, Boeretang 200, BE-2400 Mol, Belgium; kristel.mijnendonckx@sckcen.be (K.M.); katrien.hendrix@sckcen.be (K.H.); evalcke@sckcen.be (E.V.)
[2] Research Centre for Radwaste Disposal, Department of Earth and Environmental Sciences, The University of Manchester, Manchester M13 9PL, UK; joe.small.con@gmail.com
[3] ANDRA, 1-7 Rue Jean Monnet, CEDEX, FR-92298 Châtenay-Malabry, France; Achim.Albrecht@andra.fr
[4] FANC, Ravensteinstraat 36, BE-1000 Brussels, Belgium; pierre.decanniere@gmail.com (P.D.C.); maryna.surkova@fanc.fgov.be (M.S.)
[5] IRSN, BP17, FR-92262 Fontenay-aux-Roses, France; Charles.wittebroodt@irsn.fr
* Correspondence: nbleyen@sckcen.be



**Citation:** Bleyen, N.; Small, J.S.; Mijnendonckx, K.; Hendrix, K.; Albrecht, A.; De Cannière, P.; Surkova, M.; Wittebroodt, C.; Valcke, E. Ex and In Situ Reactivity and Sorption of Selenium in Opalinus Clay in the Presence of a Selenium Reducing Microbial Community. *Minerals* 2021, 11, 757. https://doi.org/10.3390/min11070757

Academic Editors: Ana María Fernández, Stephan Kaufhold, Markus Olin, Lian-Ge Zheng, Paul Wersin and James Wilson

Received: 10 May 2021
Accepted: 8 July 2021
Published: 13 July 2021





**Abstract:** $^{79}$Se is a critical radionuclide concerning the safety of deep geological disposal of certain radioactive wastes in clay-rich formations. To study the fate of selenium oxyanions in clayey rocks in the presence of a selenium reducing microbial community, in situ tests were performed in the Opalinus Clay at the Mont Terri Rock Laboratory (Switzerland). Furthermore, biotic and abiotic batch tests were performed to assess Se(VI) and Se(IV) reactivity in the presence of Opalinus Clay and/or stainless steel, in order to support the interpretation of the in situ tests. Geochemical modeling was applied to simulate Se(VI) reduction, Se(IV) sorption and solubility, and diffusion processes. This study shows that microbial activity is required to transform Se(VI) into more reduced and sorbing Se species in the Opalinus Clay, while in abiotic conditions, Se(VI) remains unreactive. On the other hand, Se(IV) can be reduced by microorganisms but can also sorb in the presence of clay without microorganisms. In situ microbial reduction of Se oxyanions can occur with electron donors provided by the clay itself. If microorganisms would be active in the clay surrounding a disposal facility, microbial reduction of leached Se could thus contribute to the overall retention of Se in clayey host rocks.

**Keywords:** selenium reduction; diffusion; sorption; Opalinus Clay; in situ; batch tests


## 1. Introduction

In several countries relying on nuclear energy, deep clay formations are studied as potential host rocks for geological disposal of intermediate-level and high-level long-lived waste (ILW and HLW). For both types of waste (mainly vitrified HLW and nitrate-bearing bituminized waste), $^{79}$Se (T$_{\frac{1}{2}}$ = 327 ka [1]) is considered as one of the main radionuclides contributing to the dose-to-man after final disposal [2–5]. As selenium is a redox sensitive element, it can exist under different oxidation states depending on the environmental conditions: mainly Se(VI), Se(IV), Se(0), Se(-I), and Se(-II). Due to the oxidizing conditions during spent fuel reprocessing, radiolysis in the waste and/or due to the presence of large amounts of nitrate in nitrate-bearing bituminized ILW, Se is expected to be released—at least in part—as Se oxyanions [3,6]. Although part of these oxyanions could sorb onto corrosion products present on the steel canisters or steel rebars [7], the remainder is expected to reach the clay host rock surrounding the waste repository.

In the clay, Se mobility is largely controlled by its oxidation state, which is affected by the prevailing redox conditions and pH. Based on thermodynamic and experimental data in soils and ground waters [3,8–11], Se(0) and metal-selenide minerals (e.g., FeSe$_2$ as ferroselite) are stable and poorly soluble (and thus immobile) at low redox potential. At slightly higher redox potentials, though still reducing conditions and at neutral pH, Se(IV) (i.e., SeO$_3{}^{2-}$ or selenite) can be found in soil pore waters. As Se(IV) can remain in solution, its mobility is higher compared to Se(0) or selenides, though it is constrained by precipitation and sorption processes. Metal-selenite minerals are rarely found in soils, though under certain pH and Eh conditions, anhydrous and hydrous metal-selenites (e.g., with Fe, Al, or Cu) have been found in near-surface environments [11,12]. Nevertheless, selenite concentrations in solution are largely controlled by its sorption onto certain minerals, such as clay minerals, pyrite, siderite, mackinawite, and iron (hydr)oxides [7,8,13–20]. Retention of selenite on redox-active clay components (e.g., Fe(II) minerals) involves surface adsorption followed by reduction to Se(0) or selenide and precipitation [16,17,20–22]. In addition, selenite can adsorb directly onto clay minerals via the formation of inner-sphere complexes [22,23]. Sorption of selenite onto clays has indeed been observed in batch tests with Toarcian argillites, Opalinus Clay, and Boom Clay [18,22,24]. Finally, under oxidizing conditions, Se(VI) (i.e., SeO$_4{}^{2-}$ or selenate) would be the most dominant species. This Se oxyanion has a high solubility and is poorly or non-sorbing (depending on the pH value of the solution with respect to the point of zero charge (pzc) of the considered mineral surface), and it can therefore remain in solution at higher concentrations. In batch sorption experiments with different clays (Boom Clay, Toarcian argillites, and Opalinus Clay), sorption of selenate could not be observed [18,24], though under certain conditions Se(VI) was able to sorb onto certain minerals (e.g., iron (oxy)hydroxides and clay minerals) [7,13,25–27]).

In addition to the pH and redox conditions, microorganisms also play an important role in Se cycling in nature and can transform Se through three distinct processes: assimilation, detoxification, and dissimilatory reduction. While assimilation and detoxification can occur under both anaerobic and aerobic conditions, dissimilatory reduction (or respiration) of Se can only occur under anoxic conditions [28]. Selenium has beneficial or toxic health properties depending on a very narrow concentration window. At low concentration, selenium is an essential micronutrient to microorganisms and is incorporated in selenoproteins necessary for enzymatic activity (e.g., for protecting cell membranes against oxidative damage) by assimilatory reduction of Se oxyanions [28–31]. However, high selenium concentrations are toxic, and detoxification is then required. Several detoxification pathways are possible, including the formation of Se(0) particles or methylation of non-volatile reduced Se species to volatile compounds [28–31]. Finally, several microbial strains are known to perform dissimilatory reduction of selenate and/or selenite. In this metabolic process, Se is used as an electron acceptor for energy production and microbial growth, while mainly organic compounds, but also inorganic compounds such as H$_2$, can act as electron donors [28]. In general, selenium oxyanions are reduced to the less soluble and less biologically available elemental Se(0). The formation of Se(0) is considered a biomineralization process and can occur either intracellularly or extracellularly [28].

To our knowledge, diffusion studies investigating the migration of Se in the clay are rather scarce, though some data is available for selenate in Boom Clay [3] and Opalinus Clay (Mont Terri Diffusion and Retention (DR) experiment [32]), and for both Se oxyanions in Callovo–Oxfordian Clay [19,33]. According to these experiments, retention of Se(VI) is very limited and its migration through the clay would be predominantly driven by diffusion of a mobile species. Reduction of Se(VI) was either not detected or not measured. On the other hand, diffusion experiments with Se(IV) in Callovo–Oxfordian Clay suggest sorption of Se(IV) on the clay, based on the decrease in Se(IV) concentration in the feed solution and on the detection of reduced Se species (though not specified) by XANES (X-ray absorption near-edge structure) along the diffusion pathway [19].

It is important to note that, most of the sorption and diffusion studies with Se oxyanions did not include microbial activity as a possible pathway for Se reduction. In other studies, microbial activity was not assessed, even though this could have contributed to the observed results. Furthermore, studies focusing on microbial Se reduction often include the addition of easily biodegradable carbon sources or electron donors, which also influences the observed processes. As microbial activity cannot be ruled out in a repository for radioactive waste, the fate and transport of Se in the near field of a radioactive waste disposal facility may not only rely on purely abiotic chemical processes but should also include microbially mediated reactivity and retention of Se. A good understanding of the diffusion, sorption, and reactivity of Se oxyanions in a clay host rock of such a disposal facility, including the effect of microorganisms, is required to support the selection of transport parameters for performance assessment calculations of the deep geological repository.

In this paper, the in situ fate of selenium oxyanions in the Opalinus Clay, including the effect of a selenium reducing microbial community, was studied for the first time. For this, in situ injection tests with selenate were performed in the borehole of the BN experiment (bitumen–nitrate–clay interaction experiment) at the Mont Terri Rock Laboratory (Switzerland), in which microbial nitrate reactivity and diffusion were previously studied [34]. Additional batch tests were performed to support the results of the in situ experiment. Both biotic and abiotic batch tests were performed in water sampled from the BN borehole and in the presence of Opalinus Clay, stainless steel, and (for the biotic tests) a microbial community, all originating from the in situ BN experiment. No additional carbon source or electron donor were provided, to ensure biogeochemical conditions as close as possible to those prevailing in situ. The acquired data from the batch tests were used to develop a model to predict the behavior of selenium in Opalinus Clay.

## 2. Materials and Methods

All preparatory work was performed and test suspensions were prepared and stored under anoxic conditions in a UV-sterilized glove box (Ar atmosphere with $O_2 < 0.001\%$). All chemicals used had a high purity ($\geq 99\%$) and were equilibrated with inert atmosphere, by bringing them in the glove box one day in advance to remove all traces of oxygen gas. Note that both the $Na_2SeO_4$ and $Na_2SeO_3$ (99% Se; Sigma Aldrich, Overijse, Belgium) used in the batch and in situ tests contained impurities: ~0.4% of the total Se content in $Na_2SeO_4$ was Se(IV) and ~10% of the total Se content in $Na_2SeO_3$ was Se(VI).

### 2.1. Design of the BN Experiment

The BN in situ experiment was performed in a borehole drilled in the Opalinus Clay at the Mont Terri Rock Laboratory (Jura, Switzerland). The Opalinus Clay is a Mesozoic shale formation in the Jura Mountains of northwestern Switzerland. A detailed overview of the Mont Terri Rock Laboratory, including the location of the in situ experiment discussed here and the characteristics of the Opalinus Clay (mineralogy and geochemistry), is provided by Bossart et al. [35] and Pearson et al. [36]. The mineral composition of the Opalinus Clay at the Mont Terri site comprises mainly quartz, illite, and mixed-layer illite–smectite, kaolinite, chlorite, biotite and muscovite, calcite, aragonite, siderite, dolomite and/or ankerite, albite and/or plagioclase, K-feldspar, pyrite, organic matter (mostly kerogen), and other trace minerals like apatite [35,36]. The BN experiment is performed in the shaly lithofacies of the Opalinus Clay, which consists of between 39 and 80 dry wt.% illite, chlorite and kaolinite, and between 5 and 20 dry wt.% of the mixed-layer illite-smectites. Furthermore, it contains between 10 and 27 dry wt.% of quartz, between 4 and 35 dry wt.% of carbonates, between 0.9 and 1.4 dry wt.% pyrite and between 0.8 and 1.4 dry wt.% organic matter [35].

The water collected in situ from the Opalinus Clay is of the NaCl type (~60–400 mM). The chloride concentration depends on the location within the Opalinus Clay, with an increased salinity from the top to the base of the formation [36]. Besides $Na^+$ and $Cl^-$, other major components of the Opalinus Clay pore water are sulfate, other cations ($Mg^{2+}$, $Ca^{2+}$, $K^+$ and $Sr^{2+}$), dissolved carbonate (corresponding to a $pCO_2$ ranging from ~$10^{-2}$ to $10^{-3}$ kPa), and dissolved organic carbon (usually below 1.7 mmol C $L^{-1}$). The concentration of other cations as well as the alkalinity are linked to the chloride content [36].

The experimental set-up and its installation have been described in detail by Bleyen et al. [34]. In short, the downhole equipment consists of three anoxic test chambers (labeled

'Interval 1 to 3') of ~90 cm in length, isolated from each other and from the gallery surface by inflatable packers to avoid both cross-contamination and oxygen ingress from the gallery atmosphere. The intervals are in contact with the surrounding clay through a cylindrical sintered stainless steel porous filter screen surrounding the central supporting stainless steel tube. Stainless steel water lines connect each of the downhole intervals to a water sampling unit (with sampling containers and a sampling port) and an on-line chemical monitoring system inside the gallery of the Rock Laboratory.

The intervals were filled under anoxic conditions with deoxygenated Opalinus Clay artificial pore water (APW; Table 1), immediately after installation of the downhole equipment. The chemical composition of APW is based on the relationship between sulfate, cations, and chloride, as determined by Pearson et al. [36]. Because of the existence of a strong chloride concentration profile in the Mont Terri anticline structure, the APW composition was also adjusted as a function of the salinity to match the in situ $Cl^-$ concentration of pore water at the exact location of the BN experiment in the rock laboratory [34]. The composition of the solutions sampled in the interval after 8 months of equilibration was close to that of the initially injected APW, except for small increases in total inorganic and organic carbon species (TIC and TOC, respectively; Table 1) and remained stable over time [34].

**Table 1.** Chemical composition of the artificial pore water (APW; target composition [34]) used to saturate the intervals and of sampled solutions from Interval 1 and 3 (one sample each) used for the batch tests (Section 2.2). The measurement uncertainty (95% confidence) on the concentrations is 4–5% ($[SO_4^{2-}]$), 5.5% ($[NO_3^-]$), 6% ($[Cl^-]$), 10% ($[Na^+]$, $[K^+]$, $[Mg^{2+}]$, $[Ca^{2+}]$, $[Sr^{2+}]$, $[CH_3COO^-]$, and TIC), 15% ($[NO_2^-]$, [total dissolved Fe]) and 30% (TOC) (95% confidence), while the uncertainty on the pH is estimated to be 0.1 pH unit (95% confidence).

| Chemical Species | Concentration (mM) | | |
| :---: | :---: | :---: | :---: |
| | APW | Interval 1 | Interval 3 |
| $Na^+$ | 162 | 181 | 177 |
| $K^+$ | 1.1 | 1.2 | 1.3 |
| $Mg^{2+}$ | 8.6 | 9.71 | 10.2 |
| $Ca^{2+}$ | 12.6 | 11.7 | 10.9 |
| $Sr^{2+}$ | 0.38 | 0.32 | 0.32 |
| Total dissolved Fe | — | 0.016 | 0.0015 |
| $Cl^-$ | 181 | 202 | 197 |
| $SO_4^{2-}$ | 12.3 | 11.9 | 11.7 |
| $NO_3^-$ | — | 0.04 | 0.06 |
| $NO_2^-$ | — | <0.1 | <0.1 |
| $CH_3COO^-$ | — | <0.02 | <0.02 |
| TIC | 2.8 | 0.53 | 2.2 |
| TOC | — | <0.83 | <0.83 |
| pH | 7.8 | 7.7 | n.a. [1] |

—: component not added to APW composition (thus initially absent); [1] n.a.: not analyzed.

The water in each of the intervals is continuously circulating from the downhole installation to the surface equipment (with water sampling and online measuring equipment) and back using a magnetically driven gear pump combined with a flow meter. This circulation results in a homogenous solution throughout the circuit of each interval and allows for monitoring of the chemical composition of the interval solutions, either by online measurements or by sampling and offline measurements. Each circuit is equipped with five water sampling containers (40 or 150 mL Whitey stainless steel cylinders from Swagelok), which can be disconnected safely when additional chemical analyses are required, without causing a perturbation to the geochemistry of the interval solution. To monitor the pH and redox potential in the intervals, pH and redox probes (pH::lyser and redo::lyser from S::can Messtechnik GmbH, Vienna, Austria, now subsidiary of Badger Meter) are installed in the water circuit inside gas-tight stainless steel flow-through cells.

In each circuit, additional removable containers ('clay loops') can be placed (one or multiple in series), which contain a piece of Opalinus Clay in contact with the interval solution through a stainless steel filter screen (i.e., simulating the borehole). These clay loops can be disconnected in the same way as the sampling containers, allowing the analysis of the solid phases without perturbing the borehole.

*2.2. Batch Tests*

$Na_2SeO_4$ (50 µM) or $Na_2SeO_3$ (10 µM) was first added to pooled solutions sampled from BN Interval 1 and Interval 3, providing part of the microbial community present in the BN borehole in the biotic tests. For the abiotic tests, these solutions were sterilized by filtration at 0.22 µm before the start of the batch tests. No natural Se background could be detected in the BN borehole water used for these tests (i.e., <1.7 µM total Se and <0.013 µM Se(VI) and Se(IV), the corresponding limits of detection). To some of the solutions, 0.1 mM $Na_2HPO_4$ was added as a source of phosphorus for microbial growth (Table 2). No additional growth nutrient or electron donor was added.

**Table 2.** Overview of the test conditions studied in the batch tests with Se species. Duplicates were prepared for each condition and were labeled with 'a' and 'b'.

| Test Code | Clay (g) | Stainless Steel (mm × mm) [1] | $Na_2HPO_4$ (mM) | Abiotic/Biotic |
|---|---|---|---|---|
| **Tests with 50 µM selenate** | | | | |
| Ab_Se(VI) | — | — | — | abiotic |
| Ab_Se(VI)-C | 0.1 g | — | — | abiotic |
| Ab_Se(VI)-F | — | 5 × 5 | — | abiotic |
| Ab_Se(VI)-CF | 0.1 g | 5 × 5 | — | abiotic |
| Bio_Se(VI)-C | 0.1 g | — | — | biotic |
| Bio_Se(VI)-CF | 0.1 g | 5 × 5 | — | biotic |
| Bio_Se(VI)-P | — | — | 0.1 | biotic |
| **Tests with 10 µM selenite** | | | | |
| Ab_Se(IV) | — | — | — | abiotic |
| Ab_Se(IV)-C | 0.1 g | — | — | abiotic |
| Ab_Se(IV)-F | — | 5 × 5 | — | abiotic |
| Ab_Se(IV)-CF | 0.1 g | 5 × 5 | — | abiotic |
| Bio_Se(IV)-C | 0.1 g | — | — | biotic |
| Bio_Se(IV)-CF | 0.1 g | 5 × 5 | — | biotic |
| Bio_Se(IV)-P | — | — | 0.1 | biotic |

[1] Thickness of the stainless steel filter = 1 mm; meaning of the test codes: Ab: abiotic; Bio: biotic; C: clay; F: filter; CF: clay and filter; P: addition of phosphate; —: component not added to the batch tests.

Secondly, small pieces of Opalinus Clay and stainless steel were prepared to investigate their impact on the Se reactivity and retention, as both are present in the BN borehole (Section 2.1). Pieces of Opalinus Clay (~0.1 g each) were cut from a larger piece (35 mm × 35 mm × 5 mm), which had been exposed to in situ conditions inside a removable container ('clay loop') in contact with the anoxic solution circulating in Interval 3 during injection of nitrate for 3 years [37]. Pieces of a stainless steel filter (5 mm × 5 mm × 1 mm) were cut from a larger filter plate, which had been in another clay loop in contact with the solution circulating in Interval 1 during injection of nitrate and $H_2$ for 2 years [34]. Both solids were chosen since they were already acclimatized to the conditions of the BN experiment and may contain a different microbial community than the interval solutions. For the abiotic batch tests, the solid pieces were autoclaved (121 °C, 2 bar, 20 min) in empty amber glass septum vials. The efficiency of the sterilization process was verified by placing a Sterikon plus Bioindicator (Sigma-Aldrich, Overijse, Belgium) in a similar glass septum vial, which was autoclaved along with the other vials.

Seventy milliliters of the Se-containing solutions was added to sterile amber glass septum vials containing pieces of Opalinus Clay (C) and/or stainless steel filter (F) according to Table 2. For the abiotic tests, the filter sterilized Se-containing solution was added immediately after autoclaving the solid pieces. Two blanks were included as negative controls, i.e., interval solution with a piece of clay (Blank_C) or with a piece of clay and filter (Blank_CF). The septa were closed for the duration of the experiment. The suspensions were sampled regularly by taking aliquots of 3 mL with a sterile needle after thoroughly shaking to homogenize the suspension. The solution was not filtered before analysis but subsampled to perform the different analyses, as described in Sections 2.4 and 2.5. The solid phases of the abiotic batch test were analyzed microscopically (Section 2.6).

### 2.3. In Situ Injection Tests

Two consecutive injections of Interval 3 with Se-containing APW were performed in 2019. The first injection of Interval 3 was performed by replacing only the solution in the surface circulation loop (volume ~0.8 L) with a freshly prepared and sterile Se-containing APW solution, while bypassing the downhole equipment (volume ~1.5 L) and thus the interval itself. For this, APW (composition given in Table 1) with 15 $\mu$M $Na_2SeO_4$ was prepared under anoxic conditions. The solution was filter sterilized (0.22 $\mu$m) and stored in a sterile vessel until injection. The detailed injection procedure is described in [34]. Care was taken to prevent $O_2$ ingress and microbial contamination during injection. After replacement of the solution in the surface equipment, the bypass was opened, and the solution was circulated at 40 mL min$^{-1}$ through the entire circuit to allow mixing with the remaining solutions in the downhole equipment. After an overnight homogenization period, the flow rate was decreased to 10 mL min$^{-1}$ for the remainder of the tests. The selenate concentration after overnight homogenization was considered the starting concentration, i.e., 5.3 $\mu$M Se(VI) and less than 0.013 $\mu$M Se(IV) for the first injection. At the time of injection, two clay loops were installed in the circuit.

The second injection of Interval 3 was performed 106 days after the first. For this, APW with 243 $\mu$M $Na_2SeO_4$ was prepared and filter sterilized (0.22 $\mu$m). Injection was done by reconnecting five sampling containers (volume 0.15 L per container) filled with this solution to the circuit of Interval 3. The solution was first circulated at 40 mL min$^{-1}$ for 4 h, to achieve a good homogenization of the added solution and a dilution with the remaining interval solution. Afterwards, a sample was taken to determine the start concentration of Se (i.e., ~60 $\mu$M Se(VI) and less than 0.013 $\mu$M Se(IV)), and the flow rate was decreased to 10 mL min$^{-1}$ for the rest of the time.

The pH and redox potential were monitored online throughout the tests. Samples were taken regularly by disconnecting sampling containers and storing them at 4 °C until further analyses (Sections 2.4 and 2.5). At the end of the second injection test, one of the installed clay loops was disconnected from the interval and scanning electron microscopy coupled with energy dispersive X-ray spectroscopy (SEM–EDX) analyses were performed (Section 2.6).

No remaining $NO_3^-$ and $NO_2^-$ from previous nitrate injection tests performed to assess the microbial nitrate reduction processes in Interval 3 were found before injection of selenate (*cf.* former nitrate injections made in the period 2012–2016; see [34,37]). Furthermore, no additional carbon source or electron donor was injected, to assess microbial Se reactivity in the borehole with natural electron donors and carbon sources.

### 2.4. Chemical Analyses

After each sampling of the batch tests, Se(VI) and Se(IV) speciation analyses were performed on one of the subsamples using LC–ICP–MS (Liquid Chromatography–Inductively Coupled Plasma–Mass Spectrometry) (method modified from [38]) after filtration to remove all particles. Liquid chromatography was performed using a Flexar LC system (Perkin Elmer, Akron, OH, USA) and was coupled to a quadrupole-based ICP–MS instrument (Nexion 300S, Perkin Elmer, Akron, OH, USA), equipped with a MicroFlow PFA-ST concentric nebulizer (Elemental Scientific Inc, Omaha, NE, USA). Note that this method implies that only dissolved Se(VI) and Se(IV) species are detected. Total Se was determined in another (not filtered) subsample by ICP-OES (inductively coupled plasma optical emission spectrometry) (according to NBN EN ISO 11885:2009) after acid digestion in aqua regia (according to NBN EN ISO 15587-1:2002). Both analyses were performed by VITO (Mol, Belgium).

The detection limits for these analyses were 1.7 $\mu$M total Se and 0.013 $\mu$M Se(VI) and Se(IV). Intermittent Se(IV) measurements were performed for the biotic batch tests by spectrophotometric analyses after reaction of selenite with 2,3-diamino-naphthalene according to [39]. The detection limit of these spectrophotometric analyses of Se(IV) was 1 $\mu$M.

The chemical composition of the sampled solutions from the BN borehole was determined as soon as possible after sampling. Total Se, Se(VI), and Se(IV) analyses were performed as described above for the abiotic batch tests. The closed sampling containers were stored at 4 °C until analysis to slow down microbial growth and reactivity. Subsamples of the solution were taken under an anoxic atmosphere (Ar atmosphere with $O_2$ < 0.001%) and analyzed at SCK CEN by ion chromatography (IC; Dionex™ Aquion™ with IonPac™ AS22 column, Thermo Fisher Scientific, Geel, Belgium) for $SO_4^{2-}$, $Cl^-$, and acetate. The detection limit for the acetate concentration was 8 $\mu$M. Cations (Na, K, Mg, Ca, and Sr) were detected by inductively coupled plasma optical emission spectroscopy (iCAP™ 7400 rad, Thermo Fisher Scientific, Geel, Belgium). Total organic and inorganic carbon (TOC and TIC, respectively) concentrations were determined using a TOC/TIC analyzer with UV persulfate digestion (IL500 TC analyzer, Hach Lange, Mechelen, Belgium).

Measurement uncertainties were calculated with 95% confidence.

### 2.5. Microbial Analyses

### 2.5.1. Microbial Cell Count

The microbial Biothema ATP Kit HS (Isogen Life Science, Utrecht, The Netherlands) was used according to the manufacturer's procedure to monitor intracellular ATP (adenosine triphosphate), which provides information on the microbial activity level.

Flow cytometry was used to count the total amount of microbial cells. To this end, samples were diluted in filter sterilized (0.22 $\mu$m) Evian potable water and stained with SYBR Green I (10,000× concentrate in 0.22 $\mu$m filtered dimethyl sulfoxide) (Thermo Fisher Scientific, Geel, Belgium) at a final concentration of 1× concentrate. All samples were stained and incubated in the dark for 20 min at 37 °C. Flow cytometry was performed using a C6 Accuri TM flow cytometer (BD Biosciences, Erembodegem, Belgium) equipped with four fluorescence detectors (530/30 nm, 585/40 nm, >670 nm, and 675/25 nm), two scatter detectors, a 50 mW 488 nm laser, and a 30 mW 640 nm laser. The flow cytometer was operated with Milli-Q water (Merck Chemicals, Overijse, Belgium) as sheath fluid. Samples were analyzed in a fixed volume mode of 50 $\mu$L, and the minimal threshold was fixed on the green fluorescence (FL1-H at a relative intensity of 1000). Data were analyzed with the R programming software version 3.6.3 with the package Phenoflow [40].

At day 7 of the abiotic batch experiment, 100 $\mu$L of the samples was spread on LB agar medium (lysogeny broth medium; Thermo Fisher Scientific, Geel, Belgium) and incubated under oxic conditions at 30 °C for 3 weeks to verify abiotic conditions. At the end of the abiotic batch experiment, cultures were diluted 1/10 in APW supplemented with 0.2 mM $NaH_2PO_4 \cdot 2H_2O$ and 1 g Na-lactate to enable growth of sulfate reducing microorganisms. The total cell number was monitored after 18 and 101 days with flow cytometry. Acetate production was measured after 109 days with the K-ACETAK kit (Megazyme, Bray, Ireland) according to the manufacturer's procedure. Lactate consumption was measured after 143 days with the K-LATE kit (Megazyme, Bray, Ireland) according to the manufacturer's procedure.

### 2.5.2. DNA-Based Analysis of the Microbial Community

DNA was extracted from sampled solutions from Interval 3 during the in situ tests. For this, the microbial community of the interval water was collected on a 0.2 $\mu$m Whatman

Nuclepore track-etched membrane (25 mm Ø; Sigma-Aldrich, Overijse, Belgium). Cell lysis was achieved by adding 850 μL of autoclaved lysis buffer containing 0.1 M Tris-HCl pH 8, 0.062 M potassium ethyl xanthogenate, 0.2 M nitrilotriacetic acid, 4% wt/v Triton QS-15, and 0.8 M ammonium acetate, followed by vigorous vortexing for 30 s and overnight incubation at 70 °C. Then, samples were vortexed for 30 s, stored on ice for 10 min, and centrifuged (12,000× $g$, 10 min). Afterwards, 750 μL of the supernatant was transferred to Eppendorf tubes containing 5 PRIME Phase Lock Gel (PLG; VWR International, Leuven, Belgium), and 750 μL of phenol:chloroform:isoamyl alcohol (25:24:1) was added. Samples were centrifuged (12,000× $g$, 10 min), and the upper phase was transferred to a new PLG tube. Then, 750 μL chloroform:isoamyl alcohol (24:1) was added, and phases were separated again by centrifugation (12,000 $g$, 10 min). The top phase was transferred to a new tube and supplemented with 75 μL LiCl (8 M) and 750 μL 2-propanol, followed by 30 min at room temperature and centrifugation (16,000× $g$, 30 min, 4 °C). The pellet was air-dried and suspended in 500 μL of Milli-Q water. This suspension was transferred to a 100 kDa Amicon Filter (Millipore, Burlington, Massachusetts, USA) and centrifuged for 10 min at 13,000× $g$. The samples were washed twice with 450 μL of Milli-Q water and centrifuged each time for 10 min at 13,000× $g$. Purified DNA was recovered by inverting the filter in a new Eppendorf tube and centrifugation for 2 min at 1000× $g$. The DNA concentration was determined with a Quantifluor dsDNA sample kit (Promega, Leiden, The Netherlands).

High-throughput amplicon sequencing of the V3–V4 hypervariable region of the 16S rRNA gene was performed with the Illumina® MiSeq platform according to the manufacturer's guidelines at BaseClear B.V (Leiden, The Netherlands). DNA sequencing data were processed using the OCToPUS pipeline (**O**ptimized **CAT**Ch, **M**othur, **IPED**, **UPARSE**, and **SPAd**es) [41], which consists of the following steps: quality filtering using the Hammer algorithm implemented in the SPAdes tool [42], merging reads using the make.contigs command from the open source software package Mothur (version 1.39.1), alignment and filtering of the merged reads following the standard operating procedure as described by the authors of the Mothur software (version 1.39.1), error correction using IPED [43], chimera identification using CATCh [44], and OTU (operational taxonomic unit) clustering using UPARSE (version 7.0) with a 97% cut-off [45]. Subsampling was performed based on the lowest number of reads obtained over the ten different samples, i.e., a coverage of 29,061 reads. Rarefaction curves indicated that this level of subsampling adequately represented the bacterial diversity in the samples (Figure S1, Supplementary Materials). The datasets generated and analyzed during the current study are available in the NCBI Sequence Read Archive (SRA) repository (PRJNA717984).

### 2.6. Microscopic Analyses

SEM–EDX analyses were performed on the stainless steel filter coupons and Opalinus Clay pieces added to the abiotic batch tests (Section 2.2) and obtained from the clay loop after removal from the in situ test (Section 2.3).

After finishing the batch tests, the filter coupons were removed from the suspensions and left to dry in an anaerobic glove box (Ar atmosphere with $O_2 < 0.001\%$). Larger pieces of clay were sampled entirely, while smaller pieces dispersed in the batch test solution (in case it was disintegrated into smaller particles) were transferred with a syringe onto a Whatman Nuclepore track-etched membrane (Sigma-Aldrich, Overijse, Belgium) placed in a Swinnex holder (EMD Millipore, Darmstadt, Germany). In addition, smaller pieces of clay and stainless steel filter coupons were cut from the clay and filter present in the clay loop and removed from the circuit of Interval 3. Care was taken not to disturb the surfaces that had been in contact with the interval solution. Both dismantling of the clay loop as well as the cutting of the samples were done in an anaerobic glove box (Ar atmosphere with $O_2 < 0.001\%$). All samples were stored in anoxic conditions until analysis.

All solid samples were analyzed with a Phenom ProX tabletop SEM equipped with an EDX element identification probe (Benelux Scientific, Nazareth, Belgium).

### 2.7. Modeling

Geochemical modeling was undertaken using the geochemical code PHREEQC [46] to aid the interpretation of the experimental data. The modeling considered various processes relevant to the batch and in situ experiments: Se speciation, anion diffusion, Se(VI) reduction kinetics, and Se(IV) sorption. Modeling was focused on the in situ experiment and considered each of these processes in turn to develop a reactive-transport model of increasing complexity. Where relevant, the effect of specific processes was tested against the batch experimental data as detailed below.

#### 2.7.1. Se Speciation

A simplified speciation model considering two species, Se(VI) and Se (IV), was used for the majority of the modeling to compare with the experimental data of these species. The two species were defined as separate components and master species in the PHREEQC input file. This approach enables modeling of reaction kinetics (Section 2.7.3) between the two main species that are confirmed to be present by the chemical analyses. Representation of selenite as a single species also enables the utilization of an existing Se(IV) sorption model (Section 2.7.4).

PHREEQC calculations were also undertaken with a full aqueous speciation model using the Thermochimie database (version 9) [47], which includes thermodynamic data for Se from Olin et al. [48] to examine the hydrolysis of Se(VI) and Se(IV) and the occurrence of Ca and Mg ion pair species under Opalinus Clay conditions. Pourbaix diagrams of the selenium system were constructed using PHREEPLOT [49], again using the Thermochimie database. These diagrams and PHREEQC calculations were performed to determine the solubility and stability of elemental Se(0), a known product of microbial selenate reduction.

#### 2.7.2. Diffusion Modeling

The diffusion of Se(VI) and Se(IV) was represented in the PHREEQC model of the in situ BN experiment by a radial diffusion model following the approach of Appelo and Wersin [50] and used to model other similar Mont Terri borehole diffusion experiments [51]. The circulating borehole fluid was represented by a single cell, which has associated with it a series of 20 'stagnant' cells representing concentric regions around the borehole. The first two stagnant cells represented the filter screen and a void space adjacent to the clay and were assumed to have a radial thickness of 2 mm and 3 mm, respectively. The first cell representing the Opalinus Clay had a radial thickness of 5 mm, and for each successive cell, the radial thickness increased by a factor of 1.3, giving a total radial distance of 3.16 m. Diffusion between the stagnant concentric cells was represented using a mixing function calculated for the above geometry and as described by Appelo et al. [50] and Tournassat et al. [51].

As only two selenium oxyanions ($SeO_4^{2-}$ and $SeO_3^{2-}$) were represented in the model, the multicomponent diffusion option of PHREEQC [46,50] was not utilized. Instead, the diffusion of the two anions was represented by a single pore diffusion coefficient ($D_p$, $m^2\ s^{-1}$) and porosity ($\varepsilon$, -), which were input parameters to the mixing function [50,51]. For simplicity and consistency with previous work [34], anion exclusion effects were represented through the pore diffusion coefficient rather than a lowering of the anion accessible porosity, and $\varepsilon$ was assumed to be equal to the total porosity of the Opalinus Clay (0.17) [36]. A similar simplification to consider the same porosity for anions, cations, and neutral species was used in modeling diffusion processes in other borehole experiments at the Mont Terri Rock Laboratory [49]. The total porosity value (0.17) was also considered in previous 1-dimensional modeling of $Br^-$ and deuterium-labelled water (HDO) non-reactive tracers in diffusion tests in Intervals 1 and 2 of the BN experiment [34]. In the present study, the previously reported $Br^-$ and HDO data [34] were refitted using the PHREEQC radial diffusion model by pore diffusion coefficients ($D_p$) of $3 \times 10^{-11}\ m^2\ s^{-1}$ and $1.8 \times 10^{-10}\ m^2\ s^{-1}$, respectively, assuming a total porosity of 0.17 for both $Br^-$ and HDO (Figure S2, Supplementary Materials). A recently performed $Br^-$ diffusion test in Interval 3 (data not shown) showed that anionic diffusion was similar for all three intervals, and the PHREEQC radial diffusion model developed for Interval 1 and 2 was also applicable to fit the data from Interval 3. The pore diffusion coefficient for $Br^-$ determined from the tracer tests undertaken in Intervals 1 and 2 was assumed to represent the diffusion of the Se(IV) and Se(VI) anions in the selenium injection tests in Interval 3.

### 2.7.3. Se(VI) Reduction Kinetics

The reduction of Se(VI) to Se(IV) was represented in PHREEQC, using the KINETICS keyword, and where one mole of the Se(VI) master species was replaced by one mole of Se(IV) master species. The PHREEQC RATES keyword was used to define a first order rate law. The first order rate constant was fitted to the Se(VI) concentration data measured in the biotic batch tests and in the circulating solution of the in situ experiment after taking account of the diffusion of Se(VI) from the borehole into the Opalinus Clay.

### 2.7.4. Se(IV) Sorption onto Opalinus Clay

Se(IV) sorption distribution coefficients ($R_d$) were calculated based on the data from the abiotic batch tests according to Equation (1):

$$R_d = \frac{C_0 - C}{C} \frac{V}{m} \tag{1}$$

where $C_0$ is the initial aqueous Se(IV) concentration (mol $L^{-1}$), C is the remaining aqueous Se(IV) concentration (mol $L^{-1}$), V is the solution volume (mL), and m is the mass of the clay (g).

Frasca et al. [18] compared the sorption of selenium oxyanions on three argillaceous rocks, including samples from a core of Opalinus Clay from a borehole at the Mont Terri Rock Laboratory. They found that while Se(VI) was non-sorbing, Se(IV) sorption behavior could be reproduced by 1- and 2-site Langmuir isotherms, with more accurate representation being obtained by the 2-site model:

$$C_s = S_{max_1} \frac{K_{L_1}C}{1 + K_{L_1}C} + S_{max_2} \frac{K_{L_2}C}{1 + K_{L_2}C} \tag{2}$$

where $C_s$ is the concentration of Se(IV) sorbed onto the solid phase (mol $kg^{-1}$), C is the remaining aqueous concentration of Se(IV) (mol $L^{-1}$), $K_L$ is the sorption potential (L $mol^{-1}$), and $S_{max}$ is the sorption capacity (mol $kg^{-1}$) of each site. Parameter values obtained by Frasca et al. [18] for Opalinus Clay are shown in Table 3. No distinction in the sorption mechanism is made (i.e., direct sorption of Se(IV) vs. reduction followed by surface precipitation).

**Table 3.** Langmuir isotherm parameters from Frasca et al. [18] used in the Se(IV) sorption model for Opalinus Clay. These parameter values are used in Equation (2) to represent Se(IV) sorption using a 2-site Langmuir isotherm. $K_L$ is the sorption potential and $S_{max}$ is the sorption capacity of each site.

| Sorption Site | $K_L$ (L $mol^{-1}$) | $S_{max}$ (mol $kg^{-1}$) |
|---|---|---|
| Site 1 | $2.1 \times 10^5$ | $1.7 \times 10^{-3}$ |
| Site 2 | $1.9 \times 10^7$ | $1.4 \times 10^{-5}$ |
| Site 1 fitted [†] | $7.5 \times 10^4$ | $7.5 \times 10^{-3}$ |

[†] 'Site 1 fitted' refers to parameters fitted to represent the enhanced sorption observed in the abiotic batch experiments.

In this study, the Langmuir isotherm was used in models to compare with experimental Se(IV) concentration data from both the batch and in situ experiments. For this, the Langmuir sorption isotherm (Equation (2)) was implemented in PHREEQC representing the two sites as surface master species and defining a Se(IV) surface species for each site. The $K_L$ and $S_{max}$ parameters for Opalinus Clay (Table 3) were used to represent Se(IV) sorption in the batch and in situ experiments. The sorption potential ($K_L$, L $mol^{-1}$) was input as the log equilibrium constant for the formation of the Se(IV) surface species. The sorption capacity ($S_{max}$, mol $kg^{-1}$) was input to PHREEQC as the number of moles of the surface master species (moles of sorption sites) per liter of water. $S_{max}$ was scaled to account for the different fluid/rock ratios of the in situ conditions of the Opalinus Clay (0.089 L $kg^{-1}$, assuming a total porosity of 0.17 and dry density of 2.3 kg $L^{-1}$) and the batch experiments (700 L $kg^{-1}$). After considering the data from the abiotic batch experiments, a second modified Langmuir isotherm was parameterized with changes to $K_L$ and $S_{max}$ for sorption site 1 ('Site 1 fitted' at the bottom of Table 3) and also used to model subsequent biotic and in situ experiments.

For the in situ experiment, sorption was taken into account in all cells representing the Opalinus Clay, although in practice significant sorption only occurred in the first cell representing the first 5 mm of clay. In addition, it was found that sorption was limited by the diffusion of Se(IV) into the model cells representing clay. To represent the direct sorption onto the borehole skin surface (borehole disturbed zone, BdZ) of the Opalinus Clay from water present in the annular space (without being limited by diffusion), it was assumed that a proportion of the sorption sites present in the first cell representing the first 5 mm of the clay could directly equilibrate with the borehole fluid, i.e., they were assigned to the cell representing a fluid filled void between the filter and the clay. In the graphical results presented in this paper, it was assumed that half of the sorption sites present in the first 5 mm of clay could equilibrate directly with the borehole fluid. The effect of other proportions of sites was also examined in variant model cases.

## 3. Results

### 3.1. Abiotic Batch Tests

#### 3.1.1. Experimental Results

Intracellular ATP measurements performed after 7, 49, and 70 days showed that all samples without a solid clay fraction, except Ab_Se(IV)-F_b, were always under the detection limit (0.016 ± 0.013 pmol ATP/mL of clay water), indicating that they remained abiotic (Figure 1). Although intracellular ATP concentrations in samples Ab_Se(VI)-C_a, Ab_Se(VI)-CF_b, Ab_Se(IV)-C_a, and Ab_Se(IV)-C_b were initially high compared to the other samples, these concentrations decreased significantly during the course of the experiments (Figure 1). This suggests that if microorganisms were still initially present, their activity can be neglected. Samples were plated on a rich growth medium (LB) under oxic conditions after 7 days to further confirm the absence of microorganisms. The growth medium has been shown in the past to be highly suitable to cultivate the most dominant species present in BN borehole water. No growth was observed in any of the samples. Furthermore, cultures were prepared at the end of the batch experiments to investigate if possible viable cells could be revived in more optimal growth conditions, i.e., in the presence of sufficient phosphate and lactate, the latter as the preferential carbon and electron source for sulfate-reducing bacteria. No increase in intracellular ATP was observed, lactate was not consumed, and acetate production was not observed. Altogether, these data indicate the absence of active microorganisms, confirming abiotic conditions in the currently discussed batch tests.

Figure 2 shows the evolution of Se(VI), Se(IV), and total Se as a function of time for all abiotic test conditions. No Se was detected in the BN borehole water used for these tests, nor in the blanks with a piece of Opalinus Clay and/or stainless steel filter (Blank_C and Blank_CF).

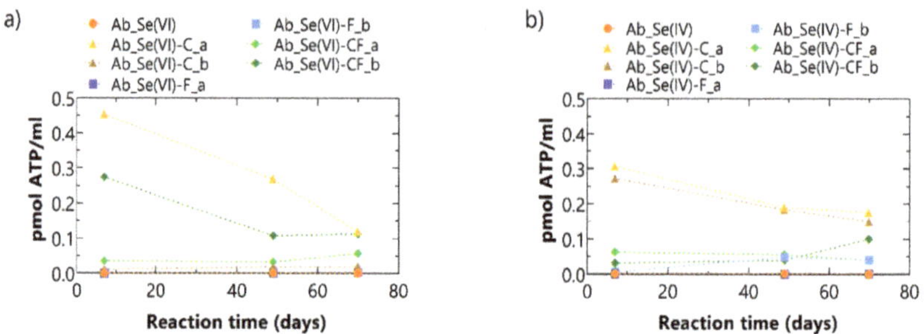

**Figure 1.** Intracellular ATP measurements for samples in the presence of (**a**) Se(VI) or (**b**) Se(IV) during the abiotic test. Details of the test conditions can be found in the legend and in Table 2.

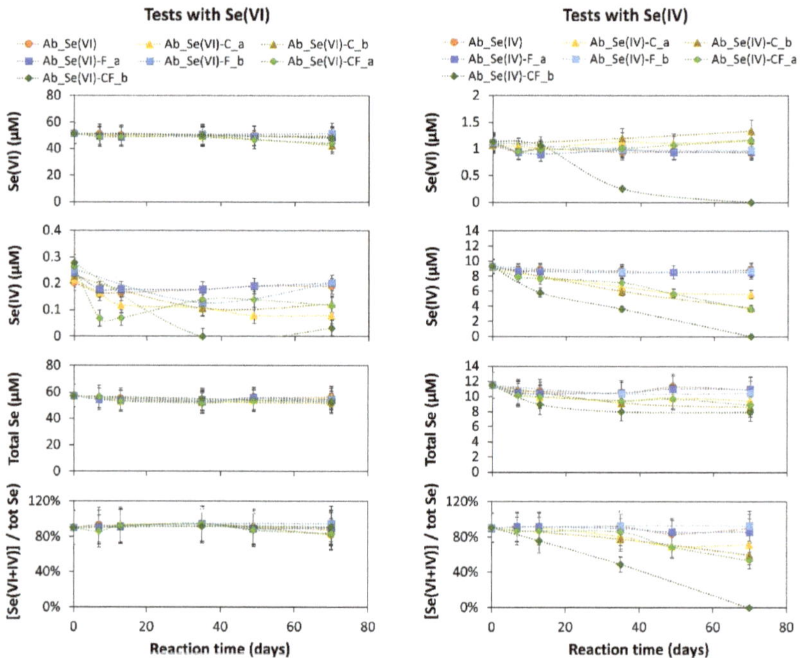

**Figure 2.** Abiotic test results for Se(VI) (left) and Se(IV) (right) reactivity in the presence of Opalinus Clay (C) and/or stainless steel filter (F). Details of the test conditions can be found in the legend and in Table 2. The error bars represent the measurement uncertainty (95% confidence).

Under abiotic conditions, Se(VI) remained stable in BN borehole water, and addition of Opalinus Clay and/or stainless steel did not have a significant impact on the Se(VI) concentration nor on the total Se concentration in solution. On the other hand, in the suspensions containing clay (Ab_Se(VI)-C and CF), the selenite concentration (present as an impurity in $Na_2SeO_4$) decreased to a significantly lower concentration compared to the initial background concentration observed in all other solutions (Figure 2). When comparing the dissolved Se species (i.e., Se(IV) and Se(VI)) to the concentration of total Se in the suspension, all Se species were accounted for. Note that a selenate impurity was also present in all tests with Se(IV). In one of these tests, a decrease in this Se(VI) background concentration could be observed (test Ab_Se(IV)-CF_b), although in all other test cases, Se(VI) remained stable as in the abiotic tests with Se(VI).

A decrease in the Se(IV) concentration could be observed in some of the batch experiments with added selenite, similar to that observed for the background selenite concentration in the test conditions with added selenate. In sterile BN borehole water without additives, the Se(IV) concentration did not decrease. When Opalinus Clay was added, a decrease in Se(IV) concentration occurred in all suspensions, with or without stainless steel, though more variation in the decrease rate was observed when stainless steel was included (Figure 2). The decrease rate ranged from $0.05 \pm 0.02$ to $0.1 \pm 0.01$ µM Se(IV) day$^{-1}$, with an average of 0.07 µM Se(IV) day$^{-1}$ for the solutions with only clay and 0.1 µM Se(IV) day$^{-1}$ for the solutions with both clay and stainless steel. In contrast, the Se(IV) concentration did not decrease in the presence of stainless steel alone (Figure 2). In all test conditions with clay, the total Se concentration decreased (2 to 4 µM), indicating that part of the Se was lost onto the solid phases. This was not the case when only stainless steel was present. Note that after introduction of the clay piece into the water, it slowly disintegrated into smaller particles. Consequently, the observed loss of total Se in the solutions with clay may be slightly underestimated as part of the smallest clay particles was sampled along with the solution, and no additional filtration was performed before total Se measurement. Comparison of the dissolved Se species (Se(VI) + Se(IV)) to the total Se concentration in the suspension shows that part of the total Se in the suspensions with clay after 70 days could not be linked to dissolved Se(VI) or Se(IV), indicating that other—more reduced—Se species were formed.

SEM–EDX analyses were performed on stainless steel filters and clay pieces after finishing the abiotic tests. The main results of sample Ab_Se(IV)-CF_a are shown in Figure 3. More results from this and other samples are shown in a supplementary dataset (Figure S3, Supplementary Materials). Based on these results, Se was present on the stainless steel filter surface in the abiotic tests with selenite only when clay was also present (Figure 3). Clear deposits of Se, more or less spherical in shape, could be observed, though the amount was limited. This is in agreement with the limited decrease in Se(IV) and total Se in the suspensions (Figure 2). EDX spot analysis shows that these spots were mostly containing Se. In these spots, oxygen was not found, while this was the case when performing EDX spot analysis on the stainless steel itself (Figure S3, Supplementary Materials). These results thus suggest that the Se deposits are not composed of precipitated or sorbed selenium oxyanions, but rather of more reduced selenium species such as elemental Se(0) or selenide, Se(-II). This is in agreement with the observed decrease in the ratio of dissolved Se species to the total Se concentration (Figure 2). When Opalinus Clay was not present, Se deposits were not found on the stainless steel filter surface (Figure S3, Supplementary Materials). SEM–EDX analysis on the clay itself did not reveal a significant amount of Se, in any of the test conditions. However, due to the disintegration of the clay pieces, the surface of the clay was quite large, and the limited amount of Se associated with the clay may have been missed by performing EDX-mapping of the clay surface.

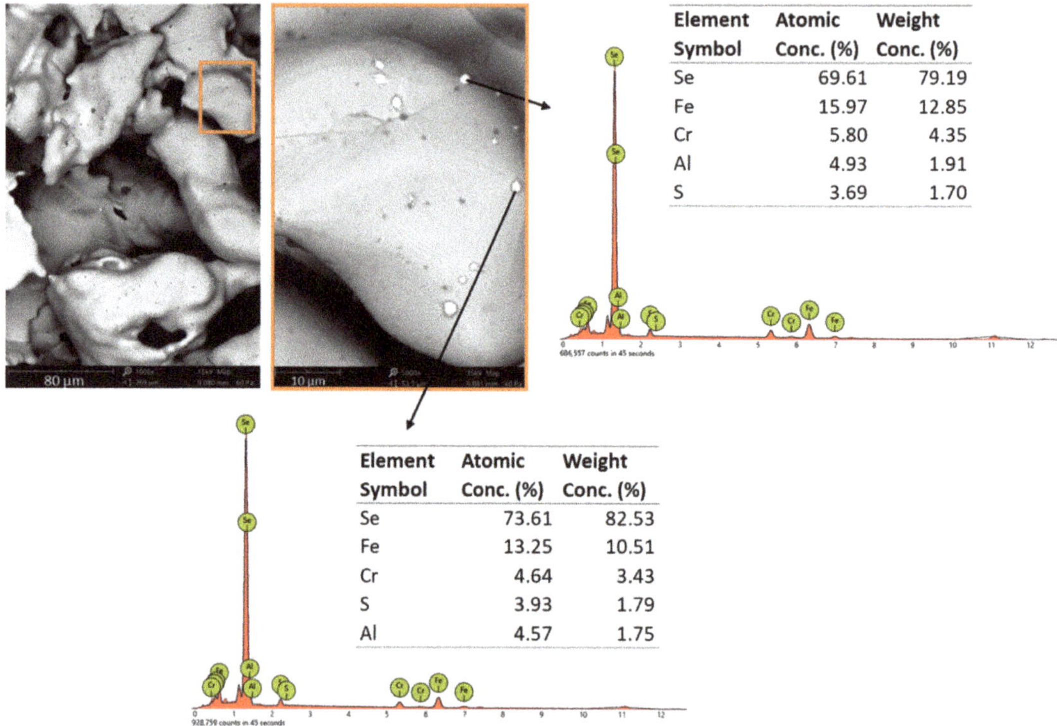

**Figure 3.** SEM–EDX images of the stainless steel filter surface of batch test Ab_Se(IV)-CF_a. EDX spot analyses were performed on distinct deposits. For EDX analysis, elements Bi, Br, Dy, I, Re, Sb, Sr, Te, Ti, V and W were disabled. The scale bar is shown in the bottom left corner of each SEM image.

### 3.1.2. Modeling Results

Modeling of the abiotic experiments was performed to examine the extent of sorption of Se(IV) onto Opalinus Clay using the 2-site Langmuir isotherm model parameterized by Frasca et al. [18]. Considering an initial Se(IV) concentration of 9.3 µM (Figure 2), the PHREEQC model of the abiotic experiments calculated that after sorption onto Opalinus Clay, the aqueous Se(IV) concentration would be 7.8 µM. This concentration was broadly consistent with the measured Se (IV) concentration in abiotic experiments with clay as the only additive (Ab_Se(IV)-C_a and Ab_Se(IV)-C_b, Figure 2), which ranged between 5.6 µM and 3.8 µM.

Figure 4 compares the sorption distribution coefficients (Rd, Equation (1)) calculated from the aqueous Se(IV) measurements from the batch tests with clay only (Ab_Se(IV)-C_a) and with clay and filter (Ab_Se(IV)-CF_a), considering a total initial concentration of 9.3 µM, with Rd measurements presented by Frasca et al. [18] for Se(IV) sorption onto Opalinus Clay (obtained from the Mont Terri Rock Laboratory). The data from Frasca et al. [18] were undertaken with an initial Se(IV) concentration of 100 µM (approximately 70 µM at equilibrium), and so the higher Rd values calculated for the present abiotic experiments were expected, given the effect of the sorption isotherm [18]. However, the Rd values calculated using the 2-site Langmuir isotherm model of Frasca et al. [18] for the range of aqueous Se(IV) concentrations (3.8 to 5.6 µM) measured in the present study were lower than (around half) the coefficients calculated from the concentration changes. This suggests that sorption of Se(IV) was enhanced in the BN batch experiments compared to what was observed by Frasca et al. [18]. The comparison also indicates that sorption processes attained a steady state more quickly in the batch experiments undertaken by Frasca et al. [18] than in our study. The slower sorption of Se(IV) apparent in our study may be attributed to the slow disintegration of the clay sample, while Frasca et al. utilized powdered Opalinus Clay material right from the start of the experiments.

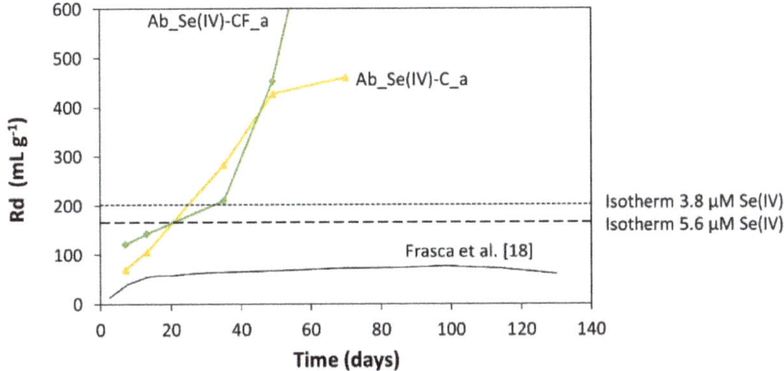


**Figure 4.** Se(IV) sorption coefficients (Rd) calculated with Equation (1) for the abiotic batch experiments Ab_Se(IV)-C_a (yellow) and Ab_Se(IV)-CF_a (green) (details on test conditions in Table 2) compared to Rd measurements from Frasca et al. [18] for Se(IV) sorption onto Opalinus Clay with initial concentration of 100 µM (black full line). The dashed lines represent the Rd calculated for the abiotic batch experiments from the 2-site Langmuir isotherm of [18] (Equations (1) and (2); Table 3) for equilibrium aqueous concentrations of 5.6 µM and 3.8 µM Se(IV), which encompass the range of concentrations measured in the batch experiments Ab_Se(IV)-C_a and Ab_Se(IV)-C_b after 70 days. Note, the Rd calculated from the Se(IV) concentration of Ab_Se(IV)-CF_a at 70 days is 1085 mL $g^{-1}$ and plots outside the scale of the figure.

Based on the sorption of Se(IV) deduced from our abiotic batch experiments, the Langmuir isotherm parameters of Frasca et al. [18] for Opalinus Clay were revised in order to use these parameters in the subsequent models of the biotic batch experiments (Section 3.2.2) and the in situ experiment (Section 3.3.3). Using sorption potential ($K_L$) and sorption capacity ($S_{max}$) values of $7.5 \times 10^4$ L mol$^{-1}$ and $7.5 \times 10^{-3}$ mol kg$^{-1}$, respectively, for the first site (see Table 3), the isotherm could be offset to represent higher Rd values than that of Frasca et al. [18], with an Rd of 460 mL g$^{-1}$ at an aqueous Se(IV) concentration of 5.6 µM representing the steady state measured concentration in test Ab_Se(IV)-C_a.

### 3.2. Biotic Batch Tests

### 3.2.1. Experimental Results

Intracellular ATP concentrations considerably increased in the first six days, and this increase was more pronounced in the presence of Se(IV) (Figure 5). Afterwards, concentrations decreased slightly or remained stable, which is typically observed in batch experiments with borehole clay water [52–54]. No intracellular ATP measurements were performed at the end of the batch experiment.

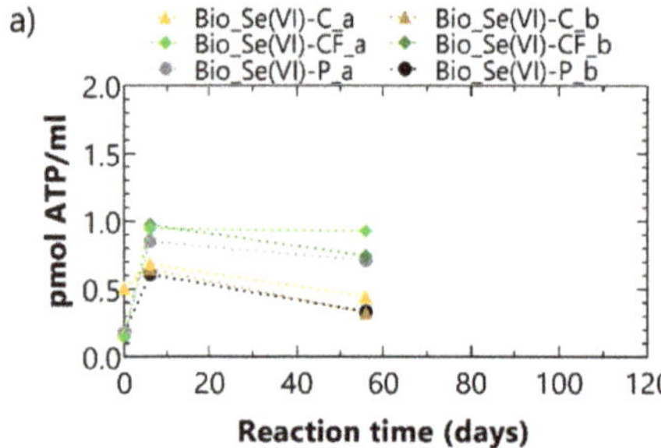
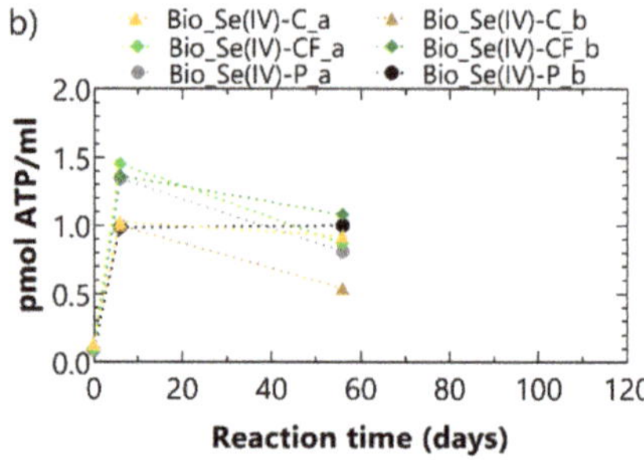

**Figure 5.** Intracellular ATP measurements for samples in the presence of (**a**) $Na_2SeO_4$ or (**b**) $Na_2SeO_3$ during the biotic test. Details on the test conditions can be found in the legend and in Table 2.

Under biotic conditions, a decrease in the concentration of Se(VI) was observed in the presence of clay with or without stainless steel filter, though at a slightly higher decrease rate for tests with clay and stainless steel compared to with clay alone (Figure 6). Overall, decrease rates of $0.2 \pm 0.07$ µM Se(VI) day$^{-1}$ and of $0.3 \pm 0.07$ µM Se(VI) day$^{-1}$ were found for tests with clay and with clay and stainless steel filter, respectively. At the end of the test, ~65% of the total decrease in Se(VI) could be linked to the production of Se(IV) in the solutions with clay alone (Bio_Se(VI)-C), while Se(IV) only made up ~6 and 47% of the Se(VI) concentration decrease in the samples with a stainless steel filter (Bio_Se(VI)-CF). In the absence of clay or stainless steel filter solid phases (Bio_Se(VI)-P), no significant decrease in Se(VI) concentration could be found. However, a small increase in selenite concentration was observed (by 0.4 to 2 µM), indicating that some selenate reduction to selenite had occurred, though at a limited rate (overall decrease rate of $0.06 \pm 0.09$ µM Se(VI) day$^{-1}$). Based on the decrease in total Se concentration (by 5 to 7 µM for Bio_Se(VI)-C and by 13 to 22 µM for Bio_Se(VI)-CF), part of the Se must be associated with the solid phase. In addition, here the observed loss of total Se may be slightly underestimated due to the partial disintegration of the clay possibly containing sorbed Se. As part of the smallest clay particles was sampled along with the solution at the end of the experiment and total Se was measured without additional filtration, the measured total Se concentration may have indeed slightly overestimated the total Se in solution (similar to the abiotic tests, see Section 3.1.1). The concentration of total Se did not decrease in the tests without solid phases (Bio_Se(VI)-P).

When comparing the concentration of dissolved Se species [Se(VI) + Se(IV)] to the concentration of total Se in the suspensions, only for tests with stainless steel filter and especially for Bio_Se(VI)-CF-b, part of the total Se could not be linked to Se(IV) or Se(VI). Indeed, the total concentration of Se(IV) and Se(VI) to the total Se concentration in suspension was 57% in sample Bio_Se(VI)-CF-b (Figure 6), indicating the formation of other (more reduced) Se species. The significant decrease in this ratio in test Bio_Se(VI)-CF-b coincided with the observed decrease in the concentration of produced selenite after reaching a maximum at day 35 (Figure 6).

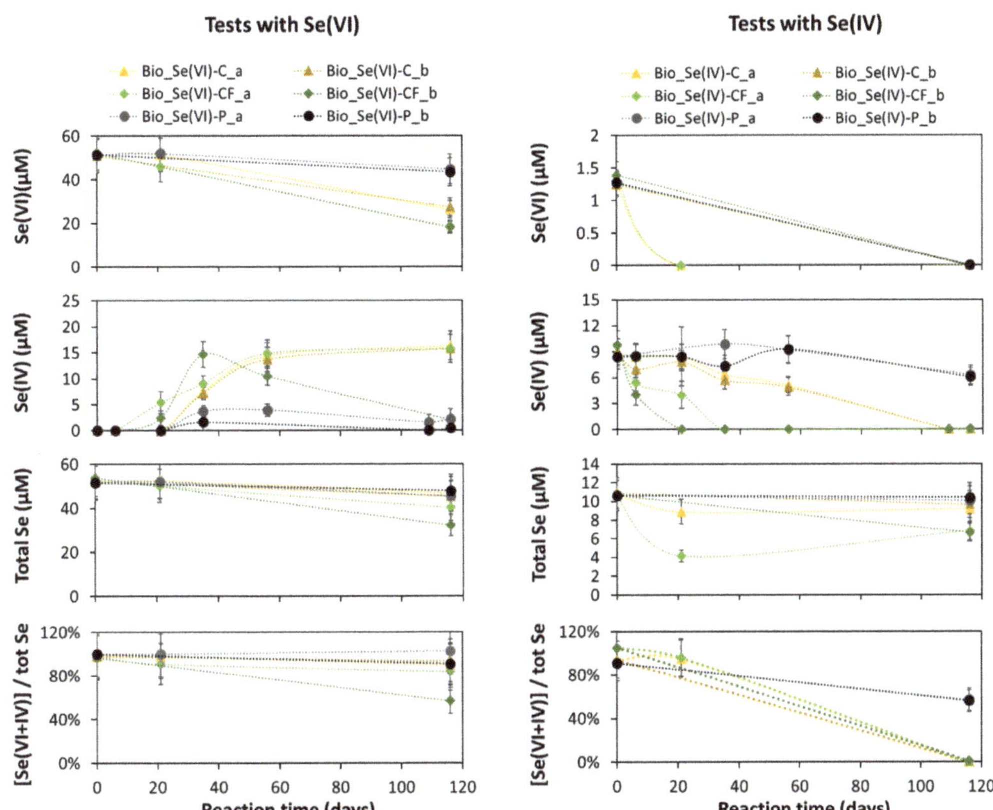

**Figure 6.** Biotic test results for Se(VI) (left) and Se(IV) (right) reactivity in the presence of Opalinus Clay (C) and/or stainless steel filter (F). Details on the test conditions can be found in the legend and in Table 2. The error bars represent the measurement uncertainty (95% confidence).

For the biotic batch tests with Se(IV), a fast decrease in the Se(IV) concentration could be observed for the tests with clay (Bio_Se(IV)-C) (Figure 6). This decrease was even enhanced when stainless steel was also present (Bio_Se(IV)-CF). Overall, a decrease rate of $0.1 \pm 0.02$ µM Se(IV) day$^{-1}$ and of 0.3 to $0.5 \pm 0.05$ µM Se(IV) day$^{-1}$ was observed for tests with clay and with clay and stainless steel filter, respectively, resulting in a complete removal of selenite from the solution. The total selenium concentrations decreased for both test conditions with clay, though for the tests with clay alone, this decrease did not exceed the analytical uncertainty, while a significant decrease was observed for the tests with clay and stainless steel. Note that the total Se concentration in Bio_Se(IV)-CF_a already decreased after 21 days and seemed to re-increase afterwards. This could be tentatively attributed to the fact that the clay sample was not yet disintegrated after 21 days (visual observation), while this was the case at the end of the experiment. Part of the Se species sorbed on the clay was thus detected in the suspension at the end of the test, which was likely not (or less) the case after 21 days. Furthermore, based on the ratio of the total concentration of Se(IV) and Se(VI) to the total Se concentration, the total Se content in tests Bio_Se(IV)-C and Bio_Se(IV)-CF measured at the end of the test was not accounted for by these two dissolved selenium species alone (Figure 3), indicating the formation of other (more reduced) Se species.

In the absence of clay or filter material (Bio_Se(IV)-P), a small decrease in Se(IV) concentration could be observed, corresponding to a Se(IV) decrease rate of $0.02 \pm 0.01$ µM

Se(IV) day$^{-1}$. On the other hand, no decrease in the total Se concentration could be observed in these solutions. Based on the ratio of the total concentration of Se(IV) and Se(VI) to the total Se, only ~57% of the total Se concentration in the batch tests without solid phases could be linked to dissolved Se species, again indicating that Se(IV) had been reduced.

### 3.2.2. Modeling Results

The biotic batch experiment Bio_Se(VI)-CF-a, which examined the reactivity of 50 µM Se(VI) with Opalinus Clay and stainless steel filter material, was selected for geochemical modeling, as it displayed clear trends in the concentration data and since its components were the most representative of the in situ tests described in Section 3.3.

Figure 7 compares the output of the PHREEQC models that include the reduction of Se(VI) to Se(IV) defined by a first order kinetic reaction with a rate constant of 0.01 day$^{-1}$. The decrease in the modeled Se(VI) concentration agreed well with the experimental data. As a result of the first order reduction process, the model predicted that the aqueous Se(IV) concentration would increase steadily, although part of the produced Se(IV) was modeled to sorb on the solid phases. In Model 1 (Figure 7), Se(IV) sorption was represented by the 2-site Langmuir isotherm of Frasca et al. [18]. Based on this, the aqueous Se(IV) would have increased to 35 µM after 120 days, with around 2 µM of the selenite sorbed onto the solid phases. Model 2 (dashed line in Figure 7) used the modified isotherm model (Table 3) to represent the increased sorption apparent in the abiotic batch tests (Section 3.1.2). This model provided a closer fit to the Se(IV) and total Se data, again indicating the differences in sorption of Se(IV) between the results from Frasca et al. [18] and our study.

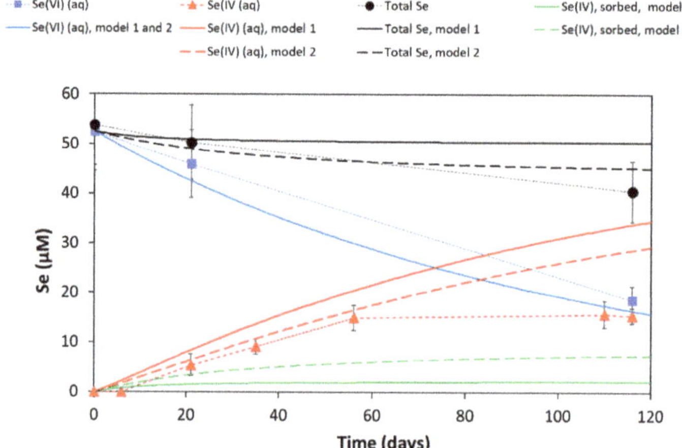


**Figure 7.** Measured aqueous Se(VI) (blue squares), Se(IV) (red triangles), and total aqueous Se (black circles) concentrations in batch experiment Bio_Se(VI)-CF-a (more details in Table 2), compared to two PHREEQC models representing first order reduction of Se(VI) and sorption of Se(IV). Model 1 (full lines) includes the 2-site Langmuir isotherm of Frasca et al. [18], while in model 2 (dashed lines) the isotherm has been adjusted to represent the Rd values for Se(IV) sorption obtained from the abiotic batch experiments. The two green curves represent the sorbed concentration of Se(IV) according to Equation (2).

During the first 60 days of the batch experiment, the two models provided a reasonable representation of the trend in measured Se(IV) concentrations, i.e., an increase to around 15 µM. However, after ~60 days, the measured Se(IV) stabilized at around 15 µM, whereas the modeled selenite concentration continued to increase, controlled by the continuing selenate reduction process and subsequent selenite sorption. The constant selenite concentration observed in the analytical data may be attributed to a solubility control by a discrete solid phase, perhaps in addition to the sorption process (see further Section 3.4). Comparison of Model 2 with the measured Se(IV) and total aqueous Se concentrations indicates that after 116 days, between 5 and 10 µM of the initially added Se would have sorbed.

### 3.3. In Situ Tests

### 3.3.1. Chemical Analyses Results

Figure 8 shows the evolution of the selenium species in the interval solution after injection with 5 µM (at t = 0) and with 60 µM (at t = 106 days) selenate. The selenite concentration is not shown, since it was below detection limit (0.013 µM) for all time points. Taking into account the analytical uncertainty, the total Se concentration was completely covered by the Se(VI) concentration, indicating that in the circulating interval solution, there was no significant production of other Se species.

During the first injection test, the pH remained rather stable at 7.6, while the selenate concentration decreased (Figure 8). A sharp pH decrease was however noticed after Se was most likely fully removed from the solution, indicating that another reaction occurred after selenium was removed from the solution. Initially, the Eh decreased only slowly, likely due to stabilization of the electrode (after being verified in air). This means that during the initial Se removal, the Eh of the interval solution remains unknown. Concomitantly with the sharp pH decrease, the Eh first decreased and then reached a stable value at −114 mV (Figure 8). This behavior again points towards an additional ongoing reaction in the solution after Se removal. Seventeen days after starting this test, both the pH and the Eh increased again and remained stable afterwards at ~7.3 and −60 mV, respectively. During the second injection test, both pH and Eh remained rather stable in time at a value of 7.3 and −31 mV, respectively.

The concentration of the main cations and anions present in the BN borehole water were stable (within the measurement uncertainty) during both injection tests (Figure 8). On the other hand, some variations in the carbon components could be observed. During the first in situ test, the TOC concentration remained more or less stable, although a slight decrease was observed in the first 13 days (when Se(VI) was removed completely). Between day 6 and 13, the acetate concentration increased from below 0.008 mM (detection limit) to 0.15 mM. Afterwards, the produced acetate was consumed, since it was no longer present after 27 days. The TIC concentration increased slightly during the initial TOC decrease, but increased more during acetate consumption, suggesting complete oxidation of acetate to $CO_2$. During the second in situ test, the TOC concentration was initially rather high (3.1 mM), especially compared to the values observed in the first test. The TOC concentration decreased significantly after 7 days to 0.2 mM C and remained constant afterwards. No changes in TIC concentration and no acetate was detected in the second in situ injection test.

SEM–EDX analyses were performed on the clay sample from the clay loop that was installed in the circuit of Interval 3 during both injection tests with Se(VI) (data not shown). However, due to the low amount of Se injected and the high surface area throughout the in situ experiment, Se was only found on the clay in only very limited amounts. This as well as the limitations of the SEM–EDX technique prevent a conclusive interpretation regarding the speciation of the sorbed or precipitated Se species.

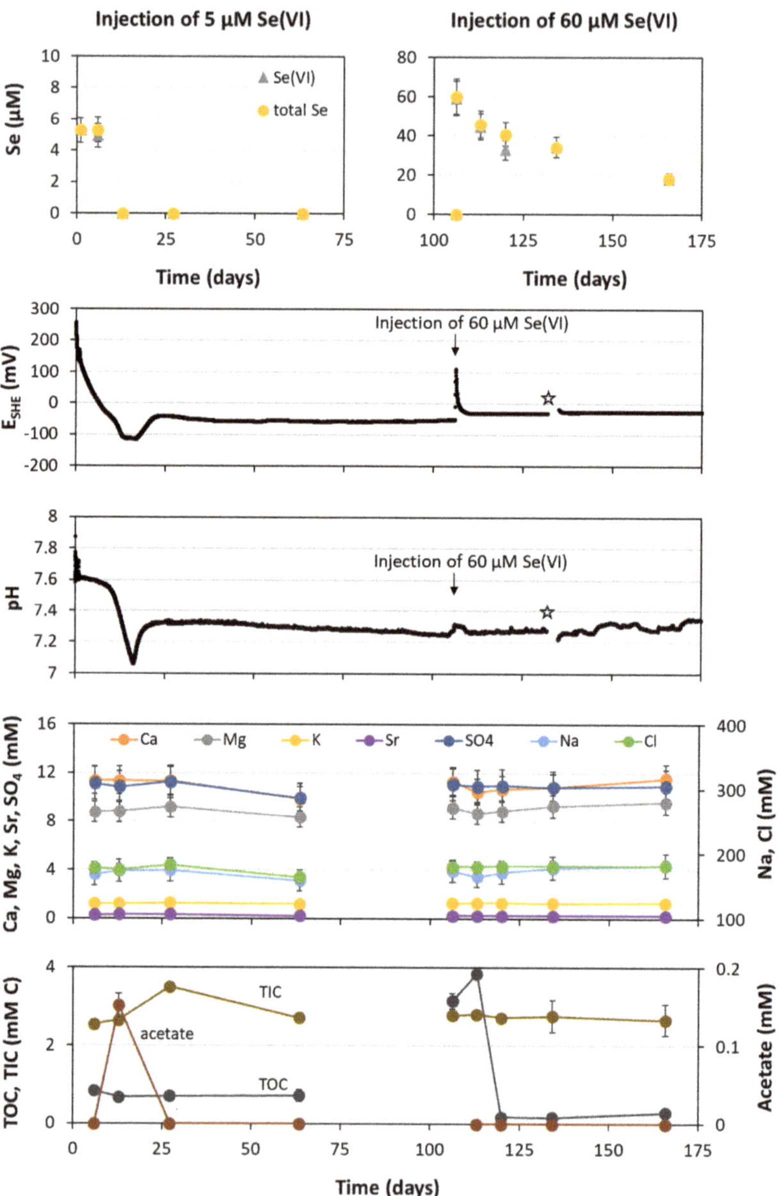



**Figure 8.** Evolution of the concentrations of Se(VI) (selenate) and total Se, the main components of the borehole pore water, organic and inorganic C, and pH and Eh as a function of time for both in situ injection tests, as shown at the top. The X-axis shows the entire experimental time frame, encompassing both injection tests with Se(VI). The experimental Se data were split in two graphs, one for each injection test, in order to visualize the ~12 times higher initial Se(VI) concentration of the second injection test. The error bars represent the measurement uncertainty (95% confidence). The combined uncertainty on the pH and Eh is 0.06 pH units and 32 mV, respectively (95% confidence interval), taking into account the measurement uncertainty observed immediately after calibration and the bias observed after ~1 year [37]. The star at ~132 days indicates an electrical power failure, resulting in a temporary lack of water circulation and online analysis.

3.3.2. Microbiological Analyses Results

Microbial presence and activity during the in situ test was monitored by flow cytometry (FC) and intracellular ATP measurements, respectively. Overall, the evolution was quite similar for both measurements (Figure 9). The first injection of selenate resulted in an increase in microbial activity and in the total number of cells followed by a decrease in both cell number (FC data) and microbial activity (ATP measurements) after 13 days (i.e., when Se was depleted). An increase in cell number and microbial activity was again observed after the injection with 60 µM selenate, although to a lesser extent. In addition, after a small decrease at day 120, microbial activity increased again until the end of the in situ test (Figures 7 and 9). The limited availability of an easily accessible carbon source can explain the observed higher increase in microbial activity compared to the total cell count.

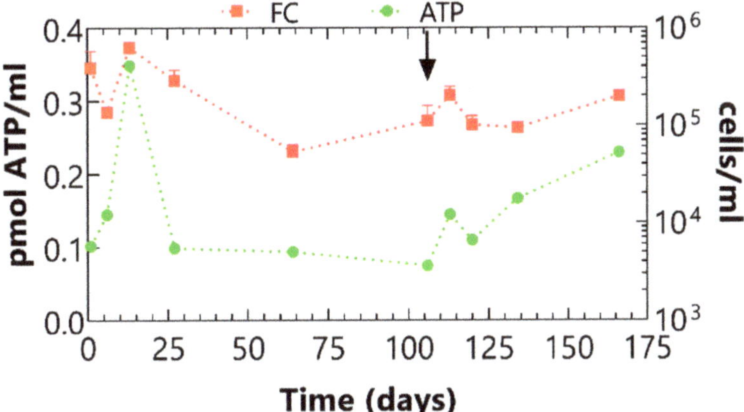



**Figure 9.** Intracellular ATP measurements (green) representative of microbial activity and total cell concentration based on flow cytometry (red) during the in situ test. ATP measurements are plotted on the left $Y$-axis, while flow cytometry results are plotted on the right $Y$-axis. The time of the second injection is indicated with an arrow.

Taxonomical characterization indicated the presence of three dominant phyla in the interval solutions: Proteobacteria, Firmicutes, and Chloroflexi (Figure 10). Relative abundance of the different phyla suggests that Proteobacteria decreased for 13 days after injection, after which a re-increase was observed. Firmicutes behaved in the opposite way. However, absolute abundances based on flow cytometry calculated according to Props et al. [55] indicate that after a decrease in the first 6 days after the first injection, Proteobacteria remained rather constant throughout the in situ test (Figure 10). On the other hand, on day 13 after injection with 5 µM selenate (i.e., when selenate was depleted), a clear growth of Firmicutes was observed, and their abundance decreased again gradually afterwards. No increase of this phylum was observed after injection with 60 µM selenate (Figure 10).

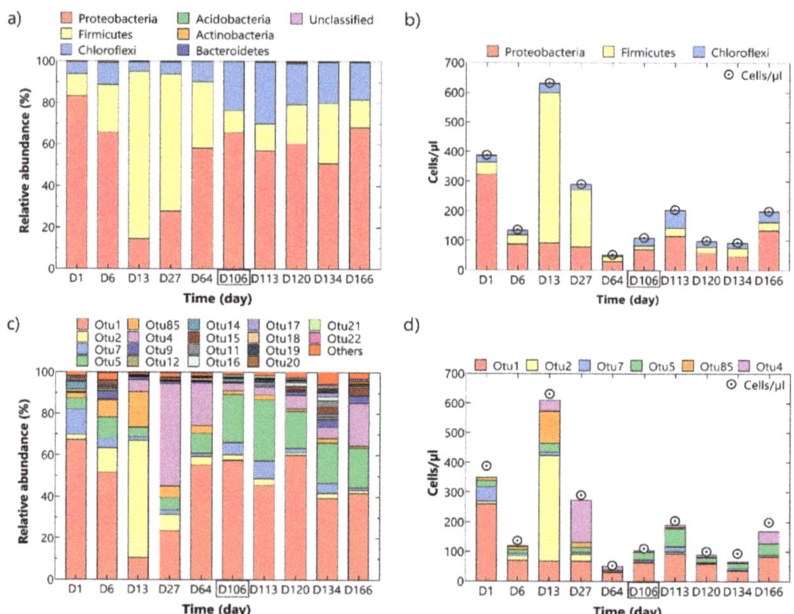

**Figure 10.** Relative (**a,c**) and absolute (**b,d**) abundance of the most dominant phyla (**a,b**), and OTUs (**c,d**) present in the borehole water after injection with 5 µM followed by an injection with 60 µM selenate. The time of the second injection is highlighted with a rectangle in the X axis (D106).

The abundance of the Proteobacteria could mainly be attributed to one OTU, i.e., OTU1, which represents a member of the genus *Pseudomonas* (Figure 10, Table S1 in Supplementary Materials). In addition, OTU2 and OTU85, both representing members of the sulfate reducing genus *Desulfosporosinus* (phylum Firmicutes), showed an actual growth between day 6 and day 13 after the injection (Figure 10, Table S1 in Supplementary Materials). Growth was also observed for OTU4, probably a member of the Peptococcaceae family (phylum Firmicutes). The abundance of OTU4 further increased up to 27 days after the injection (Figure 10, Table S1 in Supplementary Materials). Afterwards, samples were again mainly dominated by OTU1 (Figure 10, Table S1 in Supplementary Materials). Much less variation was observed after the injection with 60 µM selenate (Figure 10). The abundance of OTU4 slightly decreased from day 113 (i.e., day 7 after the second injection) up to day 134 and doubled again afterwards. The absolute abundance of OTU5, a member of the Anaerolineaceae family (phylum Chloroflexi, Table S1 in Supplementary Materials), was doubled in the sample taken on day 113, compared to the sample taken at the start of the second injection, i.e., day 106 (Figure 10). Finally, the abundance of OTU4, increased at the end of the injection test with 60 µM selenate (Figure 10).

### 3.3.3. Modeling Results

Figure 11 compares the measured concentrations of Se(VI) and total selenium in the second in situ injection test with the modeled concentrations considering diffusion as either Br$^-$ anion or deuterated water (HDO) based on previous diffusion tests undertaken in Interval 1 and Interval 2 of the BN experiment [34].

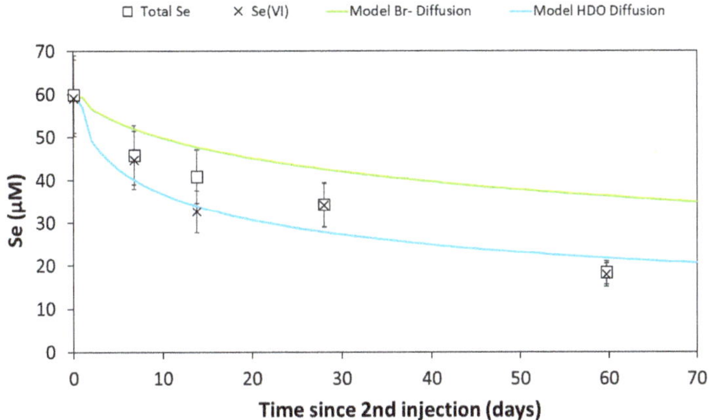



**Figure 11.** Measured concentrations (symbols) of Se(VI) and total Se in water samples after injection of the BN borehole with 60 µM Se(VI) compared to the modeled diffusion of Se(VI), using the pore diffusion coefficients of either $Br^-$ (green line) or deuterated water (HDO; blue line) to represent the diffusion of Se(VI) anions. Note that only the time after the second injection with selenate is shown.

The results show that the measured concentrations of Se(VI) decreased at a faster rate than the model representing $Br^-$ anion diffusion (Figure 11). Compared to the model for deuterated water diffusion, the higher measured Se(VI) concentration in the first few samples could be explained by pore exclusion affecting the Se(VI) anions. However, the sample collected after 60 days had a similar (but slightly lower) concentration than that modeled based on diffusion of deuterated water.

Figure 12 presents the results of the reactive transport model (in PHREEQC), where the reduction of Se(VI) to Se(IV) is represented as a first order kinetic reaction and where the diffusion of the Se(VI) and Se(IV) species is represented by the diffusion of $Br^-$. By varying the rate constant for the Se(VI) reduction reaction a satisfactory fit of the modeled Se(VI) concentration to the experimental data was obtained for a rate constant of 0.01 day$^{-1}$ (or $1.157 \times 10^{-7}$ s$^{-1}$), equivalent to a reaction half-life of around 60 days. This first order rate constant was the same as that used to model Se(VI) reduction kinetics in the biotic batch experiment (Section 3.2.2).

Se(IV) was not detected in the samples taken from the interval solution, although it was detected in the biotic batch experiments. To investigate Se(IV) behavior in the in situ experiment, the reactive transport model included Se(IV) sorption using the 2-site Langmuir isotherm model, in the first instance with parameter values obtained by Frasca et al. [18]. Figure 12 illustrates how sorption lowered the aqueous Se(IV) concentration to low concentrations, close to the Se(IV) detection limit (0.013 µM) (red curve). The modeled amount of sorbed Se(IV) (green curve) expressed as an equivalent concentration in the borehole fluid attained 19.5 µM after 60 days, whereas the modeled aqueous Se(IV) increased to a concentration of just 0.37 µM at 60 days. While this modeled aqueous Se(IV) concentration (red curve) was still higher than the level of detection, it should be noted that the 2-site Langmuir isotherm model with parameter values from Frasca et al. [18] underestimated the Rd of Se(IV) in the batch tests (Sections 3.1.2 and 3.2.2). A model of the in situ experiment parameterized with the modified 2-site Langmuir isotherm (Table 3) that represents the sorption of Se(IV) apparent in the abiotic batch experiments was found to lower the aqueous concentration to 0.23 µM at 60 days. Further uncertainty concerns the number of sorption sites present on the surface of the borehole at the interface with the borehole fluid. In the model presented in Figure 12, it was assumed that half of the sites present in the first 5 mm of clay rapidly attained equilibrium with the fluid and were not limited by diffusion. Variant models showed, however, the sensitivity of the results for a number of sorption sites. In a model where no sites were able to equilibrate directly with the borehole fluid, diffusion limits equilibration of the sorption and a high aqueous Se(IV) concentration of 10 µM was calculated to occur after 30 days, which slowly declined as the Se(IV) diffused into the clay. In a third model variant where 99% of the sites in the first cell representing the clay were considered to equilibrate, the aqueous Se(IV) concentration after 60 days was lowered to 0.12 µM.

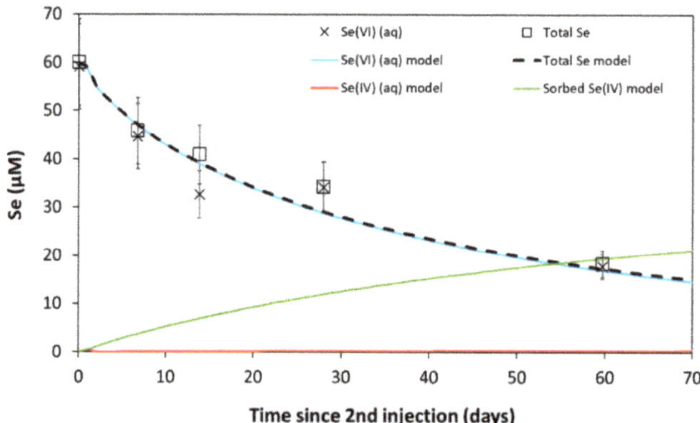


**Figure 12.** Measured concentrations of Se(VI) and total Se in water samples from the 60 µM Se(VI) in situ injection test (symbols) compared to the modeled concentrations in the circulating fluid in the BN borehole (lines). To model the Se(VI) concentration (blue line), diffusion as defined by $Br^-$ diffusion and the reduction reaction of Se(VI) to Se(IV) defined by a first order reaction with a rate constant 0.01 days$^{-1}$ is considered. Se(IV) is also subjected to diffusion as represented by $Br^-$ diffusion and to sorption onto Opalinus Clay as described by a 2-site Langmuir isotherm [18] (Equation (2)). Aqueous Se(IV) in the circulating borehole fluid is indicated by the red line, while the concentration of Se(IV) that is sorbed onto Opalinus Clay is indicated by the green line represented as an equivalent concentration in the borehole fluid. The total Se concentrations calculated by the model are indicated by the dashed black line. Note that only the time after the second injection with selenate is shown.

Overall, the extent of Se(IV) sorption in the in situ BN experiment was much higher than in the batch experiments. Several hypotheses can be envisaged to explain this difference, especially more reducing conditions (lower Eh value and also a larger redox buffer capacity) prevailing in the in situ BN experiment than in the laboratory tests, but perhaps also a higher accessibility of the sorption sites in the decompacted skin of the borehole disturbed zone compared to on the piece of clay used in the batch tests, and/or a lower fluid/rock ratio in the borehole.

### 3.4. Selenium Speciation Modeling

Additional speciation modeling results are presented here to aid the interpretation of the lab and in situ experiments. Figure 13 presents a Pourbaix diagram to illustrate the variation in the speciation of selenium under the conditions of the in situ Se injection experiment and more reducing Eh conditions measured in the Mont Terri pore water chemistry (PC) experiment [56]. The diagram shows the stability of Se(VI) ($SeO_4^{2-}$) under oxidizing conditions (Eh > +480 mV at pH 7) and the stability of Se(IV) ($HSeO_3^-$, $SeO_3^{2-}$) under less oxidizing conditions (Eh +480 mV to +230 mV at pH 7). Elemental Se(0) is stable under more reducing conditions, including the conditions of the BN in situ test and the PC experiment. For the upper (Eh) boundary of the $Se(0)_s$ field (at pH 7 and at Eh higher than around +100 mV), $Se(0)_s$ is in equilibrium with Se(IV) aqueous species ($HSeO_3^-$) and the steady-state Se(IV) concentration is expected to increase with increasing Eh (Figure S4, Supplementary Materials).

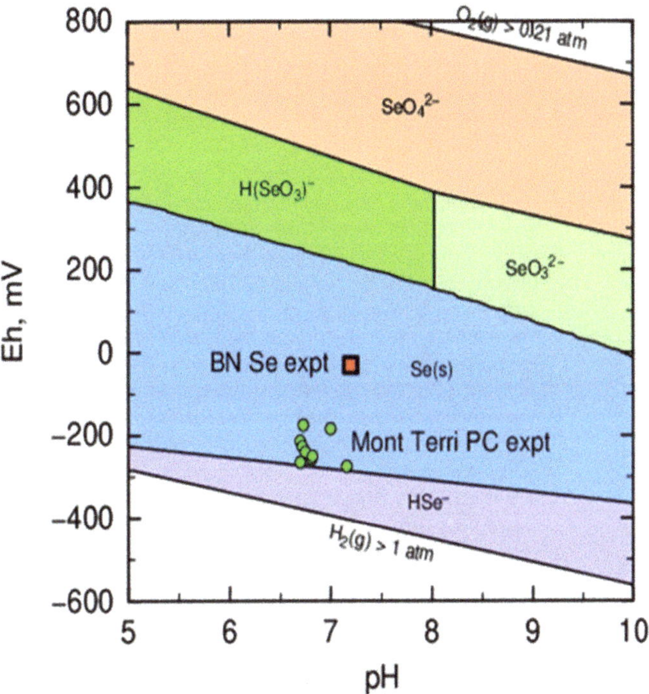


**Figure 13.** Pourbaix diagram showing the stability of Se aqueous species and more reduced Se species compared to the in situ steady state Eh and pH conditions of the BN Se injection test (red square) and measurements from the Mont Terri Pore water Chemistry (PC) experiment (green circles) [55]. Calculated for a total Se concentration of 50 µM using PHREEPLOT [49] with the Thermochimie database [47].

## 4. Discussion

### 4.1. Abiotic Se Reactivity

Under abiotic conditions, the Se(VI) concentrations in solution remained stable, indicating that selenate did not react significantly with, or sorb significantly onto, Opalinus Clay minerals or stainless steel filters in the abiotic batch tests. Although previous studies showed that Se(VI) can sorb weakly onto aluminum and iron oxides [13,25,27], and clay minerals such as kaolinite, illite, and smectites (montmorillonite) [8,25], selenate adsorption strongly decreases with increasing solution pH. Indeed, the non-specific electrostatic adsorption of Se(VI) onto Al oxides and Fe oxides decreases significantly at pH above 7 and 5 respectively, i.e., above their point of zero charge (pzc) when the number of positively charged protonated sites present at their surface strongly decreases [13,25]. Moreover, sorption on kaolinite and illite does not significantly occur, or is very low, at neutral pH, and the same can be seen for sorption of Se(VI) on soils [8,25]. The batch tests were performed at near neutral pH (i.e., pH of interval solutions between 7 and 8), which explains why a weak sorption of Se(VI) was not observed.

In contrast to Se(VI), a clear decrease in selenite concentration was observed in the presence of Opalinus Clay at a rate of 0.05 to 0.1 µM day$^{-1}$ in the abiotic batch tests. When clay and a stainless steel filter were present, Se precipitates were found on the filter surface. On the other hand, no loss of Se(IV) or precipitation of Se on the filter surface was observed when only stainless steel was present. These results indicate that clay is required for the sorption of selenite. Selenite is known to sorb, or to precipitate, on clays both through a direct adsorption on clay mineral edges (formation of inner-sphere complexes with aluminol and silanol groups located on the edges [22,23]) and through a mechanism of adsorption followed by reduction by, e.g., Fe(II) minerals and precipitation of poorly soluble reduced selenium species [16,17,20–22]. The latter process would explain the likely reduced Se species found on the filter surface by SEM–EDX in the abiotic tests with selenite, clay, and filter.

Compared to the study of Frasca et al. [18], in which sorption of Se(IV) on Opalinus Clay powder was investigated [18], the sorption of Se(IV) observed in our batch tests (in both abiotic and biotic conditions) was rather slow, though the overall extent of sorption was higher (i.e., lower steady state concentration). The slower overall sorption process can be attributed to the low initial reactive surface area and the slow disintegration of the clay piece in the solution. Differences in the extent of the sorption may be due to differences in the water to clay ratio, but also in the content of Fe(II) minerals in the Opalinus Clay samples used in the present study and by Frasca et al. [18]. Indeed, based on mineralogical and geochemical data from Pearson et al. [36], the content of Fe(II) minerals such as pyrite and siderite can vary considerably within the Opalinus Clay from the Mont Terri Rock Laboratory. As these minerals are generally considered as the main contributors to Se(IV) sorption, variations in their content would indeed lead to variations in sorption. Furthermore, speciation modeling undertaken in the present study indicates that where Se(IV) concentration is controlled by the formation of Se(0), the steady-state Se(IV) aqueous concentration is strongly sensitive to Eh (Figure S4, Supplementary Materials). In the previous study by Frasca et al. [18], Eh varied between 246 and 305 mV in the batch experiments to study sorption of Se(IV) onto Opalinus Clay. The enhanced sorption apparent in the present study could thus also reflect generally more reducing conditions in the abiotic batch tests, compared to those of Frasca et al. [18], thereby decreasing the concentration of Se(IV) in equilibrium with Se(0).

*4.2. Microbially Mediated Se Reactivity*

In the presence of an active microbial community originating from the BN borehole, both selenate and selenite were removed from the batch test solutions when Opalinus Clay (and stainless steel filter) were added as well. Note that concentrations used in our experiments were shown not to be toxic for the microbial community of the BN borehole water in previous tests (data not shown), so respiratory reduction of Se oxyanions can be assumed.

The observed decrease in selenate concentration can be completely explained by microbial selenate reduction, as no abiotic reactions or sorption were observed in the abiotic experiments. The present study thus shows that Se(VI) remains non-reactive in an abiotic, reducing environment in the presence of clay and requires microbial catalysis to transform into thermodynamically stable Se species of lower oxidation state, in the first instance to selenite. Part of the produced selenite will be sorbed onto the solid phases (clay and/or stainless steel filter), and the remainder accumulates in the solution until reaching its solubility limit and/or is reduced further, either abiotically or by microorganisms.

In the batch tests with selenite, Se(IV) concentration decreased 2 to 5 times more rapidly under biotic conditions compared to abiotic conditions, though the evolution of the total Se concentration was comparable. The latter can be explained by a slow disintegration of the clay piece and/or formation of colloidal Se(0). Subsampling of this suspension without filtration (or centrifugation) resulted in samples containing some of the dispersed solid phases. This caused some overestimation of the total Se content present in solution. Based on the evolution of the selenite concentration, however, microbial activity enhances the overall decrease in selenite concentration due to a combination of abiotic sorption phenomena and microbial selenite reduction. The final reduced Se species would likely either be Se(0), as has been observed previously in biotic batch tests [57–59], or Se(0) and/or Se(-II) from abiotic processes [16,17]. Indeed, speciation modeling indicates that Se(0) could precipitate at Eh below +230 mV at pH 7.

For both Se(IV) and Se(VI) under biotic test conditions, the presence of stainless steel combined with Opalinus Clay seems to enhance the loss of dissolved selenium species from the solution. This may be explained by the additional surface available to sorb Se reduction end products by the presence of other microbial species in the stainless steel filter samples compared to the clay samples, or by an effect on the Eh. Given the strong dependence of Se(0) solubility and thus the steady-state Se(IV) aqueous concentration on Eh (Figure S4, Supplementary Materials), it is indeed possible that significant variations in the Se(IV) concentration could result from variations in Eh between the batch experiments. These Eh variations could be caused by the presence or absence of stainless steel or clay.

Without Opalinus Clay, only small changes in the Se concentration occur under biotic test conditions, compared to the observed decrease in Se(VI) or Se(IV) concentrations in the presence of clay. This suggests that the Opalinus Clay provides additional electron donors for Se reduction. Based on the present knowledge regarding Se reducing microorganisms, organics (e.g., acetate, lactate) are the most frequently reported electron donors [28]. According to previous studies made on clay aqueous extracts [60,61], the low molecular weight organic fraction found in Opalinus Clay pore water contains variable concentrations of acetate, formate, and other organic compounds, which could have leached from the Opalinus Clay into the batch test solution and may indeed have served as electron donors. However, we cannot rule out that inorganic compounds such as $H_2$ (e.g., formed during anaerobic steel corrosion or as a fermentation product) were used as (additional) electron donors.

### 4.3. In Situ Fate of Se(VI)

Injection of selenate in the BN borehole will result in diffusion of selenate through the clay pore space, but this diffusion will be affected by pore exclusion phenomena related to the ionic radii and charge. Speciation calculations of the injected Se in the borehole water (not presented) indicate that around 75% of the Se(VI) would be present as the divalent species $SeO_4^{2-}$, with the remainder present as uncharged ion pair species $CaSeO_4$ and $MgSeO_4$. The diffusion of these three Se(VI) species will be dependent on their ionic radii and charge. Divalent $SeO_4^{2-}$, being subject to a greater anion exclusion from the electrical double layer of clay minerals than the monovalent $Br^-$ anion used as tracer, is expected to diffuse more slowly, analogous to the relative diffusion of sulfate and chloride measured in Opalinus Clay from the Benken borehole [62]. However, the aqueous Se(VI) present as uncharged ion pairs should not be subject to anion exclusion, although their diffusion relative to that of water may be slowed as a result of their large radii (i.e., steric hindrance in the smallest constricted pores). Diffusion studies of a wide range of anions and cations in the Mont Terri diffusion and retention (DR) experiment [32] conclude that diffusion of Se(VI) was slower than that of $Br^-$ or $I^-$, approximately proportional to the ratio of their diffusion coefficients in water [63]. Overall, the diffusion of Se(VI) is expected to be similar to or slower than that of the $Br^-$ tracer tests undertaken. In both injection tests of BN, however, the Se(VI) concentration decreased faster than that of the $Br^-$ tracer and was even similar to the diffusion of (deuterated) water. Based on the abiotic batch tests, no sorption of Se(VI) was expected either on the stainless steel components of the experiment or on the clay itself. On the other hand, the biotic batch tests showed the potential of the microbial community in Interval 3 to reduce selenium oxyanions. Microbial selenate reduction could thus explain the observed faster decrease in selenate concentration.

The PHREEQC model shows that the evolution of the selenate and total Se concentration can indeed be fitted well when considering both diffusion of Se(VI) (using the pore diffusion coefficient ($D_p$) obtained for $Br^-$) in combination with a reduction of Se(VI), using a reaction rate constant, which was the same for the in situ test and the biotic batch tests. This indicates that the biotic batch tests provided a good laboratory model of the ongoing in situ processes.

In contrast to the batch tests, total Se concentrations were consistent with the selenate concentrations, and selenite was never detected in the sampled interval solutions. This can be attributed to the sorption of produced Se(IV) onto the clay surrounding the borehole, similar to what was observed in the abiotic batch tests. The extent of Se(IV) sorption in the in situ BN experiment is much higher than in the batch experiments, probably as a consequence of more reducing conditions prevailing in the borehole. This would be in line with the observed preservation of selenite in the $O_2$ perturbed outer zone of a Callovo–Oxfordian Clay core during a diffusion experiment, while reduction of selenite was observed deeper in the clay core [19]. Additionally, a higher accessibility of the sorption sites in the decompacted skin of the BdZ (borehole disturbed zone) compared to on the piece of clay used in the batch tests and/or the lower in situ fluid/rock ratio may also contribute to the higher sorption observed in situ. Furthermore, based on the biotic batch tests, an additional microbial reduction of Se(IV) is expected. Speciation modeling indicates that Se(0) would be in equilibrium with aqueous Se(IV), and the steady-state Se(IV) concentration increases with increasing Eh. At the pH and Eh conditions of the in situ tests, Se(0) would be stable and the steady-state Se(IV) concentration would be several orders of magnitude below its detection limit of 0.013 µM. Thus, Se(IV) sorption, Se(0) precipitation, and microbial selenite reduction have likely all contributed to the disappearance of selenite from the borehole water.

The results of the BN injection tests with selenate show some differences to previously obtained results from the DR in situ experiment at the Mont Terri Rock Laboratory [32]. In the DR experiment, selenate was injected in a borehole at an initial concentration of 96 µM, which is comparable to the second injection of 60 µM Se(VI) in the BN experiment. On the other hand, the DR experiment had a much greater ratio of circulating water to borehole surface area than the BN experiment, and thus diffusion of the anionic tracers was rather limited. Over the course of 288 days, the slow decrease in the selenate concentration in the DR experiment was largely similar to non-reactive anions, although it appears that the Se concentration decreased slightly more in comparison to the $Br^-$ or $I^-$ tracers at the end of the experiment. Differences in behavior of selenate in the DR and BN in situ tests could be attributed (in part) to a difference in the microbial community present in the DR borehole compared to the BN borehole. In particular, the BN interval used for the Se studies has been stimulated previously by injection of nitrate.

Microbiological analyses indicate that the microbial community in the BN borehole was affected by injection of Se. The cell count and microbial activity both increased while Se was present and decreased again when Se was depleted. Overall, the microbial community is dominated by only a few OTUs, similar to what has been observed previously in boreholes in the Opalinus Clay [34,64].

Based on the taxonomical characterization of the most important OTUs and the literature available, a hypothesis regarding their contribution to the overall chemical evolution can be made. Certain dominant OTUs present during Se injection are members of the Anaerolineaceae family, which are often isolated from oil and hydrocarbon (e.g., toluene) contaminated environments [65–67]. However, their ecology and physiology remain largely unknown, as only a limited number of isolates have been characterized. Known members have a fermentative metabolism, utilizing carbohydrates and proteinaceous carbon sources under anaerobic conditions [68,69]. The second injection with 60 µM selenate seems to have resulted in a higher abundance of Anaerolineaceae compared to their presence after injection of only 5 µM selenate. Putatively, this may be linked with the higher TOC concentration at the beginning of the second injection with 60 µM selenate. Furthermore, the presence of *Pseudomonas* as one of the dominant OTUs in the BN borehole is not surprising, since it was one of the dominant species present during and after previous injection tests with nitrate in the BN experiment [34]. The abundance of *Pseudomonas* decreased in the first week after the first injection of 5 µM Se(VI), though it remained stable afterwards, at rather high amounts. Members of this genus have been found in microbial aggregates, able to reduce selenium oxyanions in anaerobic granular sludge [70]. In addition, *Pseudomonas* has been suggested to ferment organic macromolecules and to release small organic acids and $H_2$ [64]. Finally, three OTUs belonging to the Peptococcaceae family (OTU2, OTU4, and OTU85) were found abundantly during the Se injection tests as well. Members of the Peptococcaceae family have been reported with the ability to reduce Se [71,72], and as such they could also have performed Se reduction in the BN borehole. Overall, the presence of fermenting species during the in situ test with Se in the BN borehole suggests that such species could have fermented organic macromolecules in the borehole water or clay into more easily bioavailable carbon species such as acetate, as was found 13 days after the first injection with 5 µM Se(VI). Such carbon species could then be used as electron donors by other microorganisms, such as the sulfate and selenate reducing members of the Peptococcaceae family (as the phylum Firmicutes).

## 5. Conclusions

Geological disposal of HLW and ILW in deep clay formations will ultimately lead to the dissolution and slow leaching of soluble radionuclides such as $^{79}$Se, a redox sensitive radionuclide, which is expected to leach at least partially as Se oxyanions. Although part of the leached selenium could sorb on corrosion products of the steel canisters [7], the remainder will reach the surrounding clay host rock. The reducing environment in this clay host rock and the possible microbial presence might however change its speciation and thus also its transport behavior. A good understanding of the retention and reactivity of Se species under such environmental conditions and in the presence of microorganisms is therefore required and was the focus of this study.

Both lab and in situ tests were conducted, studying abiotic and biotic Se reactivity in clay water with or without Opalinus Clay and stainless steel. The pH and Eh conditions were as close as possible to unperturbed Opalinus Clay conditions (i.e., pH 7–8 and anoxic environment). Abiotically, Se(VI) remained stable in clay water with/without clay or stainless steel, i.e., no decrease in concentration of Se(VI) or total Se was observed. On the other hand, aqueous Se(IV) concentrations decreased in the batch tests with Opalinus Clay, though not with stainless steel alone. In the tests with selenite, precipitation of more reduced Se species on the solid phase was observed by SEM–EDX, though only when Opalinus Clay was present. The differences in the rate and extent of the sorption of Se(IV) observed in the present study compared to previous sorption experiments seem to be linked to the reactive surface area of the clay, to the mineralogical content, and to differences in Eh.

When a microbial community is present, with the ability to reduce Se species, both selenate and selenite were reduced, and a more extensive Se reduction and/or retention was observed in case clay and/or stainless steel filters were present (in situ or in batch tests). In the batch tests, microbial reduction of Se(VI) to Se(IV) was observed, resulting in some (intermediate) accumulation of Se(IV) in the solution. This was not observed in situ, where the decrease in Se(VI) was not accompanied by an increase in Se(IV) in the interval solution. Based on the observed selenite sorption onto the solid phases in the batch tests, the lack of Se(IV) in the in situ solution is most likely linked to more sorption of Se(IV) on the in situ clay surface and/or further reduction of Se(IV) to precipitating Se species such as elemental Se(0). Overall however, the biotic batch tests provided a good laboratory model of the ongoing in situ processes.

The present study thus suggests that in the absence of an active Se reducing microbial community, selenate leaching from radioactive waste into the surrounding clay would not be retained by sorption. In that case, non-reactive transport of Se(VI) through the clay is expected to be the main migration process. Although it remains uncertain whether microorganisms can be active in and around a repository (i.e., high pH, high ionic strength, and limited free space and water), microbial activity cannot be ruled out, especially in the excavation disturbed zone of the host rock surrounding the disposal facility. With a microbial community with Se reducing ability (locally) present, both diffusion and microbial Se(VI) reduction can be expected. Moreover, microbial reduction of Se oxyanions can occur in situ without addition of easily biodegradable nutrients and electron donors, i.e., the clay itself provides the electron donor for Se reduction. If microorganisms would be active in the clay surrounding a disposal facility, microbial reduction of Se oxyanions could thus contribute to the overall retention of Se, thereby slowing down the migration of Se out of the near field of the repository.

Future research is necessary to study the impact of high concentrations of nitrate leaching from bituminized waste on the fate of Se(VI) in a clayey host rock, both in the presence and absence of microorganisms. In addition, the speciation of Se sorbed onto the clay needs to be investigated, to further clarify the ongoing (microbial) reactions. The nature of electron donors present in Opalinus Clay and responsible for microbial Se reduction remains also to be studied along with the impact of organic electron donors from the waste and $H_2$ produced by anaerobic corrosion of metals and radiolysis.

**Supplementary Materials:** The following are available online at https://www.mdpi.com/article/10.3390/min11070757/s1, Figure S1: Rarefaction curves for the Illumina® 16S rDNA gene amplicon sequencing data showing for each sample the number of OTUs as a function of the number of reads. Figure S2: PHREEQC radial diffusion model fitted to bromide and deuterium diffusion tests in Interval 1 and Interval 2 of the BN experiment assuming a porosity of 0.17 for both tracer species (data from [34]). Figure S3: Overview of the SEM–EDX images and data of stainless steel and clay samples from the abiotic batch tests. The test codes are indicated on the top of each page and correspond to Table 2. Figure S4: Modeled solubility of Se(0)s in equilibrium with Se(IV) as a function of Eh, at pH 7. Table S1: Taxonomical classification of all OTUs with a relative abundance >1% in the water after the injections with selenate in Interval 3. Identity is 100% unless stated otherwise (between parentheses). OTU dataset: The datasets generated and analyzed during the current study are available in the NCBI Sequence Read Archive (SRA) repository (PRJNA717984).

**Author Contributions:** Conceptualization, N.B., K.H., A.A., P.D.C., C.W. and E.V.; Data curation, N.B., J.S.S., K.M. and K.H.; Formal analysis, N.B., K.M. and K.H.; Investigation, N.B., K.M. and K.H.; Methodology, N.B., K.M. and K.H.; Project administration, N.B.; Software, J.S.S.; Supervision, N.B., A.A., P.D.C., M.S., C.W. and E.V.; Validation, N.B., J.S.S., K.M. and K.H.; Visualization, N.B., J.S.S., K.M. and K.H.; Writing—original draft, N.B., J.S.S., K.M. and K.H.; Writing—review and editing, N.B., J.S.S., K.M., K.H., A.A., P.D.C., M.S., C.W. and E.V. All authors have read and agreed to the published version of the manuscript.

**Funding:** This research was performed within the framework of the BN experiment, which is funded by Andra, FANC, IRSN, and SCK CEN. No external funding was received from outside the Mont Terri Consortium.

**Data Availability Statement:** The chemical data used in this study are available from the authors upon request. Data from DNA-based analyses can be found online in the NCBI Sequence Read Archive (SRA) repository (PRJNA717984).

**Acknowledgments:** This work was undertaken in close cooperation with Swisstopo, the operator of the Rock Laboratory and the management team at the Mont Terri project. Financial support was provided by the Mont Terri Consortium. The technical assistance of Steven Smets, Veerle Van Gompel, and Wim Verwimp (SCK CEN) is greatly appreciated, as is the help in the conceptualization and interpretation of the work by Hugo Moors, Mohamed Mysara, Carla Smolders, and Natalie Leys (SCK CEN). Finally, we thank the team from the radiochemical analysis (RCA) unit at SCK CEN, and Kristof Tirez and Filip Beutels (VITO) for performing the chemical analyses.

**Conflicts of Interest:** The authors, members of the Mont Terri partner organizations funding the Consortium, or of a contracted organization, declare no conflict of interest.

## References

1. Jörg, G.; Bühnemann, R.; Hollas, S.; Kivel, N.; Kossert, K.; Van Winckel, S.; Gostomski, C.L. Preparation of radiochemically pure (79)Se and highly precise determination of its half-life. *Appl. Radiat. Isot.* **2010**, *68*, 2339–2351. [CrossRef] [PubMed]
2. Marivoet, J.; Weetjens, E. Impact of advanced fuel cycles on geological disposal. *Mater. Res. Soc. Symp. Proc.* **2009**, *1193*, 117. [CrossRef]
3. De Cannière, P.; Maes, A.; Williams, S.; Bruggeman, C.; Beauwens, T.; Maes, N.; Cowper, M. *Behaviour of Selenium in Boom Clay*; External Report SCK CEN-ER-120; SCK CEN: Mol, Belgium, 2010; p. 328.
4. Weetjens, E. *Safety Assessment Calculations for Geological Disposal of Eurobitum Bituminised Waste in Boom Clay*; Restricted Contract Report SCK CEN-R-5887; SCK CEN: Mol, Belgium, 2015; p. 83.

5. Andra. *Safety Options Report—Post-Closure Part*; Report nr CG-TE-D-NTE-AMOA-SR2-0000-15-0062; Andra: Paris, France, 2016; p. 466.

6. Bingham, P.A.; Connelly, A.J.; Cassingham, N.J.; Hyatt, N.C. Oxidation state and local environment of selenium in alkali borosilicate glasses for radioactive waste immobilisation. *J. Non-Cryst. Solids* **2011**, *357*, 2726–2734. [CrossRef]

7. Goberna-Ferrón, S.; Asta, M.P.; Zareeipolgardani, B.; Bureau, S.; Findling, N.; Simonelli, L.; Greneche, J.-M.; Charlet, L.; Fernández-Martínez, A. Influence of Silica Coatings on Magnetite-Catalyzed Selenium Reduction. *Environ. Sci. Technol.* **2021**, *55*, 3021–3031. [CrossRef] [PubMed]

8. Bar-Yosef, B.; Meek, D. Selenium sorption by kaolinite and montmorillonite. *Soil Sci.* **1987**, *144*, 11–19. [CrossRef]

9. Masscheleyn, P.H.; Delaune, R.D.; Patrick, W.H., Jr. Transformations of selenium as affected by sediment oxidation-reduction potential and pH. *Environ. Sci. Technol.* **1990**, *24*, 91–96. [CrossRef]

10. Elrashidi, M.A.; Adriano, D.C.; Workman, S.M.; Lindsay, W.L. Chemical equilibria of selenium in soils: A theoretical development. *Soil Sci.* **1987**, *144*, 141–152. [CrossRef]

11. Krivovichev, V.; Charykova, M.; Vishnevsky, A. The Thermodynamics of Selenium Minerals in Near-Surface Environments. *Minerals* **2017**, *7*, 188. [CrossRef]

12. Charykova, M.V.; Krivovichev, V.G. Mineral systems and the thermodynamics of selenites and selenates in the oxidation zone of sulfide ores—A review. *Mineral. Petrol.* **2017**, *111*, 121–134. [CrossRef]

13. Su, C.; Suarez, D.L. Selenate and selenite sorption on iron oxides an infrared and electrophoretic study. *Soil Sci. Soc. Am. J.* **2000**, *64*, 101–111. [CrossRef]

14. Missana, T.; Alonso, U.; García-Gutiérrez, M. Experimental study and modelling of selenite sorption onto illite and smectite clays. *J. Colloid Interface Sci.* **2009**, *334*, 132–138. [CrossRef]

15. Rovira, M.; Giménez, J.; Martínez, M.; Martínez-Lladó, X.; de Pablo, J.; Martí, V.; Duro, L. Sorption of selenium(IV) and selenium(VI) onto natural iron oxides: Goethite and hematite. *J. Hazard. Mater.* **2008**, *150*, 279–284. [CrossRef]

16. Breynaert, E.; Scheinost, A.C.; Dom, D.; Rossberg, A.; Vancluysen, J.; Gobechiya, E.; Kirschhock, C.E.A.; Maes, A. Reduction of Se(IV) in Boom Clay: XAS solid phase speciation. *Environ. Sci. Technol.* **2010**, *44*, 6649–6655. [CrossRef]

17. Hoving, A.L.; Münch, M.A.; Bruggeman, C.; Banerjee, D.; Behrends, T. Kinetics of selenite interactions with Boom Clay: Adsorption–reduction interplay. *Geol. Soc. Lond. Spec. Publ.* **2019**, *482*, 225–239. [CrossRef]

18. Frasca, B.; Savoye, S.; Wittebroodt, C.; Leupin, O.; Michelot, J.-L. Comparative study of Se oxyanions retention on three argillaceous rocks: Upper Toarcian (Tournemire, France), Black Shales (Tournemire, France) and Opalinus Clay (Mont Terri, Switzerland). *J. Environ. Radioact.* **2014**, *127*, 133–140. [CrossRef]

19. Savoye, S.; Schlegel, M.L.; Frasca, B. Mobility of selenium oxyanions in clay-rich media: A combined batch and diffusion experiments and synchrotron-based spectroscopic investigation. *Appl. Geochem.* **2021**, *128*, 104932. [CrossRef]

20. Scheinost, A.C.; Kirsch, R.; Banerjee, D.; Fernandez-Martinez, A.; Zaenker, H.; Funke, H.; Charlet, L. X-ray absorption and photoelectron spectroscopy investigation of selenite reduction by $Fe^{II}$-bearing minerals. *J. Contam. Hydrol.* **2008**, *102*, 228–245. [CrossRef]

21. Kang, M.; Chen, F.; Wu, S.; Yang, Y.; Bruggeman, C.; Charlet, L. Effect of pH on aqueous Se (IV) reduction by pyrite. *Environ. Sci. Technol.* **2011**, *45*, 2704–2710. [CrossRef] [PubMed]

22. Bruggeman, C.; Maes, A.; Vancluysen, J.; Vandemussele, P. Selenite reduction in Boom clay: Effect of FeS2, clay minerals and dissolved organic matter. *Environ. Pollut.* **2005**, *137*, 209–221. [CrossRef] [PubMed]

23. Peak, D.; Saha, U.K.; Huang, P.M. Selenite adsorption mechanisms on pure and coated montmorillonite: An EXAFS and XANES spectroscopic study. *Soil Sci. Soc. Am. J.* **2006**, *70*, 192–203. [CrossRef]

24. Beauwens, T.; De Cannière, P.; Moors, H.; Wang, L.; Maes, N. Studying the migration behaviour of selenate in Boom Clay by electromigration. *Eng. Geol.* **2005**, *77*, 285–293. [CrossRef]

25. Goldberg, S. Modeling selenate adsorption behavior on oxides, clay minerals, and soils using the triple layer model. *Soil Sci.* **2014**, *179*, 568–576. [CrossRef]

26. Balistrieri, L.S.; Chao, T.T. Selenium adsorption by goethite. *Soil Sci. Soc. Am. J.* **1987**, *51*, 1145–1151. [CrossRef]

27. Duc, M.; Lefèvre, G.; Fedoroff, M.; Jeanjean, J.; Rouchaud, J.C.; Monteil-Rivera, F.; Dumonceau, J.; Milonjic, S. Sorption of selenium anionic species on apatites and iron oxides from aqueous solutions. *J. Environ. Radioact.* **2003**, *70*, 61–72. [CrossRef]

28. Staicu, L.C.; Barton, L.L. Selenium respiration in anaerobic bacteria: Does energy generation pay off? *J. Inorg. Biochem.* **2021**, *222*, 111509. [CrossRef] [PubMed]

29. Stadtman, T.C. Biosynthesis and function of selenocysteine-containing enzymes. *J. Biol. Chem.* **1991**, *266*, 16257–16260. [CrossRef]

30. Heider, J.; Böck, A. Selenium metabolism in micro-organisms. *Adv. Microb. Physiol.* **1993**, *35*, 71–109. [CrossRef]

31. Eswayah, A.S.; Smith, T.J.; Gardiner, P.H.E. Microbial transformations of selenium species of relevance to bioremediation. *Appl. Environ. Microbiol.* **2016**, *82*, 4848–4859. [CrossRef] [PubMed]

32. Gimmi, T.; Leupin, O.X.; Eikenberg, J.; Glaus, M.A.; Van Loon, L.R.; Waber, H.N.; Wersin, P.; Wang, H.A.O.; Grolimund, D.; Borca, C.N.; et al. Anisotropic diffusion at the field scale in a 4-year multi-tracer diffusion and retention experiment—I: Insights from the experimental data. *Geochim. Cosmochim. Acta* **2014**, *125*, 373–393. [CrossRef]

33. Descostes, M.; Blin, V.; Bazer-Bachi, F.; Meier, P.; Grenut, B.; Radwan, J.; Schlegel, M.L.; Buschaert, S.; Coelho, D.; Tevissen, E. Diffusion of anionic species in Callovo-Oxfordian argillites and Oxfordian limestones (Meuse/Haute–Marne, France). *Appl. Geochem.* **2008**, *23*, 655–677. [CrossRef]

34. Bleyen, N.; Smets, S.; Small, J.; Moors, H.; Leys, N.; Albrecht, A.; De Cannière, P.; Schwyn, B.; Wittebroodt, C.; Valcke, E. Impact of the electron donor on in situ microbial nitrate reduction in Opalinus Clay: Results from the Mont Terri rock laboratory (Switzerland). *Swiss J. Geosci.* **2017**, *110*, 355–374. [CrossRef]

35. Bossart, P.; Bernier, F.; Birkholzer, J.; Bruggeman, C.; Connolli, P.; Dewonck, S.; Fukaya, M.; Herfort, M.; Jensen, M.; Matray, J.-M.; et al. Geologic setting and overview of key experiments in the Mont Terri rock laboratory. *Swiss J. Geosci.* **2017**, *110*, 3–22. [CrossRef]

36. Pearson, F.J.; Arcos, D.; Bath, A.; Boisson, J.Y.; Fernández, A.M.; Gabler, H.E.; Gaucher, E.; Gautschi, A.; Griffault, L.; Hernán, P.; et al. *Mont Terri Project—Geochemistry of Water in the Opalinus Clay Formation at the Mont Terri Rock Laboratory*; Geology Series No. 5; Federal Office for Water and Geology: Bern, Switzerland, 2003; p. 321.

37. Bleyen, N.; Albrecht, A.; De Cannière, P.; Wittebroodt, C.; Valcke, E. Non-destructive on-line and long-term monitoring of in situ nitrate and nitrite reactivity in a clay environment at increasing turbidity. *Appl. Geochem.* **2019**, *100*, 131–142. [CrossRef]

38. Martínez-Bravo, Y.; Roig-Navarro, A.F.; López, F.J.; Hernández, F. Multielemental determination of arsenic, selenium and chromium(VI) species in water by high-performance liquid chromatography–inductively coupled plasma mass spectrometry. *J. Chromatogr. A* **2001**, *926*, 265–274. [CrossRef]

39. Walkinson, J.H. Fluorometric determination of selenium in biological material with 2,3-diaminonaphthalene. *Anal. Chem.* **1966**, *38*, 92–97. [CrossRef] [PubMed]

40. Props, R.; Monsieurs, P.; Mysara, M.; Clement, L.; Boon, N. Measuring the biodiversity of microbial communities by flow cytometry. *Methods Ecol. Evol.* **2016**, *7*, 1376–1385. [CrossRef]

41. Mysara, M.; Njima, M.; Leys, N.; Raes, J.; Monsieurs, P. From reads to operational taxonomic units: An ensemble processing pipeline for MiSeq amplicon sequencing data. *Gigascience* **2017**, *6*, 1. [CrossRef]

42. Bankevich, A.; Nurk, S.; Antipov, D.; Gurevich, A.A.; Dvorkin, M.; Kulikov, A.S.; Lesin, V.M.; Nikolenko, S.I.; Pham, S.; Prjibelski, A.D.; et al. SPAdes: A new genome assembly algorithm and its applications to single-cell sequencing. *J. Comput. Biol.* **2012**, *19*, 455–477. [CrossRef]

43. Mysara, M.; Leys, N.; Raes, J.; Monsieurs, P. IPED: A highly efficient denoising tool for Illumina MiSeq Paired-end 16S rRNA gene amplicon sequencing data. *BMC Bioinform.* **2016**, *17*, 1–11. [CrossRef]

44. Mysara, M.; Saeys, Y.; Leys, N.; Raes, J.; Monsieurs, P. CATCh, an ensemble classifier for chimera detection in 16S rRNA sequencing studies. *Appl. Environ. Microbiol.* **2015**, *81*, 1573–1584. [CrossRef] [PubMed]

45. Edgar, R.C. UPARSE: Highly accurate OTU sequences from microbial amplicon reads. *Nat. Methods* **2013**, *10*, 996–998. [CrossRef]

46. Parkhurst, D.L.; Appelo, C.A.J. Description of input and examples for PHREEQC version 3—A computer program for speciation, batch-reaction, one-dimensional transport, and inverse geochemical calculations. In *U.S. Geological Survey Techniques and Methods*; U.S. Geological Survey: Reston, VA, USA, 2013; p. 497, Book 6, Chapter A43.

47. Giffaut, E.; Grive, M.; Blanc, P.; Vieillard, P.; Colas, E.; Gailhanou, H.; Gaboreau, S.; Marty, N.; Made, B.; Duro, L. Andra thermodynamic database for performance assessment: ThermoChimie. *Appl. Geochem.* **2014**, *49*, 225–236. [CrossRef]

48. Olin, A.; Nolang, B.; Osadchii, E.G.; Ohman, L.O.; Rosen, E. *Chemical Thermodynamics of Selenium*; NEA North Holland Elsevier Science Publishers B.V.: Amsterdam, The Netherlands, 2015; Volume 7, p. 893.

49. Kinniburgh, D.; Cooper, D. PhreePlot: Creating Graphical Output with PHREEQC. 2011. Available online: http://www.phreeplot.org (accessed on 24 June 2020).

50. Appelo, C.A.J.; Wersin, P. Multicomponent diffusion modeling in clay systems with application to the diffusion of tritium, iodide, and sodium in Opalinus Clay. *Environ. Sci. Technol.* **2007**, *41*, 5002–5007. [CrossRef]

51. Tournassat, C.; Alt-Epping, P.; Gaucher, E.C.; Gimmi, T.; Leupin, O.X.; Wersin, P. Biogeochemical processes in a clay formation in situ experiment: Part F–Reactive transport modelling. *Appl. Geochem.* **2011**, *26*, 1009–1022. [CrossRef]

52. Mijnendonckx, K.; Van Gompel, A.; Coninx, I.; Bleyen, N.; Leys, N. Water-soluble bitumen degradation products can fuel nitrate reduction from non-radioactive bituminized waste. *Appl. Geochem.* **2020**, *114*, 104525. [CrossRef]

53. Bleyen, N.; Hendrix, K.; Moors, H.; Durce, D.; Vasile, M.; Valcke, E. Biodegradability of dissolved organic matter in Boom Clay pore water under nitrate-reducing conditions: Effect of additional C and P sources. *Appl. Geochem.* **2018**, *92*, 45–58. [CrossRef]

54. Mariën, A.; Bleyen, N.; Aerts, S.; Valcke, E. The study of abiotic reduction of nitrate and nitrite in Boom Clay. *Phys. Chem. Earth* **2011**, *36*, 1639–1647. [CrossRef]

55. Props, R.; Kerckhof, F.-M.; Rubbens, P.; De Vrieze, J.; Hernandez Sanabria, E.; Waegeman, W.; Monsieurs, P.; Hammes, F.; Boon, N. Absolute quantification of microbial taxon abundances. *ISME J.* **2017**, *11*, 584–587. [CrossRef] [PubMed]

56. Wersin, P.; Leupin, O.; Mettler, S.; Gaucher, E.C.; Mäder, U.; De Cannière, P.; Vinsot, A.; Gäbler, H.; Kunimaro, T.; Kiho, K. Biogeochemical processes in a clay formation in situ experiment: Part A—Overview, experimental design and water data of an experiment in the Opalinus Clay at the Mont Terri Underground Research Laboratory, Switzerland. *Appl. Geochem.* **2011**, *26*, 931–953. [CrossRef]

57. Steinberg, N.A.; Oremland, R.S. Dissimilatory selenate reduction potentials in a diversity of sediment types. *Appl. Environ. Microbiol.* **1990**, *56*, 3550–3557. [CrossRef]

58. Oremland, R.S.; Herbel, M.J.; Blum, J.S.; Langley, S.; Beveridge, T.J.; Ajayan, P.M.; Sutto, T.; Ellis, A.V.; Curran, S. Structural and spectral features of selenium nanospheres produced by Se-respiring bacteria. *Appl. Environ. Microbiol.* **2004**, *70*, 52–60. [CrossRef] [PubMed]

59. Avendaño, R.; Chaves, N.; Fuentes, P.; Sánchez, E.; Jiménez, J.I.; Chavarría, M. Production of selenium nanoparticles in *Pseudomonas putida* KT2440. *Sci. Rep.* **2016**, *6*, 37155. [CrossRef]

60. Courdouan, A.; Christl, I.; Meylan, S.; Wersin, P.; Kretzschmar, R. Characterization of dissolved organic matter in anoxic rock extracts and in situ pore water of the Opalinus Clay. *Appl. Geochem.* **2007**, *22*, 2926–2939. [CrossRef]

61. Eichinger, F.; Lorenz, G.; Eichinger, L. *WS-H Experiment: Water Sampling from Borehole BWS-H2—Report on Physical-Chemical and Isotopic Analyses*; Mont Terri Technical Note TN 2010-49; Swisstopo: Wabern, Switzerland, 2011; p. 24.

62. Van Loon, L.R.; Leupin, O.X.; Cloet, V. The diffusion of $SO_4^{2-}$ in Opalinus Clay: Measurements of effective diffusion coefficients and evaluation of their importance in view of microbial mediated reactions in the near field of radioactive waste repositories. *Appl. Geochem.* **2018**, *95*, 19–24. [CrossRef]

63. Li, Y.-H.; Gregory, S. Diffusion of ions in sea water and in deep-sea sediments. *Geochim. Cosmochim. Acta* **1974**, *38*, 703–714. [CrossRef]

64. Bagnoud, A.; de Bruijn, I.; Andersson, A.F.; Diomidis, N.; Leupin, O.X.; Schwyn, B.; Bernier-Latmani, R. A minimalistic microbial food web in an excavated deep subsurface clay rock. *FEMS Microbiol. Ecol.* **2016**, *92*. [CrossRef]

65. Ficker, M.; Krastel, K.; Orlicky, S.; Edwards, E. Molecular characterization of a toluene-degrading methanogenic consortium. *Appl. Environ. Microbiol.* **1999**, *65*, 5576–5585. [CrossRef] [PubMed]

66. Liang, B.; Wang, L.Y.; Mbadinga, S.M.; Liu, J.F.; Yang, S.Z.; Gu, J.D.; Mu, B.Z. *Anaerolineaceae* and *Methanosaeta* turned to be the dominant microorganisms in alkanes-dependent methanogenic culture after long-term of incubation. *AMB Express* **2015**, *5*, 117. [CrossRef] [PubMed]

67. McIlroy, S.J.; Kirkegaard, R.H.; Dueholm, M.S.; Fernando, E.; Karst, S.M.; Albertsen, M.; Nielsen, P.H. Culture-independent analyses reveal novel *Anaerolineaceae* as abundant primary fermenters in anaerobic digesters treating waste activated sludge. *Front. Microbiol.* **2017**, *8*. [CrossRef] [PubMed]

68. Sun, L.; Toyonaga, M.; Ohashi, A.; Matsuura, N.; Tourlousse, D.M.; Meng, X.-Y.; Tamaki, H.; Hanada, S.; Cruz, R.; Yamaguchi, T.; et al. Isolation and characterization of *Flexilinea flocculi* gen. nov., sp. nov., a filamentous, anaerobic bacterium belonging to the class *Anaerolineae* in the phylum *Chloroflexi*. *Int. J. Syst. Evol.* **2016**, *66*, 988–996. [CrossRef]

69. Yamada, T.; Sekiguchi, Y. Cultivation of uncultured *Chloroflexi* subphyla: Significance and ecophysiology of formerly uncultured *Chloroflexi* 'subphylum I' with natural and biotechnological relevance. *Microbes Environ.* **2009**, *24*, 205–216. [CrossRef] [PubMed]

70. Gonzalez-Gil, G.; Lens, P.N.L.; Saikaly, P.E. Selenite reduction by anaerobic microbial aggregates: Microbial community structure, and proteins associated to the produced selenium spheres. *Front. Microbiol.* **2016**, *7*, 571. [CrossRef] [PubMed]

71. Stolz, J.F.; Basu, P.; Santini, J.M.; Oremland, R.S. Arsenic and selenium in microbial metabolism. *Annu. Rev. Microbiol.* **2006**, *60*, 107–130. [CrossRef] [PubMed]

72. Knotek-Smith, H.M.; Crawford, D.L.; Möller, G.; Henson, R.A. Microbial studies of a selenium-contaminated mine site and potential for on-site remediation. *J. Ind. Microbiol. Biotechnol.* **2006**, *33*, 897–913. [CrossRef] [PubMed]

 *minerals*

*Article*

# A MiniSandwich Experiment with Blended Ca-Bentonite and Pearson Water—Hydration, Swelling, Solute Transport and Cation Exchange

Katja Emmerich [1,*], Eleanor Bakker [1], Franz Königer [2], Christopher Rölke [3], Till Popp [3], Sarah Häußer [4], Ralf Diedel [4] and Rainer Schuhmann [1,2]

1    Karlsruher Institut für Technologie (KIT), Kompetenzzentrum für Materialfeuchte (IMB-CMM), 76344 Eggenstein-Leopoldshafen, Germany; eleanor.bakker@kit.edu (E.B.); rainer.schuhmann@kit.edu (R.S.)
2    Ingenieur-Gesellschaft für Sensorik in der Umwelttechnik mbH (ISU), 76185 Karlsruhe, Germany; franz.koeniger@isu-karlsruhe.com
3    Institut für Gebirgsmechanik (IfG), 04279 Leipzig, Germany; christopher.roelke@ifg-leipzig.de (C.R.); till.popp@ifg-leipzig.de (T.P.)
4    Stephan Schmidt KG, 65599 Dornburg-Langendernbach, Germany; Sarah.Haeusser@schmidt-tone.de (S.H.); ralf.diedel@schmidt-tone.de (R.D.)
*    Correspondence: katja.emmerich@kit.edu

**Citation:** Emmerich, K.; Bakker, E.; Königer, F.; Rölke, C.; Popp, T.; Häußer, S.; Diedel, R.; Schuhmann, R. A MiniSandwich Experiment with Blended Ca-Bentonite and Pearson Water—Hydration, Swelling, Solute Transport and Cation Exchange. *Minerals* **2021**, *11*, 1061. https://doi.org/10.3390/min11101061

Academic Editors: Ana María Fernández, Stephan Kaufhold, Markus Olin, Lian-Ge Zheng, Paul Wersin and James Wilson

Received: 29 July 2021
Accepted: 23 September 2021
Published: 28 September 2021

**Publisher's Note:** MDPI stays neutral with regard to jurisdictional claims in published maps and institutional affiliations.



**Abstract:** Shaft seals are geotechnical barriers in nuclear waste deposits and underground mines. The Sandwich sealing system consists of alternating sealing segments (DS) of bentonite and equipotential segments (ES). MiniSandwich experiments were performed with blended Ca-bentonite (90 mm diameter and 125 mm height) to study hydration, swelling, solute transport and cation exchange during hydration with A3 Pearson water, which resembles pore water of Opalinus Clay Formation at sandy facies. Two experiments were run in parallel with DS installed either in one-layer hydrate state (1W) or in air-dry two-layer hydrate (2W) state. Breakthrough at 0.3 MPa injection pressure occurred after 20 days and the fluid inlet was closed after 543 days, where 4289 mL and 2984 mL, respectively, passed both cells. Final hydraulic permeability was 2.0–2.7 $\times$ 10$^{-17}$ m$^2$. Cells were kept for another 142 days before dismantling. Swelling of DS resulted in slight compaction of ES. No changes in the mineralogy of the DS and ES material despite precipitated halite and sulfates occurred. Overall cation exchange capacity of the DS does not change, maintaining an overall value of 72 $\pm$ 2 cmol(+)/kg. Exchangeable Na$^+$ strongly increased while exchangeable Ca$^{2+}$ decreased. Exchangeable Mg$^{2+}$ and K$^+$ remained nearly constant. Sodium concentration in the outflow indicated two different exchange processes while the concentration of calcium and magnesium decreased potentially. Concentration of sulfate increased in the outflow, until it reached a constant value and chloride concentration decreased to a minimum before it slightly increased to a constant value. The available data set will be used to adapt numerical models for a mechanism-based description of the observed physical and geochemical processes.

**Keywords:** anion distribution; CEC; exchangeable cations; hydration; MiniSandwich; sandwich sealing system; solute transport; swelling pressure

## 1. Introduction

Shaft and drift-sealing systems for a nuclear waste repository and in deep underground mines limit the fluid inflow from the adjacent rock in the early stage after closure of the repository or a mine. In mines, partial active areas are protected from sudden fluid inflow, and in repositories and mines the release of possibly contaminated fluids at later stage are prevented, or at least delayed [1]. Current German concepts of shaft seals of nuclear waste repositories contain the hydraulic Sandwich sealing system as a component of the lower seal in the host rock [2].

The Sandwich sealing system consists of alternating sealing segments (DS) of bentonite and equipotential segments (ES), which are characterized by a higher hydraulic conductivity. Within ES, fluid is evenly distributed over the cross section of the seal. Water bypassing the seal via the excavation damaged zone (EDZ) or penetrating the seal inhomogeneously is contained and a more homogeneous hydration and swelling of the DS is obtained.

Proof of functionality of the system has been provided in semi-technical scale experiments [3,4] for different host rocks. In 2019, an in-situ large-scale experiment was launched at the Mont Terri Rock Laboratory (Mont Terri RL, Switzerland) in the sandy facies of the Opalinus clay to address the interaction between the sealing system and a potential host rock [5]. The sandy facies of the Opalinus clay at Mont Terri RL corresponds to the clay formation in the generic site model SOUTH [6] of the potential clay host rock in Germany.

Additional semi-technical scale experiments and extensive laboratory tests are carried out for material development and material parametrization to supplement the in-situ experiment by modeling. The current study covered MiniSandwich experiments [7] with blended Ca-bentonite to study hydration, swelling, solute transport and cation exchange during hydration with Pearson water.

The pore water of the Opalinus clay at Mont Terri Rock Laboratory (Mont Terri RL) (Pearson water) contains 150–250 mmol·L$^{-1}$ NaCl. The concentration of divalent cations and anions is lower than 15% of the cations and anions, respectively [8–10].

The current German concept for engineered barriers does not specify a bentonite but two national bentonites from Bavaria (Calcigel, Clariant) and Westerwald (Secursol UHP, Stephan Schmidt KG) would be available [5]. Both bentonites are Ca-bentonites. Cation exchange in the interlayer of the swellable clay minerals [11], together with anion exchange, dissolution, and precipitation of secondary phases will take place during the hydration of the sealing system with a pore fluid, mainly containing sodium cations until geochemical equilibrium is obtained. Thereby, alteration of the hydration of the bentonites as well as hydraulic conductivity and swelling pressure is expected.

Cation exchange and reactive solute transport during the early stage of the hydration of engineered barriers was observed at elevated temperatures from the Long-Term Test of Buffer Material (LOT) and the Alternative Buffer Material (ABM) test at Äspö Hard Rock Laboratory (Äspö HRL) (e.g., [12–15]), but pore fluid composition at Äspö HRL and Mont Terri RL differ in composition and sealing systems will not be exposed to elevated temperatures. The objective of the current study was to enhance the understanding of hydromechanical behavior and reactive solute transport during hydration of a Sandwich-sealing system composed of Ca-bentonite under the inflow of A3 Pearson water, which resembles Opalinus Clay pore water at sandy facies [8,9].

## 2. Materials and Methods

### 2.1. Materials and Experimental Setup

Fine sand N45 (Nivelsteiner Sandwerke und Sandsteinbrüche GmbH, Herzogenrath, Germany) [11] was used for ES. The bentonite Secursol UHP (Ruppach, Lower Saxony, Germany) by the Stephan Schmidt group (Langendernbach, Germany) with a smectite content of about 80% (Table 1) from a batch of granular material, also used in a semi-technical scale experiment (HTV-6), was used in a material blend for the DS. The specific density ($\rho_s$) of Secursol UHP of 2.77 g/cm$^3$ calculated from the phase content was confirmed by measurement after drying at 200 °C with a He pycnometer (Pycnomatic ATC Porotec, Hofheim, Germany).

**Table 1.** Phase content determined by XRD and specific grain density * of Secursol UHP.

| Phase | [wt.%] 2W [#] | [wt.%] Anhydrous | $\rho_s$ [g/cm$^3$] * Anhydrous |
|---|---|---|---|
| Smectite (di) | 79 | 75 | 2.75 |
| Quartz | 10 | 12 | 2.66 |
| Kaolinite | 2 | 2 | 2.6 |
| Mica (di) | 3 | 4 | 2.75 |
| Feldspar (orthoclase) | 2 | 2 | 2.61 |
| Anatase | 3 | 4 | 3.78 |
| Traces [§] | 1 | 1 | 3.16 |
| Total | 100 | 100 | 2.77 |

Note: * $\rho_s$ calculated based on crystal structure. # air dried at ambient conditions with a 2W hydration state and water content of about 15.4% (measured at 200 °C). § apatite, rutile, hematite, maghemite (all detectable by XRD) and carbonates (only detectable by STA).

The Westerwald bentonite is from the Miocene age and consists of highly altered tuffs from the late Oligocene age. Secursol UHP is characterized by a very high swelling pressure of >4 MPa at a dry density of about 1.55 g/cm$^3$, which even exceeds 10 MPa at higher dry densities [5]. Therefore, the bentonite was blended with N45 to reduce its smectite content, and thus its swelling pressure, to avoid damage of the oedometer cells.

Pearson water [16] close to A3 composition [8,17] which resembles the pore fluid of the sandy facies of Opalinus clay at Mont Terri URL [9] (Table 2), perfused the MiniSandwich columns. The ionic strength was 0.5 M. The Pearson water was prepared in two batches.

**Table 2.** Pearson water composition at sandy facies of Opalinus Clay [8], pH and liquid density ($\rho_l$).

| Title | Na$^+$ | Ca$^{2+}$ | Mg$^{2+}$ | K$^+$ | HCO$_3{}^-$ | Cl$^-$ | SO$_4{}^{2-}$ | Cl$^-$/SO$_4{}^{2-}$ | pH | $\rho_l$ |
|---|---|---|---|---|---|---|---|---|---|---|
| | | | [mmol/L] | | | | | [–] | [–] | [g/cm$^3$] |
| Target [1] | 164 | 11.9 | 9.2 | 2.55 | 0.54 | 160 | 24 | 6.67 | 7.9 | 1.01 |
| Batch 1 [2] | 153 | 11.6 | 8.7 | 2.5 | n.d. | 150 | 24 | 6.25 | 7.7 | 1.01 |
| Batch 2 | 230 | 15.9 | 11.37 | 4.06 | n.d. | 230 [3] | 30 [3] | 7.67 | n.d. | n.d. |
| | | | [mg/L] | | | | | [–] | | |
| Target [1] | 3770 | 477 | 224 | 100 | 33 | 5672 | 2305 | 2.46 | | |
| Batch 1 [2] | 3517 | 465 | 211 | 98 | n.d. | 5318 | 2305 | 2.31 | | |
| Batch 2 | 5288 | 637 | 276 | 159 | n.d. | 8154 | 2882 | 2.83 | | |

Note: [1] [9,16]. [2] [5]. [3] estimated from cation content. n.d. not determined.

The MiniSandwich specimens consisted of three ES sandwiched by two DS of blended bentonite (Figure 1b), separated by filter papers (manufacturer: Macherey-Nagel, material: cellulose, pore size: 7.0–12.0 µm). The diameter of the columns was 90 mm and a total sample height of 125 mm was envisaged. Actual sample heights of 130 and 131 mm, respectively, were obtained (Table 3). Sample heights decreased to 129.2 and 129.8 mm, respectively, by applying the axial load prior to the start of the hydration.

Two experiments were run in parallel with the bentonite in DS, installed either in air-dry state (cell 10) or after conditioning by drying over a concentrated H$_2$SO$_4$ solution atmosphere for several months (cell 9) to study different initial hydration states of the smectite (Table 4). The water content of the bentonite determined at (105 °C) 200 °C was (13.3%) 16.5% in air dry state and (0.9%) 3.9% after drying at very low relative humidity. Only the water content determined at 200 °C will be used for further calculations.

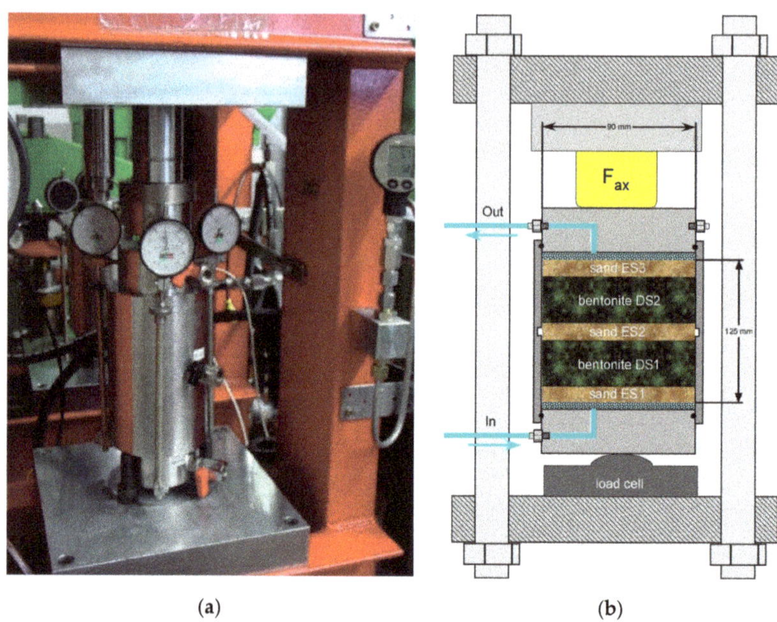



(**a**)  (**b**)

**Figure 1.** Experimental setup. (**a**) Hydraulic testing rig with oedometer cell. Note the three strain gauges, measuring the axial displacement over the sample lengths; (**b**) Schematic sketch of the oedometer cell.

**Table 3.** Installation bulk ($\rho_b$) and dry density ($\rho_d$), porosity (n), pore volume ($V_p$), volume of water in pores during installation ($V_{w,i}$ i) and saturation water content ($w_s$) of cells 9 and 10.

| Cell 9 | m | h | $\rho_b$ | $\rho_d$ | n | $V_p$ | $V_{w,i}$ | $w_s$ |
|---|---|---|---|---|---|---|---|---|
|  | [g] | [mm] | [g/cm³] | [g/cm³] | [%] | [mL] | [mL] | [%] |
| ES3 | 158.79 | 16 | 1.56 | 1.56 | 41.4 | 42.1 | 0 | 26.77 |
| DS2 | 357.2 | 42 | 1.34 | 1.30 | 52.8 | 141.1 | 10.6 | 41.33 |
| ES2 | 158.79 | 16 | 1.56 | 1.56 | 41.4 | 42.1 | 0 | 26.77 |
| DS1 | 357.3 | 41 | 1.37 | 1.33 | 51.7 | 134.7 | 10.6 | 39.25 |
| ES1 | 158.79 | 16 | 1.56 | 1.56 | 41.4 | 42.1 | 0 | 26.77 |
| Sum | 1190.87 | 131 |  |  | 48.3 | 402.2 | 21.2 |  |
| **Cell 10** | m | h | $\rho_b$ | $\rho_d$ | n | $V_p$ | $V_{w,i}$ | $w_s$ |
|  | [g] | [mm] | [g/cm³] | [g/cm³] | [%] | [mL] | [mL] | [%] |
| ES3 | 158.80 | 16 | 1.56 | 1.56 | 41.3 | 42.1 | 0 | 26.77 |
| DS2 | 370.31 | 41 | 1.42 | 1.26 | 54.3 | 141.5 | 42.3 | 43.58 |
| ES2 | 158.80 | 16 | 1.56 | 1.56 | 41.3 | 42.1 | 0 | 26.77 |
| DS1 | 391.42 | 41 | 1.50 | 1.33 | 51.7 | 134.7 | 44.7 | 39.25 |
| ES1 | 158.80 | 16 | 1.56 | 1.56 | 41.3 | 42.1 | 0 | 26.77 |
| Sum | 1238.13 | 130 |  |  | 48.7 | 401.7 | 87.0 |  |

**Table 4.** Water content of DS, hydration state of smectite and gas permeability during installation and final swelling pressure and fluid permeability of cell 9 and cell 10.

| Parameter | Unit | Cell 9 | Cell 10 |
|---|---|---|---|
| Water content, W (200 °C) | [%] | 3.1 | 12.9 |
| Hydration state of smectite | | 1W | 2W |
| Gas permeability | [m$^2$] | $8.7 \times 10^{-12}$ | $5.6 \times 10^{-12}$ |
| Swelling pressure | [MPa] | 0.7 | 0.6 |
| Fluid permeability | [m$^2$] | $2.0 \times 10^{-17}$ | $2.7 \times 10^{-17}$ |

One aliquot of bentonite (dry weight) was mixed with 0.28 aliquots of N45 (water content <0.1%) resulting in a starting water content in DS of 3.1% (cell 9) and 12.9% (cell 10). The specific density of the blend was 2.75 g/cm$^3$. The bentonite sand mixture could be compacted only to a dry density ($\rho_d$) between 1.26 and 1.33 g/cm$^3$ (Table 3), and a very low resulting EMDD (Dixon et al. 1985) between 0.95 and 1.02 g/cm$^3$. The N45 in ES was installed with a $\rho_d$ of 1.56 g/cm$^3$.

For the MiniSandwich experiments modified oedometer cells with two movable pistons and filter plates for adjusting well-defined saturation conditions [18] were located in hydraulic load frames (Figure 1a). The axial strain/displacement was measured by three gauges (rotated by 120°) and was regularly reset to zero by increasing or decreasing the axial load measured by a load cell. Thus, during the tests, axial expansion due to swelling of the sample has been reset recompacting the material to the initial volume to keep the MiniSandwich under quasi-constant volume conditions.

After installation, the initial gas permeability defined the starting conditions. It was determined with nitrogen by average of five measurements of the gas pressure difference to atmospheric pressure (0.12–0.18 MPa) at gas flow between 200 and 1000 mL/min (Bronkhorst F-231M flow controller). The injection pressure of Pearson water was 0.1 MPa and increased to 0.3 MPa after 4 days. After 543 days, the fluid inlet was closed and the experiments were run for another 142 days. During saturation, the development of the swelling pressure and the fluid volume balance (in- and outflow) were monitored and the stationary fluid permeability was calculated. After saturation and development of steady-state flow conditions, the outflow fluid was collected to determine its ion content for studying ion transport and cation exchange processes. Both columns were dismantled after 685 days and analyzed for water content and chemical/mineralogical changes.

### 2.2. Methods

### 2.2.1. Sampling

Immediately after unloading of the oedometer cells the upper pistons and filter plates were removed. In a stepwise dismantling procedure from top to bottom, each of the different sandwich segments was uncovered and probed. In both experiments, the upper filter plates were found to be nearly saturated with the fluid. Two samples were taken in each segment, one for determination of the water content (about 50–100 g each) and the other one for determination of the chemical and mineralogical composition/changes (about 50 g of each ES and about 250 g of each DS). Samples for determination of water content were immediately placed in the oven after initial weighing. Samples for chemical and mineralogical analyses were shrink-wrapped in foil until further analysis.

The fabric of the dismantled DS in both cells was still characterized by recognizable bentonite granules and fine sand grains. The following analyses were performed with homogenized material and separated bentonite granules were studied if additional information was expected.

### 2.2.2. Water Content

The water content and moisture of the materials were determined mostly in duplicate after heating at 105 °C and 200 °C for 24 h or until constant weight ($\Delta m < 0.1\%$) was obtained [19]. Heating to 200 °C is necessary to dehydrate swellable clay minerals [20] in bentonites. The water content was calculated with respect to dry mass of the sample after heating, while moisture was calculated in relation to the initial mass of the sample.

### 2.2.3. X-ray Diffraction Analysis (XRD)

Mineralogical quantification of Secursol UHP was performed by X-ray diffraction (XRD) measurements on powder samples < 500 μm. Samples were milled using the McCrone Micronizing mill with zirconium oxide cylindrical grinding elements (McCrone Microscopes and Accessories, Westmont, IL, USA). A Bruker D8 Advance A25 diffractometer (Bruker AXS GmbH, Karlsruhe, Germany) equipped with a LYNXEYE XE Detector (opening degree 2.94° and 192 channels) was used. Patterns were recorded between 2 and 80° 2θ with CuKα radiation, a counting time of 2 s and a step size of 0.02° 2θ, a fixed slit of 0.18°, Soller collimator of 2.5° (primary and secondary side) and an automatic knife edge for powder samples. Rietveld software PROFEX 4.3.2 (Nicola Döbelin, Solothurn, Switzerland) was utilized for quantitative analysis [21]. Quantitative phase content was converted in chemical composition and compared with chemical composition from X-ray fluorescence analysis (Supplementary Materials).

### 2.2.4. Cation Exchange Capacity (CEC), Exchangeable Cations (EC) and Cations in Pearson Water and Outflow Liquid

CEC was measured by the modified Cu-trien method according to Meier and Kahr [22]. Approximately 50 mg of ground sample were dispersed in a 15 mL centrifuge tube after adding 10 mL of Millipore water and 5 mL of a copper-triethylenetetramine (Cu-trien) solution of 0.01 mol/L by shaking for 3 h on a vibrating table. After centrifugation at 4500 rpm for 10 min (Multifuge 3S-R, Heraeus Holding GmbH, Hanau, Germany), the absorbance of the supernatants was measured at a wavelength ($\lambda$) of 580 nm with a UV–Vis spectrophotometer (Genesys 10 UV, Thermo Electron Corporation, Waltham, MA, USA) using polystyrene microcuvettes (Lab logistics Groups GmbH, Meckenheim, Germany) with a path length of 1 cm. The concentrations of Cu-trien in the supernatants were determined from a calibration curve and CEC [cmol(+)/kg] was calculated on the basis of the depletion of the supernatant by uptake of Cu-Trien by the smectite. The concentrations of exchangeable cations were analyzed from supernatant after dilution (if required) and acidification with $HNO_3$ (1M Suprapur) by inductively-coupled plasma–optical emission spectrometry (ICP-OES) (Optima 8300 DV, Perkin Elmer Inc., Waltham, MA, USA). Cations in Pearson water and outflow liquid were measured in the same way by ICP-OES.

### 2.2.5. Aqueous Leachate/Soluble Ions and Anions in Pearson Water and Outflow Liquid

Approximately 2 g of the sample were weighed into 50 mL centrifuge tube and dispersed in 40 mL of Millipore water by shaking for 24 h on a vibrating table. Thereafter, the samples were centrifuged at 4500 rpm for 10 min. Conductivity and temperature were measured in the supernatant using conductometer WTW LF 318 (Xylem Analytics Germany Sales GmbH and Co. KG, Weilheim in Oberbayern, Germany). The pH was measured with pH strips (Merck, pH 4.0–7.0 Special indicator).

The supernatant was filtered through a 0.45 μm filter (syringe filter, cellulose acetate, d = 25 mm, LLG-Labware GmbH, Meckenheim, Germany). Cations were analyzed by ICP-OES method and anions by ionic chromatography (Dionex Aquion IC System with Autosampler by Thermo Fisher Scientific GmbH, Dreieich, Germany). Anions in Pearson water and outflow liquid were measured in the same way by IC.

### 3. Results

#### 3.1. Geotechnical Parameters

Both cells were installed with identical dry densities and thus resulting porosity (Table 3). Due to the different hydration state and water content of the DS material 5.3% of the total porosity and 7.7% of the porosity of both DS in cell 9 were filled with water, while 21.7% of the total porosity and 31.5% of the porosity of both DS in cell 10 were already filled with water after installation.

Right at the start of the hydration/re-saturation 44.2 mL of liquid entered cell 9, and 45.4 mL liquid went into cell 10 within 5 min of starting the hydration. Thereby, ES1 in both cells were saturated. Fluid uptake and inflow permeability decreased strongly in both cells (Figures 2 and 3) until a fast breakthrough after 14 d was observed for both cells. The fluid breakthrough in cell 9 occurred after a fluid uptake (difference inflow/outflow fluid volume) of 359 mL, that corresponded to a saturation of 94.5%. The breakthrough in cell 10 was accompanied with a fluid uptake of 309 mL corresponding to a saturation of 98%. Actual fluid uptake was 4.5% and 2.5% higher than calculated (Tables 3 and 5). After the breakthrough 4289 mL and 2984 mL, respectively, Pearson water passed the outflow of cell 9 and 10 until termination of the experiment.

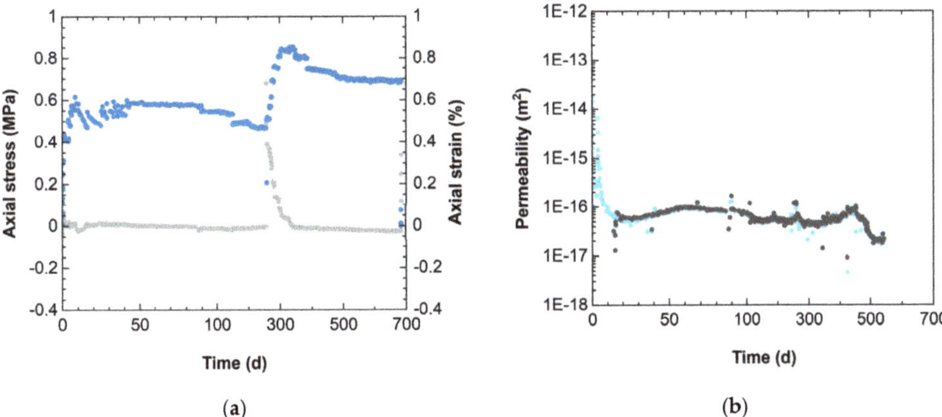

**Figure 2.** (**a**) Axial strain (grey) and axial stress (blue) and (**b**) inflow (blue) and outflow (grey) permeability cell 9.

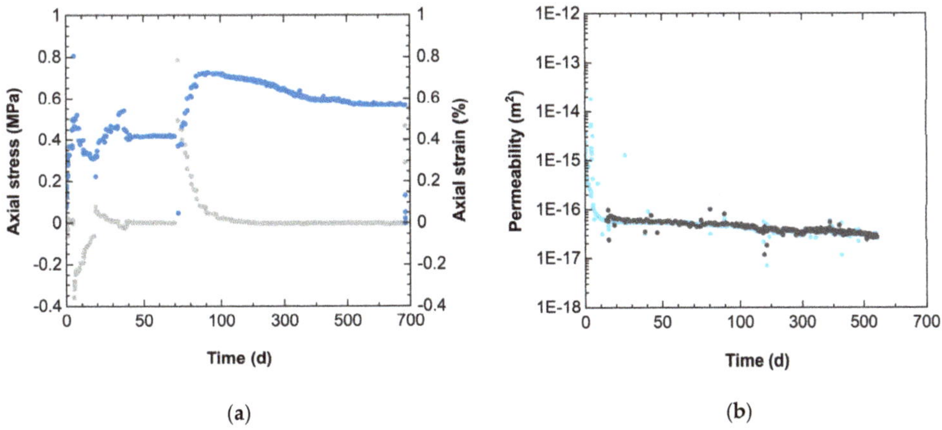

**Figure 3.** (**a**) Axial strain (grey) and axial stress (blue) and (**b**) inflow (blue) and outflow (grey) permeability cell 10.

**Table 5.** Water content (200 °C) ($w_e$), volume of water ($V_W$), dry density and volume (height) change after dismantling of cell 9 and 10.

| Cell 9 | $w_e$ | $V_w$ | $\rho_d$ | $\Delta V$ or $\Delta h$ |
|---|---|---|---|---|
| | [%] | [mL] | [g/cm$^3$] | [%] |
| ES3 | 22.81 | 36.2 | 1.66 | −6.5 |
| DS2 | 44.15 | 153.0 | 1.25 | 3.7 |
| ES2 | 22.81 | 36.2 | 1.66 | −6.5 |
| DS1 | 44.65 | 154.8 | 1.24 | 6.6 |
| ES1 | 24.77 | 39.3 | 1.61 | −3.2 |
| sum | | 419.6 | | |

| Cell 10 | $w_e$ | $V_w$ | $\rho_d$ | $\Delta V$ or $\Delta h$ |
|---|---|---|---|---|
| | [%] | [mL] | [g/cm$^3$] | [%] |
| ES3 | 22.25 | 35.3 | 1.68 | −7.5 |
| DS2 | 43.27 | 141.9 | 1.26 | −0.4 |
| ES2 | 23.78 | 37.8 | 1.64 | −4.8 |
| DS1 | 44.94 | 155.8 | 1.24 | 7.0 |
| ES1 | 25.12 | 39.9 | 1.60 | −2.6 |
| sum | | 410.7 | | |

Inflow permeability was higher in cell 9 than in cell 10 until breakthrough. Afterwards hydraulic permeability decreased very slowly to 2–2.7 × 10$^{-17}$ m$^2$ until the experiment was terminated (Figures 2 and 3).

A pressure loss in the hydraulic storage medium and thus a power loss in the hydraulic cylinder that ensures the constant volume of the MiniSandwich occurred in cell 9 at day 258 and in cell 10 at day 72 which resulted in free swelling of the MiniSandwich. After compensation of the pressure loss the volume was constant again (axial strain 0%) and an increased swelling pressure was observed, which decreased only slightly during 200–250 d until it was close to constant in both cells. The total axial stress was slightly higher in cell 9 than in cell 10 (Table 4). Despite of the diameter/height ratio of 1:1.5 friction effects of the material at the cell walls were negligible because of the limited axial displacement (volume constancy). Thus, the measured total axial stress corresponds to the overall swelling pressure.

### 3.2. Volume Changes Due to Swelling of DS

Under the assumption of full saturation, the volume change and change in dry density of each DS and ES were calculated from the water content during dismantling (Table 5). DS in cell 9 and cell 10 expanded up to 7%, while expansion was equal in DS1 and DS2 of cell 9 and only DS1 expanded in cell 10. Thereby, the fine sand in all ES was compacted to a dry density of 1.66 g/cm$^3$ which is slightly above its measured $\rho_{d,max}$ (1.644 g/cm$^3$) and in the range of measurement uncertainty.

### 3.3. Cation Exchange Capacity (CEC) and Exchangeable Cations (EC)

Secursol UHP is a Ca,Mg-bentonite with a CEC of 93 cmol(+)/kg (Table 6). By blending the bentonite the smectite content in DS was 58.6% and the CEC of the DS material was 73 cmol(+)/kg. The apparent pH of dispersed Secursol UHP was 4.7. After dismantling, a slightly lower average CEC was measured in DS of cell 9 and cell 10 (Table 7), but deviation is in the range of measurement uncertainty. The exchangeable Na$^+$ strongly increased from 0.7 to 24–29 cmol(+)/kg while exchangeable Ca$^{2+}$ decreased by about 1/3 from 42 to 25–30 cmol(+)/kg with lower values in DS1 of cell 9 and cell 10. Exchangeable Mg$^{2+}$ was nearly constant in DS of both cells compared to the initial Mg$^{2+}$ of the installed bentonite blend. Exchangeable K$^+$ was very low in DS of both cells but increased slightly. Exchangeable Fe$^{3+}$ was slightly higher in DS1 of cell 10 but in any other DS was close to the value of the initial blend. Sum of exchangeable cations now equaled the measured CEC (Table 7).

**Table 6.** CEC, exchangeable cations, conductivity of suspension and soluble cations and anions of Secursol UHP (granular material HTV-6 [5]).

| CEC | Na$^+$ | Ca$^{2+}$ | Mg$^{2+}$ | K$^+$ | Fe$^{3+}$ | Sum | Cl$^-$ | SO$_4{}^{2-}$ |
|---|---|---|---|---|---|---|---|---|
| | [cmol(+)/kg] | | | | | | [cmol(-)/kg] | |
| | 0.9 | 54 | 19 | 1.3 | 0.03 | 75 | - | - |
| 93 | [mg/g] | [mg/g] | [mg/g] | [mg/g] | [mg/g] | [mg/g] | [mg/g] | [mg/g] |
| | 0.2 | 12.9 | 2.8 | 0.6 | 0.01 | 17 | - | - |
| Conductivity | Na$^+$ | Ca$^{2+}$ | Mg$^{2+}$ | K$^+$ | Fe$^{3+}$ | Sum | Cl$^-$ | SO$_4{}^{2-}$ |
| [µS/cm] | [cmol(+)/kg] | | | | | | [cmol(-)/kg] | |
| | 0.34 | 1.11 | 0.81 | 0.14 | 0.83 | 3.2 | 0.2 | 0.3 |
| 36.8 | [mg/g] | [mg/g] | [mg/g] | [mg/g] | [mg/g] | [mg/g] | [mg/g] | [mg/g] |
| | 0.08 | 0.22 | 0.10 | 0.05 | 0.15 | 0.6 | 0.05 | 0.13 |

**Table 7.** CEC, exchangeable cations, electric conductivity of supernatant suspension and soluble cations and anions (granular material HTV-6 [5]).

| | | CEC | Na$^+$ | Ca$^{2+}$ | Mg$^{2+}$ | K$^+$ | Fe$^{3+}$ | Sum |
|---|---|---|---|---|---|---|---|---|
| | | [cmol(+)/kg] | | [cmol(+)/kg] | | | | |
| Blend * | | 73 | 0.7 | 42 | 15 | 1.0 | 0.02 | 59 |
| Cell 9 | DS2 | 74 | 29 | 30 | 14 | 1.7 | 0.01 | 75 |
| Cell 9 | DS1 | 70 | 28 | 25 | 14 | 2.0 | 0.01 | 69 |
| Cell 10 | DS2 | 71 | 24 | 30 | 12 | 1.3 | 0.03 | 67 |
| Cell 10 | DS1 | 70 | 25 | 27 | 14 | 1.9 | 0.06 | 69 |

| | | Conductivity | Na | Ca | Mg | K | Fe | Sum * | Cl$^-$ | SO$_4{}^{2-}$ |
|---|---|---|---|---|---|---|---|---|---|---|
| | | [mS/cm] | | | [cmol(+)/kg] | | | | [cmol($-$)/kg] | |
| Blend * | | 0.0287 | 0.3 | 0.9 | 0.6 | 0.1 | 0.7 | 2.6 | 0.16 | 0.23 |
| Cell 9 | DS2 | | 14 | 7 | 4 | 0.7 | 0.1 | 25 | 2.6 | 4.1 |
| Cell 9 | DS1 | | 15 | 7 | 5 | 0.8 | 0.1 | 27 | 2.9 | 4.2 |
| Cell 10 | DS2 | | 11 | 5 | 3 | 0.5 | 0.2 | 20 | 2.5 | 4.3 |
| Cell 10 | DS1 | | 14 | 7 | 5 | 0.8 | 0.2 | 26 | 2.3 | 3.5 |

Note: * calculated.

### 3.4. Cations in Outflow and Soluble Cations (SC)

Sodium concentration in the outflow of both cells increased with a steep linear slope during the first 100 days. Afterwards, the Na concentration increased further with a linear but less steep slope up to 200 days (cell 9) and 300 days (cell 10), respectively, until it reached a nearly constant value of about 157 and 150 mmol/L (Figures 4 and 5) corresponding to the inflow concentration. For cell 10 a slight decrease in Na concentration was observed around day 100 and 300 of the experiment at the beginning of each change in slope of outflow concentration (Figure 5). The concentration of Ca, Mg and K decreased potentially ($x^{-1/2}$) approaching asymptotically a concentration close to the concentration of each cation in the inflow Pearson water (Figures 4 and 5). Thereby the calcium concentration was 16 and 20 mmol/L for cell 9 and 10, respectively, and thus slightly higher than 11.6 mmol/L in the inflow. The asymptotic concentrations of Mg and K were 7.2/8.7 mmol/L and 1.6/1.5 mmol/L, respectively, for cell 9 and 10.

The cumulative amount of Na in the volume of Pearson water that passed the MiniSandwich cells was 1.29 (cell 9) and 0.93 (cell 10) times of the initial CEC of the DS and the cumulative amount of Na in the outflow amounted 1.15 and 0.74 times the initial CEC of the DS, respectively, in cell 9 and cell 10. Corresponding values were 0.20/0.14 and 0.48/0.40 (Ca), 0.15/0.11 and 0.20/0.16 (Mg) and 0.02/0.02 and 0.01/0.01 (K). The resulting ratio of cations in the outflow to cation input into each MiniSandwich (cell 9/cell 10) to was 0.89/0.80 (Na), 2.44/2.88 (Ca), 1.36/1.50 (Mg) and 0.54/0.52 (K).

After dismantling, in ES of both cells 56–104% of the expected soluble Na was found with lowest values in the outer ES1 and ES3 of cell 9 and highest values in the outer ES1 and ES3 of cell 10. Both Ca and Mg showed a similar distribution pattern in ES of both cells (Table 8). In contrast, less than 56% of expected soluble K was found after dismantling in ES of both cells, but, again, the distribution pattern was the same as for the other cations. No soluble $Fe^{3+}$ was measured in ES of both cells.

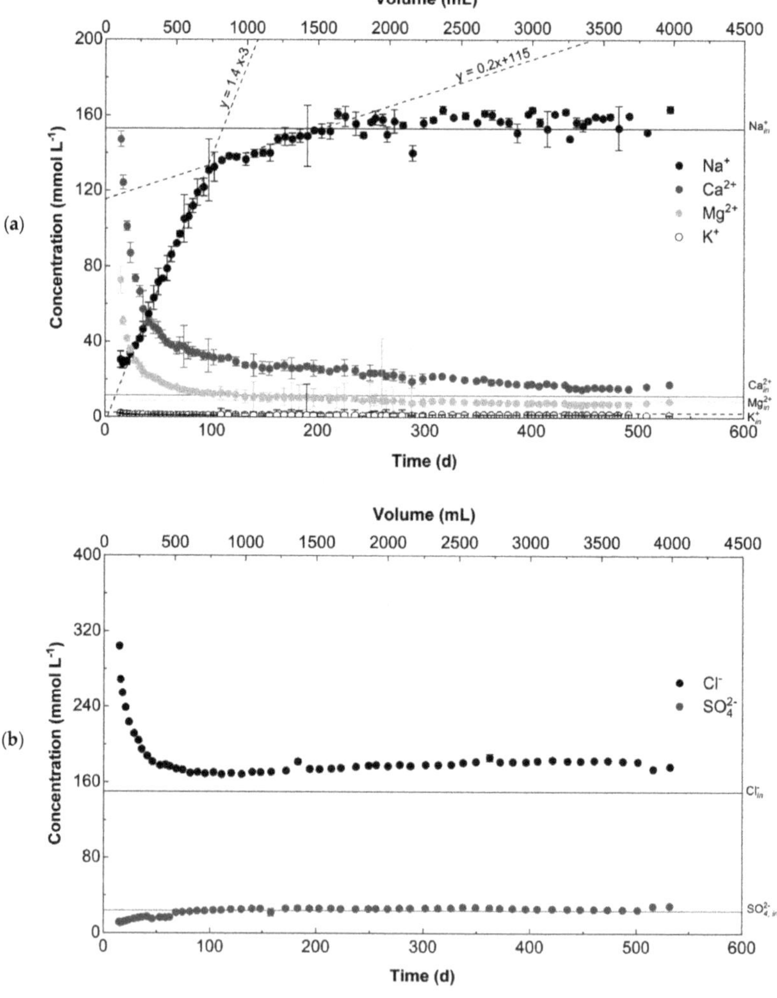



**Figure 4.** Cation concentration (**a**) and anion concentration in outflow (**b**) of cell 9.

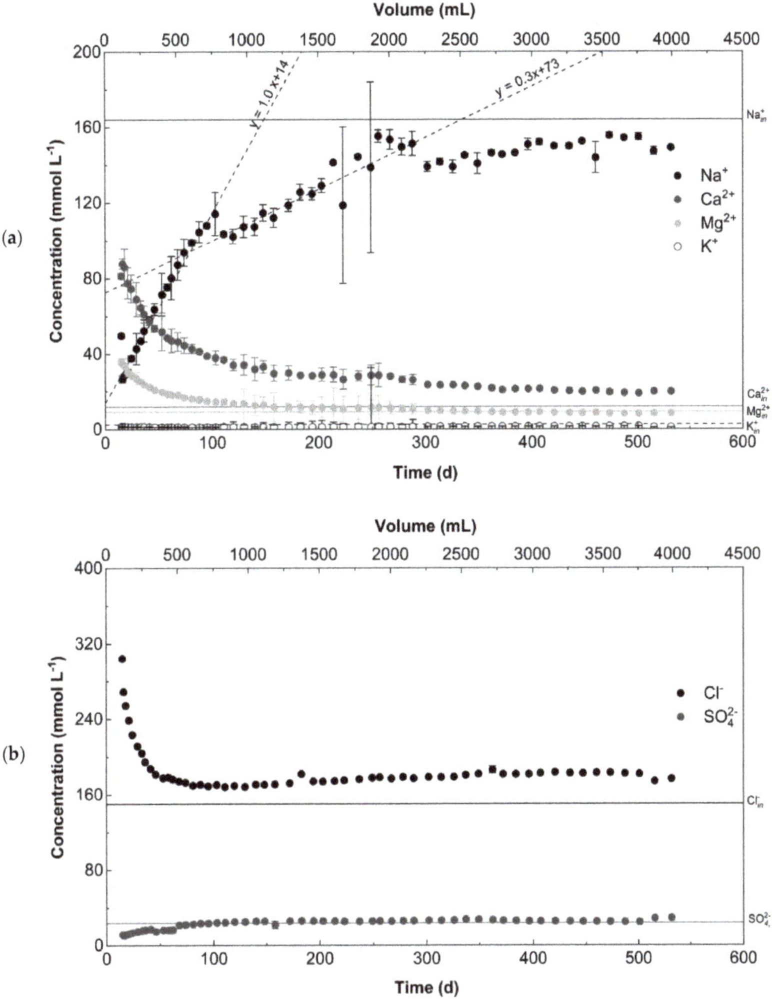

**Figure 5.** Cation concentration (**a**) and anion concentration in outflow (**b**) of cell 10.

In contrast the soluble Na in the DS of both cells was twice as high as expected and the concentrations of soluble Ca, Mg and K were five to seven times higher than expected from water content (Table 8). About 0.02 to 0.04 mg/g Fe were found in DS of both cells. The molar ratio of soluble cations to soluble anions was about 3.7 (Table 7) in DS of both cells while it was about 6.4 in the raw Secursol UHP (Table 6) and 0.64–0.82 in the ES of both cells after dismantling.

**Table 8.** Cation concentration in DS and ES after dismantling.

| Cell 9 | 1<br>Na<br>[mg/g] | 2<br>Ca<br>[mg/g] | 3<br>Mg<br>[mg/g] | 4<br>K<br>[mg/g] | 5<br>Fe<br>[mg/g] | 6<br>Na<br>[mg/g] | 7<br>Ca<br>[mg/g] | 8<br>Mg<br>[mg/g] | 9<br>K<br>[mg/g] | 10<br>Fe<br>[mg/g] | 11<br>Na<br>c1:c6 | 12<br>Ca<br>c2:c7 | 13<br>Mg<br>c3:c8 | 14<br>K<br>c4:c9 |
|---|---|---|---|---|---|---|---|---|---|---|---|---|---|---|
| ES3 | 0.698 | 0.096 | 0.036 | 0.010 | 0.000 | 0.804 | 0.106 | 0.048 | 0.022 | 0.00 | 0.87 | 0.90 | 0.74 | 0.46 |
| DS2 | 3.182 | 1.339 | 0.496 | 0.281 | 0.022 | 1.556 | 0.206 | 0.094 | 0.043 | 0.00 | 2.04 | 6.51 | 5.30 | 6.50 |
| ES2 | 0.837 | 0.114 | 0.042 | 0.012 | 0.000 | 0.804 | 0.106 | 0.048 | 0.022 | 0.00 | 1.04 | 1.07 | 0.87 | 0.56 |
| DS1 | 3.384 | 1.310 | 0.564 | 0.311 | 0.023 | 1.574 | 0.208 | 0.095 | 0.044 | 0.00 | 2.15 | 6.30 | 5.96 | 7.12 |
| ES1 | 0.842 | 0.102 | 0.044 | 0.013 | 0.000 | 0.873 | 0.115 | 0.052 | 0.024 | 0.00 | 0.96 | 0.88 | 0.84 | 0.55 |
| **Cell 10** | Na<br>[mg/g] | Ca<br>[mg/g] | Mg<br>[mg/g] | K<br>[mg/g] | Fe<br>[mg/g] | Na<br>[mg/g] | Ca<br>[mg/g] | Mg<br>[mg/g] | K<br>[mg/g] | Fe<br>[mg/g] | Na<br>c1:c6 | Ca<br>c2:c7 | Mg<br>c3:c8 | K<br>c4:c9 |
| ES3 | 0.695 | 0.132 | 0.034 | 0.008 | 0.000 | 0.784 | 0.104 | 0.047 | 0.022 | 0.00 | 0.89 | 1.27 | 0.73 | 0.36 |
| DS2 | 2.619 | 1.011 | 0.356 | 0.213 | 0.040 | 1.525 | 0.202 | 0.092 | 0.042 | 0.00 | 1.72 | 5.02 | 3.89 | 5.03 |
| ES2 | 0.469 | 0.071 | 0.026 | 0.007 | 0.004 | 0.838 | 0.111 | 0.050 | 0.023 | 0.00 | 0.56 | 0.64 | 0.52 | 0.31 |
| DS1 | 3.191 | 1.373 | 0.576 | 0.300 | 0.030 | 1.584 | 0.209 | 0.095 | 0.044 | 0.00 | 2.01 | 6.56 | 6.05 | 6.81 |
| ES1 | 0.790 | 0.106 | 0.040 | 0.012 | 0.000 | 0.885 | 0.117 | 0.053 | 0.025 | 0.00 | 0.89 | 0.91 | 0.76 | 0.49 |

Note: Column 1–4: measured in supernatant of conductivity measurement (eluted). Column 5–6: calculated from water content after dismantling based on ion content of Pearson water batch 1; c: column.

*3.5. Anions in Outflow and Soluble Anions (SA)*

The $Cl^-/SO_4^{2-}$ weight ratio was 2.31 (molar ratio 6.25, Table 2) in the inflow Pearson water and 2.55 (molar ratio 6.9) in the outflow of cell 9 and 2.66 (molar ratio 7.2) in the outflow of cell 10 (Figures 4 and 5). In contrast, the $Cl^-/SO_4^{2-}$ weight ratio in the DS was 0.51/0.47 in cell 9 and 0.50/0.43 in cell 10 (Table 9). In ES the $Cl^-/SO_4^{2-}$ weight ratio varied between 0.42 and 0.69 with the lowest ratio in ES3 of cell 9 and ES2 in cell 10 and the highest ratio in ES1 of both cells (Table 9). Thereby, the $SO_4^{2-}$ concentration was significantly higher than expected from water content and $SO_4^{2-}$ concentration in the Pearson water while the $Cl^-$ concentration was only about 39–77% of the expected concentration in DS of both cells (Table 9).

**Table 9.** Anion concentration in DS and ES after dismantling.

| | 1 | 2 | 3 | 4 | 5 | 6 | 7 | 8 | 9 |
|---|---|---|---|---|---|---|---|---|---|
| **Cell 9** | $Cl^-$ [mg/g] | $SO_4^{2-}$ [mg/g] | sum [mg/g] | c1:c2 | $Cl^-$ [mg/g] | $SO_4^{2-}$ [mg/g] | Sum [mg/g] | c1:c5 | c2:c6 |
| ES3 | 0.78 | 1.32 | 2.10 | 0.59 | 1.22 | 0.53 | 1.74 | 0.64 | 2.51 |
| DS2 | 0.93 | 1.97 | 2.89 | 0.47 | 2.35 | 1.02 | 3.37 | 0.39 | 1.93 |
| ES2 | 0.93 | 1.41 | 2.34 | 0.66 | 1.22 | 0.53 | 1.74 | 0.77 | 2.67 |
| DS1 | 1.01 | 2.01 | 3.02 | 0.51 | 2.38 | 1.03 | 3.41 | 0.43 | 1.94 |
| ES1 | 0.96 | 1.39 | 2.36 | 0.69 | 1.32 | 0.57 | 1.89 | 0.73 | 2.44 |
| **Cell 10** | $Cl^-$ [mg/g] | $SO_4^{2-}$ [mg/g] | sum [mg/g] | c1:c2 | $Cl^-$ [mg/g] | $SO_4^{2-}$ [mg/g] | Sum [mg/g] | c1:c5 | c2:c6 |
| ES3 | 0.78 | 1.28 | 2.06 | 0.61 | 1.19 | 0.51 | 1.70 | 0.66 | 2.49 |
| DS2 | 0.95 | 2.23 | 3.18 | 0.43 | 2.31 | 1.00 | 3.30 | 0.41 | 2.23 |
| ES2 | 0.53 | 1.27 | 1.80 | 0.42 | 1.27 | 0.55 | 1.82 | 0.42 | 2.31 |
| DS1 | 0.92 | 1.86 | 2.78 | 0.50 | 2.39 | 1.04 | 3.43 | 0.39 | 1.79 |
| ES1 | 0.89 | 1.33 | 2.22 | 0.67 | 1.34 | 0.58 | 1.92 | 0.66 | 2.29 |

Note: Column 1–4: measured in supernatant of conductivity measurement (eluted). Column 5–6: calculated from water content after dismantling based on ion content of Pearson water batch 1. c: column.

*3.6. Charge Balance of Cations and Anions in the Outflow*

The nearly constant concentrations of ions in the outflow added up to 204 mmol(+)/L cations and 217 mmol(−)/L anions in cell 9 and 210 mmol(+)/L cations and 217 mmol(−)/L anions in cell 10. In contrast, the cumulative amount in the total outflow volume was 924 mmol(+) cations and 894 mmol(−) anions in cell 9 and 631 mmol(+) cations and 582 mmol(−) anions in cell 10.

*3.7. Phase Content*

No changes in the mineral phase content were observed by XRD, but changed hydration state of the air-dry bentonite in DS blend material was observed by decrease of the basal space from 14.9 Å to <14 Å prominently detectable from the d001 reflection (Figure S1, Supplementary Materials). Samples from ES showed no alteration to the original mineral phases of the N45 sand. Reflections corresponding to halite and gypsum appeared in ES samples from both cells, however, the amount was below the limit of quantification (Figure S2, Supplementary Materials).

STA curves of evolved water (Figure S3, Supplementary Materials) of air-dry bentonite in DS blend material showed a single dehydration peak compared to the characteristic dehydration peak with a shoulder at higher temperatures of the air-dry raw Secursol UHP with mainly exchangeable divalent $Ca^{2+}$ and $Mg^{2+}$. Only small changes were observed for the $CO_2$/carbonate traces in the bentonite after dismantling, where the shoulder between 200 and 300 °C and a small peak between 850 and 950 °C disappeared for the bentonite granules (Figure S3, Supplementary Materials) but were still visible in the blended DS material (data not shown). MS curves of evolved $SO_2$ of the bentonite granules showed decomposition of sulfates between 400 and 700 °C, 800 and 1000 °C and above 1000 °C, that precipitated during drying of the samples (Figure S3, Supplementary Materials) even more pronounced in the blended DS material (data not shown).

ES material showed evolved water between 100–200 °C which corresponded to dehydration of sorbed water, not present in the original N45 material. Limited changes occur to the carbonate peak centered on 370 °C in N45 material, which was broadened for all ES and shifted to around 350 °C in ES3 in both cells. A sharp carbonate peak also appeared at 607 °C in ES3 of cell 10. Several ES also showed sulfate decomposition occurring between close to 900 and 1000 °C and above 1000 °C (Figure S4, Supplementary Materials).

## 4. Discussion

### 4.1. Hydromechanical Behavior

Fast breakthrough in both cells occurred very quickly due to the low dry density in DS during installation and the fabric resulting from blending of bentonite granules with fine sand. Thereby breakthrough occurred at saturation slightly below 100%. Due to the low dry density and the reduced smectite content, the initial hydration state of the smectite had no influence on the breakthrough time and only a small influence on the axial stress (i.e., the swelling pressure). In cell 9, the smectite underwent a volume change by increasing of the unit cell in c direction of 26.5% due to the transition from 1W into 2W hydration. In cell 10, the bentonite in the DS was already in 2W hydration state and thus swelling lead only to small changes in the interlayer height (i.e., volume of unit cell). The swelling pressure in cell 9 was diminished by the larger inter-aggregate pores in the bentonite/sand blend in the DS where the swelling granules could expand.

The 1W hydration state of the DS in cell 9 resulted in larger inflow permittivity at the beginning of the experiments and about 45% higher volume of Pearson water that went through the MiniSandwich compared to cell 10 until both cells were terminated.

Permeability in cell 9 showed three separate increases of about a half order of magnitude in the course of the experiment while the underlying trend was a continuous slow decrease after saturation and breakthrough. First, the breakthrough might have enabled discrete flow paths that caused a temporary faster fluid flow. These paths slowly closed and the permeability decreased again. The second and third increase of the permeability might have been similarly caused by local changes in the discrete fluid pathways due to the more pronounced swelling of the bentonite granules in cell 9 and a rearrangement of the inter-aggregate pores in the DS as described for pure bentonites in Meleshyn, et al. [23].

Accurate determination of swelling pressures under consideration of the heterogeneity of the MiniSandwich setup with different material segments is a challenge, because the axial effect of swelling DS is partly compensated by compacting the sand in ES. However, the selected experimental MiniSandwich setup corresponds to the arrangement used in real shaft conditions, which facilitates a realistic upscaling. As already mentioned, the measured axial stress necessary to ensure volume constancy during the test corresponds to the swelling pressure of the whole sandwich. Thereby, it is worth to noting that the applied maximum hydraulic pressure to ensure saturation over the experimental time is about half of the maximum swelling pressure of the MiniSandwich (swelling pressure for cell 9: 0.7 MPa; for cell 10: 0.6 MPa). Only in the beginning of the experiment was a stepwise increase of the axial strain required to ensure the water access to the material (Figures 2a and 3a). In the later state of the experiments, the axial load remained nearly constant, which indicated that stationary conditions for swelling and fluid flow were reached.

The initial gas permeability was also influenced by the hydration state of the bentonite and was slightly higher in cell 9 than in cell 10, while the resulting fluid permeability was similar in both cells after saturation at a very low value of $2.0 \times 10^{-17}$ m$^2$ (Table 4). The initial gas permeability was in the range of the gas permeability of pure Secursol UHP, around $\times 10^{-12}$ m$^2$, installed with $\rho_d$ = 1.31 g/cm$^3$ [5]. Deviation is again explained by the blending of the bentonite in the DS with fine sand.

*4.2. Mineralogical/Geochemical Behavior*

No transformation of the smectite into another clay mineral in the DS of both cells in the course of the experiment could be detected by XRD, but XRD and STA measurements showed a different hydration behavior of the smectite equilibrated at ambient conditions (40–60% r.h. and 22–25 °C) after dismantling that was induced by the $Na^+$ exchange into the smectite interlayer observed by the CEC/EC measurements.

Furthermore, no changes occurred in the carbonate traces of the bentonite in DS of both cells after dismantling but MS curves of evolved $SO_2$ showed decomposition of sulfates, while the raw Secursol UHP did not contain any sulfate. No crystalline sulfates were detectable by XRD. Thus, their concentration was very low in the bentonite samples and, in addition, they were precipitates that are expected to have a very fine crystal/grain size. Both low concentrations and small size lower the decomposition temperature during STA strongly up to 200 K compared to tabulated values [24] which makes assignment of the observed peaks in the MS curves of the evolved $SO_2$ tenuous. In any case, decomposition temperature would increase in the order $Na_2SO_4 < MgSO_4 < CaSO_4$ and thus, the three intervals observed for evolved $SO_2$ indicate all three sulfates precipitated either in the course of the experiment and/or mainly during drying of the samples after dismantling. Both processes cannot be distinguished without geochemical modelling or microfluidic experiments, like those described by Poonoosamy, et al. [25,26]. It seemed that pores in bentonite granules and inter-granular pores in DS blended materials formed different microreactors for dissolution/precipitation reactions.

The deficit between the sum of exchangeable cations and the CEC of the raw Secursol UHP could be explained by protons as the pH of the dispersed bentonite was slightly acidic. Protons are not detectable by ICP-OES in the supernatant of the CEC measurement. In the course of the experiment, these protons are only partially replaced by cations of the through flowing Pearson water. The sum of exchangeable cations in DS of both cells after dismantling equaled the CEC but a strong increase of soluble cations (Table 8) added to the sum of exchangeable cations. The charge imbalance of soluble cations and anions is explained by $HCO_3{}^-$.

Excess soluble sulfate in the DS and ES after dismantling indicate precipitation of sulfates in the course of the experiment and not only during drying of the samples after dismantling, although an enrichment in soluble cations was only observed in the DS of both cells. Thereby, Fe was observed only as soluble cation in DS but not in ES.

Imbalance of cation and anion concentrations in the solids and in the fluids indicate that neither ES nor DS materials were already in equilibrium with the Pearson water and observed ion concentration gradients in ES and DS of both cells have to be modelled for the throughflow experiment that was followed by a batch experiment with adapted numerical models for a mechanism-based description of the observed physical and geochemical processes.

## 5. Summary and Conclusions

The Sandwich sealing system, which consists of sealing segments (DS) of bentonite and equipotential segments (ES) of higher hydraulic conductivity, is a component in the German concept of shaft seals for nuclear waste deposits. Functionality was proved on semi-technical scale experiments. An in-situ large-scale demonstration experiment is running at the Mont Terri rock laboratory (Switzerland) that addresses the interaction between the Sandwich sealing system and the Opalinus clay. Prediction and evaluation of the in-situ behavior of the Sandwich sealing system require numerical models for a mechanism-based description of the observed physical and geochemical processes. These models can be validated with MiniSandwich experiments as performed in the current study.

Current MiniSandwich experiments were performed with blended Ca-bentonite (Secursol UHP) and Pearson water. Two experiments were run in parallel with DS installed at the same dry density but either in 1W hydration state or in air-dry (2W) state. Breakthrough and and almost complete saturation occurred at 0.3 MPa injection pressure very fast after 20 days due to the modified fabric of the compacted bentonite blend and a reduced EMDD in the DS. While the influence of the hydration state of the smectite on the break-through was masked by blending the bentonite with a coarser material 4289 mL and 2984 mL, respectively, passed the cell with DS installed in 1W and 2W state within 543 d. Influence of the hydration state of DS during installation on initial and final hydraulic permeability ($2.0$–$2.7 \times 10^{17}$ m$^2$) was also masked by the blending of the DS material and hydraulic permeability in both cells was nearly constant, but still decreased slightly.

Swelling of DS resulted in slight compaction of ES, up to 7%. There were no changes in the mineralogy of the DS and ES material, despite precipitated halite and sulfates occurring, and no changes in CEC of the bentonite blend were observed. While halite precipitated during drying after dismantling sulfates also could have precipitated in the course of the experiments. Exchangeable Na$^+$ strongly increased while exchangeable Ca$^{2+}$ decreased. Exchangeable Mg$^{2+}$ and K$^+$ remained nearly constant. Sodium concentration in the outflow indicated two different exchange processes, while the concentration of calcium and magnesium decreased potentially. Concentration of sulfate$^-$ increased in the outflow until it reached a constant value, and concentration of chloride decreased to a minimum before it slightly increased to a constant value. While cations and anions in the outflow are nearly balanced with respect to their charge, the imbalance of cation and anion concentrations in the solids and in the cumulated outflow volume indicated that neither ES nor DS materials were already in equilibrium with the Pearson water.

**Supplementary Materials:** The following are available online at https://www.mdpi.com/article/10.3390/min11101061/s1, X-ray fluorescence (XRF), Table S1: Chemical composition of Secursol UHP, Simultaneous Thermal Analysis (STA), Figure S1: XRD pattern of powdered DS samples, Figure S2: XRD patterns of powdered ES samples, Figure S3: Mass spectrometer curves for DS; Figure S4: mass spectrometer curves for ES.

**Author Contributions:** Conceptualization, K.E., F.K., R.S., C.R. and T.P.; methodology, K.E., E.B. and C.R.; analysis, C.R., E.B. and K.E.; resources, S.H. and R.D.; data curation, K.E.; writing—original draft preparation, K.E. and E.B.; writing—review and editing, K.E., E.B., C.R., T.P., R.S., F.K., S.H. and R.D.; visualization, K.E., C.R. and E.B.; project administration, K.E. and R.S.; funding acquisition, K.E. and R.S. All authors have read and agreed to the published version of the manuscript.

**Funding:** This research was funded by the German Federal Ministry for Economic Affairs and Energy (BMWi) under 02 E 11587 and 02 E 11799.

**Data Availability Statement:** Not Applicable.

**Acknowledgments:** The authors would like to thank Laure Delavernhe, Peter Bohac (former members of CMM) and Nadja Werling (IMB-CMM, KIT) for preparation of Pearson water, preliminary bentonite characterization, analysis of ions in outflow liquid and STA measurements. We thank Maya Denker and Elisabeth Eiche (AGW, KIT) for several IC measurements while our equipment was broken, Marita Heinle (IFG, KIT) for ICP-OES measurements and Silke Berberich (IMB-CMM, KIT) for many IC and ICP-OES measurements. Thanks also to Thomas Wilsnack and his colleagues (IBeWa) for measurement of the specific density of the bentonite and Matthias Schellhorn (SSKG) who made us aware of the bentonite Secursol UHP. Wiebke Baille and Michael Skubisch (RUB) are thanked for the determination of maximum dry density of N45 according to DIN 18126. We acknowledge support by the KIT-Publication Fund of the Karlsruhe Institute of Technology.

**Conflicts of Interest:** The authors declare no conflict of interest. The funders had no role in the design of the study; in the collection, analyses, or interpretation of data; in the writing of the manuscript, or in the decision to publish the results.

# References

1. Gruner, M. Utilization of bentonite in wase disposal mining; Einsatz von Bentonit im Entsorgungsbergbau. *Bergbau* **2010**, *61*, 394–403.
2. Kudla, W.; Herold, P. *Zusammenfassender Abschlussbericht für das Verbundvorhaben Schachtverschlüsse für Endlager für Hochradio-Aktive Abfälle (ELSA—Phase 2): Konzeptentwicklung für Schachtverschlüsse und Test von Funktionselementen von Schachtverschlüssen*; Technische Universität Bergakademie Freiberg; BGE Technology GmbH: Freiberg, Germany, 2021.

3. Königer, F.; Emmerich, K.; Kemper, G.; Gruner, M.; Gaßner, W.; Nüesch, R.; Schuhmann, R. Moisture spreading in a multi-layer hydraulic sealing system (HTV-1). *Eng. Geol.* **2008**, *98*, 41–49. [CrossRef]
4. Schuhmann, R.; Emmerich, K.; Kemper, G.; Königer, F. *Verschlusssystem mit Äquipotenzialsegmenten für die Untertägige Entsorgung (UTD und ELA) Gefährlicher Abfälle zur Sicherherstellung der Homogenen Befeuchtung der Dichtelemente und zur Verbesserung der Langzeitstabilität: Schlussbericht*; Karlsruher Institut für Technologie: Karlsruhe, Germany, 2009.
5. Emmerich, K.; Schuhmann, R.; Königer, F.; Bohac, P.; Delavernhe, L.; Wieczorek, K.; Czaikowski, O.; Hesser, J.; Shao, H. *Joint Project: Vertical Hydraulic Sealing System Based on the Sandwich Principle—Preproject (Sandwich-VP): Joint Final Report*; Karlsruher Institut für Technologie: Karlsruhe, Germany, 2020.
6. Jobmann, M.; Bebiolka, A.; Jahn, S.; Lommerzheim, A.; Maßmann, J.; Meleshyn, A.; Mrugalla, S.; Reinhold, K.; Rübel, A.; Stark, L.; et al. *Sicherheits- und Nachweismethodik für ein Endlager im Tongestein in Deutschland—Synthesebericht*; BGE Technology GmbH: Peine, Germany, 2017; p. 137.
7. Roelke, C.; Emmerich, K.; Shao, H.; Popp, T.; Koeniger, F.; Schuhmann, R. *The MiniSANDWICH: Fully Controlled Lab Testing and HM Modelling of a Bentonite Based SANDWICH Shaft Seal System*; CMM: Karlsruhe, Germany, 2019; pp. 153–165.
8. Pearson, F.J. *Artificial Waters for Use in Laboratory and Field Experiments with Opalinus Clay: Status June 1998*; Mont Terri Technical Note 99-31; Paul Scherrer Institut: Villigen, Switzerland, 1998.
9. Pearson, F.J.; Arcos, D.; Bath, A.; Boisson, J.Y.; Fernández, A.M.; Gäbler, H.E.; Gaucher, E.; Gautschi, A.; Griffault, L.; Hernán, P.; et al. *Mont Terri Project: Geochemistry of Water in the Opalinus Clay Formation at the Mont Terri Rock Laboratory*; Office Fédéral des Eaux et de la Géologie OFEG: Bern, Switzerland, 2003; Volume 5.
10. Van Loon, L.; Soler, J.; Bradbury, M. Diffusion of HTO, $^{36}$Cl$^-$ and $^{125}$I$^-$ in Opalinus Clay samples from Mont Terri: Effect of confining pressure. *J. Contam. Hydrol.* **2003**, *61*, 73–83. [CrossRef]
11. Emmerich, K.; Kemper, G.; Königer, F.; Schlaeger, S.; Gruner, M.; Gaßner, W.; Hofmann, M.; Nüesch, R.; Schuhmann, R. Saturation Kinetics of a Vertical Multilayer Hydraulic Sealing System Exposed to Rock Salt Brine. *Vadose Zone J.* **2009**, *8*, 332–342. [CrossRef]
12. Karnland, O.; Olsson, S.; Dueck, A.; Birgersson, M.; Nilsson, U.; Hernan-Håkansson, T.; Pedersen, K.; Nilsson, S.; Eriksen, T.E.; Rosborg, B. *Long Term Test of Buffer Material at the Äspö Hard Rock Laboratory, LOT Project: Final Report on the A2 Test Parcel*; TR-09-29; Svensk Kärnbränslehantering AB: Oskarshamn, Sweden, 2009.
13. Dohrmann, R.; Kaufhold, S. Characterization of the Second Package of the Alternative Buffer Material (ABM) Experiment; II Exchangeable Cation Population Rearrangement. *Clays Clay Miner.* **2017**, *65*, 104–121. [CrossRef]
14. Dohrmann, R.; Olsson, S.; Kaufhold, S.; Sellin, P. Mineralogical investigations of the first package of the alternative buffer material test—II. Exchangeable cation population rearrangement. *Clay Miner.* **2013**, *48*, 215–233. [CrossRef]
15. Wallis, I.; Idiart, A.; Dohrmann, R.; Post, V. Reactive transport modelling of groundwater-bentonite interaction: Effects on exchangeable cations in an alternative buffer material in-situ test. *Appl. Geochem.* **2016**, *73*, 59–69. [CrossRef]
16. Wersin, P.; Mazurek, M.; Waber, N.; Mäder, U.; Gimmi, T.; Rufer, D.; De Haller, A. *Rock and Porewater Characterisation on Drillcores from the Schlattingen Borehole*; NAGRA: Wettingen, Switzerland, 2013; p. 343.
17. Pearson, F.J. *Artificial Waters for Use in Laboratory and Field Experiments with Opalinus Clay*; Internal Report; Paul Scherrer Institut: Villigen, Switzerland, 1998; pp. 44–98.
18. Popp, T.; Rölke, C.; Salzer, K. Hydromechanical properties of bentonite–sand block assemblies with interfaces in engineered barrier systems. In *Gas Generation and Migration in Radioactive Waste Repositories*; Shaw, R., Ed.; Geological Society: London, UK, 2015; Volume 415, pp. 19–33.
19. DIN52102. *Test Methods for Aggregates—Determination of Dry Bulk Density by the Cylinder Method and Calculation of the Ratio of Density*; German Institute for Standardization: Berlin, Germany, 2013.
20. Emmerich, K.; Giraudo, N.; Schuhmann, R.; Schnetzer, F.; Kaden, H.; Thissen, P. On the Prediction of Water Contents in Na-Saturated Dioctahedral Smectites. *J. Phys. Chem. C* **2018**, *122*, 7484–7493. [CrossRef]
21. Doebelin, N.; Kleeberg, R. Profex: A graphical user interface for the Rietveld refinement program BGMN. *J. Appl. Crystallogr.* **2015**, *48*, 1573–1580. [CrossRef] [PubMed]
22. Meier, L.P.; Kahr, G. Determination of the cation exchange capacity (CEC) of clay minerals using the complexes of copper (II) ion with triethylenetetramine and tetraethylenepentamine. *Clays Clay Miner.* **1999**, *47*, 386–388. [CrossRef]
23. Meleshyn, A.Y.; Zakusin, S.V.; Krupskaya, V.V. Swelling Pressure and Permeability of Compacted Bentonite from 10th Khutor Deposit (Russia). *Minerals* **2021**, *11*, 742. [CrossRef]
24. Smykatz-Kloss, W. *Differential Thermal Analysis, Application and Results in Mineralogy*; Springer: Berlin, Germany, 1974; p. 188.
25. Poonoosamy, J.; Klinkenberg, M.; Deissmann, G.; Brandt, F.; Bosbach, D.; Mäder, U.; Kosakowski, G. Effects of solution supersaturation on barite precipitation in porous media and consequences on permeability: Experiments and modelling. *Geochim. Cosmochim. Acta* **2020**, *270*, 43–60. [CrossRef]
26. Poonoosamy, J.; Westerwalbesloh, C.; Deissmann, G.; Mahrous, M.; Curti, E.; Churakov, S.V.; Klinkenberg, M.; Kohlheyer, D.; von Lieres, E.; Bosbach, D. A microfluidic experiment and pore scale modelling diagnostics for assessing mineral precipitation and dissolution in confined spaces. *Chem. Geol.* **2019**, *528*, 119264. [CrossRef]

*Article*

# Interaction of Corroding Iron with Eight Bentonites in the Alternative Buffer Materials Field Experiment (ABM2)

Paul Wersin [1], Jebril Hadi [1,*], Andreas Jenni [1], Daniel Svensson [2], Jean-Marc Grenèche [3], Patrik Sellin [2] and Olivier X. Leupin [4]

[1] Institute of Geological Sciences, University of Bern, 3012 Bern, Switzerland; paul.wersin@geo.unibe.ch (P.W.); andreas.jenni@geo.unibe.ch (A.J.)

[2] Swedish Nuclear Fuel and Waste Management Co., 169 03 Solna, Sweden; Daniel.Svensson@skb.se (D.S.); patrik.sellin@skb.se (P.S.)

[3] Institut des Molécules et Matériaux du Mans IMMM UMR CNRS 6283, Le Mans Université, 72085 Le Mans, France; jean-marc.greneche@univ-lemans.fr

[4] National Cooperative for the Disposal of Radioactive Waste, 5430 Wettingen, Switzerland; olivier.leupin@nagra.ch

* Correspondence: jebril.hadi@geo.unibe.ch

**Abstract:** Bentonite, a common smectite-rich buffer material, is in direct contact with corroding steel in many high-level radioactive waste repository designs. The interaction of iron with the smectite-rich clay may affect its swelling and sealing properties by processes such as alteration, redox reactions and cementation. The chemical interactions were investigated by analysing the Fe/clay interfaces of eight bentonite blocks which had been exposed to temperatures up to 130 °C for five years in the ABM2 borehole at the Äspö Hard Rock Laboratory managed by the Swedish Nuclear Fuel and Waste Management Co (SKB). Eleven interface samples were characterised by high spatial resolution methods, including scanning electron microscopy coupled with energy dispersive X-ray spectroscopy and μ-Raman spectroscopy as well as by "bulk" methods X-ray diffraction, X-ray fluorescence and $^{57}$Fe Mössbauer spectrometry. Corrosion induced an iron front of 5–20 mm into the bentonite, except for the high-Fe bentonite where no Fe increase was detected. This Fe front consisted mainly of ferric (oxyhydr)oxides in addition to the structural Fe in the smectite fraction which had been partially reduced by the interaction process. Fe(II) was also found to extend further into the clay, but its nature could not be identified. The consistent behaviour is explained by the redox evolution, which shifts from oxidising to reducing conditions during the experiment. No indication of smectite alteration was found.

**Keywords:** bentonite; iron; in situ experiment; interface

**Citation:** Wersin, P.; Hadi, J.; Jenni, A.; Svensson, D.; Grenèche, J.-M.; Sellin, P.; Leupin, O.X. Interaction of Corroding Iron with Eight Bentonites in the Alternative Buffer Materials Field Experiment (ABM2). *Minerals* **2021**, *11*, 907. https://doi.org/10.3390/min11080907

Academic Editor: Andrey G. Kalinichev

Received: 30 June 2021
Accepted: 18 August 2021
Published: 22 August 2021

**Publisher's Note:** MDPI stays neutral with regard to jurisdictional claims in published maps and institutional affiliations.



## 1. Introduction

Compacted bentonite is foreseen as a buffer material in many concepts for high-level radioactive waste repositories [1–3]. This is due to its favourable sealing properties, which include high swelling capacity and low permeability. In this function the bentonite buffer will be in contact with other materials in the engineered barrier system (EBS) such as cement or steel, which will interact with the clay [4]. Additionally, the buffer will be exposed to heat arising from the radioactive decay of the waste inside the metal canister. In some concepts, the canister consists of carbon steel which will corrode and result in oxidised iron species. These species may interact with the clay by sorption [5], precipitation [6] and complex redox processes [7,8]. This interaction process may affect the functionality of the bentonite barrier, for example by weathering and transformation of smectite, but the details of this process are not yet fully understood.

A considerable amount of experimental work on Fe-bentonite interaction has been conducted, however this has been dedicated to highly simplified batch-type systems (e.g., [9,10] and references therein). Less work has focussed on parameters that are more representative of repository conditions, such as low liquid/solid and low Fe/bentonite ratios, variable redox conditions, higher temperatures (up to 140 °C), realistic dimensions and longer timescales. Examples of such studies are the FEBEX mock-up test [11], the FEBEX in situ test at the Grimsel Test Site [12–15] or the ABM in situ tests at the Äspö Hard Rock Laboratory [16–23]. In general, similar patterns of neo-formed iron minerals at the Fe/bentonite interface have been observed, notably magnetite, ferric (oxyhydr)oxides (e.g., hematite, goethite, lepidocrocite) and siderite. In contrast to many small-scale batch experiments, little or no smectite alteration has been found. However, minor amounts of trioctahedral smectite were distinguished in some of the blocks in contact with the iron heater contact in the ABM1 experiment [17,18] and the ABM2 experiment [21,22] identified an iron-bearing saponite at the interface from a FEBEX bentonite block in the ABM2 experiment. Corrosion of the steel surface in all samples induced an iron front of variable thickness reaching into the bentonite. Despite these numerous studies the mechanisms of iron transfer, from the corroding steel into the clay and its subsequent diffusion within, are not understood in detail.

This study aims at improving the knowledge base of Fe-bentonite interaction in a repository-type setting. This was done by investigating the changes in iron speciation of eight interface samples of seven bentonite materials with different mineralogical and chemical characteristics. These samples had been exposed to similar conditions at maximum temperatures of 130 °C. The methodology including high spatial resolution profiling developed in a previous study [20] was extended by the use of $^{57}$Fe Mössbauer spectrometry. Our work complements previous "bulk" studies on ABM2 samples of [21,22,24].

## 2. Materials and Experimental Methods

### 2.1. Description of the Emplaced Bentonite Materials

Eight doughnut shaped blocks with seven different bentonite materials from the ABM2 in situ test (see below) were studied. Details of the origin of the materials are provided by [16].

Granular MX-80 bentonite: MX-80 is a natural Na-rich bentonite mined in Wyomimg (American Colloid Company, Colony, WY, USA). The raw material with a mesh size of 16–200 was purchased from American Colloid Company. The granular bentonite material, which was emplaced in the ABM2 borehole (see following section), consisted of highly compacted MX-80 with very low moisture content (<5 wt %) and a bimodal grain size distribution (8–12 mm and 0.6–3.1 mm) [25]. The smectite content determined from XRD Rietveld analysis is 80.5 ± 3.6 wt % [26]. The cation exchange capacity (CEC) is 0.85 eq/kg [16]. The Fe content determined from XRF is 3.1 wt % (Table 1).

**Table 1.** Fe content and iron species in raw samples.

| Bentonite | Total Fe | | Distribution (% of Total Fe) | | |
|---|---|---|---|---|---|
| | wt % | Fe$_{str}$ in Smectite | Fe in Oxy-Hydroxides | Fe in Pyrite | Fe in Other |
| MX80 | 3.1 | >90 | 1 < 5 (goe) | <5 | <5 (ilm) |
| Ibecoseal | 3.0 | >91 | <5 (goe/hem/mag) | <5 | <5 (mar) |
| IKosorb | 2.1 | >92 | <5 (goe) | <5 | <5 (mar, mic) |
| Kunigel | 1.5 | >75 | <2 (goe) | <25 | |
| Rokle | 11.6 | 40 | 60 (goe) | | |
| Deponit | 3.8 | >75 | <5 (goe) | <20 | |

Granular MX-80 bentonite/quartz mixture: This bentonite was the same granular material as that described above to which quartz with a grain size of 0.1–0.5 mm [25] was added to yield a homogeneous MX-80/quartz (70/30) mixture. The smectite content determined from XRD Rietveld analysis is 63.7 ± 2.7 wt % [26].

Ibeco Seal M-90 (Ibecoseal): This natural Na bentonite is mined by S&B Industrial Minerals in the Askana region in Georgia/CIS. It is characterised by a high montmorillonite content, as indicated from the high CEC of 0.88 eq/kg [16]. The Fe content determined by XRF is 3.0 wt % (Table 1).

Ikosorb: This natural white Ca, Mg-rich bentonite from Morocco has a high montmorillonite content, as indicated from the measured CEC (0.90 eq/kg) [16] and a rather low Fe content (2.1 wt %) (Table 1).

Kunigel VI: This natural Na-bentonite is produced by Kunimine Industries Co. It has a comparatively low montmorillonite content (~55 wt %) [27] and a CEC of 0.61 eq/kg [16]. It displays the lowest Fe content (1.3 wt %) (Table 1) of all studied materials.

Rokle: This natural bentonite originates from the Rokle deposit in the Kadan basin of the Czech Republic. It has a moderately high smectite content as indicated from the measured CEC value of 0.74 eq/kg [16]. The Fe content is 11.6 wt % which the highest of all studied materials.

Deponit CAN (Deponit): This natural Ca bentonite is mined on the island of Milos in the South Aegean (Greece). It has a moderately high smectite content of 72 wt % [28] and a CEC of 0.84 eq/kg [16]. The Fe content is rather high, a value of 3.8 wt % was obtained from XRF analyses (Supplementary Materials S3).

### 2.2. The ABM2 Test Package, Excavation and On-Site Sampling

The ABM (Alternative Buffer Materials) test is an internationally supported in situ experiment conducted by SKB (Swedish Nuclear Fuel and Waste Management Co., Solna, Sweden) in the Äspö Hard Rock Laboratory (Äspö HRL), Sweden. The main objective of the ABM test is to access the stability of different bentonites under adverse, but representative conditions of the near-field of high-level radioactive waste repositories [16]. Within the ABM test, three test packages (ABM1, ABM2, and ABM3) consisting of various bentonite materials stacked upon each other as blocks were emplaced in three boreholes and heated via a central steel tube composed of common carbon steel, P235TR1 [29]. The layout of the ABM2 test package and the blocks sampled in this study are illustrated in Figure 1.

The "block" materials #11, #12, #13, #24 and #26 consisted of pre-compacted blocks (height 100 mm and diameter 300 mm) emplaced in direct contact with the steel heater. The granular materials of blocks #8, #25 and #26 were inserted in a prefabricated iron-based cage on-site. This circular cage was made of a cylindrical inner steel ring, steel frames, and steel fibre cloth wrapped around the cages. The installation procedure is detailed in [29]. The description of the experiment and its dismantling are presented in [22] and [23]. In short: the experiment was saturated with natural Äspö groundwater for one year via a sand filter previously emplaced between the blocks and the borehole wall. The Äspö groundwater at the location of the ABM test package is of Na-Ca-Cl type with ~180 mM $Cl^-$, ~100 mM $Na^+$ and ~50–60 mM $Ca^{2+}$ including some $SO_4^{2-}$ (~5 mM), $Mg^{2+}$ (0.3–2.4 mM), $K^+$ (~0.3 mM) and $HCO_3^-$ (~0.4 mM) [16,30]. The package was then heated to temperatures of 120–140 °C at the heater contact for a period of three years before allowing subsequent cooling for an additional year. This was followed by retrieval through overcoring and uplifting. Samples were removed in situ from the heater steel tube, unfortunately in some areas the Fe-bentonite interface was damaged during this process. The cages were simply slid off the tube and were thus kept intact during in situ sampling. This resulted in the Fe-bentonite interface being generally better preserved in the caged samples.

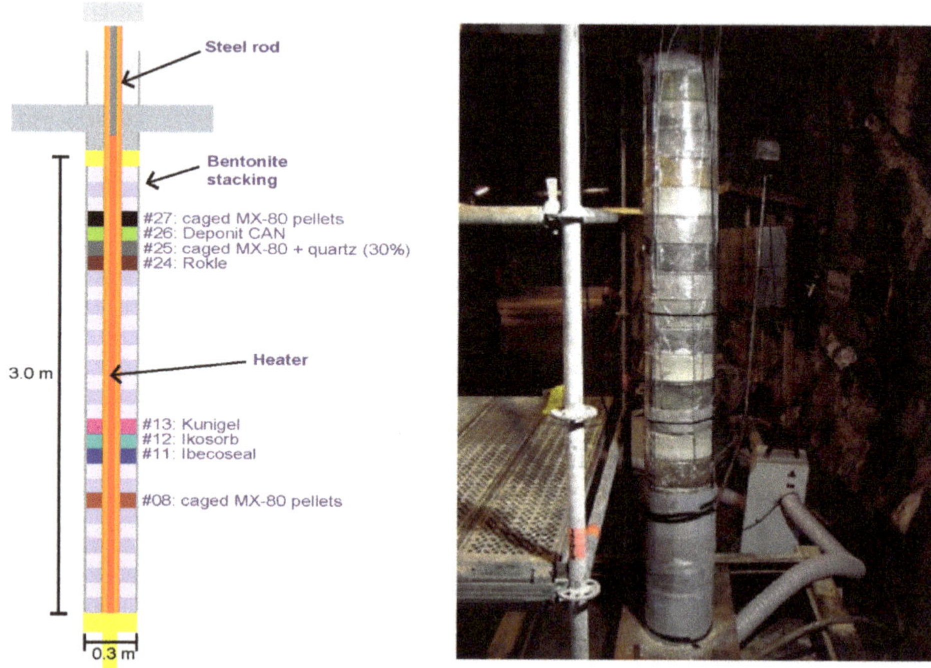


**Figure 1.** ABM2 borehole with stacked blocks and caged pellets surrounding the central steel heater [29].

### 2.3. Analysis of Fe-Clay Contact Zone

Sample preparation: Two types of samples were prepared (Figure 2). The first type underwent preparation for microscopic analysis at high spatial resolution (SEM/EDX, μ-Raman). A subsample comprising the Fe-bentonite interface was cut out, freeze-dried, vacuum-embedded in epoxy resin and finally polished (using petroleum). The resulting polished section (Figure 2A) was subsequently stored in a desiccator until further microscopic analyses. The second type consisted of powdered subsamples at different distances from the contact zone (Figure 2B), which were used for bulk analysis (Mössbauer spectrometry, XRD and XRF). These were prepared anaerobically in the glovebox ($N_2/H_2$ mixture with Pd catalyst), including separation, drying, milling and storage.

SEM/EDX analysis: The uncoated sample surface was examined in a SEM (EVO-50 XVP, Carl Zeiss AG, Jena, Germany) equipped with an EDAX® Sapphire light-element detector in low vacuum mode (10–20 Pa) with a beam acceleration of 20 kV, a sample current of 500 pA, and a working distance of 8.5 mm. The beam current was adjusted to yield a dead time of 8–15% for EDX analysis. EDX element maps with a resolution of 128 × 100 pixels were acquired using a dwell time of 200 μs/pixel. Mappings were conducted with a magnification of 80, which results in pixel size of ~11 μm$^2$ and maps of ~1.4 mm × 1.1 mm. Mapped elements generally included C, O, Na, Mg, Al, Si, P, S, Cl, K, Ca, Ti, and Fe but only Fe, Ca, S and Al data are reported here. The total grid dimension was usually 30–40 maps along the x axis perpendicular to the interface and 8–10 maps along the y axis parallel to the interface. Given the parameters of analysis (resolution, dwell time), the acquisition time per block was ~12 h. Output data from the operating software (Smartsem® by ZEISS for the SEM part and Genesis® by AMETEK for the EDX) were collected and treated with a MATLAB in-house algorithm in order to establish chemical profiles, large scale elemental mappings, and backscatter images.

**Figure 2.** (**A**) Polished samples obtained from bentonite blocks/cages used for SEM and μ-Raman analysis: (**a**) Ibecoseal #11, (**b**) Ikosorb #12, (**c,d**) Kunigel #13, (**e**) MX-80+quartz #25, (**f**) Deponit #26, (**g**) MX-80 with steel part #27. (**B**) Cross section of upper part of caged pellets #25 and powdered samples obtained neighbouring Deponit material. Yellow arrow shows interface with heater, black arrow shows attached Deponit bentonite with interesting shades of red, green and blue. (**G**) grey, (**R**) red, (**B**) blue.

The main data obtained from the SEM-EDX survey are presented as "Al-normalized" chemical profiles of the major elements, representing the atomic ratio of a given element over Al as a function of the distance to the interface [15]. The so-called "Al-normalized" values were computed directly from the quantification results [31,32]. For a given element, each point of the profile represents the average ratio of the content in this element over the Al content of a given column parallel to the interface of the analysis grid. Ratios over Al are not sensitive to variations in other elements. Al is assumed to be immobile, based on the lack of clay alteration (major Al carrier) and therefore, the very low Al mobility. This was confirmed by Al profiles showing constant Al perpendicular to the interface (Supplementary Materials S1). The error bars account for twice the standard deviation. Raw EDX data were corrected using individual Standard Element Coefficients (SEC) factors for each element. These factors were determined from the EDX analysis of six different raw bentonites (MX-80, Ibecoseal, Ikosorb, Deponit, Rokle and Kunigel) of very similar composition for which reference XRF data were also available [16].

μ-Raman spectroscopy: Raman spectroscopy was performed with a Jobin Yvon LabRAMHR800 instrument consisting of a BX41 confocal microscope (Olympus) coupled to an 800 mm focal length spectrograph. A non-attenuated He–Ne laser (20 mW, polarized 500:1) with an excitation wavelength of 632.817 nm (red) was focused on the sample surface and the Raman signal was collected in reflection mode. The sampled volume was a few $\mu m^3$ using a 100x objective lens. Spectra were measured in Raman shift intervals of 150 to 1400 $cm^{-1}$ in five steps of 250 $cm^{-1}$. Acquisition time for each step was $2 \times 15$ s, i.e., 2.5 min in total. Acquisition time was doubled for some analyses in the clay matrix. The spectra were recorded with Labspec V4.14 software (HORIBA Scientific). Identification of the species was done using the spectra library included in the HORIBA Edition of the KnowItAll®. The spectra presented in this report indicate the name(s) of identified species and corresponding reference number(s) in the library, which actually combines several entries for inorganics, minerals, and gemstones from Minlab v3 or RRUFF [33].

XRF analyses: Glass pellets were made by fusing a 1:10 mixture of sample powder and Li-tetraborate at 1150 °C. XRF analyses of major elements were performed on a PW 2400 Philips spectrometer and corrected with the internal Philips software ×40 on the basis of a set of international rock standards. Loss on ignition (LOI) was determined by mass difference before and after fusing. Water content was determined during the same process (105 °C for 2 h).

$^{57}$Fe Mössbauer spectrometry: Mössbauer spectrometry was employed to measure the iron reduction level and to identify Fe bearing phases present in the sample. The spectra were recorded at room temperature (RT, 300 K) and at 77 K using a constant acceleration transducer and a $^{57}$Co source dispersed in a Rh matrix. Velocity calibrations were carried out using an $\alpha$-Fe foil at RT. The values of the hyperfine parameters were refined using a least-square fitting procedure (MOSFIT in-house unpublished program) with a discrete number of independent quadrupolar doublets and magnetic sextets composed of Lorentzian lines. The values of isomer shift (I.S.) are reported relative to that of the $\alpha$-Fe spectrum obtained at RT. The fractions of each Fe species are proportional to the relative spectral area. Indeed, the f-Lamb-Mössbauer factors which correspond to the fraction of gamma rays emitted and absorbed without recoil are assumed to be identical for the different phases present in the samples and for the different Fe species present in the same phase [34–36]. The fitting strategy consisted in the use of a minimal number of components (quadrupolar doublet or magnetic sextet) for discriminating between high spin octahedral $Fe^{3+}$ (HS-oct-Fe(III)), high spin octahedral $Fe^{2+}$ (HS-oct-Fe(II)), low spin octahedral $Fe^{2+}$ (pyrite) and magnetically ordered species (goethite and hematite). In addition, it is important to emphasize that the quadrupolar component may in part be attributed to superparamagnetic species originating from very fast relaxation phenomena: it is, therefore, necessary to compare Mössbauer spectra recorded at different temperatures. Goethite and hematite in particular display hyperfine structures which are strongly dependent on the crystalline grain size, the distance between close grains, and temperature [37]. In the present study, this temperature dependency was used to discriminate between "large" grains (or aggregates) of goethite or hematite (~>30 nm, magnetically ordered at room temperature and 77 K), "medium-sized" grains (~5–30 nm, paramagnetic at room temperature, but magnetically ordered at 77 K, i.e., superparamagnetic), and "small" grains (~<5 nm, paramagnetic at both temperatures, thus it is much more difficult to discriminate from other species such as the clay structural Fe(III)). In addition, the extent of the hyperfine magnetic field ($B_{hf}$) and of quadrupolar shift ($2\varepsilon$) also enables discriminating goethite from hematite (the latter exhibits a higher $B_{hf}$).

XRD analyses: Studies were conducted using an Anton Paar domed sample holder (Anton Paar Austria) for air-sensitive materials equipped with a polycarbonate dome. The powdered samples were loaded on the sample holder in the anaerobic chamber, the surface was flattened with a glass slide and the dome was closed before the samples were removed from the chamber. The raw bentonites were also analysed without the dome. The samples were analysed with a X'Pert PRO X-ray and recorded using Cu K$\alpha$ radiation with a wavelength of 1.54 Å and an X-ray tube operated at 40 mA and 40 kV. The samples were scanned from 5 to 60° 2$\theta$ using a step size of 0.0167° 2$\theta$ and a time of 10 s per step, with automated divergence slits. Samples were spun during the measurement, at a rate of one revolution every 8 s.

## 3. Results

### 3.1. Macroscopic Observations

The caged blocks which had been filled with compacted granular MX-80 and granular MX-80/quartz (Figure 3 left) were transformed to a homogeneous mass without any visible trace of the original pellet texture (Figure 3 right). This homogenization had been induced by swelling processes during water saturation.

The second obvious feature was the severe corrosion of the steel frame, which also affected the adjacent bentonite (Figure 3 right). The surfaces of the blocks appeared brown-orange contrasting with the grey colour of the unaffected MX-80 material. The impact on the upper caged blocks (#25 and #27) was stronger than that on the lower block (#08) where the grey colour remained visible. The neighbouring blocks were also affected by corrosion of the steel frame of the cage, observable as a brownish rusty front penetrating several mm into the neighbouring blocks.

**Figure 3.** Example of caged pellets before installation (**left**) and after dismantling (**right**) [29].

The extent of the rusty front is larger in the caged granular materials than in the adjacent blocks. It is worth noting that the granular materials have a lower density compared to the "normal" blocks according to data on ABM1 [20]. Cracks filled with iron oxides and pronounced corrosion halos deep inside the caged blocks were observed in caged blocks. The origin of this phenomenon is discussed in Section 4.3.

The upper blocks (#24–#27) had a number of contrasting features compared to the lower blocks. The latter were still humid, had a homogeneous texture with rare fractures, presumably having occurred after sampling and storage, and rather well preserved interfaces to the metal pieces. The former, on the other hand, appeared more altered, drier, with many more fractures as well as poorer preserved interfaces with the heater. From the polished sections (Figure 2A) it can be concluded that the impact of corrosion was more extensive in the upper blocks. A 5–10 mm wide concentric rim of white precipitate, identified as anhydrite, was observed on the surface and fracture walls of the upper caged blocks. The reason for these different features in the upper blocks can be attributed to a boiling event, which probably occurred during the experiment in this area where the maximum temperatures occurred as a result of local pressure release [22,24]. This is possibly related to a local transmissive fracture in the surrounding rock.

### 3.2. Quantitative EDX Profiles

The distribution of Fe relative to the adjacent steel surface was studied by quantitative area surface measurements (see Methods section). The total area was divided into 8–10 rows and 30–40 columns, which resulted in 240–400 cells overall. Each cell corresponds to one EDX area measurement. The average EDX analyses of 8–10 individual measurements (referred to "area measurements" below) were calculated for each column. Error bars represent twice the standard deviation. The complete set of profiles is shown in the Supplementary Materials S1).

The Fe content (shown as Fe/Al ratios) in the blocks shows an increase toward the Fe-clay interface. The MX-80 materials exhibit a similar shape of the Fe front (Figure 4). A sharp Fe increase at the contact in a narrow zone (<5 mm) is followed by a zone of more gradual change in the clay (~5–10 mm). The actual steel crust consisting of various corrosion products without bentonite is not part of the profiles. The large variation observed in some locations reflects the local presence of accessory iron minerals, whose distribution is more or less inhomogeneous as indicated from SEM analysis, and which were also observed in the reference material. Blocks #25 (MX80/qz) and #27(MX80) from the upper part of the package, where the boiling event occurred (see below), display slightly higher Fe enrichment and a larger front compared to block #08 (MX80) from the lower part.

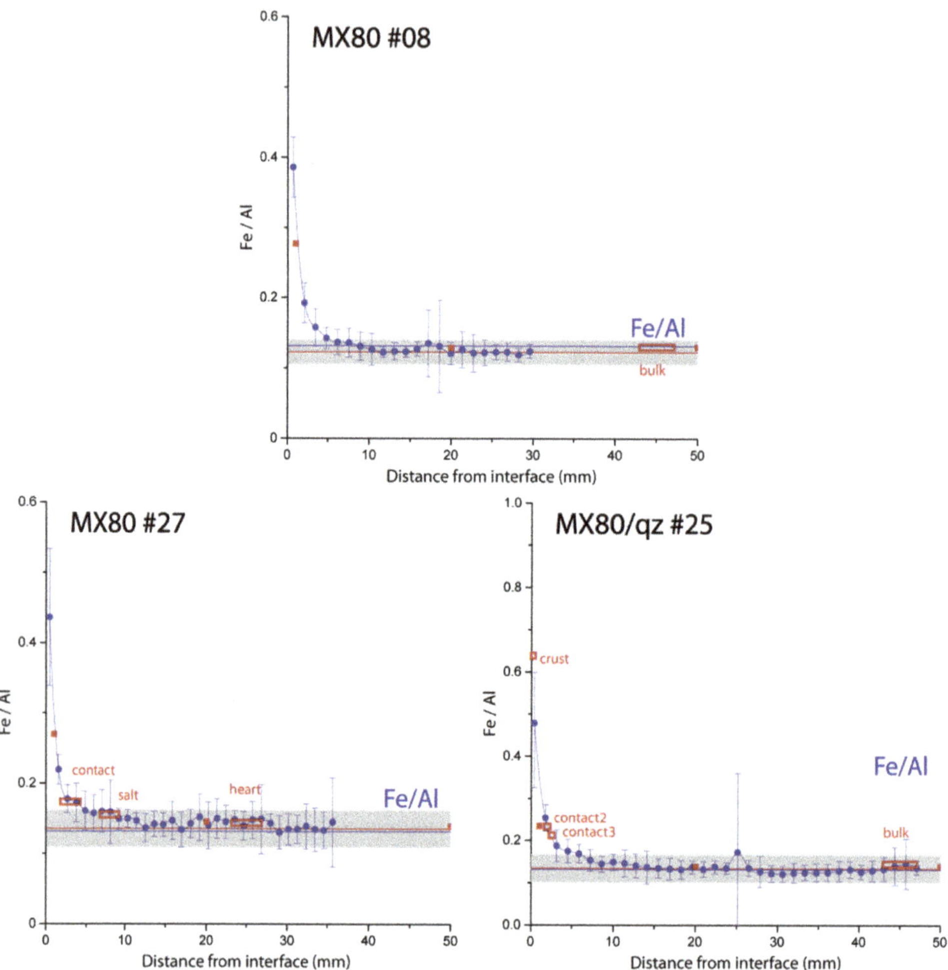



**Figure 4.** Al-normalised Fe profile perpendicular to steel-clay contact for blocks with MX-80. Points: EDX measurements, squares: XRF bulk data from [22], empty rectangles: XRF measurements on powdered samples, grey area: range of bulk.

The other materials display similar Fe behaviour perpendicular to the Fe-clay contact (Figure 5). An exception is the Rokle bentonite which contains by far the highest Fe content. In this material, there is no indication of an increase in Fe compared to the bulk material. The variations are large which is explained by the inhomogeneous distribution of Fe oxides also observed in the reference materials (see below). In some blocks (Ikosorb, Kunigel) there seems to be an indication of a second front further inside the clay. In the latter material, this front is close to the uncertainty of the measurements.

In summary, the different blocks show similar Fe profiles towards the contact with steel, regardless of the type of material (except for the high-Fe Rokle bentonite). The maximum increase relative to the bulk material varies between a factor of 2.5 to 4. The position in the package does not have a significant effect. Comparing the same materials (MX80) however, it appears that in the upper part where the boiling event occurred, slightly more Fe from the corroding steel migrated into the clay. Using the adequate calibration (see Methods section) the Fe content determined from EDX agrees with that obtained from XRF within the analytical error.

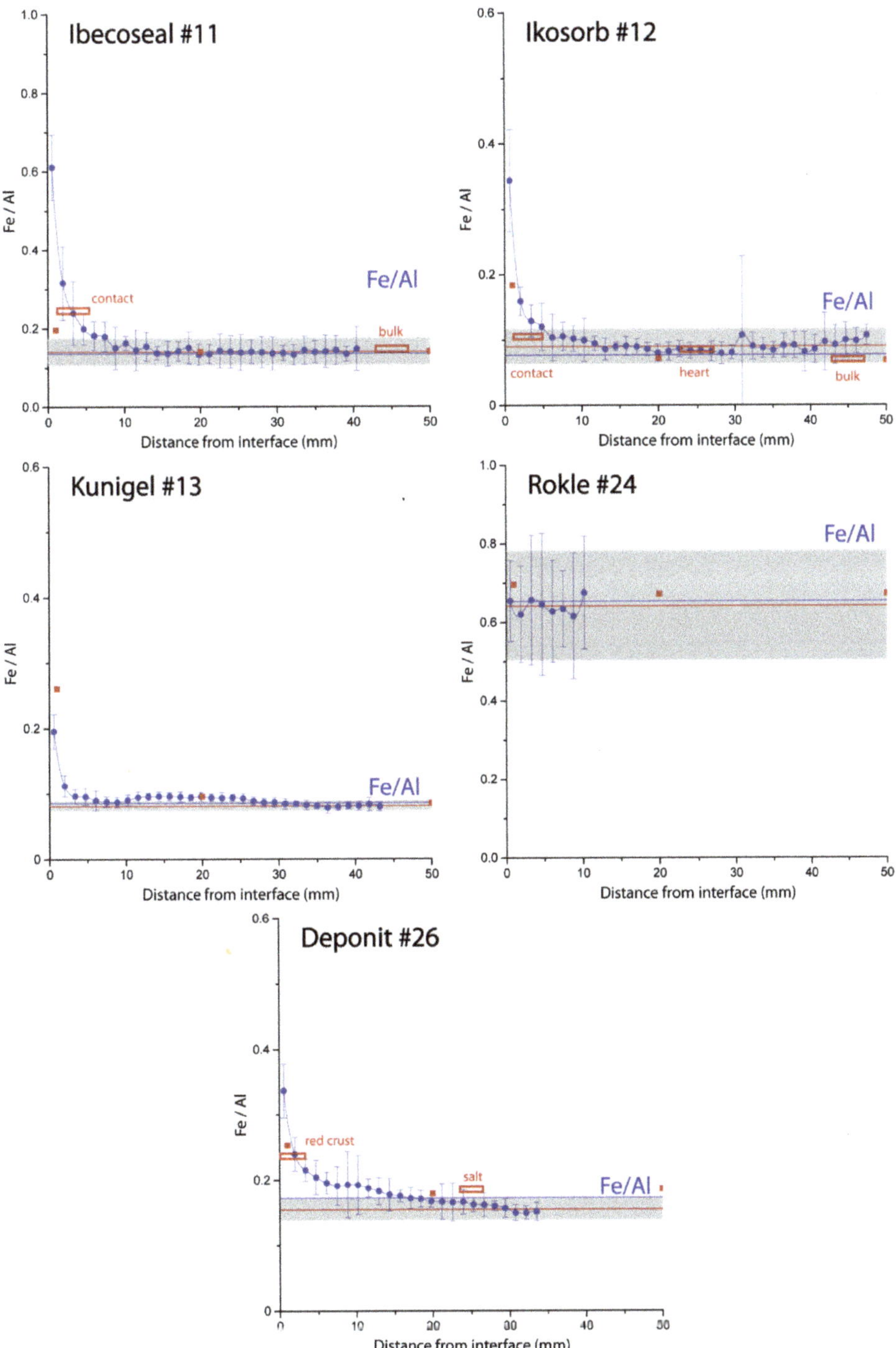

**Figure 5.** Al-normalised Fe profile perpendicular to steel/clay contact for other blocks. Points: EDX measurements, squares: XRF bulk data from [22], empty rectangles: XRF measurements on powdered samples; grey area: range of bulk.

### 3.3. Identification of Iron Phases

The identification of iron phases in relation to their location relative to the corroding steel source(s) was performed using combined information from SEM/EDX, µ-Raman, XRD and Mössbauer spectrometry. The entire datasets are presented in Supplementary Materials S2 (µ-Raman), S3 (XRD) and S4 (Mössbauer).

#### 3.3.1. Pre-Existing Iron Mineral Phases

The goal here was to determine the mineral phases both in the raw materials and the bulk samples, which were located far from the metal source in order to facilitate the interpretation of the "interface" samples affected by corrosion phenomena. It should be highlighted that the main iron pool is structural Fe in the smectite except in case of the high Fe Rokle sample.

MX-80: Bulk samples from blocks #08, 25 and #27 revealed pyrite, ilmenite and to a lesser extent Fe(III) oxides as accessory minerals according to combined SEM/EDX and µ-Raman spectroscopy (Figure 6). The grain size distribution is broad, ranging from µm to mm scale.

From XRD diffractograms (full data in Supplementary Materials S3) pyrite, goethite and hematite could be identified (Table 1), thus partly supporting data from µ-Raman.

XRF analyses of the raw sample and the bulk ABM sample do not show difference in Fe content (Supplementary Materials S3). From the Mössbauer spectrometry data obtained at room temperature (Figure 7) (full data in Supplementary Materials S4), only the octahedral structural iron in the clay could be identified in the MX-80 raw and the bulk sample (distant from the heater with no interaction with Fe) from block #08. This indicates that the contents of accessories identified by SEM/EDX, µ-Raman and XRD are too low to be captured by Mössbauer spectrometry. Imposing the presence of pyrite in the fitting procedure did not lead to an improvement of the fits. The two MX-80 samples show reduction levels of 18% and 26% for the raw sample and bulk sample of block #08, respectively. The same difference between raw and bulk sample was observed for the MX-80/quartz block #25. The presence of structural Fe(II) at such levels is not uncommon for MX-80 [38,39]. The reason for the difference in reduction levels between the two samples is not clear, but may be due to the natural variation in the samples.

Ibecoseal: This material contains various Fe accessory minerals which could be identified by combined SEM/EDX and µ-Raman (Supplementary Materials S2). Most phases occur as small aggregates (<10 µm) unevenly distributed in the matrix. According to EDX mapping combined with µ-Raman, marcasite appears to be the main Fe-bearing accessory phase, followed by mixtures of magnetite with either goethite or hematite. The latter two minerals form coatings around magnetite grains. This suggests an oxidative alteration process related to the original feature of the Ibecoseal bentonite.

From XRD diffractograms (Supplementary Materials S3) marcasite and hematite (but not goethite) can be deduced from their main reflection, confirming SEM/µ-Raman data.

The bulk sample of block #11 exhibits a slightly higher Fe content (by 8%) than the reference sample (Table 2). From Mössbauer spectrometry data obtained at room temperature and 77 K (S4), the reference sample only contains structural Fe while the bulk sample also contains some hematite and goethite (together ~9%). This would be consistent with the Fe$_{tot}$ increase, but SEM/EDX and µ-Raman analysis suggests that not all of Fe(III) oxides determined in the bulk sample are additional phases. The reduction level of the raw sample is about 31%, thus higher than that of MX-80. This may be attributed to the larger amount of illite/mica in Ibecoseal as suggested from the XRD data (Figure S3). The reduction level of the bulk sample is 34%, in the same range as the reference sample.

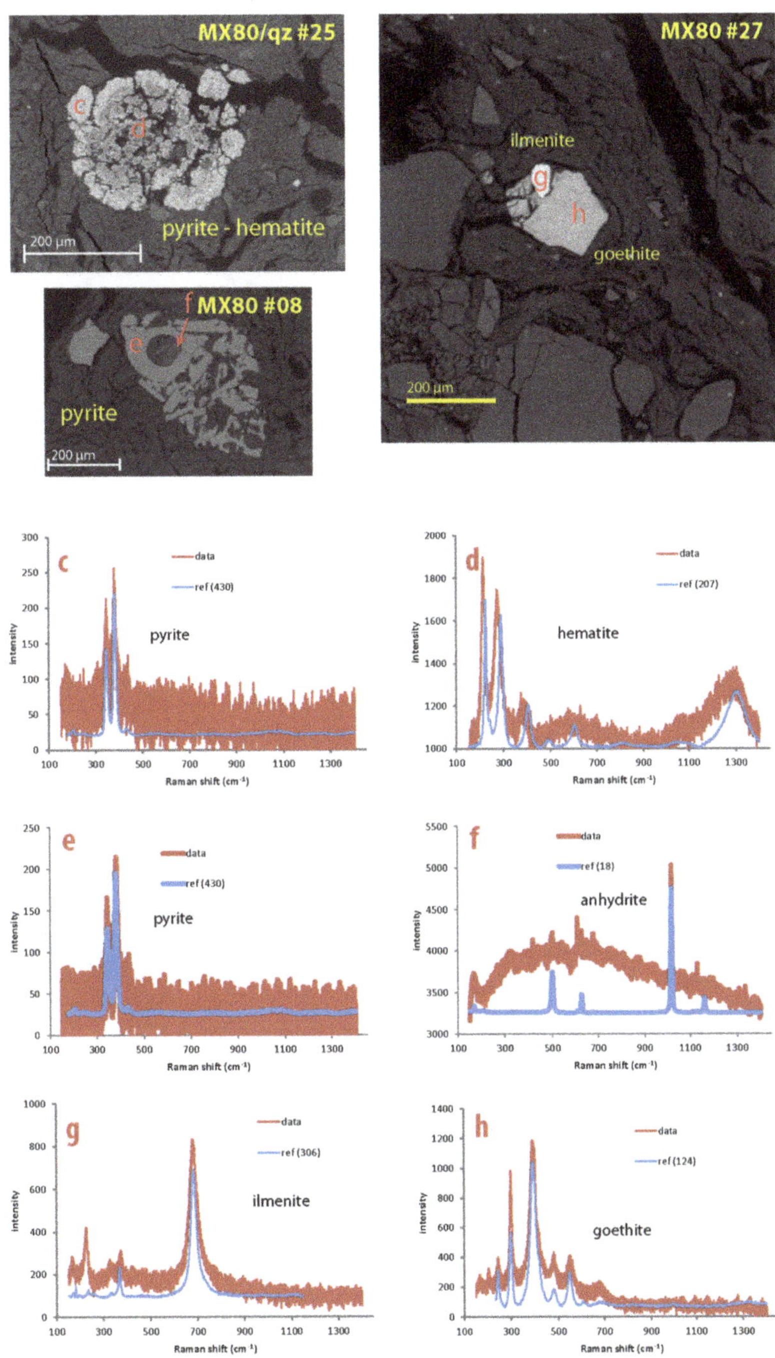



**Figure 6.** SEM micrographs (**upper**) and µ-Raman spectra (**lower**) at indicated locations of unaffected MX-80 bentonite blocks #8 and #27 and MX-80/quartz block #25. Note that further µ-Raman spectra are shown in Figure S2(1).

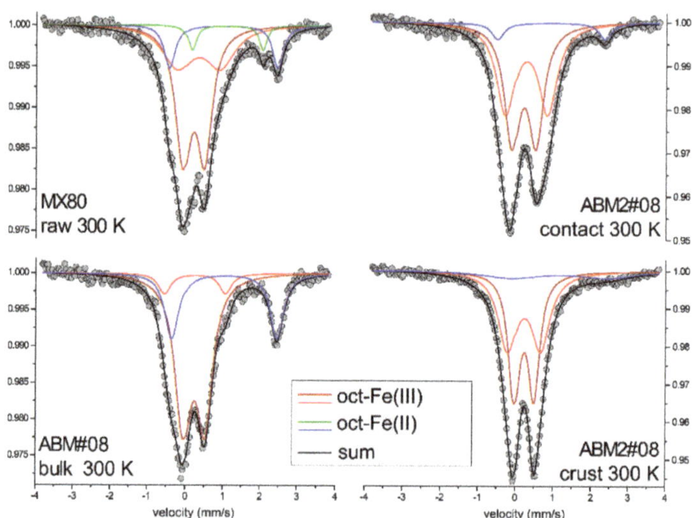



**Figure 7.** Room temperature Mössbauer spectra of raw, reacted and bulk MX80 from block #08. The refined values of hyperfine parameters are listed in Tables of Supplementary Materials S4.

Ikosorb: This bentonite has the lowest amount of Fe-bearing accessory minerals of all the studied bentonites as indicated from SEM/EDX, μ-Raman and XRD and XRF analysis (Supplementary Materials S1–S3). Note that because of the limited amount of material no Mössbauer analysis was conducted. Rare large grains (~mm range) of sulphide (marcasite altered from pyrite) and smaller amounts of Fe oxides (probably goethite according to its appearance) were identified.

Kunigel: This bentonite has the lowest Fe content (1.4 wt %) of all studied materials with a significant fraction of pyrite. The pyrite pool is estimated to represent ~20% of the total Fe based on the Fe/S proportion from SEM/EDX maps and XRF (Figures S1 and S3). Pyrite grain sizes are <50 μm in diameter and appear to be homogeneously distributed in the clay matrix. Some rare small grains of goethite were observed. Due to the limited sample amounts no Mössbauer and XRD analyses were carried out.

Rokle: This is the bentonite with the highest Fe content (12 wt %) of all studied materials. A large proportion (~50%) of Fe is contained in Fe(III) oxides (mainly goethite) (Figure S2(9,10)) which display a large grain size distribution (μm–mm scale) and a heterogeneous distribution in the clay material.

Deponit: The main accessory Fe-bearing mineral observed was pyrite which was estimated to constitute up to 19% of the total Fe from EDX profiles and XRF data (Figures S1 and S3). Occasionally, small grains of goethite were found by SEM (Figure S2(11,12). These findings are supported by XRD and Mössbauer data (Figures S3 and S4), the latter indicating that the pyrite fraction represents at most 20% of total Fe.

Summary: From the ensemble of the multi-method data the relative amounts of the different Fe-bearing phases were estimated (Table 2). The major Fe fraction consists of structural Fe in the smectite in all bentonite materials except Rokle where the main fraction is goethite. All samples contain Fe oxyhydroxides (mainly goethite) and pyrite.

### 3.3.2. Newly Formed Iron Mineral Phases

The identification of neo-formed Fe phases in the contact zone is not straightforward in view of (i) inherent limitations of the applied analytical methods, (ii) the (at least potentially) microcrystalline character of the precipitates (e.g., Fe hydroxides) [15,20] and (iii) the mechanically disturbed contact zone with very limited amounts of sample material. Nevertheless, valuable information could be obtained from combined Mössbauer-XRF analysis on powdered bulk samples, which was complemented by SEM/μ-Raman analysis on polished samples from the caged blocks.

The well-defined contact areas of the cage frames of blocks #08, #25 and #27 were selected for detailed analysis of iron-bearing phases. The SEM backscatter images and μ-Raman spectra are shown in Figure 8. All interfaces exhibit a similar layered structure: an inner compact layer (50–100 μm) with magnetite and goethite and an outer mechanically disturbed zone (100–500 μm) of a mixture of corrosion products (goethite, lepidocrocite and magnetite) and clay aggregates (Figure 8). The proportion of ferric (oxyhydr)oxides (mainly goethite) increases towards the clay in parallel with the decrease of magnetite. In the case of block #08, siderite crystals were identified in fractures in the outer part of the corrosion layer (Supplementary Materials S2).

Further information was obtained from Mössbauer spectrometry in combination with XRF on powdered samples of clay adjacent to the corrosion layer from blocks #08 (MX-80), #11 (Ibecoseal), #12 (Ikosorb), #25 (MX-80/qz), #26 (Deponit) and #27 (MX-80) Table 2.

**Table 2.** Fe content, reduction level and $Fe^{2+}/Fe^{3+}$ ratio of raw, bulk and interface samples (crust: direct contact to Fe/clay interface; contact: sample $\geq$0.3 mm from interface; red/green/blue: samples with reddish/greenish and blueish appearance, respectively; salt: sample containing white precipitate; goe: goethite, hem: hematite; n.a.: not analysed). Error on reduction level from Mössbauer spectrometry is about 4%.

| Block # | Material | Sample Type | Distance Interface | Total Fe | Increase Total Fe | Reduction Level | Fe(II)/Fe(III) | Fe Oxides from 77 K Mössbauer |
|---|---|---|---|---|---|---|---|---|
| | | | (mm) | mmol/kg | % | % | (-) | |
| #08 | MX-80 | raw | raw | 490 | | 18 | 0.22 | n.a. |
| | | contact | 0.3–5 | 610 [a] | 25 | 13 | 0.14 | n.a. |
| | | bulk | >45 | 485 | −1 | 26 | 0.36 | n.a. |
| #11 | Ibecoseal | raw | raw | 476 | | 31 | 0.45 | n.a. |
| | | crust | <0.3 | 2128 [a] | 347 | 24 | 0.31 | goe |
| | | contact | 0.3–5 | 856 | 80 | 33 | 0.48 | goe, hem |
| | | bulk | >45 | 512 | 8 | 34 | 0.51 | goe, hem |
| #12 | Ikosorb | raw | raw | 321 | | | | n.a. |
| | | crust | <0.3 | 1420 [a] | 343 | 9 | 0.10 | goe, hem |
| | | contact | 0.3–5 | 436 | 36 | 10 | 0.11 | goe, hem |
| | | heart | 15–25 | 353 | 10 | 21 | 0.27 | |
| | | bulk | >45 | 290 | −10 | 10 | 0.11 | |
| #13 | Kunigel | raw | raw | 232 | | | | n.a. |
| | | crust | <0.3 | 290 | 25 | | | n.a. |
| #25 | MX80+quartz | | raw | 343 | | 18 | 0.22 | n.a. |
| | | crust | <0.3 | 1989 | 480 | 19 | 0.23 | goe |
| | | contact 1 | 1.5 | 1658 | 384 | 21 | 0.27 | goe, (hem) |
| | | contact 2 | 3 | 604 | 76 | 22 | 0.28 | goe |
| | | contact 3 | 4.5 | 552 [a] | 61 | 21 | 0.27 | goe, (hem) |
| | | bulk | 45 | 378 | 10 | 27 | 0.37 | goe, hem |
| #26 | Deponit | | raw | 581 | | 9 | 0.13 | n.a. |
| | | red | n.a. | 1231 | 112 | 8 | 0.10 | goe |
| | | green | n.a. | 694 | 20 | 17 | 0.25 | goe |
| | | blue | n.a. | 752 | 29 | 22 | 0.38 | goe |
| #27 | MX-80 | | raw | 490 | | 18 | 0.22 | n.a. |
| | | crust | <0.3 | 645 | 32 | 15 | 0.18 | n.a. |
| | | salt | 7 | 576 | 18 | 11 | 0.12 | goe |
| | | bulk | 25 | 533 [a] | 9 | | | n.a. |

[a] Amount of material insufficient for XRF analysis, thus inferred from the EDX chemical profiles.

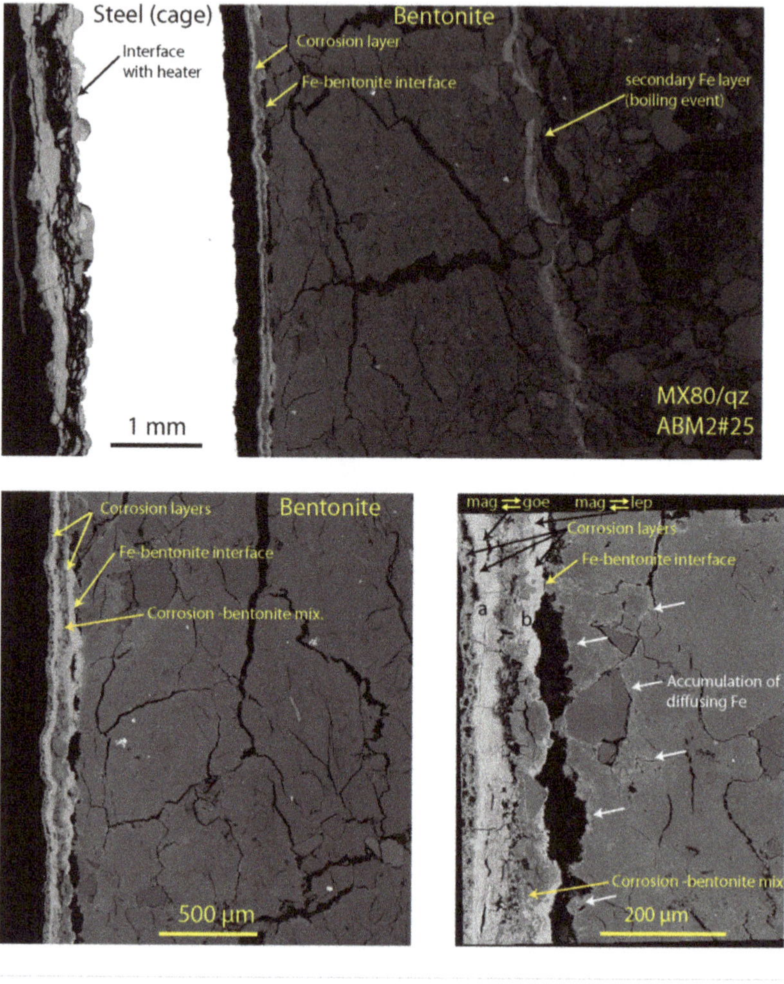



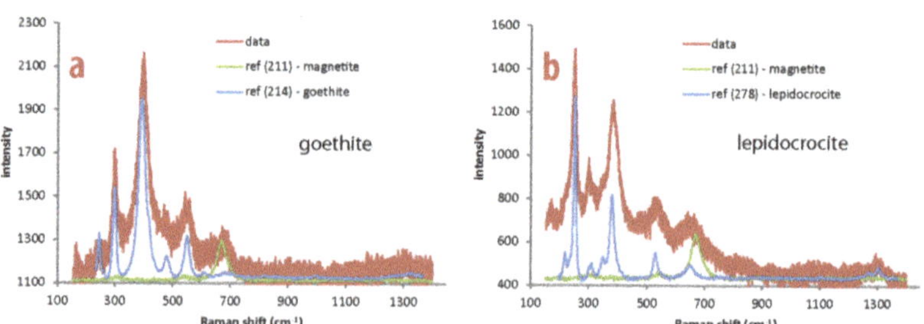



**Figure 8.** SEM micrographs (**upper**) of steel-clay interface of block #25 (MX-80/qz) and µ-Raman spectra of two areas (**lower, a,b**).

The Fe enrichment in the samples closest to the interface of blocks #8, #11, #27 and, to a lesser extent, also #26 parallels the increase in paramagnetic $Fe^{3+}$ leading to lower reduction levels compared to the raw and bulk samples. According to the Mössbauer data, the increase in $Fe^{3+}$ is mainly related to goethite precipitation at the interface and some hematite precipitation identified in blocks #11, #12 and #25. It is important to note that, despite the lower reduction levels, $Fe^{2+}$ content is also increased in the contact zone, which can be attributed to reduced structural Fe in the smectite fraction. In the case of the contact samples from blocks #12 and #25, Fe enrichment is accompanied by a similar increase in paramagnetic $Fe^{3+}$ and $Fe^{2+}$, thus no net change in reduction level. The corresponding Mössbauer spectra in these two samples could be reproduced by assuming the additional presence of goethite and of structural $Fe^{2+}$ in equal amounts.

Further away from the contact, where the Fe content is still above that of the bulk material, the reduction level increases relative to the sample(s) closer to the contact, as indicated for the samples of blocks #11, #12, #25 and #26. This suggests a higher proportion of $Fe^{2+}$ compared to the immediate contact area. In the case of block #25, this trend is very weak. In the case of block #26, the "blue" sample exhibits a slight increase of Fe compared to the bulk and an even higher reduction level, thus a higher proportion of $Fe^{2+}$ compared to the other corresponding samples from blocks #08, #11.

### 3.4. Profiles of Mg, Ca and S

All samples show an accumulation of Mg in the clay in the first mm from the interface, as illustrated by the increase in the Mg/Al ratio (Figure 9). Thus, the Mg/Al ratio increases by a factor of 1.4–2.5 relative to the background level. The amount of Mg is higher in the upper part, indicated for example in the granular MX-80 blocks (Figure 9). In the case of block #26, (Deponit), the Mg accumulation extends over >20 mm. In this zone, magnesium sulphates were identified by SEM/EDX besides accumulations of gypsum/anhydrite (Figure 10) (see below). This is where the centre of the boiling event is suspected.

Although the Fe content is also increased at the interface, there is no clear correlation between the Fe and Mg profiles. The latter generally exhibit a narrower accumulation front and a "monophasic" shape. An exception is the Deponit block which displays a large front, but a comparatively moderate maximum increase in the Mg/Al ratio. The different shapes and extents indicate that different processes control the profiles of Fe and Mg. While the Fe distribution is obviously related to the corrosion of the adjacent steel, the Mg distribution is related to internal processes and the large temperature gradient in the clay. The distribution of exchangeable cations was clearly affected during the experiment as shown by [24]. Notably, these authors identified a general depletion of exchangeable Mg in the bentonite blocks in parallel with the enrichment close to the heater. This loss was more notable in the upper part which was interpreted as being possibly related to the boiling event.

EDX profiles of Ca and S illustrate the significant accumulation of $CaSO_4$ in the upper part (Figure 10), confirming the macroscopic observations, XRD and µ-Raman spectroscopy data which indicated the neo-formation of anhydrite close to the contact, especially in pre-existing voids (e.g., Figure S2(2)). In the lower part, neo-formation of $CaSO_4$ close to the contact also occurred, but in much smaller amounts.

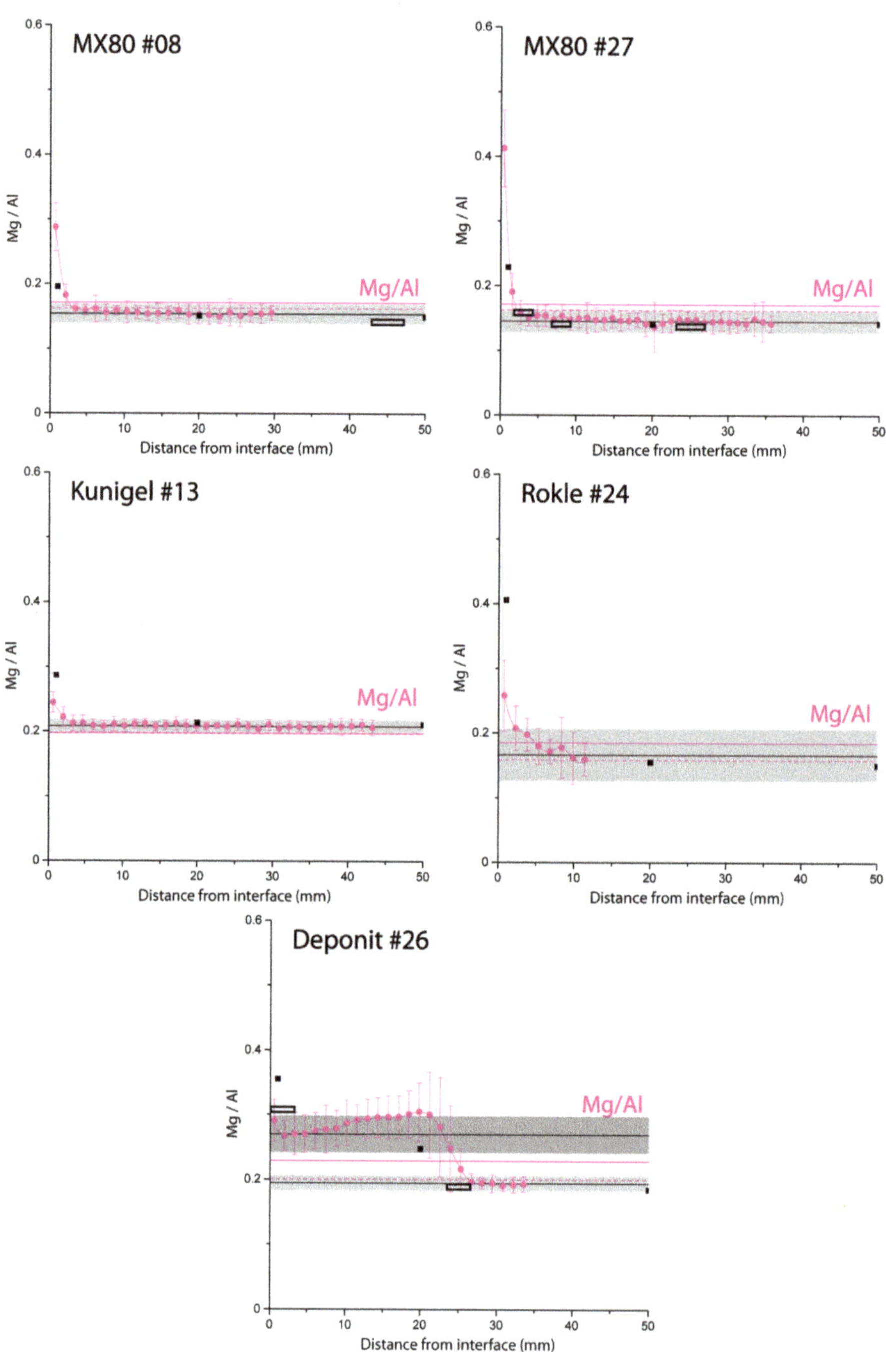



**Figure 9.** Al-normalised Mg profile perpendicular to steel/clay contact for five different blocks. Points: EDX measurements, squares: XRF bulk data from [22], empty rectangles: XRF measurements on powdered samples; grey area: range of bulk.

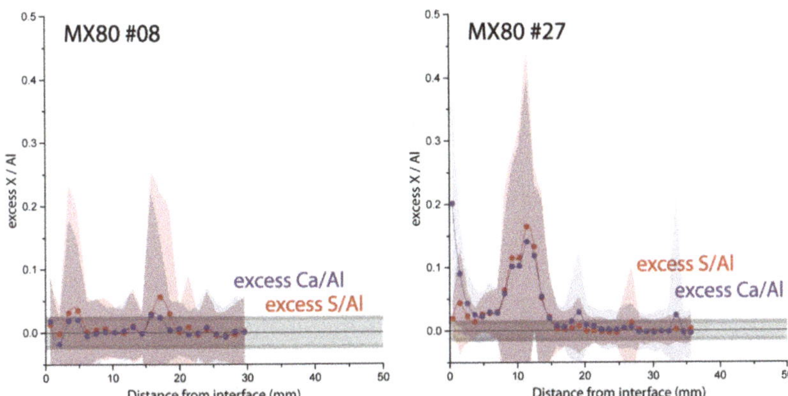

**Figure 10.** Al-normalised excess Ca and S profiles (bulk content has been subtracted) perpendicular to steel/clay contact for blocks with MX-80. EDX measurements with range, grey horizontal area: range of bulk. Light purple area: range of Ca/Al values. Red shaded area: range of S/Al values.

## 4. Discussion

### 4.1. Corrosion Layer and Fe-Clay Interaction Zone

The microscopic and spectroscopic data yield a consistent picture with regards to the structure and the mineralogical patterns in the metal-clay interface area. The steel surface is covered by a corrosion layer (up to several hundreds of μm) consisting of magnetite, goethite, lepidocrocite and, at least partly, of siderite. Note that siderite was also identified in other samples from MX80 blocks #08, #17, #30 and #31 with differential thermal analysis-mass spectrometry by [22]. The corrosion layer is affected to a variable degree by mechanical disturbances resulting in an outer sublayer of corrosion products mixed with clay material. The corrosion layer was not included in the Fe/Al profiles, which aim at the characterisation of Fe migration in the clay matrix. The same corrosion layer features, although less pronounced, were observed in the ABM1 package which was exposed to similar conditions for a shorter period [20]. An analogous pattern at the Fe-bentonite interface was described for the FEBEX in situ experiment at the Grimsel Test Site [13,15]. The layered structure of the interface area is schematically illustrated in Figure 11.

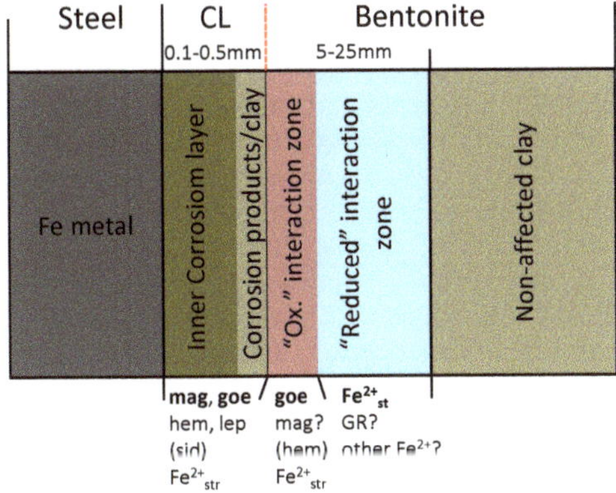


**Figure 11.** Sketch of steel-bentonite interface with neo-formed iron phases (not to scale). CL: corrosion layer. mag: magnetite, hem: hematite, lep: lepidocrocite, sid: siderite, GR: green rust, $Fe^{2+}_{str}$: structural $Fe^{2+}$ in smectite.

The presence and spatial distribution of neo-formed magnetite and siderite besides ferric (oxyhydr)oxides (goethite, hematite, lepidocrocite) in the corrosion layer indicates that redox conditions shifted from oxidising to reducing during the experiment. The increase in the proportion of magnetite relative to ferric (oxyhydr)oxides towards the metal surface suggests that reducing conditions started at the metal contact. It should be noted that magnetite was likely present in the original thin corrosion layer, which was at least partly oxidised to ferric (oxyhydr)oxide during initial the aerobic phase [40]. Moreover, partial oxidation of reduced iron phases during dismantling and sampling cannot be ruled out.

The iron accumulation at the interface on the clay side, which is visible in some areas as reddish crust, consists predominantly of Fe(III) oxides (mainly goethite, less hematite) as deduced from Mössbauer spectrometry. An increase of paramagnetic $Fe^{2+}$ was also identified by this method which was interpreted as structural $Fe^{2+}$ of the clay. The presence of neo-formed magnetite in the Fe-enriched clay could not be confirmed, but its presence in small quantities cannot be ruled out on the basis of Mössbauer spectrometry. It is noteworthy that the identification of newly formed Fe oxides by μ-Raman in polished samples is difficult due to the high background fluorescence of the clay matrix. The Fe front extending further into the clay (further than about 10 mm from contact), is characterised by a higher reduction level compared to the contact zone according to Mössbauer spectrometry, such as indicated from samples of blocks #12 and #26).

The additional $Fe^{3+}$ originates mainly from goethite, but the nature of neo-formed Fe(II) phases is uncertain. The diffusion of soluble $Fe^{2+}$ released from the corrosion layer may undergo different pathways in smectites according to previous studies. $Fe^{2+}$, may sorb to edge sites of the montmorillonite surface or sorb via cation exchange in the interlayers [41,42]. Simultaneously, redox reactions with structural Fe may occur. This could lead to the reduction of structural $Fe^{3+}$ and precipitation of Fe(III) oxide or mixed Fe(II)/Fe(III) oxide (green rust) phases [7]. The reduction of structural Fe is confirmed by Mössbauer spectrometry, but no other neo-formed Fe(II) could be identified. This reduction process appears to be most extreme for block #26 (Deponit), where a blue zone reaches far into the clay (~30 mm). In this zone, the reduction level is 30% compared to 12% in the reference sample. It should be noted that in this area the centre of the boiling event is suspected (see above) which may have enhanced the corrosion and Fe-bentonite interaction process.

No indications of montmorillonite alteration other than partial reduction of structural Fe were found. Thus, no neo-formed non-swelling clay phases, such as Fe-rich 1:1 clay minerals (e.g., berthierine, cronstedite) or chlorite minerals [14,43] could not be detected. The absence of notable amounts of such phases is supported by the constant Al/Si ratio (EDX) perpendicular to the Fe/clay contact evidenced by all samples (Supplementary Materials S1). It should be noted that Svensson (2015) identified neo-formed Fe saponite at the direct contact Fe/clay contact by XRD analysis in another ABM2 block (#9) consisting of FEBEX bentonite. The presence of saponite was partly confirmed by XRD and IR analysis by [22]. In our study, no saponite at the Fe/clay interface was detected with the applied methods; its presence in minor amounts, however, cannot be ruled out.

A comparison of the granular MX-80 block in ABM1 [20] with the corresponding one in ABM2 (block #08) reveals a very similar Fe front in both bentonite materials. This suggests on the one hand that the same Fe diffusion and Fe-bentonite interaction processes were active. On the other hand, it suggests very slow further migration of the front in ABM2 which lasted about twice as long as ABM1.

Similar relationships regarding the iron speciation in the bentonite affected by iron corrosion were also deduced from the studied profile in the FEBEX experiment [15]. The Fe front, however, was larger (>140 mm) which was explained by the longer duration of this experiment (18 years) and, in particular, by the longer duration of oxidising conditions.

### 4.2. A Model of the Fe Diffusion Process

The transfer of corroded Fe to the clay is strongly dependent on the redox conditions which affect both the steel corrosion and the diffusion of Fe in the clay. From the previous discussion it can be inferred that the ABM2 package was exposed to oxidising conditions that shifted to reducing conditions over the course of the experiment. The $Fe^{2+}/Fe^{3+}$ ratio decreases from the metal surface towards the interface, which is followed by an increase of the ratio on the clay side. This detailed analysis of the spatial distribution of the iron species enables a period of variable redox conditions to be distinguished. In this transient period anaerobic corrosion occurred while oxidising conditions still prevailed in the bentonite. In the following, we propose a phenomenological model that integrates the observations in the interface area of the different blocks and separates the evolution into three phases:

Initial state: Steel is coated with a thin iron oxide layer (magnetite and ferric (oxyhydr)oxides) having resulted from atmospheric corrosion [40] which is in contact with unsaturated bentonite containing structural Fe(III) in smectite as main iron pool.

Phase 1: Aerobic corrosion of steel and oxidation of the magnetite layer leads to the formation of Fe(III) oxides. Depending on the moisture content, either the formation of hematite or goethite/lepidocrocite is favoured. $O_2$ and $H_2O$ transfer to steel diminishes as the corrosion proceeds and the corrosion layer thickens.

Phase 2: Anaerobic corrosion of the steel within the corrosion layer leads to the generation of $Fe^{2+}$ and the formation of magnetite and siderite in the corrosion layer. In addition, (fast) electron transfer across the corrosion layer occurs [44,45], generating Fe(II) at the corrosion layer/bentonite interface. This diffusing $Fe^{2+}$ reacts with the remaining $O_2$ present in the bentonite to further produce Fe(III) oxides accumulating at the vicinity of the interface. Diffusion of $O_2$ in the bentonite may be slowed down by sorption processes [46].

Phase 3: Anaerobic conditions throughout: continued anaerobic corrosion of the steel, production of Fe(II) (and magnetite/siderite) and (fast) electron transfer across the corrosion layer. Diffusion of $Fe^{2+}$ and accumulation of $Fe^{2+}$ is observed in the clay. The mechanism of the diffusion and interaction process still needs to be established. It possibly involves a redox reaction with structural $Fe^{3+}$ leading to an increase of structural $Fe^{2+}$ and the precipitation of other Fe(III) oxides or green rust phases at the surface of smectite. Figure 12 illustrates this proposed model as simplified scheme.

### 4.3. A Phenomenological Description of Caged Granular Material

As noted in Section 3.1, iron oxides filling former voids between pellet surfaces surrounded by reddish halos (extending several mm into the clay) were observed in the granular materials (MX80 and MX80/qz) which had otherwise formed a homogeneous mass during the experiment (Figure 3). It should be noted that this feature was limited to the granular materials in the metal cages. Based on the previously described above, a sequence of events is proposed in Figure 13.

The geometrical configuration of the fractures mimics the original shapes of the pellets. This suggests that the corrosion-derived iron oxides were formed in the voids between the pellets during the saturation process. This also implies that anaerobic corrosion releasing $Fe^{2+}$ started when the voids between the pellets were still present. During transport in voids $Fe^{2+}$ reacted with $O_2$ in the partially saturated bentonite. Transport of $O_2$ in this material was presumably attenuated by sorption to the clay [46]. This led to a "corrosion" halo around the voids and contributed to the gradual depletion of $O_2$ in the clay. The Fe oxide filled voids provided preferential pathways for further transport of $Fe^{2+}$ away from the steel/clay interface deeper into the clay.

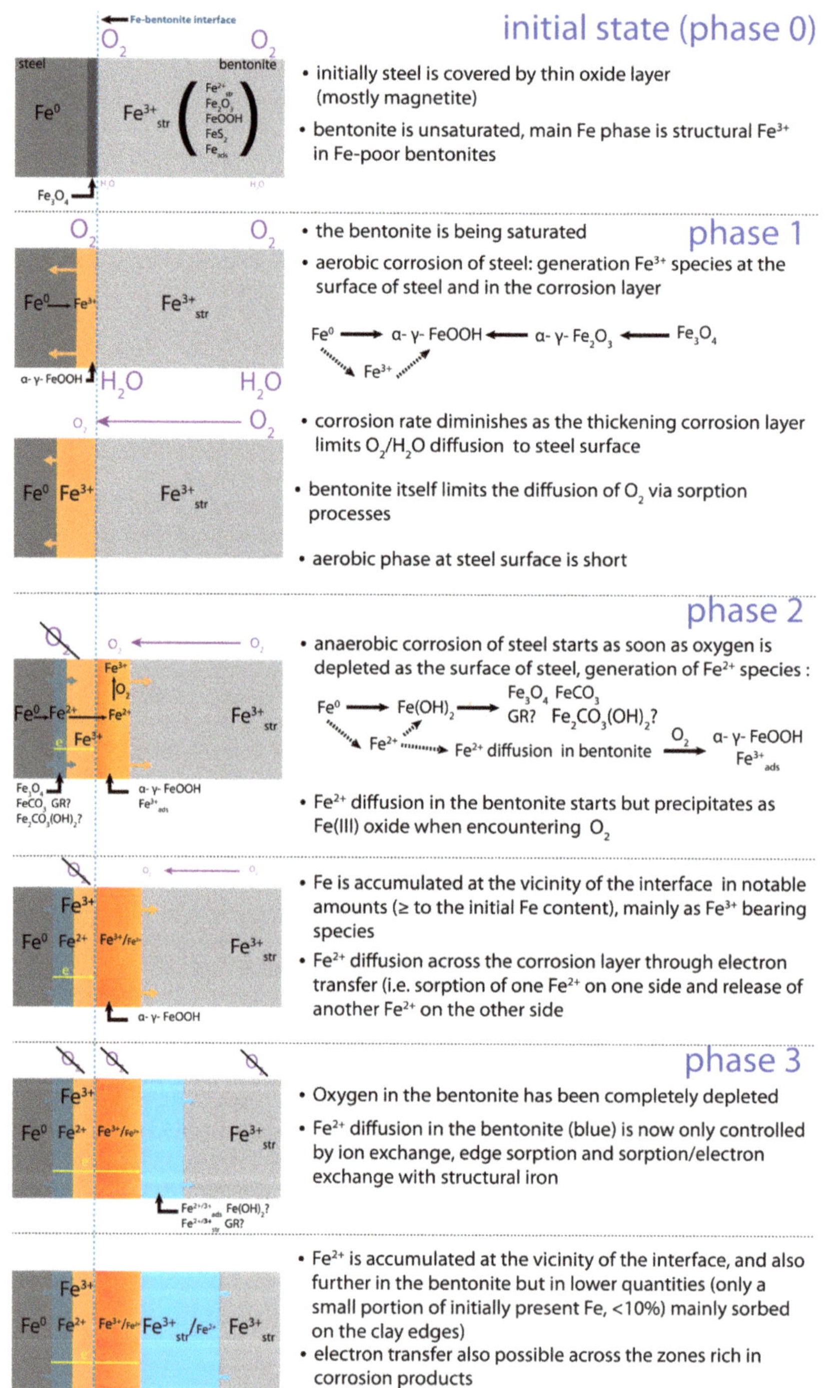



- initially steel is covered by thin oxide layer (mostly magnetite)
- bentonite is unsaturated, main Fe phase is structural $Fe^{3+}$ in Fe-poor bentonites

- the bentonite is being saturated
- aerobic corrosion of steel: generation $Fe^{3+}$ species at the surface of steel and in the corrosion layer

- corrosion rate diminishes as the thickening corrosion layer limits $O_2/H_2O$ diffusion to steel surface
- bentonite itself limits the diffusion of $O_2$ via sorption processes
- aerobic phase at steel surface is short

- anaerobic corrosion of steel starts as soon as oxygen is depleted as the surface of steel, generation of $Fe^{2+}$ species :

- $Fe^{2+}$ diffusion in the bentonite starts but precipitates as Fe(III) oxide when encountering $O_2$

- Fe is accumulated at the vicinity of the interface in notable amounts ($\geq$ to the initial Fe content), mainly as $Fe^{3+}$ bearing species
- $Fe^{2+}$ diffusion across the corrosion layer through electron transfer (i.e. sorption of one $Fe^{2+}$ on one side and release of another $Fe^{2+}$ on the other side

- Oxygen in the bentonite has been completely depleted
- $Fe^{2+}$ diffusion in the bentonite (blue) is now only controlled by ion exchange, edge sorption and sorption/electron exchange with structural iron

- $Fe^{2+}$ is accumulated at the vicinity of the interface, and also further in the bentonite but in lower quantities (only a small portion of initially present Fe, <10%) mainly sorbed on the clay edges)
- electron transfer also possible across the zones rich in corrosion products

**Figure 12.** Conceptual model of corrosion and Fe-clay interaction process in the ABM2 in situ experiment [15].

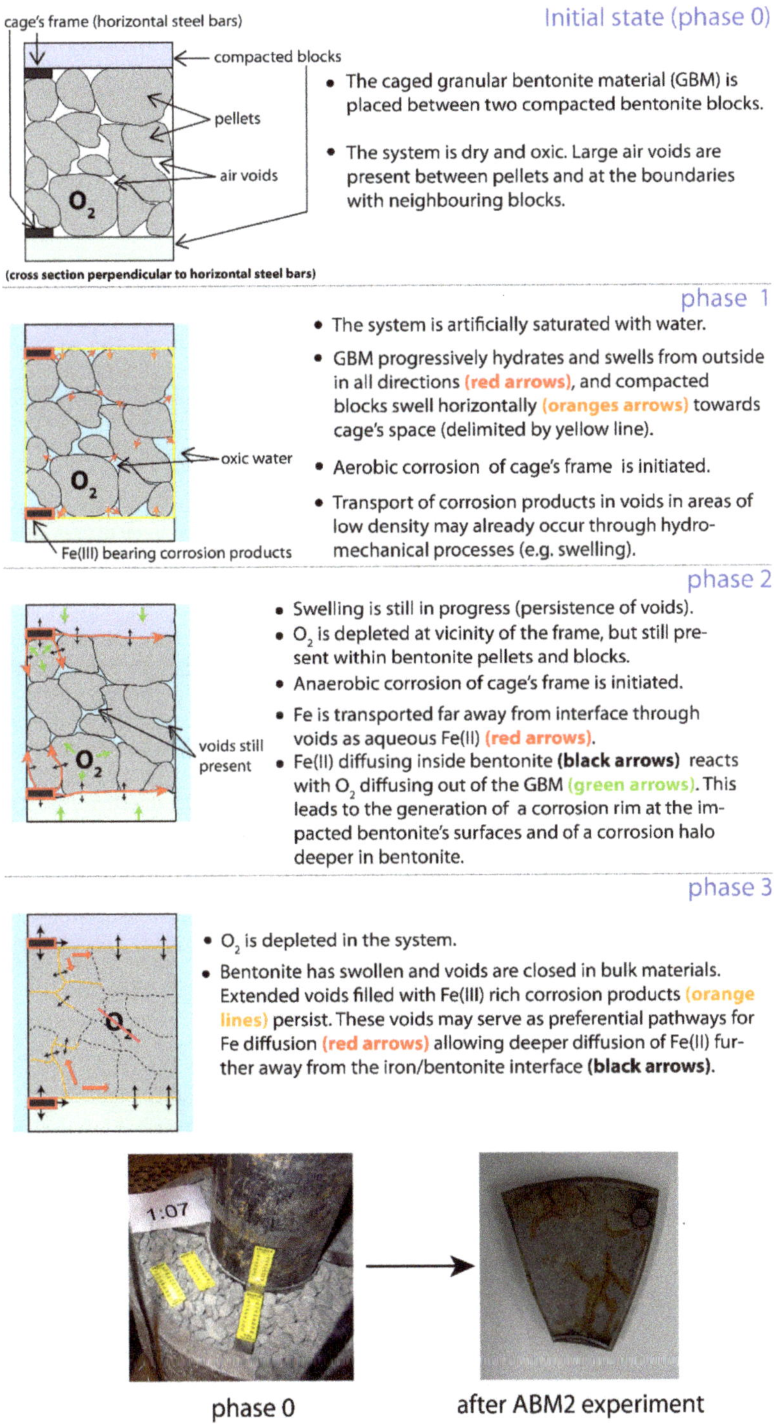

**Initial state (phase 0)**

cage's frame (horizontal steel bars)

compacted blocks

pellets

air voids

$O_2$

(cross section perpendicular to horizontal steel bars)

- The caged granular bentonite material (GBM) is placed between two compacted bentonite blocks.

- The system is dry and oxic. Large air voids are present between pellets and at the boundaries with neighbouring blocks.

**phase 1**

oxic water

$O_2$

Fe(III) bearing corrosion products

- The system is artificially saturated with water.

- GBM progressively hydrates and swells from outside in all directions **(red arrows)**, and compacted blocks swell horizontally **(oranges arrows)** towards cage's space (delimited by yellow line).

- Aerobic corrosion of cage's frame is initiated.

- Transport of corrosion products in voids in areas of low density may already occur through hydro-mechanical processes (e.g. swelling).

**phase 2**

$O_2$

voids still present

- Swelling is still in progress (persistence of voids).
- $O_2$ is depleted at vicinity of the frame, but still present within bentonite pellets and blocks.
- Anaerobic corrosion of cage's frame is initiated.

- Fe is transported far away from interface through voids as aqueous Fe(II) **(red arrows)**.
- Fe(II) diffusing inside bentonite **(black arrows)** reacts with $O_2$ diffusing out of the GBM **(green arrows)**. This leads to the generation of a corrosion rim at the impacted bentonite's surfaces and of a corrosion halo deeper in bentonite.

**phase 3**

- $O_2$ is depleted in the system.

- Bentonite has swollen and voids are closed in bulk materials. Extended voids filled with Fe(III) rich corrosion products **(orange lines)** persist. These voids may serve as preferential pathways for Fe diffusion **(red arrows)** allowing deeper diffusion of Fe(II) further away from the iron/bentonite interface **(black arrows)**.



**Figure 13.** Proposed sequence of processes inside caged pellets based on macroscopic observations.

## 5. Conclusions

The chemical interaction between carbon steel and bentonite was investigated by analysing the Fe/clay interfaces of eight bentonite blocks that had been exposed to temperatures of up to 130 °C for five years in the ABM2 borehole at the Äspö HRL. High spatial resolution methods (SEM/EDX, μ-Raman) were applied in combination with "bulk" methods (XRD, XRF, $^{57}$Fe Mössbauer spectrometry) to determine the Fe front and to unravel processes occurring in the clay as a result of steel corrosion.

Corrosion induced an iron front of 5–20 mm into the bentonite, except for the high-Fe bentonite where no Fe increase was detected. The Fe fronts consisted mainly of ferric (oxyhydr)oxides in addition to the structural Fe in the smectite fraction which had been partially reduced by the interaction process. Additional Fe(II) extended further into the clay, but its nature could not be identified. Mg was also found to be enriched at the Fe/clay interface, presumably because of the elevated temperatures and the temperature gradient within the clay. The Mg accumulation was larger in the upper part of the test package where a boiling event had probably occurred during the experiment. In this upper part, precipitates $MgSO_4$ salts as well as large amounts of anhydrite/gypsum were formed. In lower parts of the test package smaller amounts of $CaSO_4$ in the contact zone were also observed.

The corrosion and Fe-clay interaction process is strongly linked to the redox evolution in the borehole. A conceptual model with three phases is proposed, partly based on similar findings in previous in situ experiments. Initially, when conditions were oxidising, aerobic corrosion at the steel surface occurred producing a ferric (oxyhydr)oxide corrosion layer. After depletion of $O_2$ within the corrosion layer, anaerobic corrosion led to the formation of magnetite and some siderite in the inner part of the corrosion layer. However, ferric (oxyhydr)oxides continued to be formed at the contact to the clay where aerobic conditions still prevailed. Once $O_2$ was depleted in this area, $Fe^{2+}$ diffused into the bentonite and induced a complex interaction process with the clay. The interaction mechanism is not understood in detail but may include $Fe^{2+}$ sorption, reduction of structural Fe together with the formation of ferric (oxyhydr)oxides and/or mixed Fe(III)/Fe(III) phases (green rust).

The blocks, consisting of granular bentonite material contained in metal cages, exhibited specific corrosion features that were not observed in the other blocks. Such features included iron (oxyhydr)oxides filling former voids along with reddish halos extending into the clay. A phenomenological model for the formation of these features is proposed. The model includes the preferential transport of corrosion-derived $Fe^{2+}$ through the voids and a still occurring reaction of $O_2$ in the partially saturated bentonite.

No indications of montmorillonite alteration other than the reduction of structural iron were found. It should be noted that the formation of minor amounts of trioctahedral smectite, which was identified at the Fe/clay contact in previous studies in other blocks of the ABM2 experiment, cannot be ruled out.

**Supplementary Materials:** The following are available online at https://www.mdpi.com/article/10.3390/min11080907/s1, Supplementary Materials S1: SEM-EDX chemical profiles, Supplementary Materials S2: μ-Raman spectroscopy data of reference materials and contact zone, Supplementary Materials S3: XRD and XRF data, Supplementary Materials S4: Mössbauer spectrometry data.

**Author Contributions:** Conceptualization, P.W., A.J., D.S., P.S. and O.X.L.; methodology, P.W., J.H. and A.J.; software, J.H.; validation, J.H., P.W. and A.J.; formal analysis, J.H. and J.-M.G.; investigation, J.H.; resources, J.H.; data curation, J.H.; writing—original draft preparation, P.W.; writing—review and editing, P.W. and J.H.; visualization, J.H. and P.W.; supervision, P.W.; project administration, P.W.; funding acquisition, P.W. All authors have read and agreed to the published version of the manuscript.

**Funding:** This research was partially funded by Nagra (Switzerland) and ṢKB (Sweden).

**Data Availability Statement:** The data used in this study is available from the corresponding author upon request.

**Acknowledgments:** We would like to acknowledge Nadine Lötscher, Thomas Aebi and Stephan Brechbühl (University of Bern) for their cautious preparation of the polished surfaces. We thank Stephan Kaufhold and Reiner Dohrmann from BGR (Germany) for providing the large CEC/XRF data set on the ABM2 samples. The whole team involved in ABM2 is thanked for fruitful discussions. Josh Richards (University of Bern) is acknowledged for editing of language and style. The reviews of two anonymous reviewers significantly helped to improve the manuscript.

**Conflicts of Interest:** The authors declare no conflict of interest.

# References


1. Nagra. *Project Opalinus Clay: Safety Report. Demonstration of Disposal Feasibility for Spent Fuel, Vitrified High-Level Waste and Long-Lived Intermediate-Level Waste (Entsorgungsnachweis)*; Nagra Technischer Bericht NTB 02-05; Nagra: Wettingen, Switzerland, 2002.
2. SKB. *Long-Term Safety for the Final Repository for Spent Nuclear Fuel at Forsmark. Main Report of the SR-Site Project. Volume II*; SKB Tecnical Reports TR 11-01; SKB: Stockholm, Sweden, 2011.
3. Posiva. *Safety Case for the Disposal of Spent Nuclear Fuel at Olkiluoto—Synthesis 2012*; Posiva Reports 2012-12; Posiva: Olkiluoto, Finland, 2013.
4. Jenni, A.; Wersin, P.; Mäder, U.K.; Gimmi, T.; Thoenen, T.; Baeyens, B.; Hummel, W.; Gimmi, T.; Ferrari, A.; Marschall, P.; et al. *Bentonite Backfill Performance in a High-Level Waste Repository: A Geochemical Perspective*; Nagra Technischer Bericht NTB 19-03; Nagra: Wettingen, Switzerland, 2019.
5. Muurinen, A.; Tournassat, C.; Hadi, J.; Greneche, J.M. *Sorption and Diffusion of Fe(II) in Bentonite*; Posiva Working Reports WR 2014-04; Posiva: Olkiluoto, Finland, 2014.
6. Carlson, L.; Karnland, O.; Oversby, V.; Rance, A.; Smart, N.; Snellman, M.; Vähänen, M.; Werme, L. Experimental studies of the interactions between anaerobically corroding iron and bentonite. *Phys. Chem. Earth.* **2007**, *32*, 334–345. [CrossRef]
7. Latta, D.E.; Neumann, A.; Premaratne, W.A.P.J.; Scherer, M.M. Fe(II)–Fe(III) Electron Transfer in a Clay Mineral with Low Fe Content. *ACS Earth Space Chem.* **2017**, *1*, 197–208. [CrossRef]
8. Hadi, J.; Tournassat, C.; Ignatiadis, I.; Greneche, J.M.; Charlet, L. Modelling CEC variations versus structural iron reduction levels in dioctahedral smectites. Existing approaches, new data and model refinements. *J. Colloid Interface Sci.* **2013**, *407*, 397–409. [CrossRef] [PubMed]
9. Wersin, P.; Birgersson, M.; Olsson, S.; Karnland, O.; Snellman, M. *Impact of Corrosion-Derived Iron on the Bentonite Buffer within the KBS-3H Disposal Concept The Olkiluoto Site as Case Study*; Posiva Reports 2007-11; Posiva: Olkiluoto, Finland, 2007.
10. Bradbury, M.; Berner, U.; Curti, E.; Hummel, W.; Kosakowski, G.; Thoenen, T. *The Long Term Geochemical Evolution of the Nearfield of the HLW Repository*; Nagra Technical Reports NTB 12-01; Nagra: Villingen, Switzerland, 2014.
11. Martin, P.-L.; Barcala, J.M.; Huertas, F. Large-scale and long-term coupled thermo-hydro-mechanic experiments with bentonite: The FEBEX mock-up test. *J. Iber. Geol.* **2006**, *32*, 259–282.
12. Fernandez, A.M.; Villar, M.V. Geochemical behaviour of a bentonite barrier in the laboratory after up to 8 years of heating and hydration. *Appl. Geochem.* **2010**, *25*, 809–824. [CrossRef]
13. Wersin, P.K.; Kober, F. (Eds.) *FEBEX-DP. Metal Corrosion and Iron-Bentonite Interaction Studies*; Nagra Arbeitsbericht NAB 16-16; Nagra: Wettingen, Switzerland, 2017.
14. Fernandez, A.M.; Kaufhold, S.; Sanchez-Ledesma, D.M.; Rey, J.J.; Melon, A.; Robredo, L.M.; Fernandez, S.; Labajo, M.A.; Clavero, M.A. Evolution of the THC conditions in the FEBEX in situ test after 18 years of experiment: Smectite crystallochemical modifications after interactions of the bentonite with a C-steel heater at 100 degrees C. *Appl. Geochem.* **2018**, *98*, 152–171. [CrossRef]
15. Hadi, J.; Wersin, P.; Serneels, V.; Greneche, J.M. Eighteen years of steel-bentonite interaction in the FEBEX "in situ" test at the Grimsel Test Site in Switzerland. *Clays Clay Miner.* **2019**, *67*, 111–131. [CrossRef]
16. Svensson, D.; Dueck, A.; Nilsson, U.; Olsson, S.; Sandén, T.; Lydmark, S.; Jägerwall, S.; Pedersen, K.; Hansen, S. *Alternative Buffer Material—STATUS of the Ongoing Laboratory Investigation of Reference Materials and Test Package 1*; SKB Technical Reports TR 11-06; SKB: Stockholm, Sweden, 2011.
17. Svensson, P.D.; Hansen, S. Redox chemistry in two iron-bentonite field experiments at Äspö Hard Rock Laboratory, Sweden: An XRD and Fe k-edge XANES study. *Clays Clay Miner.* **2013**, *61*, 566–579. [CrossRef]
18. Kaufhold, S.; Dohrmann, R.; Sandén, T.; Sellin, P.; Svensson, D. Mineralogical investigations of the first package of the alternative buffer material test—I. Alteration of bentonites. *Clay Miner.* **2013**, *48*, 199–213. [CrossRef]
19. Dohrmann, R.; Olsson, S.; Kaufhold, S.; Sellin, P. Mineralogical investigations of the first package of the alternative buffer material test—II. Exchangeable cation population rearrangement. *Clay Miner.* **2013**, *48*, 215–233. [CrossRef]
20. Wersin, P.; Jenni, A.; Mäder, U.K. Interaction of corroding iron with bentonite in the ABM1 experiment at Äspö, Sweden: A microscopic approach. *Clays Clay Miner.* **2015**, *63*, 51–68. [CrossRef]
21. Svensson, D. The Bentonite Barrier—Swelling Properties, Redox Chemistry and Mineral Evolution. Ph.D. Thesis, Lund University, Lund, Sweden, 2015.
22. Kaufhold, S.; Dohrmann, R.; Götze, N.; Svensson, D. Characterization of the Second Parcel of the Alternative Buffer Material (ABM) Experiment—I Mineralogical Reactions. *Clays Clay Miner.* **2017**, *65*, 27–41. [CrossRef]

23. Hadi, J.; Wersin, P.; Jenni, A.; Greneche, J.M. *Redox Evolution and Fe-Bentonite Interaction in the ABM2 Experiment, Äspö Hard Rock Laboratory*; Nagra Technischer Bericht NTB 17-10; Nagra: Wettingen, Switzerland, 2017.

24. Dohrmann, R.; Kaufhold, S. Characterization of the second package of the Alternative Buffer Material (ABM) experiment—II Exchangeable cation population rearrangement. *Clays Clay Miner.* **2017**, *65*, 104–121. [CrossRef]

25. Wollenberg, R.; Schroeder, H. *Herstellung und Charakterisierung von Bentonitsystemen für den Einsatz als Versiegelungsmaterial*; Nagra Arbeitsbericht NAB 06-20; Nagra: Wettingen, Switzerland, 2006.

26. Nagra. *Alternative Buffer Material–Status Report*; Nagra Arbeitsbericht NAB 11-19; Nagra: Wettingen, Switzerland, 2011.

27. Leal Olloqui, M. *A Study of Alteration Processes in Bentonite*; University of Bristol: Bristol, UK, 2019. Available online: https://research-information.bris.ac.uk/en/studentTheses/a-study-of-alteration-processes-in-bentonite (accessed on 10 October 2020).

28. Kumpulainen, S.; Kiviranta, L. *Mineralogical, Chemical and Physical Study of Potential Buffer and Backfill Materials from ABM Test Package 1*; Posiva Working Reports WR 2011-41; Posiva: Olkiluoto, Finland, 2011.

29. Eng, A.N.U.; Svensson, D. *Äspö Hard Rock Laboratory—Alternative Buffer Material—Installation Report*; SKB International Progress Report IPR-07-15; SKB: Stockholm, Sweden, 2007.

30. Muurinen, A. *Chemical Conditions in the AO Parcel of the Long-Term Test of Buffer Material in Äspö (LOT)*; Posiva Working Report WR 2003-32; Posiva: Olkiluoto, Finland, 2003.

31. Lábár, J.L.; Török, S. A peak-to-background method for electron probe X-ray micro-analysis applied to individual small particles. *X-ray Spectrom.* **1992**, *21*, 183–190. [CrossRef]

32. Trincavelli, J.; Limandri, S.; Bonetto, R. Standardless quantification methods in electron probe microanalysis. *Spectrochim. Acta Part B At. Spectrosc.* **2014**, *101*, 76–85. [CrossRef]

33. Lafuente, B.; Drowns, R.T.; Yang, H.; Stone, N. The Power of Databases: The RRUFF Project. In *Highlights in Mineralogical Chrystallography*; Armbruster, T., Danisi, R.M., Eds.; De Gruyter: Berlin, Germany, 2015; pp. 1–30.

34. Mössbauer, R.L. Kernresonanzabsorption von Gammastrahlung in Ir191. *Naturwissenschaften* **1958**, *45*, 538–539. [CrossRef]

35. Tzara, C. Diffusion des photons sur les atomes et les noyaux dans les cristaux. *J. De Phys. Radium* **1961**, *22*, 303–307. [CrossRef]

36. Gütlich, P.; Bill, E.; Trautwein, A.X. *Mössbauer Spectroscopy and Transition Metal Chemistry*; Springer: Berlin Heidelberg, Germany, 2011; p. 569.

37. Vandenberghe, R.; De Grave, E. Application of Mössbauer Spectroscopy in Earth Sciences. In *Mössbauer Spectroscopy—Tutorial Book*; Yoshida, Y., Langouche, G., Eds.; Springer: Berlin/Heidelberg, Germany, 2013; pp. 91–185.

38. Grim, R.E.; Güven, N. *Bentonites—Geology, Mineralogy, Properties and Uses*; Elsevier Scientific Publishing Company: Amsterdam, The Netherland, 1978; Volume 24, p. 256.

39. Sitek, J.; Araiova, B.; Toth, I. Reduction of Fe in montmorillonite. *Acta Phys. Slovaca* **1995**, *45*, 67–70.

40. King, F. *Corrosion of Carbon Steel under Anaerobic Conditions in a Repository for SF and HLW in Opalinus Clay*; Nagra Technischer Bericht NTB 08-12; Nagra: Wettingen, Switzerland, 2008.

41. Tournassat, C. *Interactions cations—Argiles: Le cas du Fe(II). Application au contexte de stockage profond des déchets radioactifs*; Université Joseph Fourier: Saint-Martin-d'Hères, France, 2003.

42. Soltermann, D.; Fernandes, M.M.; Baeyens, B.; Dahn, R.; Joshi, P.A.; Scheinost, A.C.; Gorski, C.A. Fe(II) uptake on natural montmorillonites. I. Macroscopic and spectroscopic characterization. *Environ. Sci. Technol.* **2014**, *48*, 8688–8697. [CrossRef]

43. Mosser-Ruck, R.; Cathelineau, M.; Guillaume, D.; Charpentier, D.; Rousset, D.; Barres, O.; Michau, N. Effects of Temperature, pH, and Iron/Clay and Liquid/Clay Ratios on Experimental Conversion of Dioctahedral Smectite to Berthierine, Chlorite, Vermiculite, or Saponite. *Clays Clay Miner.* **2010**, *58*, 280–291. [CrossRef]

44. Yanina, S.V.; Rosso, K.M. Linked reactivity at mineral-water interfaces through bulk crystal conduction. *Science* **2008**, *320*, 218–222. [CrossRef] [PubMed]

45. Handler, R.M.; Frierdich, A.J.; Johnson, C.M.; Rosso, K.M.; Beard, B.L.; Wang, C.; Latta, D.E.; Neumann, A.; Pasakarnis, T.; Premaratne, W.A.P.J. Fe(II)-Catalyzed recrystallization of goethite revisited. *Environ. Sci. Technol.* **2014**, *48*, 11302–11311. [CrossRef] [PubMed]

46. Leupin, O.X.; Smart, N.R.; Zhang, Z.; Stefanoni, M.; Angst, U.; Papafotiou, A.; Diomidis, N. Anaerobic corrosion of carbon steel in bentonite: An evolving interface. *Corros. Sci.* **2021**, *187*, 109523. [CrossRef]

*Article*

# Thermally Induced Bentonite Alterations in the SKB ABM5 Hot Bentonite Experiment

Ritwick Sudheer Kumar [1,*], Carolin Podlech [1], Georg Grathoff [1], Laurence N. Warr [1] and Daniel Svensson [2]

[1] Institute of Geography and Geology, University of Greifswald, 17489 Greifswald, Germany; carolin.podlech@uni-greifswald.de (C.P.); grathoff@uni-greifswald.de (G.G.); warr@uni-greifswald.de (L.N.W.)

[2] Department of Research and Safety Assessment, SKB—Swedish Nuclear Fuel and Waste Management Company, SE-102 40 Stockholm, Sweden; Daniel.Svensson@skb.se

[*] Correspondence: sudheerkur@uni-greifswald.de

**Abstract:** Pilot sites are currently used to test the performance of bentonite barriers for sealing high-level radioactive waste repositories, but the degree of mineral stability under enhanced thermal conditions remains a topic of debate. This study focuses on the SKB ABM5 experiment, which ran for 5 years (2012 to 2017) and locally reached a maximum temperature of 250 °C. Five bentonites were investigated using XRD with Rietveld refinement, SEM-EDX and by measuring pH, CEC and EC. Samples extracted from bentonite blocks at 0.1, 1, 4 and 7 cm away from the heating pipe showed various stages of alteration related to the horizontal thermal gradient. Bentonites close to the contact with lower CEC values showed smectite alterations in the form of tetrahedral substitution of $Si^{4+}$ by $Al^{3+}$ and some octahedral metal substitutions, probably related to ferric/ferrous iron derived from corrosion of the heater during oxidative boiling, with pyrite dissolution and acidity occurring in some bentonite layers. This alteration was furthermore associated with higher amounts of hematite and minor calcite dissolution. However, as none of the bentonites showed any smectite loss and only displayed stronger alterations at the heater–bentonite contact, the sealants are considered to have remained largely intact.

**Keywords:** bentonite; HLRW; ABM test; smectite alteration; SEM-EDX; repository; high temperatures

**Citation:** Sudheer Kumar, R.; Podlech, C.; Grathoff, G.;Warr, L.N.; Svensson, D. Thermally Induced Bentonite Alterations in the SKB ABM5 Hot Bentonite Experiment. *Minerals* **2021**, *11*, 1017. https://doi.org/10.3390/min11091017

Academic Editors: Ana María Fernández, Stephan Kaufhold, Markus Olin, Lian-Ge Zheng, Paul Wersin and James Wilson

Received: 11 August 2021
Accepted: 14 September 2021
Published: 18 September 2021

**Publisher's Note:** MDPI stays neutral with regard to jurisdictional claims in published maps and institutional affiliations.

## 1. Introduction

Bentonite barriers are currently being considered for the sealing of high-level radioactive waste (HLRW) repositories to encapsulate the waste safely and prevent the circulation of hydrous fluids. Bentonites are types of clays formed commonly as an alteration product of volcanic ash, with a high content (typically >60%) of smectite minerals [1]. Smectites have the unique property of swelling in the presence of water, which leads to extremely low permeabilities (values in $10^{-13}$ m·s$^{-1}$ for highly compacted bentonites [2]), ideal for sealing underground cavities. Therefore, bentonites are likely to be installed in future engineered radioactive waste sites that use the multibarrier storage concept [3].

An important part of evaluating the suitability of bentonite barriers is to test their performance in pilot repository sites where materials are emplaced under constrained conditions for several years [3–6]. It is also important to determine how the bentonite properties respond to more extreme HLRW conditions, in particular when subjected to elevated temperatures [7,8]. High-temperature conditions are known to increase the rate of smectites alterations significantly [9,10], and may therefore lead to the eventual breakdown of the clay seal in HLRW repositories. The SKB (Swedish Nuclear Fuel and Waste Management Company, Stockholm, Sweden) has been testing the performance of bentonite barriers for many years in a granite formation at a depth of 500 m using the currently favored KBS-3 (kärnbränslesäkerhet, nuclear fuel safety) concept at a test site in the Äspö Hard Rock Laboratory [5]. The SKB conducted a series of six ABM (alternative buffer material) experiments between 2006 and 2017 using a range of compacted bentonites simulating different temperatures and water saturation conditions [11–15], where the ABM3, ABM4 and ABM6 tests are still running and are expected to be excavated in 2024.

The ABM5 test, which is the subject of this study, was installed in 2012 and excavated in June 2017. This experiment was placed in a tunnel at a depth of approximately 420 m as part of the Äspö Hard Rock Laboratory site. It contained twelve compacted bentonites, of which the following five were the subject of this study: MX-80, FEBEX, Asha-NW BFL-L, Rokle and Ibeco SEAL M-90. ABM5 differed from previous tests in that notably high temperatures were reached (150–250 °C) compared to the 80–130 °C conditions of the ABM1–2 tests. As a result, it represents one of the hottest bentonite experiments yet conducted in an underground rock laboratory [16]. In addition, the ABM5 series also suffered conditions related to fracturing of the host rock, uncontrolled water inflow and boiling [16] and, as a result, possibly experienced steam–bentonite interactions. All these conditions make ABM5 of particular interest for studying one of the worst-case scenarios of an HLRW repository. Previous studies of the ABM test materials have revealed localized bentonite alterations induced by thermal, mineralogical and geochemical conditions in the underground experimental environment [11–16]. A notable decrease in swelling pressure was documented in some of the Na-bentonites in the ABM1 test, attributed to interlayer cation exchange of $Ca^{2+}$ replacing $Na^+$ [11]. A decrease in exchangeable $Na^+$ and $Mg^{2+}$ cations and an increase in $Ca^{2+}$ cations were also documented in some ABM2 investigations [13,15]. Based on differences in the Na/Mg cation ratio, Dohrmann and Kaufhold [15] suggested that cation exchange was influenced by possible water loss caused by the pressure drop recorded in the ABM2 experiment. Moreover, a reduction in cation exchange capacity (CEC) by 5.5 cmol·$kg^{-1}$ near the heating tube–bentonite contact zone was documented by Dohrmann et al. [13] in some bentonites following underground alterations in the ABM2 experiment.

Investigations of some ABM1 and ABM2 materials also revealed localized increases in the $Fe_2O_3$ content due to corrosion of the heating tube [12,14], the occasional accumulation of organic carbon and anhydrite [12] as well as minor dissolution of clinoptilolite and cristobalite [12]. In addition, the formation of a tri-octahedral smectite phase, possibly saponite, was identified in some samples from the bentonite–heater contact [12,14].

The hot bentonite ABM5 experimental setup contained ring-shaped bentonite blocks inserted through an iron tube of 3 m height and 10 cm diameter, which contained three 1000 W heaters inside. The three heaters were positioned as follows: (1) along the whole test length (main heater), (2) at a depth of 0 to 1 m (top heater) and (3) at a depth of 2 to 3 m (bottom heater), in order to achieve an equal distribution of temperature [5]. Each compacted bentonite block had an average thickness of 10 cm, an inner diameter of 11 cm and an outer diameter of 27.7 cm. A pressure of 100 MPa corresponding to a load of 5089 kN was applied to the bentonites for compaction [5]. The outermost slot between the bentonite blocks and the surrounding rock was filled with gravel and titanium tubes of 6 mm diameter used for artificial saturation. The chemistry of these artificial fluids was assumed to be Na-Ca-Cl-dominated groundwater [17]. There was also an additional water inflow to the experimental system through a fracture located 0.8 m beneath the floor [16]. The gravel around the bentonite blocks distributed water slowly and evenly through the system. Thermocouples were installed in five blocks (3rd, 9th, 15th, 21st and 27th block) to measure the internal temperature variations. Within each of the five bentonite blocks (Table 1) three thermocouples were installed at 0.5 cm, 4 cm and 7 cm from the bentonite–heater interface and, in the same five block positions, a thermocouple was installed in the heating tube's interior [5].

**Table 1.** Depth and block position of the compacted bentonites installed in the ABM5 experiment [5] with maximum thermocouple readings from inside the heating tube. The studied bentonites are marked in green.

| Depth (m) | Block Number | Compacted Bentonite Blocks | Maximum Thermocouple Reading Inside the Heating Tube (°C) |
|---|---|---|---|
| 0.1 | 30 | MX-80 | - |
| 0.2 | 29 | MX-80 | - |
| 0.3 | 28 | Asha 505 | - |
| 0.4 | 27 | Calcigel | 188 |
| 0.5 | 26 | Deponit CAN | - |
| 0.6 | 25 | FEBEX | - |
| 0.7 | 24 | GMZ | - |
| 0.8 | 23 | Ibeco SEAL M-90 | - |
| 0.9 | 22 | Ikosorb | - |
| 1.0 | 21 | Kunigel V1 | 240 |
| 1.1 | 20 | MX-80 | - |
| 1.2 | 19 | Asha NW BFL-L | - |
| 1.3 | 18 | Rokle | - |
| 1.4 | 17 | Saponite | - |
| 1.5 | 16 | Asha 505 | - |
| 1.6 | 15 | MX-80 | 251 |
| 1.7 | 14 | Rokle | - |
| 1.8 | 13 | FEBEX | - |
| 1.9 | 12 | Saponite | - |
| 2.0 | 11 | Ibeco SEAL M-90 | - |
| 2.1 | 10 | Calcigel | - |
| 2.2 | 9 | Asha NW BFL-L | 251 |
| 2.3 | 8 | MX-80 | - |
| 2.4 | 7 | Ikosorb | - |
| 2.5 | 6 | GMZ | - |
| 2.6 | 5 | Kunigel V1 | - |
| 2.7 | 4 | Deponit CAN | - |
| 2.8 | 3 | Asha NW BFL-L | 156 |
| 2.9 | 2 | MX-80 | - |
| 3.0 | 1 | MX-80 | - |

In ABM5, the starting temperature was regulated to 50 °C to avoid any boiling due to the increased water pressure in the surrounding gravel filter. The temperature was later increased to 150 °C and then to 250 °C for almost six months in 2016 before final excavation of the materials for investigation (Table 1) [16]. As a result, boiling occurred in the experiment as the pressure could not be maintained. The first mineralogical results published by Kaufhold et al. [16] described reactions similar to those occurring in previous experiments. Dissolution and precipitation of carbonate and sulfur phases were reported. Notable Fe corrosion and an increase in the total amount of $Mg^{2+}$ at the heater contact zone were also reported.

In this contribution, the nature of the mineral alterations under these extreme HLRW repository conditions is further documented and discussed. The prime aim of the study was to establish more specifically the changes in smectite composition occurring in the ABM5 bentonite blocks subjected to enhanced thermal conditions and to relate these changes to the more general alteration features.

## 2. Materials and Methods

The following set of five bentonite samples were obtained: (1) Asha NW BFL-L(ANB) from the depth of 2.7 to 2.8 m (3rd block), (2) Ibeco SEAL M-90 (IBS) from the depth of 1.9 to 2 m (11th block), (3) FEBEX (FEB) from the depth of 1.7 to 1.8 m (13th block), (4) Rokle (ROK) from the depth of 1.6 to 1.7 m (14th block) and (5) MX-80 from the depth of 0 to 0.1 m (30th block) (Table 1). The origin of the MX-80, IBS, ROK, ANB and FEB bentonites were from Wyoming (USA), Askana (Georgia), the Kadan Basin (Prague, Czech Republic), Kutch (India) and Almeria (Spain), respectively [5]. The assumed temperatures for each of the five bentonites were selected based on the location of thermocouples (Table 1). The samples of the ABM5 test package were cut down into arc-shaped slices vertically and then sliced horizontally into two halves. The bentonite blocks were tightly sealed in air-tight sealed bags and stored in this state prior to analysis. Machine cutting marks were visible on sample surfaces when the bags were opened and sampled in the laboratory. All the instruments used for analyses are housed in the mineralogical laboratories of the University of Greifswald, Germany.

### 2.1. Sampling Strategy

For detailed analysis of each bentonite, four specimens were extracted horizontally from each section. Vertical sampling was not possible as only one of the blocks was obtained from SKB for the study, but the largest alteration was assumed to occur horizontally due to the thermal gradient. Each arc-shaped bentonite block obtained had a 9 to 10 cm radius and a vertical height of 4.5 to 5.5 cm (approximately half of the total bentonite block's height). The top or bottom of each block represented either a bentonite–bentonite contact zone or a machine-cut face. A 1-cm-thick surface portion of the sample material was removed to avoid impurities from the bentonite–bentonite contacts and areas of minor deformation associated with machine cutting. The 8 to 10 cm portion was also removed from each block due to gravel impurities present in the outer circle. Samples were extracted from the bentonite block–heating tube contact zone (0.1 cm) and at 1, 4 and 7 cm away from the heating pipe in order to study material subjected to the horizontal thermal gradient (Figure 1). The relative humidity of the laboratory during the sampling and further analyses deviated between 45–50% at 23–25 °C.

**Figure 1.** Sampling locations in a bentonite block (0.1, 1, 4 and 7 cm) from heater contact to outer rim in a bentonite block, e.g., sample MX-80.

### 2.2. X-ray Diffraction (XRD) Analysis

Bulk (whole rock) powder mineral assemblages were determined by XRD analysis using a Bruker D8 Advance diffractometer with CoKα-radiation (40 kV, 30 mA). The samples were micronized using a Glen Creston McCrone micronizing mill to a particle size of <10 µm. The micronized samples were prepared as random powders without any preferred particle orientations by placing them in sample holders using the side-loading technique [18]. Random powders were measured from 3° to 100° 2θ using a step size of 0.02° 2θ with a divergence slit set to 0.5° and a scanning rate of 1° 2θ/minute. The software EVA (Bruker) and the Profex database [19] were used for mineral identification.

For detailed clay mineral analysis, 1 g of each sample was sieved (<63 µm) and the <2 µm size fraction separated gravitationally using Stokes' law [18]. The <2 µm clay suspension was oriented on a glass slide (45 mg/cm$^2$) and then measured in an air-dried condition, after ethylene glycol saturation (24 h) and after heating up to 550 °C for one hour to identify the clay mineral assemblages [18]. The oriented slides were measured from 4° to 40° 2θ using a step size of 0.02° 2θ with a divergence slit set to 1° and a scanning rate of 1° 2θ/minute.

The random powder samples were quantified by Rietveld refinement using the software Profex/BGMN (Version 3.14.3, Nicola Doebelin, RMS Foundation, Bettlach, Switzerland) [19]. Two separate smectite structure files (a one-water layer, Na-smectite, and a two-water layer, Ca-smectite) were applied for quantifying these phases in the program (Figure 2). This procedure obtained a better fit, considering the smectites have varying interlayer compositions and amounts of adsorbed water [20]. For these two structure files, all settings other than the interlayer parameter and the layer factor parameter were refined globally to avoid distinct refinement factors such as b1l and k2l (profile-shaping parameters) in the quantification [21]. The mineral species quantified at <1% were considered uncertain unless distinct reflections were observed over the entire XRD pattern. The level of error involved in the quantitative analyses is considered to be similar to those reported by Kemp et al. [22]. Using similar methods, they measured errors of +1% for >50 wt.%, ±5% for 50 to 20 wt.% and ±10% for <10 wt.%.

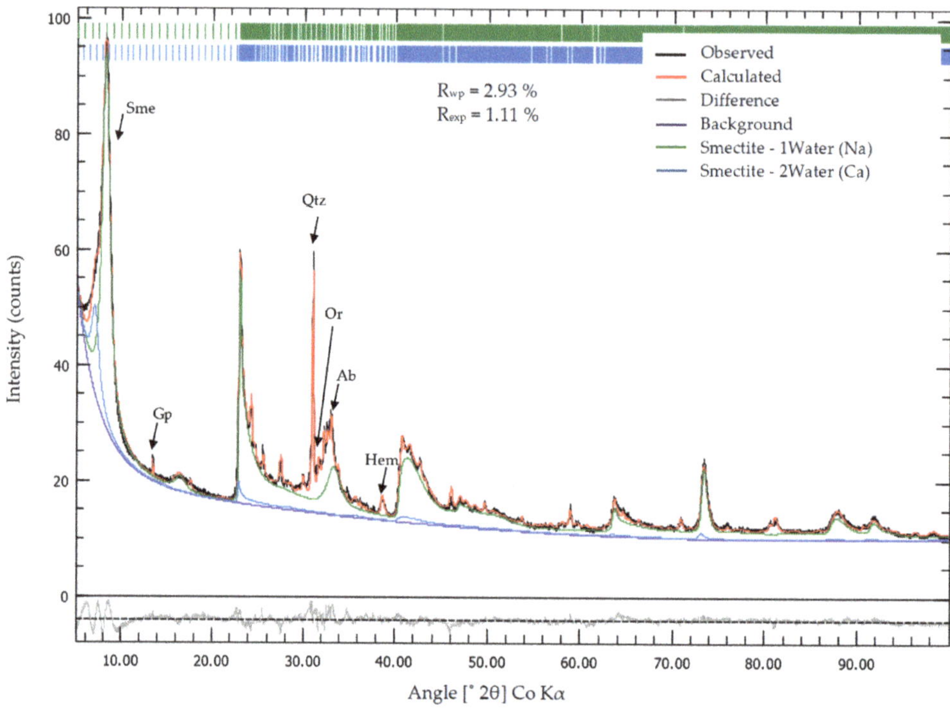



**Figure 2.** X-ray diffraction pattern of the MX-80 bentonite (0.1 cm sample) showing Rietveld refinement using 1-water and 2-water layer smectite structures. (Hem, hematite; Sme, smectite; Qz, quartz; Gp, gypsum; Or, orthoclase; Ab: albite).

### 2.3. Cation Exchange Capacity (CEC) and Exchangeable Cations (EC)

The CEC was determined using the Cu-trien method [23,24] by adding 150 mg of micronized sample to 50 mL of ultrapure water and ultrasonically dispersing it. For each sample, the CEC was determined twice to check the reproducibility of the results. The measurements showed an average standard deviation of $\pm 5.7$ cmol·kg$^{-1}$.

The suspension obtained after CEC measurement was used to quantify the exchangeable cations by atomic absorption spectroscopy (AAS). The relative $Ca^{2+}$, $Na^{+}$, $Mg^{2+}$ and $K^{+}$ cation concentrations were determined using the calibration curve method [25].

### 2.4. Smectite Purification and Energy-Dispersive X-ray (EDX) Spectroscopy

The chemical composition of purified smectites was measured by energy-dispersive X-ray spectroscopy (EDX). Purification and preparation of the samples was undertaken using the methodology described by Podlech et al. [26]. The degree of purity of the separates was controlled by transmission electron microscopy examination using a JEOL JEM-2100Plus instrument. For this study, a ZEISS EVO MA10 scanning electron microscope (SEM) with an EDAX element analyzer was used. The measurements were performed at a 10.5 mm working distance, with an accelerating voltage of 15 kV, a 6000× magnification and a dwell time of 100 μs. Thirty frames per map were measured for three EDX sections per sample, and the average values (n = 3) were used to quantify the smectite's composition. The calculated element-specific mean deviations in weight percentage (wt.%) from the EDX mapping were 0.25 for Na (0–5 wt.%), 0.14 for K (0–5 wt.%), 0.35 for Ca (0–5 wt.%), 0.16 for Mg (0–5 wt.%), 0.40 for Al (0–16 wt.%), 0.72 for Si (0–35 wt.%) and 0.36 for Fe (0–10 wt.%). The obtained oxide chemistries, the EDX analyses were normalized to 100%.

The composition of the smectites was calculated using the structural formula method (SFM) [27,28]. The calculations were done on the basis of 11 oxygen atoms equivalent per half unit cell (e·phuc$^{-1}$). While evaluating the EDX maps, traces of impurities such as chlorine, sulfur and titanium were identified in most smectites. The sulfur and titanium impurities were neglected in calculating the smectite oxide compositions and structural formulae. The atomic percentage of $Cl^-$ and associated $Na^+$ was also removed, assuming minor NaCl impurities to be present. The di-octahedral smectite structure was used for the SFM calculations by restricting the octahedral occupancy to a maximum of 2. The SFM method has its limitations in calculating octahedral charges when $Fe^{3+}$ and $Fe^{2+}$ are not separately measured. In this study, the Fe content was assumed to be $Fe^{3+}$ as it is considered to be the dominant form in di-octahedral smectites, based on published studies [29]. The presence of minor impurities in purified mineral specimens may also lead to erroneous SFM calculations. For example, the ROK smectite fraction was found to contain traces of hematite and therefore considered unsuitable for study.

### 2.5. pH Measurements

The relative pH values for all samples were determined with a pH meter. For this, 50 mg of each bentonite micronized powder was dispersed in 40 mL of double-distilled water using an ultrasonic homogenizer (clay to water ratio 1:800). The suspensions were kept on a shaking table at 110 rpm for one week before measurement.

## 3. Results

### 3.1. Sample Observations

When removed from the air-tight bags, most bentonite blocks, except the ANB material, quickly disintegrated when extracting portions for study, revealing that significant changes in physical properties occurred following alteration after ABM5. All the bentonites, except ANB, were dry at the contact zone and up to 4–6 cm away, but were moist in the deeper sections. The ANB sample was comparatively rigid and moist throughout and could withstand the pressure of a hand drill while sampling. A dark coloration was observed in all of the bentonite blocks in the contact zone region. The MX-80 bentonite block displayed a noticeable color change when visibly examined, ranging from dark-orangish brown at the contact zone to a light-yellowish gray in the outer circle (Figure 3a,b). The FEB and IBS bentonite blocks did not show any visible color variations except minor darkening of the area in the contact zone with the heating tube.

**Figure 3.** (**a**) Image showing the contact between MX-80 bentonite and the heating tube. (**b**) Image showing visible color variations in MX-80 bentonite samples. Numbers in the photograph indicate distance in cm from the bentonite–heat interface.

### 3.2. XRD Random Powder and Oriented Preparations

The random powder patterns of MX-80 bentonite sections showed mineralogical differences throughout the sample. Hematite was only observed in the 0.1 cm sample whereas gypsum occurred in samples up to 4 cm from the contact, but was not observed at 7 cm (Table 2). A 0.303 nm XRD peak, representative of calcite, was detected at 7 cm. The smectite XRD reflections at 1.24 nm in the 0.1 cm sample shifted to between 1.27 nm and 1.48 nm at 4 and 7 cm distance from the heated tube. This change is considered to represent different hydration states due to cation exchange reactions within the MX-80 smectite. All other bentonite samples did not show significant differences in XRD random powder patterns, and hence mineral content, within the horizontal sections.

**Table 2.** List of minerals identified in specific sample sections. (BDL: below detection limit.)

| Bentonites | Contact Zone (0.1 cm) | 1 cm | 4 cm | 7 cm |
| --- | --- | --- | --- | --- |
| MX-80 | Sme, Qz, Hem, Gp, Py, Sa, Ab, Ant, Cal, 1.0 nm mica. | Same as 0.1 cm, Hem is BDL | Same as 0.1 cm, Hem is BDL | Same as 0.1 cm, Hem, Py and Gp are BDL |
| ANB | Sme, Qz, Hem, Cal, Kln, Sd, Rt and Ant | Same as 0.1 cm | Same as 0.1 cm | Same as 0.1 cm |
| IBS | Sme, Qz, Or, Ab, Py, 1.0 nm mica. | Same as 0.1 cm | Same as 0.1 cm | Same as 0.1 cm |
| FEB | Sme, Qz, Or, Ab, Crs, Cal, 1.0 nm mica. | Same as 0.1 cm | Same as 0.1 cm | Same as 0.1 cm |
| ROK | Sme, Qz, Kln, Cal, Ant, Py, Or, Hem, Sd, 1.0 nm mica. | Same as 0.1 cm | Same as 0.1 cm | Same as 0.1 cm |

Hem, hematite; Cal, calcite; Sme, smectite; Qz, quartz; Gp, gypsum; Py, pyrite; Sd, siderite; Crs, cristobalite; Kln, kaolinite; Ann, annite; Or, orthoclase; Sa, sanidine; Ab, albite; Ant, anatase; Rt, rutile. IMA-CNMNC-approved mineral symbols [30].

Quantitative analysis of the MX-80 bentonite showed recognizable variations in the abundance of hematite, gypsum and calcite across the sections, despite being low in abundance (<1.3 wt.%; Table 3). The MX-80 bentonite showed a consistent trend of increasing amounts of calcite away from the contact. The calcite content increased, between 0.1 cm and 7 cm, from 0.1 to 1.3 wt.%. In contrast, only the 0.1 cm sample of MX-80 bentonite showed the presence of hematite. A similar calcite gradient was observed in ANB bentonite, as described in the MX-80 section. However, all the other minerals in ANB bentonite showed no change across the sampled block. The ROK bentonite showed a steady decrease in the quantity of hematite away from the contact zone (from 2.6 wt.% to 0.5 wt.%). However, no other specific mineral abundance trends were observed in the ROK bentonite section.

Based on mineral quantifications, the formation of calcite and hematite appears to occur in only two of the five bentonites and none of the bentonites showed significant indication of any smectite loss. The IBS and FEB bentonites revealed no specific differences in the XRD patterns following their refinement (Table 3).

**Table 3.** Rietveld refinement quantifications of the five bentonites from the ABM5 experiment (data in wt.%).

| Minerals | MX80 0.1 | MX80 1 | MX80 4 | MX80 7 | ANB 0.1 | ANB 1 | ANB 4 | ANB 7 | ROK 0.1 | ROK 1 | ROK 4 | ROK 7 | IBS 0.1 | IBS 1 | IBS 4 | IBS 7 | FEB 0.1 | FEB 1 | FEB 4 | FEB 7 |
|---|---|---|---|---|---|---|---|---|---|---|---|---|---|---|---|---|---|---|---|---|
| Hem | 1.0 | - | - | - | 2.7 | 2.6 | 3.1 | 2.5 | 2.6 | 1.3 | 1.0 | 0.5 | - | - | - | - | - | - | - | - |
| Cal | 0.1 | 0.5 | 0.6 | 1.3 | 0.9 | 1.8 | 2.0 | 5.0 | 1.2 | 1.0 | 0.7 | 1.0 | - | - | - | - | 1.1 | 0.9 | 1.6 | 1.0 |
| Sme | 81.0 | 85.4 | 82.3 | 81.3 | 90. | 89.5 | 88.5 | 87.3 | 82.0 | 82.8 | 83.5 | 82.2 | 92.3 | 92.2 | 91.5 | 93.9 | 89.1 | 86.7 | 85.8 | 85.4 |
| Qz | 5.6 | 6.0 | 4.8 | 6.1 | 2.7 | 3.0 | 3.0 | 1.7 | 3.3 | 4.0 | 4.5 | 4.5 | 1.0 | 1.0 | 1.0 | 0.9 | 2.4 | 2.2 | 1.9 | 2.0 |
| Gp | 0.5 | 0.8 | 0.9 | 0.3 | - | - | - | - | - | - | - | - | - | - | - | - | - | - | - | - |
| Py | 0.3 | 0.1 | 0.2 | - | - | - | - | - | 0.2 | 0.3 | 0.2 | 0.3 | 0.1 | 0.1 | - | - | - | - | - | - |
| Sd | - | - | - | - | 0.2 | 0.3 | 0.3 | 0.4 | 0.8 | 0.9 | 0.8 | 0.6 | - | - | - | - | - | - | - | - |
| Crs | - | - | - | - | - | - | - | - | - | - | - | - | - | - | - | - | 0.3 | - | 0.3 | - |
| Kln | - | - | - | - | 2.2 | 1.9 | 2.1 | 2.4 | 2.2 | 2.1 | 1.9 | 2.3 | - | - | - | - | - | - | - | - |
| Ann | - | - | - | - | - | - | - | - | - | - | - | - | 0.8 | 0.8 | 0.7 | 0.5 | - | - | - | - |
| Or | - | - | - | - | - | - | - | - | 3.6 | 3.6 | 3.5 | 3.8 | 1.1 | 1.7 | 1.1 | 2.1 | 2.7 | 2.5 | 1.7 | 1.6 |
| Sa | 4.8 | 3.6 | 5.1 | 4.2 | - | - | - | - | - | - | - | - | - | - | - | - | - | - | - | - |
| Ab | 6.4 | 3.4 | 5.9 | 6.6 | - | - | - | - | - | - | - | - | 4.7 | 4.2 | 5.6 | 2.7 | 4.3 | 7.8 | 8.5 | 10.0 |
| Ant | 0.4 | 0.2 | 0.2 | 0.3 | 0.6 | 0.5 | 0.5 | 0.5 | 4.1 | 4.1 | 4.0 | 5.0 | - | - | - | - | - | - | - | - |
| Rt | - | - | - | - | 0.5 | 0.3 | 0.5 | 0.3 | - | - | - | - | - | - | - | - | - | - | - | - |
| Total | 100 | 100 | 100 | 100 | 100 | 100 | 100 | 100 | 100 | 100 | 100 | 100 | 100 | 100 | 100 | 100 | 100 | 100 | 100 | 100 |

Hem, hematite; Cal, calcite; Sme, smectite; Qz, quartz; Gp, gypsum; Py, pyrite; Sd, siderite; Crs, cristobalite; Kln, kaolinite; Ann, annite; Or, orthoclase; Sa, sanidine; Ab, albite; Ant, anatase; Rt, rutile. IMA-CNMNC-approved mineral symbols are used [30]; 0.1, 1, 4 and 7 indicate distance in cm from the heating tube.

### 3.3. Relative pH Measurements

The MX-80 bentonite showed a distinct pattern of decreasing pH from slightly basic to acidic when moving toward the heater contact zone from the outer rim. All other bentonites did not show statistically relevant pH variations (Figure 4).

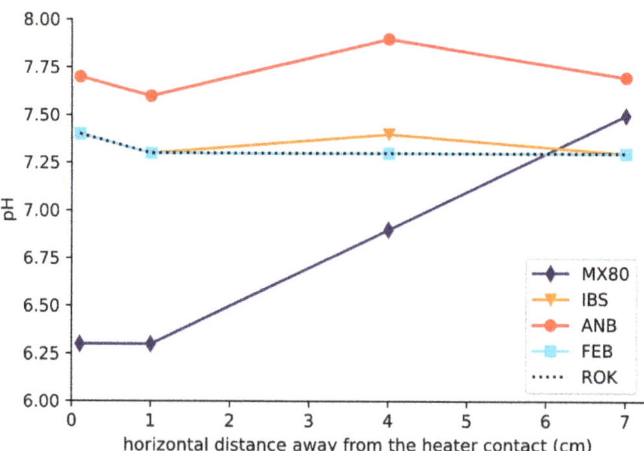


**Figure 4.** Plot showing the pH values for all bentonite clay water suspensions at different distances to the heater.

### 3.4. Cation Exchange Capacity and Exchangeable Cations

In the MX-80, FEB and IBS bentonites, the CEC was seen to decrease towards the contact zone. In contrast, ROK and ANB showed no changes in the CEC (Table 4).

**Table 4.** Exchangeable cations (in ppm) and average CEC values (in $cmol \cdot kg^{-1}$), with an average standard deviation of $\pm 5.6$ $cmol \cdot kg^{-1}$ based on two measurements for each specimen taken from the five analyzed bentonite blocks. (CO: contact zone at 0.1 cm).

| Samples | Na | K | Ca | Mg | Total Cation Charges (= Na + K + (2 × Ca) + (2 × Mg)) | Average CEC |
|---|---|---|---|---|---|---|
| MX-80 CO | 29.7 | 3.7 | 7.2 | 1.4 | 50.5 | 74 |
| MX-80 1 | 32.4 | 3.3 | 15.5 | 1.4 | 69.6 | 81 |
| MX-80 4 | 35.6 | 3.3 | 15.0 | 1.5 | 71.9 | 85 |
| MX-80 7 | 37.1 | 3.3 | 10.9 | 1.5 | 65.1 | 91 |
| IBS CO | 24.0 | 3.7 | 21.4 | 2.2 | 75.0 | 72 |
| IBS 1 | 21.7 | 3.5 | 16.6 | 2.2 | 62.8 | 75 |
| IBS 4 | 22.7 | 3.4 | 18.2 | 2.3 | 67.3 | 81 |
| IBS 7 | 26.4 | 3.6 | 23.9 | 2.8 | 83.2 | 80 |
| ROK CO | 14.8 | 2.8 | 12.2 | 3.5 | 49.0 | 104 |
| ROK 1 | 15.3 | 2.6 | 13.4 | 3.5 | 51.6 | 102 |
| ROK 4 | 16.2 | 2.5 | 15.1 | 3.7 | 56.2 | 106 |
| ROK 7 | 16.0 | 2.2 | 13.9 | 3.4 | 52.8 | 109 |
| ANB CO | 19.9 | 2.2 | 38.0 | 0.7 | 99.5 | 102 |
| ANB 1 | 19.1 | 2.0 | 36.0 | 0.7 | 94.6 | 95 |
| ANB 4 | 20.3 | 2.1 | 41.1 | 0.7 | 106.1 | 97 |
| ANB 7 | 21.6 | 1.6 | 40.1 | 0.6 | 104.7 | 113 |
| FEB CO | 17.2 | 3.2 | 21.3 | 5.4 | 73.7 | 94 |
| FEB 1 | 17.8 | 3.2 | 16.7 | 5.5 | 65.3 | 100 |
| FEB 4 | 18.5 | 3.1 | 17.7 | 5.3 | 67.5 | 99 |
| FEB 7 | 21.1 | 3.7 | 20.8 | 6.1 | 78.7 | 108 |

A relative decrease in exchangeable $Na^+$ cations towards the contact zone was observed in the MX-80, ANB, FEB and ROK bentonites. In addition, a minor increase in exchangeable $K^+$ cations towards the contact zone were also observed in the MX-80, ANB and ROK bentonites. In contrast, the IBS samples showed no significant differences in the exchangeable cations measured (Table 4).

### 3.5. Smectite Purification and Energy-Dispersive X-ray Spectroscopy (EDX)

Several differences regarding the smectite layer charge distribution (tetrahedral vs. octahedral charges) and the metal content of the sites were observed in different bentonite samples.

The amount of tetrahedral Si content decreased towards the contact in the MX-80, IBS and FEB bentonites (Figure 5a, Tables 5 and 6). This was accompanied by an increase in tetrahedral Al towards the contact in these bentonites (Table 6), which led to an increase in tetrahedral charges close to the heating tube. In addition, an increase in the total structural Fe content towards the contact was also observed in the octahedral sheets of the MX-80 and IBS bentonites (Figure 5b). The deviations in these metals led to significant changes in both tetrahedral and octahedral charges of the ABM5 sections close to the heating tube compared to the less altered bentonite located 7 cm from the contact (Table 6). Tetrahedral charges increased toward the contact, whereas octahedral charges decreased in the same direction in the case of MX-80 and IBS bentonites. In contrast, the ANB bentonite showed a decrease in both octahedral and tetrahedral charges, whereas the FEB material showed an increase in both. The overall change in the total layer charge for all studied smectites therefore showed no consistent pattern of alteration. The decrease in tetrahedral Si and increase in octahedral Fe was notably high between the 0.1 cm and 1 cm samples for all of the studied bentonites, except ROK.

**Table 5.** Chemical compositions of the purified smectites from the EDX measurements, shown as oxide percentages (average oxide percentages of the three mapped areas for each sample were used (Section 2.4) (CO: contact zone at 0.1 cm)).

| Samples | $SiO_2$ | $Al_2O_3$ | $Fe_2O_3$ | MgO | CaO | $Na_2O$ | $K_2O$ | Total |
|---|---|---|---|---|---|---|---|---|
| MX-80 CO | 62.4 | 22.6 | 9.2 | 2.4 | 0.3 | 2.8 | 0.3 | 100 |
| MX-80 1 | 64.1 | 22.8 | 8.1 | 2.1 | 0.4 | 2.5 | 0.1 | 100 |
| MX-80 4 | 64.4 | 23.1 | 6.3 | 2.4 | 0.6 | 3.1 | 0.2 | 100 |
| MX-80 7 | 66.0 | 23.6 | 4.7 | 2.5 | 0.9 | 2.0 | 0.2 | 100 |
| IBS CO | 64.8 | 20.7 | 6.8 | 5.1 | 1.4 | 0.5 | 0.7 | 100 |
| IBS 1 | 66.3 | 21.1 | 4.6 | 4.8 | 1.3 | 1.3 | 0.6 | 100 |
| IBS 4 | 65.4 | 20.9 | 4.5 | 5.3 | 1.7 | 1.4 | 0.8 | 100 |
| IBS 7 | 67.8 | 20.9 | 3.9 | 4.9 | 1.2 | 0.9 | 0.3 | 100 |
| ANB CO | 58.8 | 22.0 | 9.7 | 3.3 | 2.3 | 3.4 | 0.3 | 100 |
| ANB 1 | 60.9 | 20.8 | 8.5 | 4.0 | 2.1 | 3.5 | 0.2 | 100 |
| ANB 4 | 58.4 | 20.9 | 9.7 | 4.3 | 1.1 | 5.3 | 0.3 | 100 |
| ANB 7 | 57.6 | 20.7 | 9.7 | 4.1 | 2.8 | 4.9 | 0.2 | 100 |
| FEB CO | 62.7 | 20.1 | 5.2 | 5.4 | 1.5 | 4.4 | 0.6 | 100 |
| FEB 1 | 63.3 | 20.2 | 4.8 | 5.1 | 2.1 | 4.0 | 0.6 | 100 |
| FEB 4 | 63.7 | 20.6 | 4.5 | 5.1 | 1.6 | 3.8 | 0.7 | 100 |
| FEB 7 | 63.8 | 20.8 | 4.9 | 5.2 | 1.4 | 3.3 | 0.6 | 100 |

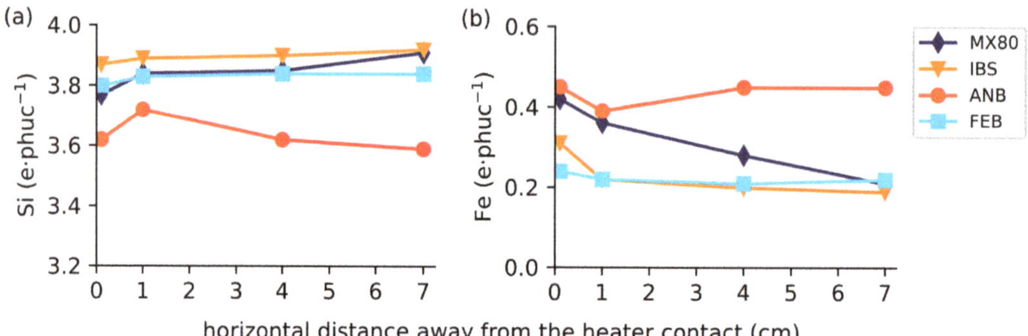


**Figure 5.** Elemental distribution of (**a**) Si in the tetrahedral sheet and (**b**) total Fe content in the octahedral sheet of the smectites.

**Table 6.** Calculated mineral formulae of smectites, including the charges for the tetrahedral (TET) sheet and the octahedral (OCT) sheet, as well as the total layer charges. (CO: contact zone at 0.1 cm.) (Sheet and layer charges are given in e·phuc$^{-1}$.)

| Samples | Tetrahedral (Max 4) | | Octahedral (Max 2) | | | Interlayer Cations | | | | TET Charge | OCT Charge | Total Layer Charge | Interlayer Cation Charge |
|---|---|---|---|---|---|---|---|---|---|---|---|---|---|
| | Si | Al | Al | Fe | Mg | K | Na | Ca | Mg | | | | |
| MX80 CO | 3.77 | 0.23 | 1.38 | 0.42 | 0.20 | 0.02 | 0.30 | 0.02 | 0.02 | −0.23 | −0.20 | −0.42 | 0.40 |
| MX80 1 | 3.84 | 0.16 | 1.45 | 0.36 | 0.18 | 0.01 | 0.28 | 0.02 | 0.00 | −0.16 | −0.19 | −0.35 | 0.34 |
| MX80 4 | 3.85 | 0.15 | 1.48 | 0.28 | 0.21 | 0.01 | 0.33 | 0.04 | 0.00 | −0.15 | −0.29 | −0.44 | 0.42 |
| MX80 7 | 3.91 | 0.09 | 1.55 | 0.21 | 0.23 | 0.01 | 0.22 | 0.06 | 0.00 | −0.09 | −0.26 | −0.35 | 0.35 |
| IBS CO | 3.87 | 0.13 | 1.33 | 0.31 | 0.36 | 0.05 | 0.06 | 0.09 | 0.10 | −0.13 | −0.36 | −0.49 | 0.49 |
| IBS 1 | 3.89 | 0.11 | 1.37 | 0.22 | 0.41 | 0.06 | 0.18 | 0.10 | 0.04 | −0.11 | −0.41 | −0.52 | 0.52 |
| IBS 4 | 3.90 | 0.10 | 1.37 | 0.20 | 0.42 | 0.06 | 0.16 | 0.11 | 0.04 | −0.10 | −0.43 | −0.53 | 0.53 |
| IBS 7 | 3.92 | 0.08 | 1.38 | 0.19 | 0.42 | 0.05 | 0.16 | 0.10 | 0.05 | −0.08 | −0.43 | −0.52 | 0.52 |
| ANB CO | 3.62 | 0.38 | 1.22 | 0.45 | 0.29 | 0.03 | 0.29 | 0.15 | 0.01 | −0.38 | −0.39 | −0.77 | 0.63 |
| ANB 1 | 3.72 | 0.28 | 1.22 | 0.39 | 0.37 | 0.02 | 0.29 | 0.14 | 0.00 | −0.28 | −0.42 | −0.70 | 0.58 |
| ANB 4 | 3.62 | 0.38 | 1.14 | 0.45 | 0.39 | 0.03 | 0.53 | 0.08 | 0.00 | −0.38 | −0.43 | −0.81 | 0.71 |
| ANB 7 | 3.59 | 0.41 | 1.10 | 0.45 | 0.38 | 0.01 | 0.44 | 0.19 | 0.00 | −0.41 | −0.58 | −0.99 | 0.83 |
| FEB CO | 3.80 | 0.20 | 1.24 | 0.24 | 0.50 | 0.05 | 0.42 | 0.10 | 0.00 | −0.20 | −0.57 | −0.76 | 0.67 |
| FEB 1 | 3.83 | 0.17 | 1.27 | 0.22 | 0.46 | 0.05 | 0.35 | 0.14 | 0.00 | −0.17 | −0.62 | −0.79 | 0.68 |
| FEB 4 | 3.84 | 0.16 | 1.31 | 0.21 | 0.45 | 0.05 | 0.36 | 0.10 | 0.00 | −0.16 | −0.55 | −0.70 | 0.61 |
| FEB 7 | 3.84 | 0.16 | 1.32 | 0.22 | 0.46 | 0.05 | 0.26 | 0.09 | 0.00 | −0.16 | −0.47 | −0.62 | 0.49 |

## 4. Discussion

### 4.1. Thermal Gradient and Boiling

The five-year-long ABM5 test represents an exceptional underground rock experiment that underwent extreme temperatures and boiling conditions. Furthermore, the temperature was not distributed equally in all of the bentonites, and, as a result, a diverse range of thermal conditions occurred (Figure 6). According to the temperature measurements of the interior of the tube, the ROK, FEB and IBS bentonites even reached maximum temperatures approaching 250 °C, whereas ANB and MX-80 bentonites experienced lower temperature readings of 156 °C and 188 °C, respectively. Peak temperatures in the 3rd, 9th and 15th block positions were also reached in January 2017, while the peak temperatures in the 21st and 27th blocks occurred later in March 2017. These readings therefore indicate that the vertical thermal distribution in the upper and lower parts of the ABM5 package was not uniform. As a result of the complex thermal history, a fair comparison between the five studied bentonites is therefore not easy to draw. The high-temperature (∼250 °C) readings in the 9T, 15T and 21T (T: thermocouples) are also likely to reflect the boiling event that affected this part of the experiment setup (Figure 6).

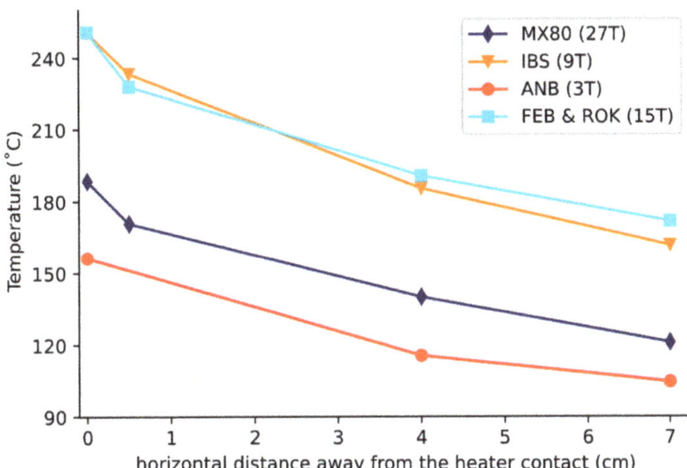

**Figure 6.** Plot showing maximum thermocouple readings closest to the studied bentonites.

### 4.2. Mineralogical Alteration of the Bentonite Blocks and Reaction Mechanisms

The Rietveld refinement and EDX-SEM analyses indicate that more significant mineralogical alterations of the ABM5 bentonites occurred at the contact with the heating tube as a response to the high temperatures and possible oxidative environment. Significant differences in smectite composition and the CEC occurred as well as minor differences in the abundance of hematite and calcite.

#### 4.2.1. Smectite Alteration

Extreme temperatures have been shown to cause a decrease in the silica content of smectites in a number of previous studies [31,32]. One of the possible reasons for the silica loss in ABM5 can be explained by the selective dissolution of tetrahedral Si and substitution by Al, where the amount of reaction increases significantly with temperature [32]. Moreover, steam formed as part of the boiling event in the ABM5 may have also enhanced the dissolution of silica in the bentonites. The possibility of steam–silica interactions and silica dissolution in this type of material have been documented in a study by Heuser et al. [33]. The ABM5 study confirmed this specific type of temperature-related reaction with the highest degree of silica substitution observed in the contact zone of the IBS, MX-80 and FEB bentonites. These changes in tetrahedral metals led to an increase in the negative tetrahedral charge and were commonly accompanied by the substitution of octahedral $Al^{3+}$ and $Mg^{2+}$ by $Fe^{3+}$ that resulted in a decrease in octahedral layer charges. As there was also an increase in the Fe content of the smectites at the contact (Figure 7, Table 6), it is likely that this metal was derived from corrosion of the heating tube or by the oxidation of pyrite. The exact reasons behind these substitutions remain unclear and require further study.

The slight increase in interlayer $K^+$ (Table 4) and total layer charges (Table 6) towards the heater contact zone observed in two of the ABM5 bentonites (MX80 and FEB) is of particular interest. Similar increases in smectite layer charge and $K^+$ fixation were observed in laboratory experiments that studied the smectite to illite conversion [34], and are generally considered to reflect the initial stages of illitization. However, based on these experiments as well as the available kinetic models [9,10] for the smectite to illite conversion, it is evident that the degree of smectite alteration was actually quite low for temperatures that reached 250 °C, as in the FEB and ROK bentonites at the heater contact. At this specific location, the temperatures were equivalent to the experimental conditions used by Huang et al. [9], where significant amounts of illitization occurred during a time period of just 30 days compared to the 5 years of reaction time characterizing the ABM5 test. Although there were probably differences in the clay to fluid ratio's, the slow reaction progress is best explained by a low activity of $K^+$ in the conditions of the repository test, and this highlights the importance of using bentonites with low concentrations of $K^+$ by avoiding those with significant quantities of K-bearing minerals. In this context, all of the bentonites of this investigation contained <5 wt.% K-feldspar content and no micaceous minerals (Table 3). However, the reason why the MX80 and FEB bentonites showed an increase in total layer charge, compared to the other bentonites (IBS and ANB) that showed a decrease, is not yet apparent, but it does appear to be related to the type of octahedral metal substitutions that occurred (Table 6, Figure 7).

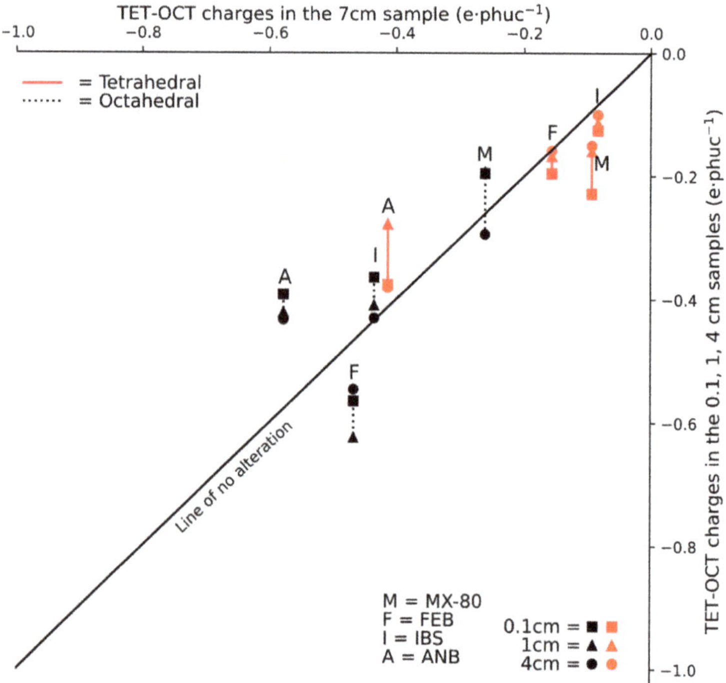


**Figure 7.** Plot showing smectite alterations in 0.1, 1 and 4 cm samples compared to less altered material at 7 cm distance from the heated tube. TET, tetrahedral; OCT, octahedral.

### 4.2.2. Mineral Abundance Variations in Hematite and Calcite

The higher hematite content in the contact zone has been attributed to iron corrosion in a number of previous studies [12,14,16]. The source of iron in the ABM5 experimental setup could have been derived from the heating tube, which would indicate the mobility of iron. The heating tube used in the ABM5 test was manufactured using P235TR1 carbon steels [5].

The ABM5 experiment was not a completely closed experiment. Kaufhold et al. [16] mentioned a fracture in the host rock 0.8 m beneath the floor. As internal fractures were observed in the experimental setup of ABM5, there is a possibility that groundwater flowed in the system carrying oxygen within.

However, hematite occurring at the heater contact was not common to all five bentonites. Only ROK and MX-80 bentonites showed a specific increase in hematite abundance. A further explanation may be the oxidation of the pyrite present in both of these bentonites. Pyrite oxidation results in the formation of sulfuric acid and $Fe^{2+}$, creating an acidic environment that favors the corrosion of the carbon steel. Such environments can also lead to the dissolution of carbonate minerals such as calcite, forming $CO_2$ as a byproduct [35]. $CO_2$ can later dissolve in the interstitial water creating carbonic acid, further increasing the system's acidity [35]. The $Fe^{2+}$ cations formed as a byproduct of pyrite oxidation then later oxidize to the $Fe^{3+}$ oxidation state, resulting in the formation of hematite [35,36]. Verron et al. [37] studied bentonite mixtures with pyrite and carbonates at 100 °C and observed the dissolution of calcite and the formation of minerals such as hematite, anhydrite and beidellite, although the latter two minerals were not investigated in the ABM5 investigation.

The decrease in calcite abundance observed at the heater contact in the MX-80 and ANB bentonites also indicate enhanced dissolution under more acidic conditions. This calcite dissolution can also be best explained as a byproduct of pyrite oxidation [37]. The pH measurements of the MX-80 bentonite showed slightly acidic pH in the samples near the heating tube and slightly basic pH towards the 7 cm section. Calcite dissolution and precipitation in some bentonites were also reported in previous ABM studies by Kaufhold et al. [14,16].

### 4.3. Cation Exchange Capacities and Exchangeable Cations

The analyzed CEC pattern from this study agrees with Dohrmann et al. [13], who described CEC values for the ABM2 bentonites, which were, on average, 5.5 cmol·kg$^{-1}$ lower than the reference materials. The current study shows an average of 10.8 cmol·kg$^{-1}$ decrease in CEC for 0.1 cm samples compared to the 7 cm samples. The high-temperature conditions of the heating tube–bentonite interface clearly negatively influenced the CEC of bentonites at this location.

One of the possible reasons for the decrease in the CEC of smectite could be a decrease in layer charge as documented by a number of studies [24,38]. The low layer charges of smectite observed in the 0.1 cm samples of ANB and IBS bentonites support this idea and the correlation fit between CEC and the total layer charges does indicate that a general trend exists between the two parameters (Figure 8).

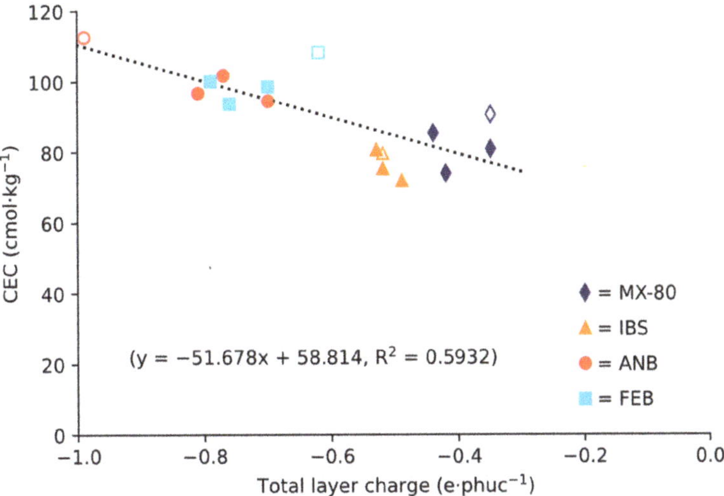


**Figure 8.** Total layer charge (e·phuc$^{-1}$) (Table 6) plotted against CEC values (cmol·kg$^{-1}$) (Table 4). (Solid markers = values of the 0.1, 1 and 4 cm samples; open markers = values of the 7 cm sample.)

Another reason for the decrease in CEC can be explained by the generation of steam during the ABM5's boiling event. Hot water vapor leading to a reduction in the CEC of Na-smectites has been experimentally demonstrated by Heuser et al [33], and the ABM2 experiment, which shows a similar CEC pattern, also suffered a similar boiling event but at a lower temperature range.

## 5. Conclusions

(1) The SKB ABM5 in situ test provided an important opportunity to investigate bentonite stability under extreme repository conditions that included elevated temperatures, fractured host rock, groundwater inflow and boiling. Mineral alterations and CEC variations occurred mainly at the contact with the heater tube, whereas little effect was detected at 7 cm distance.

(2) Smectite alteration occurred by the substitution of tetrahedral $Si^{4+}$ by $Al^{3+}$ and possible substitution of $Al^{3+}$ and $Mg^{2+}$ by $Fe^{3+}$ were detected. These substitution reactions most commonly resulted in an increase in tetrahedral charges and a decrease in octahedral charges.

(3) The increase in octahedral Fe in the smectite near the contact zone was sourced from corrosion of the carbon steel heating tube and/or by the oxidation of pyrite, which locally led to increased acidity.

(4) Minor dissolution of calcite and neoformation of hematite at the bentonite contacts with the heater tube was also related to the high-temperature, acidic and partly oxidizing conditions associated with the extreme conditions of fracturing and boiling occurring in the ABM5 test.

**Author Contributions:** Conceptualization; R.S.K., G.G., L.N.W. and C.P.; methodology, R.S.K., G.G., L.N.W. and C.P.; validation, R.S.K., C.P., G.G. and L.N.W.; formal analyses, R.S.K.; investigation, R.S.K. and C.P.; resources, D.S., L.N.W. and G.G.; writing—original draft preparation, R.S.K., L.N.W., G.G. and C.P.; writing—review and editing, all authors.; visualization, R.S.K.; supervision, G.G. and L.N.W.; project administration, G.G. and L.N.W.; funding acquisition, L.N.W. All authors have read and agreed to the published version of the manuscript.

**Funding:** This project was conducted within the framework of the UMB1 program on reaction mechanisms in bentonite barriers, which was financed by the Federal Ministry of Economics and Technology (BMWi). Project number 02E11344C. We also acknowledge the support for the Article Processing Charge from the DFG (German Research Foundation, 393148499) and the Open Access Publication Fund of the University of Greifswald.

**Data Availability Statement:** The data presented in this study are partially available on request from the corresponding author.

**Acknowledgments:** The authors are grateful to Markus Peltz (UG) for his suggestions with Rietveld refinement and SEM-EDX analyses. We would also like to thank Gabriele Wiederholt (UG), Robert Mrotzek (UG) and Florian Krebs (UG) for their help with the analytical work. This study used DFG-funded XRD and TEM instrumentation (project numbers 108031954 and 428027021) and an EFRE-funded SEM (project numbers GHS-15-0006).

**Conflicts of Interest:** The authors declare no conflict of interest.

## References

1. Christidis, G.E. Origin of the Bentonite Deposits of Eastern Milos, Aegean, Greece: Geological, Mineralogical and Geochemical Evidence. *Clays Clay Miner.* **1995**, *43*, 63–77. [CrossRef]
2. Pusch, R. *Permeability of Highly Compacted Bentonite*; SKBF/KBS TR 80-16; Svensk Kärnbränslehantering AB: Stockholm, Sweden, 1980.
3. Sellin, P.; Leupin, O.X. The Use of Clay as an Engineered Barrier in Radioactive-Waste Management—A Review. *Clays Clay Miner.* **2014**, *61*, 477–498. [CrossRef]
4. Karnland, O.; Olsson, S.; Dueck, A.; Birgersson, M.; Nilsson, S.; Erikson, T.E.; Rosburg, B. *Long Term Test of Buffer Material at the Äspö Hard Rock Laboratory, LOT Project—Final Report on the A2 Test Parcel*; TR-09-29; Svensk Kärnbränslehantering AB: Stockholm, Sweden, 2009.
5. Sandén, T.; Nilsson, U.; Andersson, L.; Svensson, D. *ABM45 Experiment at Äspö Hard Rock Laboratory—Installation Report*; P-18-20; Svensk Karnbransleforsorjning AB: Stockholm, Sweden, 2018.
6. Torres, E.; Turrero, M.J.; Moreno, D.; Sánchez, L.; Garralón, A. FEBEX In-Situ Test: Preliminary Results of the Geochemical Characterization of the Metal/Bentonite Interface. *Procedia Earth Planet. Sci.* **2017**, *17*, 802–805. [CrossRef]
7. Karnland, O.; Olsson, S.; Nilsson, U. *Mineralogy and Sealing Properties of Various Bentonites and Smectite-Rich Clay Materials*; TR-06-30; Svensk Kärnbränslehantering AB: Stockholm, Sweden, 2006.

8. Kaufhold, S.; Dohrmann, R. Distinguishing between more and less suitable bentonites for storage of high-level radioactive waste. *Clay Miner.* **2016**, *51*, 289–302. [CrossRef]
9. Huang, W.-L. An Experimentally Derived Kinetic Model for Smectite-to-Illite Conversion and Its Use as a Geothermometer. *Clays Clay Miner.* **1993**, *41*, 162–177. [CrossRef]
10. Elliott, W.C.; Matistoff, G. Evaluation of Kinetic Models for the Smectite to Illite Transformation. *Clays Clay Miner.* **1996**, *44*, 77–87. [CrossRef]
11. Svensson, D.; Dueck, A.; Nilsson, U.; Olsson, S.; Sandén, T.; Lydmark, S.; Jägerwall, S.; Pedersen, K.; Hansen, S. *Alternative Buffer Material—Status of the Ongoing Laboratory Investigation of Reference Materials and Test Package 1*; TR-11-06; Svensk Karnbransleforsorjning AB: Stockholm, Sweden, 2011.
12. Kaufhold, S.; Dohrmann, R.; Sandén, T.; Sellin, P.; Svensson, D. Mineralogical investigations of the first package of the alternative buffer material test—I. Alteration of bentonites. *Clay Miner.* **2013**, *48*, 199–213. [CrossRef]
13. Dohrmann, R.; Olsson, S.; Kaufhold, S.; Sellin, P. Mineralogical investigations of the first package of the alternative buffer material test—II. Exchangeable cation population rearrangement. *Clay Miner.* **2013**, *48*, 215–233. [CrossRef]
14. Kaufhold, S.; Dohrmann, R.; Götze, N.; Svensson, D. Characterization of the second parcel of the alternative buffer material (ABM) Experiment—I mineralogical reactions. *Clays Clay Miner.* **2017**, *65*, 27–41. [CrossRef]
15. Dohrmann, R.; Kaufhold, S. Characterization of the Second Package of the Alternative Buffer Material (ABM) Experiment—II Exchangeable Cation Population Rearrangement. *Clays Clay Miner.* **2017**, *65*, 104–121. [CrossRef]
16. Kaufhold, S.; Dohrmann, R.; Ufer, K.; Svensson, D.; Sellin, P. Mineralogical Analysis of Bentonite from the ABM5 Heater Experiment at Äspö Hard Rock Laboratory, Sweden. *Minerals* **2021**, *11*, 669. [CrossRef]
17. Dueck, A.; Johannesson, L.-E.; Kristensson, O.; Olsson, S. *Report on Hydro-Mechanical and Chemical-Mineralogical Analyses of the Bentonite Buffer in Canister Retrieval Test*; TR-11-07; Svensk Kärnbränslehantering AB: Stockholm, Sweden, 2011.
18. Moore, D.M.; Reynolds, R.C. *X-ray Diffraction and the Identification and Analysis of Clay Minerals*, 2nd ed.; Oxford University Press: Oxford, UK, 1997; ISBN 9780195087130.
19. Doebelin, N.; Kleeberg, R. Profex: A graphical user interface for the Rietveld refinement program BGMN. *J. Appl. Crystallogr.* **2015**, *48*, 1573–1580. [CrossRef]
20. Ufer, K.; Stanjek, H.; Roth, G.; Dohrmann, R.; Kleeberg, R.; Kaufhold, S. Quantitative phase analysis of bentonites by the Rietveld method. *Clays Clay Miner.* **2008**, *56*, 272–282. [CrossRef]
21. Ufer, K.; Kleeberg, R. Parametric Rietveld refinement of coexisting disordered clay minerals. *Clay Miner.* **2015**, *50*, 287–296. [CrossRef]
22. Kemp, S.J.; Smith, F.W.; Wagner, D.; Mounteney, I.; Bell, C.P.; Milne, C.J.; Gowing, C.J.B.; Pottas, T.L. An Improved Approach to Characterize Potash-Bearing Evaporite Deposits, Evidenced in North Yorkshire, United Kingdom. *Econ. Geol.* **2016**, *111*, 719–742. [CrossRef]
23. Meier, L.P.; Kahr, G. Determination of the Cation Exchange Capacity (CEC) of Clay Minerals Using the Complexes of Copper(II) Ion with Triethylenetetramine and Tetraethylenepentamine. *Clays Clay Miner.* **1999**, *47*, 386–388. [CrossRef]
24. Środoń, J.; MaCarty, D.K. Surface area and layer charge of smectite from CEC and EGME/$H_2O$-retention measurements. *Clays Clay Miner.* **2008**, *56*, 155–174. [CrossRef]
25. García, R.; Báez, A.P. Atomic Absorption Spectrometry (AAS). In *Atomic Absorption Spectroscopy*; Akhyar Farrukh, M., Ed.; InTech: Rijeka, Croatia, 2012; ISBN 978-953-307-817-5.
26. Podlech, C.; Matschiavelli, N.; Peltz, M.; Kluge, S.; Arnold, T.; Cherkouk, A.; Meleshyn, A.; Grathoff, G.; Warr, L.N. Bentonite Alteration in Batch Reactor Experiments with and without Organic Supplements: Implications for the Disposal of Radioactive Waste. *Minerals* **2021**, *11*, 932. [CrossRef]
27. Ross, C.; Hendricks, S.B. *Minerals of the Montmorillonite Group, Their Origin and Relation to Soils and Clays*; Professional Paper 205-B; USGS: Reston, VA, USA, 1945.
28. Stevens, R.E. A system for calculating analyses of micas and related minerals to end members. *U.S. Geol. Surv. Bull.* **1946**, *950*, 101–119.
29. Wolters, F. Classification of Montmorillonites. Ph.D. Thesis, Universität Karlsruhe, Karlsruhe, Baden-Württemberg, Germany, 2005.
30. Warr, L.N. IMA–CNMNC approved mineral symbols. *Mineral. Mag.* **2021**, *85*, 291–320. [CrossRef]
31. Olsson, S.; Karnland, O. Mineralogical and chemical characteristics of the bentonite in the A2 test parcel of the LOT field experiments at Äspö HRL, Sweden. *Phys. Chem. Earth Parts A/B/C* **2011**, *36*, 1545–1553. [CrossRef]
32. Leupin, O.X.; Birgersson, M.; Karnland, O.; Korkeakoski, P.; Sellin, P.; Mäder, U.; Wersin, P. *Technical Report 14-12, Montmorillonite Stability under Near-Field Conditions*; National Cooperative for the Disposal of Radioactive Waste: Wettingen, Switzerland, 2014.
33. Heuser, M.; Weber, C.; Stanjek, H.; Chen, H.; Jordan, G.; Schmahl, W.W.; Natzeck, C. The interaction between bentonite and water vapor. I: Examination of physical and chemical properties. *Clays Clay Miner.* **2014**, *62*, 188–202. [CrossRef]
34. Sato, T.; Murakami, T.; Watanabe, T. Change in Layer Charge of Smectites and Smectite Layers in Illite/Smectite During Diagenetic Alteration. *Clays Clay Miner.* **1996**, *44*, 460–469. [CrossRef]
35. Crusset, D.; Deydier, V.; Necib, S.; Gras, J.-M.; Combrade, P.; Féron, D.; Burger, E. Corrosion of carbon steel components in the French high-level waste programme: Evolution of disposal concept and selection of materials. *Corros. Eng. Sci. Technol.* **2017**, *52*, 17–24. [CrossRef]

36. Schorr, J.R.; Everhart, J.O. Thermal Behavior of Pyrite and Its Relation to Carbon and Sulfur Oxidation in Clays. *J. Am. Ceram. Soc.* **1969**, *52*, 351–354. [CrossRef]
37. Verron, H.; Sterpenich, J.; Bonnet, J.; Bourdelle, F.; Mosser-Ruck, R.; Lorgeoux, C.; Randi, A.; Michau, N. Experimental Study of Pyrite Oxidation at 100 °C: Implications for Deep Geological Radwaste Repository in Claystone. *Minerals* **2019**, *9*, 427. [CrossRef]
38. Maes, A.; Stul, M.; Cremers, A. Layer Charge-Cation-Exchange Capacity Relationships in Montmorillonite. *Clays Clay Miner.* **1979**, *27*, 387–392. [CrossRef]

*Article*

# Mineralogical Analysis of Bentonite from the ABM5 Heater Experiment at Äspö Hard Rock Laboratory, Sweden

Stephan Kaufhold [1,*], Reiner Dohrmann [1,2], Kristian Ufer [1], Daniel Svensson [3] and Patrik Sellin [3]

[1]   Department of Technical Mineralogy and Clay Mineralogy, BGR—Bundesanstalt für Geowissenschaften und Rohstoffe, Stilleweg 2, D-30655 Hannover, Germany; Reiner.Dohrmann@bgr.de (R.D.); Kristian.Ufer@bgr.de (K.U.)

[2]   Department of Technical Mineralogy, Sedimentology, LBEG—Landesamt für Bergbau, Energie und Geologie, Stilleweg 2, D-30655 Hannover, Germany

[3]   Department of Research and Safety Assessment, SKB—Svensk Kärnbränslehantering AB, P.O. Box 5864, S-102 40 Stockholm, Sweden; Daniel.Svensson@skb.se (D.S.); patrik.sellin@skb.se (P.S.)

*   Correspondence: s.kaufhold@bgr.de; Tel.: +49-511-6432765

**Abstract:** The present study reports on the analysis of all blocks of the ABM5 test, which is a medium scale bentonite buffer deposition test. In contrast to similar tests, the ABM5 was conducted at higher temperature (up to 250 °C). The aim of the study was to characterize the chemical and mineralogical reactions and to identify the effect of the extraordinarily high temperature. Reactions observed were similar to those observed in previous and/or similar tests covering cation exchange, anion inflow, dissolution and precipitation of C- and S-phases, Fe corrosion, and Mg increase at the heater. Neither the type nor the extent of the different reactions could be related to the significantly higher temperature. However, due to the absence of lubricant used between heater and bentonite, it could be proved that the calcite previously present was dissolved and precipitated as siderite at the contact, pointing towards the importance of the presence of carbonate when considering different Fe corrosion products. Moreover, for the first time, a decrease of the Mg content at the heater was observed, which was probably because a Mg-rich clay was used. The reasons for Mg increase or decrease are still not completely understood.

**Keywords:** bentonite; technical barrier; Äspö; ABM-test; smectite alteration

**Citation:** Kaufhold, S.; Dohrmann, R.; Ufer, K.; Svensson, D.; Sellin, P. Mineralogical Analysis of Bentonite from the ABM5 Heater Experiment at Äspö Hard Rock Laboratory, Sweden. *Minerals* **2021**, *11*, 669. https://doi.org/10.3390/min11070669

Academic Editor: Francisco Franco

Received: 15 April 2021
Accepted: 17 June 2021
Published: 23 June 2021

**Publisher's Note:** MDPI stays neutral with regard to jurisdictional claims in published maps and institutional affiliations.



## 1. Introduction

Bentonite is currently investigated as a geotechnical barrier to enclose metal canisters containing high-level radioactive waste (HLRW) [1,2], particularly in combination with crystalline host rocks. Bentonites are mined worldwide. Materials from different deposits often show specific properties [3]. Bentonites are used in many different applications [4]. For each application, more suitable bentonites and less suitable ones exist. The aim of the bentonite industry is to identify the optimum bentonite for a specific application, to provide superior products compared to their competitors. Bentonite selection is commonly based on empirical tests in which specific properties of the bentonites are compared. Using bentonite as an HLRW barrier material, although tested for decades in experiments, from the point of view of the bentonite mining industry, is a new application. Well accepted quality determining parameters, therefore, do not yet exist. Kaufhold and Dohrmann [5] discussed ten parameters which could be used to compare the suitability of bentonites as an HLRW barrier material. However, to date, there is no agreement about one or more quality determining parameters. This may also result from the fact that the bentonite requirements differ amongst the different concepts. A comparison of the performance of different bentonites is therefore an interesting task.

Comparative tests can be conducted both in clay laboratories and in underground rock laboratories as medium to large scale tests. In 2006, SKB started an experiment series in which different bentonites are compared directly [6] at Äspö Hard Rock Laboratory (HRL) with granite as host rock. Altogether, 30 blocks (rings) made of different bentonites (diameter 30 cm, diameter of central hole for the heater 10 cm, height 10 cm) were placed on each other without any protection between them. Two of the three packages were retrieved, analysed, and reported [7–15]. In 2013, three additional packages were installed, partly with different materials (e.g., saponite, a Mg-rich smectite [16]). The target temperature of the tests was 130 °C for all tests. The temperature of ABM5, however, was low at the beginning (50 °C) because of problems increasing the water pressure in the surrounding sand filter to avoid boiling at the high temperature. In 2016, the temperature was increased up to 250 °C for about six months and in 2017 ABM5 was excavated. Temperature profiles are shown in Figure 1. To date, the ABM5 is the hottest bentonite test conducted in an underground rock laboratory (URL). In the deposition hole drilled for the ABM5-test, a fracture was observed which provided some water inflow. In the installation report, it is mentioned that "most of the water seemed to come from one fracture located 0.8 m down from the floor".

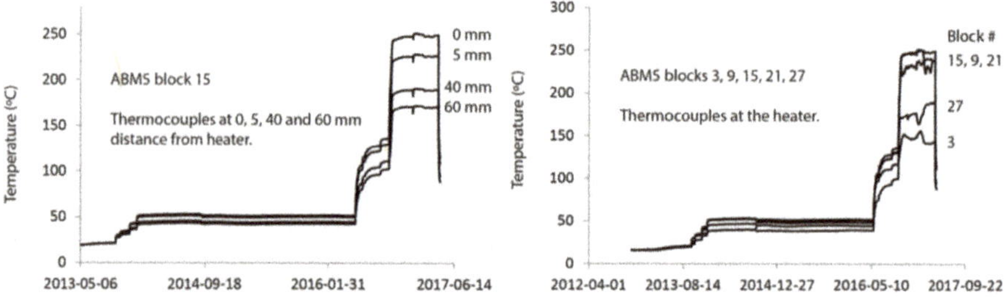


**Figure 1.** Temperature profiles (**left**: temperature in block 15 at different distances to the heater; **right**: temperatures at the heater of blocks 3, 9, 15, 21, 27).

The temperature reached 250 °C at the contacts of the samples to the heater, which were in the central part of the column (9, 15, 21). At the contact of the upper- and lowermost blocks to the heater, lower temperatures were observed (about 150 °C for blocks 3 and 27). In the hottest blocks (e.g., #15, Figure 1, left), more than 150 °C was reached at a distance of 6 cm from the heater.

To date, a couple of mineralogical reactions were identified in the different medium to large scale tests. In the LOT experiment, a redistribution of sulphate (gypsum, anhydrite) within the blocks was found [17,18]. The gypsum which was initially present in the bentonite was not evenly distributed after the test, but showed larger concentrations in the center of the blocks. Extensive cation exchange was observed as well [15,19]. Corrosion was observed both in case of Cu [17,20] and Fe heaters [12,13]. Corrosion products could only be found at the interface. In addition, a Mg enrichment at the contact of bentonite and metal was found in the case of both Cu [17] and Fe [11–13,21,22]. Samples with larger Mg increase showed the formation of trioctahedral phyllosilicate minerals or domains which could be detected by XRD. The mechanism behind the Mg enrichment is not yet clear [11,12,23]. In ABM1 and ABM2, organic material was found at the contact to the heater which could be explained by the lubricant (molykote) used for installation. Moreover, siderite formation was reported as well as the dissolution of cristobalite and/or clinoptilolite [12].

The aim of the present study is to characterize the mineralogical reactions that occurred in the ABM5 test and to compare them with the identified reactions. Another aim is to identify possible differences of the mineralogical reactions in terms of temperature effects, because the ABM5 test had the highest temperature yet applied in a medium- or large-scale test.

## 2. Materials and Methods

After dismantling the vertically installed ABM5 test in the Äspö HRL, Figeholm, Sweden, samples were collected and sent to BGR. Figure 2 illustrates the setup and dimensions of the experiment, including the location of the sand filter and iron heater (diameter 10 cm) in the holes of the bentonite rings (diameter 30 cm, height 10 cm). The sand filter was installed to support rapid water saturation. At the beginning of the heating period, the water pressure remained unexpectedly low. In order to further support saturation, the temperature was set lower at the beginning. The temperature profile is shown in Figure 1. Experimental details can be found elsewhere [6,9,16].

**Table 1.** List and state of samples sent to BGR as well as samples taken at BGR from the blocks (numbers = cm distanced from heater; "x" distance to heater could not be determined). The heater was not investigated in the present study. The numbers correspond to the block numbers in ABM5. Block 1 was the lowermost block and the others were placed on top of each other.

| Block no. | Material | Remark | Direction | Comment | 0.1 | 1 | 5 | 8 | x |
|---|---|---|---|---|---|---|---|---|---|
| 30 | MX80 | | 30 S | | ✓ | ✓ | ✓ | ✓ | |
| 29 | MX80 | | 29 N | | ✓ | ✓ | ✓ | ✓ | |
| 28 | Asha 505 | | 28 S | | ✓ | ✓ | ✓ | ✓ | |
| 27 | Calcigel | Termoelement | | Fragments | | | | ✓ | ✓ |
| 26 | Deponit CAN | | | Fragments | ✓ | ✓ | ✓ | ✓ | |
| 25 | Febex | | 25 W | | ✓ | ✓ | ✓ | ✓ | |
| 24 | GMZ | | | Fragments | | ✓ | ✓ | ✓ | |
| 23 | Ibeco SEAL | | 23 W | | ✓ | ✓ | ✓ | ✓ | |
| 22 | Ikosorb | | 22 W | | ✓ | ✓ | ✓ | ✓ | |
| 21 | Kunigel V1 | Termoelement | | Fragments | ✓ | ✓ | ✓ | ✓ | ✓ |
| 20 | MX80 | Copper | 20 W | | ✓ | ✓ | ✓ | ✓ | |
| 19 | Asha NW BFL-L | | 19 W | | ✓ | ✓ | ✓ | ✓ | |
| 18 | Rokle | | 18 W | | ✓ | ✓ | ✓ | ✓ | |
| 17 | Saponite | | 17 W | | ✓ | ✓ | ✓ | ✓ | |
| 16 | Asha 505 | Copper | 16 NW | | ✓ | ✓ | ✓ | ✓ | |
| 15 | MX80 | Termoelement + Titanium | 15 W | | ✓ | ✓ | ✓ | ✓ | |
| 14 | Rokle | | 14 NW | | ✓ | ✓ | ✓ | ✓ | |
| 13 | Febex | | 13 W | | ✓ | ✓ | ✓ | ✓ | |
| 12 | Saponite | | | Fragments | ✓ | | | ✓ | ✓ |
| 11 | Ibeco SEAL | | 11 S | | ✓ | ✓ | ✓ | ✓ | |
| 10 | Calcigel | | | Fragments | ✓ | ✓ | ✓ | ✓ | |
| 9 | Asha NW BFL-L | Termoelement | 9 N | Partial | ✓ | | ✓ | | ✓ |
| 8 | MX80 | | 8 S | | ✓ | ✓ | ✓ | ✓ | |
| 7 | Ikosorb | | 7 S | | ✓ | ✓ | ✓ | ✓ | |
| 6 | GMZ | | | Fragments | | | | | ✓ |
| 5 | Kunigel V1 | | | Fragments | ✓ | ✓ | ✓ | ✓ | |
| 4 | Deponit CAN | | | Fragments | ✓ | ✓ | ✓ | | |
| 3 | Asha NW BFL-L | Termoelement + Titanium | 3 N | | ✓ | ✓ | ✓ | | |
| 2 | MX80 | | 2 SE | | ✓ | ✓ | ✓ | ✓ | |
| 1 | MX80 | | 1 N | | ✓ | ✓ | ✓ | ✓ | |

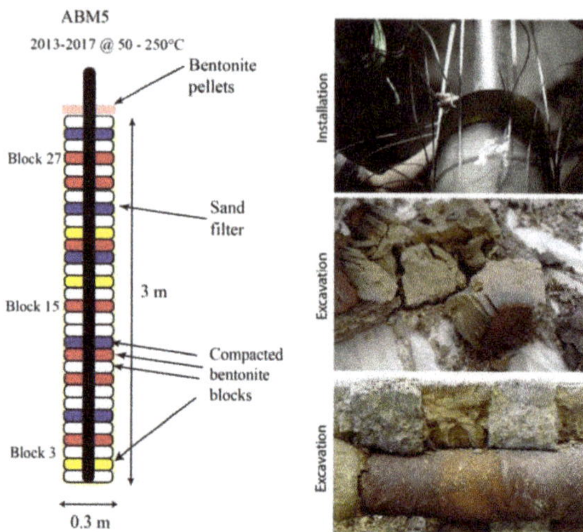


**Figure 2.** Schematic presentation of the ABM-5 test (**left**) and photographs of a part of the ABM5 experiment before and after dismantling (**right**). The sequence of the different blocks is given in Table 1.

Table 1 lists all samples sent to BGR including information about the state of the sample. Test packages ABM-4–6 are composed of bentonite material only. All bentonites known from previous test packages ABM-1–3 were used again. However, the materials FRI (Friedland clay), COX (Callo-Oxfordian clay), as well as bentonite–quartz mixtures made of MX80 were no longer used in ABM-4–6. Instead, three new bentonite raw materials were used for the first time: GMZ, the Chinese reference material, Asha NW (Ashapura, Mumbai, India), and 'Saponite', a bentonite named after a trioctahedral mineral of the smectite group: saponite. In the present paper, this bentonite is called 'Saponite' to distinguish the raw material from the mineral.

As in the case of previous analyses of ABM-blocks [12,13], each block should be sampled at different distances to the contact. In some cases, only fragments were present, which made it impossible to obtain suitable samples from defined distances. Sampling of the blocks was undertaken at four different distances to the contact (0.1, 1, 5 and 8 cm). A minimum of 2 g was required for a comprehensive characterization. The actual depths at which the 0.1 cm sample was taken, therefore, depended on the available area of the contact. The contact area ranged from 10 to 20 $cm^2$ and 2 g were collected which, assuming a density of 2 $g/cm^3$, corresponds to a sampling depth of 0.05–0.1 cm. The contact sample was collected with a sharp knife and the other samples were collected by hand drilling. The samples were then dried at 60 °C for three days and ground using a mortar mill.

The chemical analysis was performed by ACTLABs® using the analytical tools 4B1 QOP Total (Total Digestion using different acids, analysis by ICP-OES), 4C QOP XRF Fusion (Whole Rock Analysis-XRF), and 4F-Cl QOP INAA—Short Lived (INAA). The effect of different water contents on the elemental composition was eliminated by normalizing all values not considering the loss on ignition.

The organic carbon ($C_{org}$) content was measured by combustion with a LECO CS-444-Analysator after dissolution of the carbonates. Carbonates were removed by treating the samples several times at 80 °C with HCl until no further gas evolution could be observed. Samples of 170–180 mg of the dried material were used to measure the total carbon ($C_{total}$) content. Total inorganic carbon ($C_{inorg}$) was calculated by the difference of $C_{total}$ and $C_{org}$. The samples were heated in the device to 1800–2000 °C in an oxygen atmosphere and the $CO_2$ and $SO_2$ were detected by an infrared detector. The device was built by LECO (3000 Lake Avenue, St. Joseph, MI, USA).

Thermoanalytical investigations were performed using a Netzsch 449 F3 Jupiter thermobalance equipped with a DSC/TG sample holder linked to a Netzsch QMS 403 C Aeolus mass spectrometer (MS). Then, powdered material (100 mg) previously equilibrated at 53% relative humidity (RH) was heated from 25 to 1150 °C with a heating rate of 10 K/min. The devices were manufactured by Netzsch (Gebrüder-Netzsch-Straße 19, Selb, Germany).

XRD patterns were recorded using a PANalytical X'Pert PRO MPD $\Theta$–$\Theta$ diffractometer (Cu-K$_\alpha$ radiation generated at 40 kV and 40 mA), equipped with a variable divergence slit (20 mm irradiated length), primary and secondary sollers, Scientific X'Celerator detector (active length 0.59°), and a sample changer (sample diameter 28 mm). The samples were investigated from 2° to 85° 2$\Theta$ with a step size of 0.0167° 2$\Theta$ and a measuring time of 20 s per step. For specimen preparation, the back loading technique was used.

For the detailed clay mineralogical investigation, texture slides of the <2 μm fraction were prepared. Oven-dried clay fractions <2 μm of the contact samples (if available) were dispersed using ultrasound. A suspension containing approximately 60 mg solid clay fraction was sucked through porous ceramic tiles of 27 mm diameter using a vacuum-filter in order to orient the clay minerals (alumosilicates) preferentially parallel to their basal planes. The final solid density was approximately 15 mg/cm$^2$. These so-called oriented aggregates were x-rayed from 2.5 to 40° 2$\Theta$ (air-dried and ethylene glycol solvated) with a step size of 0.03° 2$\Theta$. The measuring time wasa 6 s per step. The scans were run on a PANalytical X'Pert PRO MPD $\Theta$–$\Theta$ diffractometer (Co-K$\alpha$ radiation generated at 40 kV and 40 mA), equipped with a variable divergence slit (20 mm irradiated length), a primary and secondary soller, a proportional counter, and a secondary monochromator.

A Zeiss Sigma 300 V P FEG scanning electron microscope operating at 15 kV was used to evaluate samples on the micro scale using the high-vacuum mode. The microscope was equipped with the following detectors: Bruker Xflash® 6/60 EDX detector, high-definition backscattered electron detector (HDBSD), secondary electron detector (SE2), variable pressure secondary electron detector (VPSE), and an InLens detector for detection of secondary and backscattered electrons, respectively.

For measuring mid (MIR) infrared spectra, the KBr pellet technique (1 mg sample/200 mg KBr) was applied. Spectra were collected on a Thermo Nicolet Nexus FTIR spectrometer (MIR beam splitter: KBr, detector DTGS TEC; FIR beam splitter: solid substrate, detector DTGS PE). The resolution was adjusted to 2 cm$^{-1}$. Measurements were conducted before and after drying of the pellets at 150 °C in a vacuum oven for 24 h. The spectrometer was built by Nicolet Instruments, Verona Road, Madison, WI, USA.

## 3. Results and Discussion

All data is summarized in Table S1 (Supplementary Materials).

### 3.1. Geochemical Profiles

In the LOT experiment, which was also conducted at temperatures higher than 90 °C, a redistribution of sulphate phases (gypsum) was detected [17]. Gypsum was obviously dissolved both near the crystalline host rock and near the heater and precipitated in the center of the blocks. In the ABM1 and ABM2 tests, different S-profiles were measured [12,13], which indicates that locally different conditions (water content and temperature) can lead to different gypsum/anhydrite redistribution profiles. In the LOT experiment, which was setup with MX80 blocks only, this gypsum redistribution was found in many different blocks and at different depths (from 0 to 3 m). Most of the blocks of ABM5 (present study) showed a slight increase of the S content in the central part and some dissolution at the inner and outer parts (Figure 3). This behavior was similar to the results found in the LOT A2 experiment but much less pronounced. The amount of S which precipitated in the central part of the LOT A2 blocks was larger. S concentrations below 0.1 mass% are close to detection limit and hence not discussed further. Most of the samples, however, showed S contents ranging from 0.2 to 0.4 mass%. Three samples showed an untypical behavior. In block 1 (lowermost sample), a significant increase of the S content directly at the heater was observed. The top block (#30) showed S enrichment at a distance of 1 cm. In both cases, an increase of approximately 0.2 up to 0.5 mass% S was observed. The extraordinary behavior of these two blocks (#1, #30) may result from the special hydraulic properties caused by the fact that there were no other blocks above and below. In the case of block #30, cement pore waters derived from the concrete seal above could have provided different cations and anions (compare [12]). In the case of block 1, however, no concrete was near the block, but the sand base could have promoted water inflow. Blocks 26 and 4, both Deponit CAN, showed the highest initial S content but slightly different profiles. For both samples, slightly lower values were found in the block after the experiment compared to the reference, which confirms that gypsum is at least partly soluble and transported inside the barrier.

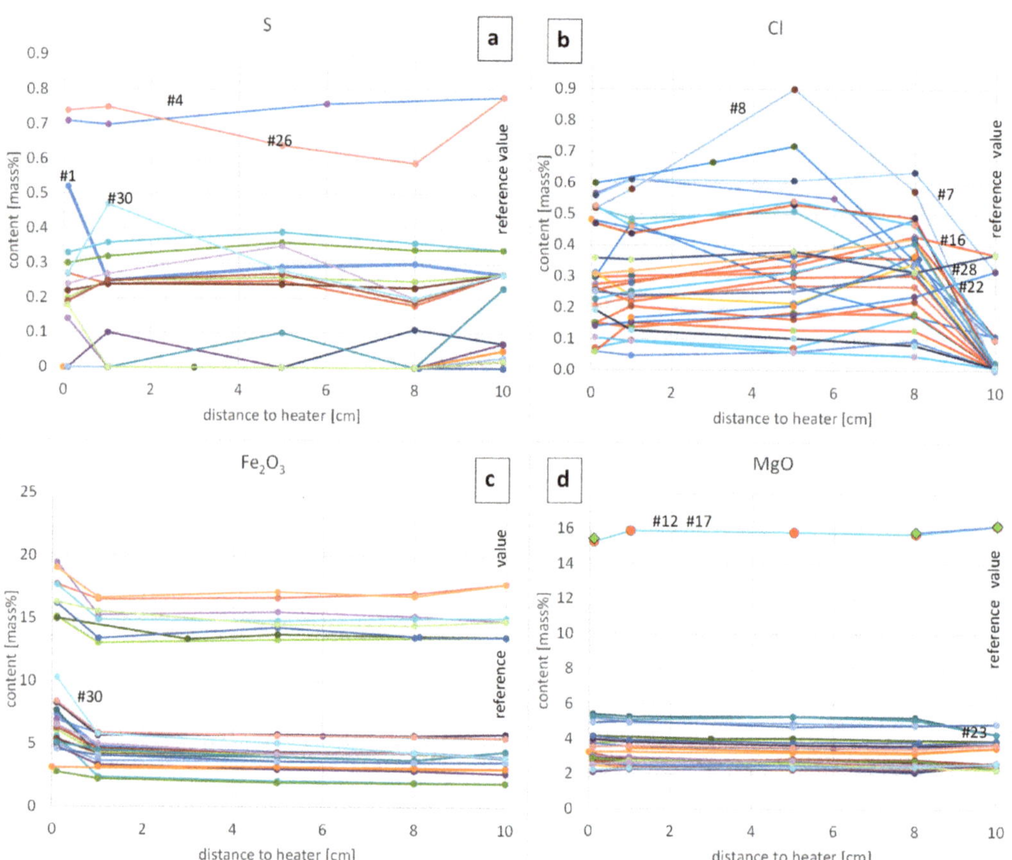

**Figure 3.** Geochemical profiles of S (**a**), Cl (**b**), Fe$_2$O$_3$ (**c**), and MgO (**d**) of all blocks.

Most samples showed an initial Cl content of ≤0.1 mass% and an increase in the reacted blocks which can only be explained by Cl derived from the rock water which migrated into the blocks. In most blocks Cl even reached the heater because the contents measured there were higher compared to the reference materials. The Cl concentrations were not higher in the contact samples of the upper- and lowermost blocks, which suggests that the Cl did not migrate to the heater and then vertically along the bentonite/heater interface, but was transported parallel to the thermal gradient (horizontally). The Cl concentrations of the ASHA 505 and IKO blocks (#7, #16, #22, #28, Figure 3b) indicate that material properties were less relevant with respect to the resulting Cl profiles than local conditions. Blocks #22 and #28 had a larger initial Cl concentration, and this decreased towards the heater. Blocks #7 and #16, however, showed the opposite effect. IKO showed Cl increase in block #7 and decrease in block #22. Asha 505 showed an increase in block #16 and decrease in block #28. The Cl migration into the blocks apparently did not depend on the material but on local differences which is discussed later.

The Fe contents were interesting because they may reflect canister corrosion. Most of the blocks showed an increase of the $Fe_2O_3$ content at the contact (0.1 mm sample) of more than 1 mass%. The largest increase was found for the top block (#30: +6.3 mass%). These values can hardly be compared with those of the other studies because of different sampling areas (as explained above). In blocks #20–#30 a slight increase of the $Fe_2O_3$ content could even be detected in the 1 cm sample. The increase ranged from 0.3 mass% (compared with the reference) up to 2 mass% (#30) and it apparently increased with block number. One may conclude that Fe diffusion into the barrier was more pronounced in the upper blocks, which could result from temperature or water content (discussed later).

Mg enrichment close to the heater was observed in many large-scale tests, regardless of whether Fe or Cu was used. This enrichment is of particular interest because it points towards the neoformation of phases such as trioctahedral smectite and/or brucite (FEBEX experiment) at the expense of dioctahedral smectite, which in turn could result in a reduction of the swelling capacity of the barrier at the heater.

In the ABM2 test [13], MgO enrichments at the contact of 0.0–3.0 mass% were observed. The most significant MgO increase was observed at the face of the FEBEX experiment (+6 mass% MgO [24]) which at least partly can be explained by the fact that a large sampling area was available (about 200 cm$^2$) and hence the very top could be sampled (<0.1 mm). Using IR spectrometry, the formation of brucite was found, and by XRD, the formation of trioctahedral domains was observed. Fernandez et al. [22] could also observe sepiolite. In the ABM2 experiment, only the formation of trioctahdral phyllosilicate domains was observed (no brucite), commonly associated with a significant MgO increase. In the previous ABM tests and in the present one, the MgO increase, however, was not typical for a special type of bentonite. As an example, the two blocks of Deponit CAN and ROK showed either 0.0–0.3 mass% or 1.8–2.8 mass% MgO increase. These values, therefore, seem to depend more on the local conditions (temperature, water content, type of solution diffusing through the bentonite) than on the type of bentonite. The significantly larger temperature of ABM5 could have affected the formation of Mg-rich phases. Kaufhold et al. [23] showed that the temperature affects the ratio of Mg/Si which can be dissolved from smectites in such a way that it decreases with increasing temperature, which means that the solubility of Si increases more than that of Mg at higher temperature. Accordingly, one could expect Si phases (or Si rich Mg-silicates) to be precipitated at the heater. The data are summarized in Supplementary Table S1 and plotted in Figure 3d. Overall, only a slight increase of the MgO value was observed in the ABM5 test. The largest values (+1 mass% MgO) were observed for the two IBECO Seal bentonites, while all others showed lower values. Notably, a slight decrease or increase of the MgO content compared to the reference value (see Figure 3d) can sometimes be explained by cation exchange (cation exchange data will be discussed in detail in a follow up study). For the first time, a trioctahedral smectite (saponite) was tested in a deposition test (in the 'Saponite' blocks). Interestingly, a slight decrease of the MgO content at the heater was observed for both 'Saponite' samples. The curve of #12, however, is not complete because it could only be sampled at the contact. The three points shown in Figure 3d, however, are identical to those measured for block #17 which was also made out of 'Saponite'. Both 'Saponite' blocks, therefore, show an MgO decrease at the heater which can be explained by cation exchange and Fe increase. Sample #17 contained 36 meq/100 g exchangeable Mg before and only 4 meq/100 g after the test which explains part of the MgO decrease. The missing MgO decrease can be explained by the increase of

$Fe_2O_3$ caused by corrosion which adds to the contact sample hence relatively decreasing the other elements.

For blocks #11 and #23 (both IBECO) more MgO was found in the entire blocks compared to the reference material. This neither can be explained by the small amount of carbonate dissolution nor based on cation exchange. The reason for this difference cannot be explained yet.

### 3.2. X-ray Diffraction

In bulk XRD, almost all samples showed a shift of the $d_{001}$ peak position which can be explained by cation exchange due to its impact on the bentonite water content after drying in typical laboratory conditions. Differences other than $d_{001}$ shifting such as varying muscovite/feldspar/quartz peak intensities were also observed, but these do not necessarily point towards actual mineral reactions. Differences of these intensities can also be explained by preferred orientation. Moreover, the reference samples were recorded with slightly higher intensity which translates into a better signal/noise ratio. Generally, XRD is not a method to unambiguously detect minor mineral changes. Varying peak intensities may point to actual mineral reactions, but these should be validated based on chemical data. All observed changes detected by XRD are summarized in Figure 4.

Most of the mineralogical changes observed were related to the dissolution/precipitation of C- and S-phases. As an example, gypsum was dissolved in MX80 blocks #8, #15, and #20 whereas gypsum remained stable in the bottom and top blocks made of the same material. Calcite was dissolved in both Ibeco Seal (IBE) blocks #11 and #23 (partly). Dolomite intensities were lower in both Calcigel blocks (#10, #26), indicating partial dissolution. Results of previous large-scale tests showed that the (partly) soluble phases can be dissolved anywhere in the barrier and precipitated elsewhere. In the ABM1 test [12], the dissolution of cristobalite and clinoptilolite was observed. In the ABM5 test, cristobalite intensities were lower only in the Deponit CAN blocks (#4, #26), possibly indicating dissolution. No cristobalite or clinoptilolite dissolution was observed in the other blocks of GMZ, Febex, and MX80. In laboratory experiments with NaOH solutions Karnland et al. [25] reported evidence for cristobalite dissolution in Wyoming bentonite samples. However, pH conditions were obviously lower in ABM5 as the observed dissolution was much lower. The identification of traces of siderite is particularly challenging because of peak coincidences with apatite and partly muscovite. Apatite was observed in Asha NW and ROK. The unambiguous identification of siderite, therefore, is not possible in these samples and other methods such as simultaneous thermal gas (STA) have to be used. No evidence for anhydrite precipitation was observed in ABM5, in contrast to ABM2.

Special attention was paid towards changes of the $d_{060}$ reflexes which in previous tests indicated the presence of trioctahedral phases at the heater or in outer parts of the blocks. All but one material consist of dioctahedral smectite. Such trioctahedral phases were found at the contact to the heater before [11–13,24] based on XRD and IR, supported by an increase in MgO concentration (XRF). In the ABM5 samples, nearly no such changes of the $d_{060}$ XRD intensities were observed, although some MgO enrichment was observed. The extent of the MgO enrichment at the heater, however, was small comparted with yet published results. The largest MgO increase was found in both Ibeco Seal blocks (#11: +1.1 mass% MgO, #23: +1.0 mass% MgO). Moreover, some MgO enrichment was found in block #16 (0.7 mass%) and block #21 (0.6 mass%). However, no changes of the $d_{060}$ reflection were found for any of these blocks with MgO increase (Figure 5b). The 'Saponite' samples (#12, #17) on the other hand showed some decrease of the MgO content at the heater. This bentonite raw material contains trioctahedral smectite (saponite). However, no changes of the $d_{060}$ reflection could be observed either (Figure 5d). An indication for the presence of trioctahedral phases was found in a #7 (Ikosorb) (Figure 5c) and #27 (Calcigel), not shown. In contrast to former ABM experiments, such trioctahedral phases are indicated not only at the contact to the heater, but also in outer parts of the blocks (marked by arrows in Figure 5c).

**Figure 4.** Summary of mineralogical changes detected by XRD powder diffraction (X = samples not available, "+" = clearly identified, "(?)" = could be present, difficult to decide) and additional techniques for trace phases (carbonate group minerals, sulphates, pyrite). MX80 samples are marked yellow.

distances from heater (cm)

| Block | Material/abbreviation | REF | d(060) 0.1 | new (0-1) |
|---|---|---|---|---|
| 30 | MX80 | 1.498 | ✓ | |
| 29 | MX80 | 1.498 | ✓ | |
| 28 | Asha 505 | 1.500 | ✓ | |
| 27 | Calcigel (CAL) | 1.498 | | no contact sample |
| 26 | Dep. CAN | 1.498 | ✓ | |
| 25 | Febex | 1.499 | ✓ | |
| 24 | GMZ | 1.498(5) | | no contact sample |
| 23 | Ibeco Seal (IBE) | 1.498 | ✓ | |
| 22 | Ikosorb (IKO) | 1.495 | ✓ | |
| 21 | Kunigel V1 (JNB) | 1.498 | ✓ | |
| 20 | MX80 | 1.498 | ✓ | |
| 19 | Asha-NW BFL-L | 1.500 | ✓ | |
| 18 | Rokle (Rawra) | 1.505 | ✓ | |
| 17 | Saponite | 1.512 | ✓ | |
| 16 | Asha 505 | 1.500 | ✓ | |
| 15 | MX80 | 1.498 | ✓ | |
| 14 | Rokle (Rawra) | 1.505 | ✓ | |
| 13 | Febex | 1.499 | ✓ | |
| 12 | Saponite | 1.512 | ✓ | |
| 11 | Ibeco Seal (IBE) | 1.498 | ✓ | |
| 10 | Calcigel (CAL) | 1.498 | ✓ | |
| 9 | Asha-NW BFL-L | 1.500 | ✓ | |
| 8 | MX80 | 1.498 | ✓ | |
| 7 | Ikosorb (IKO) | 1.495 | ✓ | 1.54(5) |
| 6 | GMZ | 1.498(5) | | no contact sample |
| 5 | Kunigel V1 (JNB) | 1.498 | ✓ | |
| 4 | Dep. CAN | 1.498 | ✓ | |
| 3 | Asha-NW BFL-L | 1.500 | ✓ | |
| 2 | MX80 | 1.498 | ✓ | |
| 1 | MX80 | 1.498 | ✓ | |

sulphate phases

| Block | Material | gypsum REF | 0.1 | 1 | 5 | 8 | pyrite XRD REF | 0.1 | 1 | 5 | 8 |
|---|---|---|---|---|---|---|---|---|---|---|---|
| 30 | MX80 | + | (?) | + | + | | + | | | | ? |
| 29 | MX80 | + | + | (?) | | | + | (?) | ? | (?) | ? |
| 28 | Asha 505 | (?) | (?) | | | | (?) | (?) | (?) | | |
| 27 | Calcigel (CAL) | | X | X | X | X | | X | X | X | X |
| 26 | Dep. CAN | + | + | + | ? | (?) | + | (?) | (?) | (?) | (?) |
| 25 | Febex | | | | | | | | | | |
| 24 | GMZ | | X | X | X | X | | X | X | X | X |
| 23 | Ibeco Seal (IBE) | | | | | | | | | | |
| 22 | Ikosorb (IKO) | | | | | | | | | | |
| 21 | Kunigel V1 (JNB) | | | | | | + | ? | ? | ? | ? |
| 20 | MX80 | + | | | | | + | ? | (?) | (?) | ? |
| 19 | Asha-NW BFL-L | | | | | | | | | | |
| 18 | Rokle (Rawra) | | | | | | | | | | |
| 17 | Saponite | | | | | | | | | | |
| 16 | Asha 505 | (?) | (?) | | | | | | | | |
| 15 | MX80 | + | | | | | + | (?) | (?) | (?) | (?) |
| 14 | Rokle (Rawra) | | | | | | | | | | |
| 13 | Febex | | | | | | | | | | |
| 12 | Saponite | | X | X | X | X | | X | X | X | X |
| 11 | Ibeco Seal (IBE) | | | | | | | | | | |
| 10 | Calcigel (CAL) | | | | | | | | | | |
| 9 | Asha-NW BFL-L | | X | X | X | X | | X | X | X | X |
| 8 | MX80 | + | | | | | + | (?) | (?) | (?) | (?) |
| 7 | Ikosorb (IKO) | | | | | | | | | | |
| 6 | GMZ | | X | X | X | X | | X | X | X | X |
| 5 | Kunigel V1 (JNB) | | | | | | + | ? | ? | ? | ? |
| 4 | Dep. CAN | + | + | (?) | (?) | X | + | (?) | (?) | (?) | X |
| 3 | Asha-NW BFL-L | | X | X | X | X | | X | X | X | X |
| 2 | MX80 | + | (?) | (?) | (?) | | + | (?) | (?) | (?) | (?) |
| 1 | MX80 | + | + | ? | + | + | + | ? | ? | ? | ? |

inorganic carbon phases

| Block | Material | calcite REF | 0.1 | 1 | 5 | 8 | dolomite/ankerite REF | 0.1 | 1 | 5 | 8 | siderite REF | 0.1 | 1 | 5 | 8 |
|---|---|---|---|---|---|---|---|---|---|---|---|---|---|---|---|---|
| 30 | MX80 | | | | | | | | | | | | | | | |
| 29 | MX80 | | | | | | | | | | | | | | | |
| 28 | Asha 505 | (?) | + | + | + | + | | | | | | | | | | |
| 27 | Calcigel (CAL) | + | X | X | X | X | ++ | X | X | X | + | | X | X | X | X |
| 26 | Dep. CAN | ++ | + | ++ | ++ | ++ | + | + | + | + | + | | (?) | (?) | (?) | (?) |
| 25 | Febex | + | + | + | + | + | | | | | | | | | | |
| 24 | GMZ | | X | X | X | X | | X | X | X | X | | X | X | X | X |
| 23 | Ibeco Seal (IBE) | + | + | (?) | | ? | + | | | | | | | | | |
| 22 | Ikosorb (IKO) | + | + | + | + | + | | | | | | | | | | |
| 21 | Kunigel V1 (JNB) | + | + | + | + | + | | | | | | | (?) | (?) | (?) | (?) |
| 20 | MX80 | + | + | + | + | + | | | | | | | | | | |
| 19 | Asha-NW BFL-L | ++ | + | + | + | + | | | | | | + | ? | + | + | + |
| 18 | Rokle (Rawra) | ? | | | | | | | | | | | | | | |
| 17 | Saponite | + | + | + | + | + | + | ? | | + | ? | | | | | |
| 16 | Asha 505 | (?) | + | + | + | | | | | | | (?) | (?) | | | |
| 15 | MX80 | ? | | | | | | | | | | | | | | |
| 14 | Rokle (Rawra) | ? | | | | | | | | | | | | | | |
| 13 | Febex | + | + | + | + | + | | | | | | | | | | |
| 12 | Saponite | + | + | X | X | + | + | (?) | X | X | + | | X | X | X | X |
| 11 | Ibeco Seal (IBE) | + | (?) | | | | + | (?) | | | | | | | | |
| 10 | Calcigel (CAL) | + | (?) | (?) | (?) | (?) | ++ | + | + | + | + | | (?) | + | + | + |
| 9 | Asha-NW BFL-L | ++ | + | X | + | X | | X | X | X | X | + | + | X | + | X |
| 8 | MX80 | + | + | + | + | + | | | | | | | | | | |
| 7 | Ikosorb (IKO) | + | + | + | + | + | | | | | | | | | | |
| 6 | GMZ | | X | X | X | X | | X | X | X | X | | X | X | X | X |
| 5 | Kunigel V1 (JNB) | + | + | + | + | + | + | + | + | + | + | | (?) | (?) | (?) | (?) |
| 4 | Dep. CAN | ++ | ++ | ++ | ++ | X | + | + | + | + | X | | (?) | (?) | (?) | X |
| 3 | Asha-NW BFL-L | ++ | + | + | + | X | | | | | | + | (?) | (?) | (?) | |
| 2 | MX80 | + | | | | | | | | | | | | | | |
| 1 | MX80 | ++ | | | | | | | | | | | | | | |

| Block | Material | cristobalite REF | 0.1 | 1 | 5 | 8 | goethite REF | 0.1 | 1 | 5 | 8 | clinoptilolite REF | 0.1 | 1 | 5 | 8 |
|---|---|---|---|---|---|---|---|---|---|---|---|---|---|---|---|---|
| 30 | MX80 | + | + | + | + | + | | | | | | | | | | |
| 29 | MX80 | + | + | + | + | + | | | | | | | | | | |
| 28 | Asha 505 | | | | | | + | + | + | + | + | + | + | + | + | + |
| 27 | Calcigel (CAL) | | X | X | X | X | | X | X | X | X | | X | X | X | X |
| 26 | Dep. CAN | + | (?) | (?) | (?) | (?) | | | | | | | | | | |
| 25 | Febex | + | + | + | + | + | | | | | | | | | | |
| 24 | GMZ | | X | X | X | X | | X | X | X | X | | X | X | X | X |
| 23 | Ibeco Seal (IBE) | | | | | | | | | | | | | | | |
| 22 | Ikosorb (IKO) | + | ++ | ++ | ++ | ++ | + | ++ | ++ | ++ | ++ | | | | | |
| 21 | Kunigel V1 (JNB) | | | | | | | | | | | + | + | + | + | + |
| 20 | MX80 | + | + | + | + | + | | | | | | | | | | |
| 19 | Asha-NW BFL-L | | | | | | | | | | | | | | | |
| 18 | Rokle (Rawra) | | | | | | ? | ? | ? | ? | ? | | | | | |
| 17 | Saponite | | | | | | | | | | | | | | | |
| 16 | Asha 505 | + | + | + | + | + | + | + | + | + | + | + | + | + | + | + |
| 15 | MX80 | + | + | + | + | + | | | | | | | | | | |
| 14 | Rokle (Rawra) | | | | | | ? | ? | ? | ? | ? | | | | | |
| 13 | Febex | + | + | + | + | + | | | | | | | | | | |
| 12 | Saponite | | X | X | X | X | | X | X | X | X | | X | X | X | X |
| 11 | Ibeco Seal (IBE) | | | | | | | | | | | | | | | |
| 10 | Calcigel (CAL) | | | | | | | | | | | | | | | |
| 9 | Asha-NW BFL-L | | X | X | X | X | | X | X | X | X | | X | X | X | X |
| 8 | MX80 | + | + | + | + | + | | | | | | | | | | |
| 7 | Ikosorb (IKO) | + | ++ | ++ | ++ | ++ | + | ++ | ++ | ++ | ++ | | | | | |
| 6 | GMZ | | X | X | X | X | | X | X | X | X | | X | X | X | X |
| 5 | Kunigel V1 (JNB) | | | | | | | | | | | + | + | + | + | + |
| 4 | Dep. CAN | + | (?) | (?) | (?) | X | | | | | | | | | | |
| 3 | Asha-NW BFL-L | + | + | + | ++ | X | | X | X | X | X | | X | X | X | X |
| 2 | MX80 | + | + | + | ++ | + | | | | | | | | | | |
| 1 | MX80 | + | + | + | + | + | | | | | | | | | | |

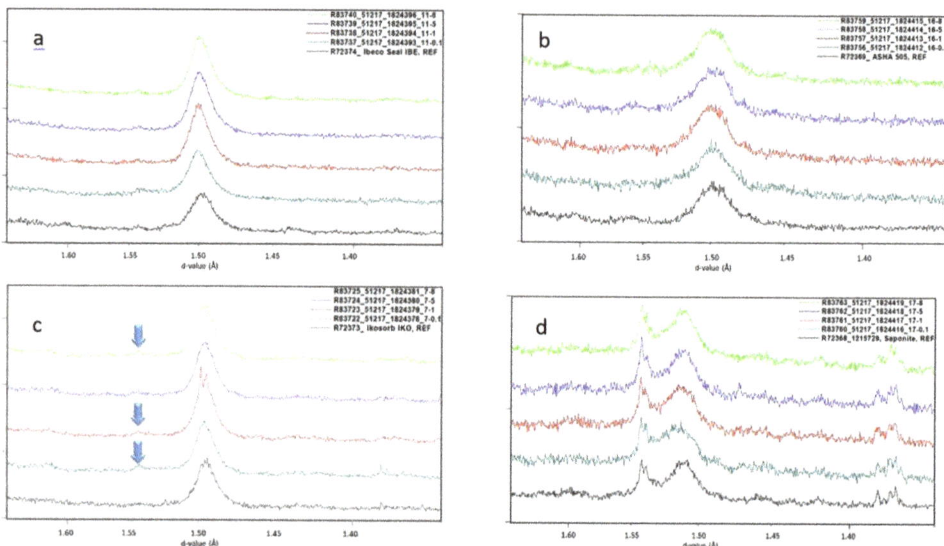

**Figure 5.** XRD analysis of the $d_{060}$ region of dioctahedral smectites in (**a**) #11 (Ibeco Seal), (**b**) #16 (Asha 505), both showing no evidence for presence of trioctahedral phases; whereas presence of trioctahedral phases could be indicated in (**c**) #7 (Ikosorb). In the only block with trioctahedral smectites (**d**) #17 ('Saponite') no evidence for presence of (additional) dioctahedral phases was observed.

XRD analyses of clay fractions of all bentonites from the samples at the contact to the heater (0.1 cm) or of fragments—if such samples were not available—were performed to verify if smectites became interstratified during the ABM5 test. Results indicated no additional interstratification of smectites for any block in comparison with the reference materials ('Saponite' and GMZ data are shown as an example in Figures 4 and 6a,b). Dioctahedral smectites of GMZ (Figure 6b) remained fully expandable (i.e., no illitization) and all bentonites showed the same behavior. The saponite clay minerals in bentonite 'Saponite' initially shows a broad peak at ca. 10.5 Å, which remained unchanged after the experiment (Figure 6a). This intensity can be explained as smectite with collapsed/non expandable layers. 'Saponite' also contains traces of chlorite, serpentine, and talc.

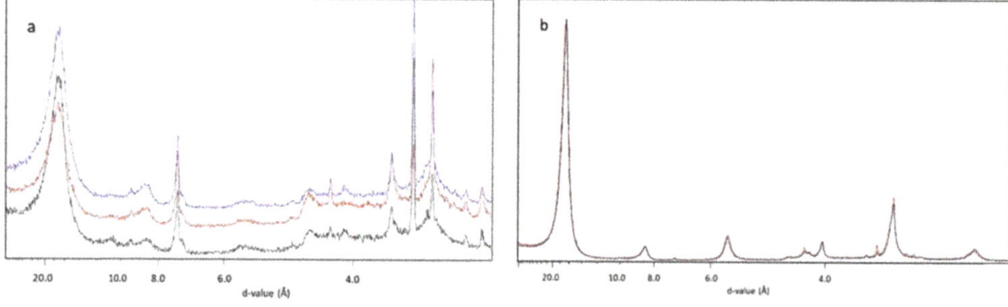

**Figure 6.** XRD peak positions for clay fractions of (**a**) 'Saponite' from the contact to the heater (0.1 cm) of blocks #12 (red) and #17 (blue, upper scan) and of (**b**) GMZ block #6 (fragments with unknown distance to the heater) and the of the according REF samples (both in black) after EG solvation.

### 3.3. Simultaneous Thermal Analysis (STA) and Elemental C-/S-Analysis

STA was performed to investigate the C- and S-phase changes at the interface. As discussed before, carbonates and sulphates can be dissolved precipitating elsewhere and pyrite can be oxidized. The STA detection limit of theses phases is low because of the sensitive mass spectrometer (MS). Pyrite, which is detected by a peak of the MS-$SO_3$ curve between 450 and 500 °C, was found in the bentonites used for blocks #4 and #5 as an example (Figure 7). Interestingly, the pyrite at the contact of block #4 was preserved (could still be detected in the sample taken after the test) but it was oxidized at the contact of block #5. Both blocks were close together, but the different results indicate that the redox potential and/or hydraulic conditions differed inside the blocks. In some samples (#4, #5, #10 in Figure 7), a decrease of the carbonate peak and an increase of the siderite peak (between 500–600 °C) was found.

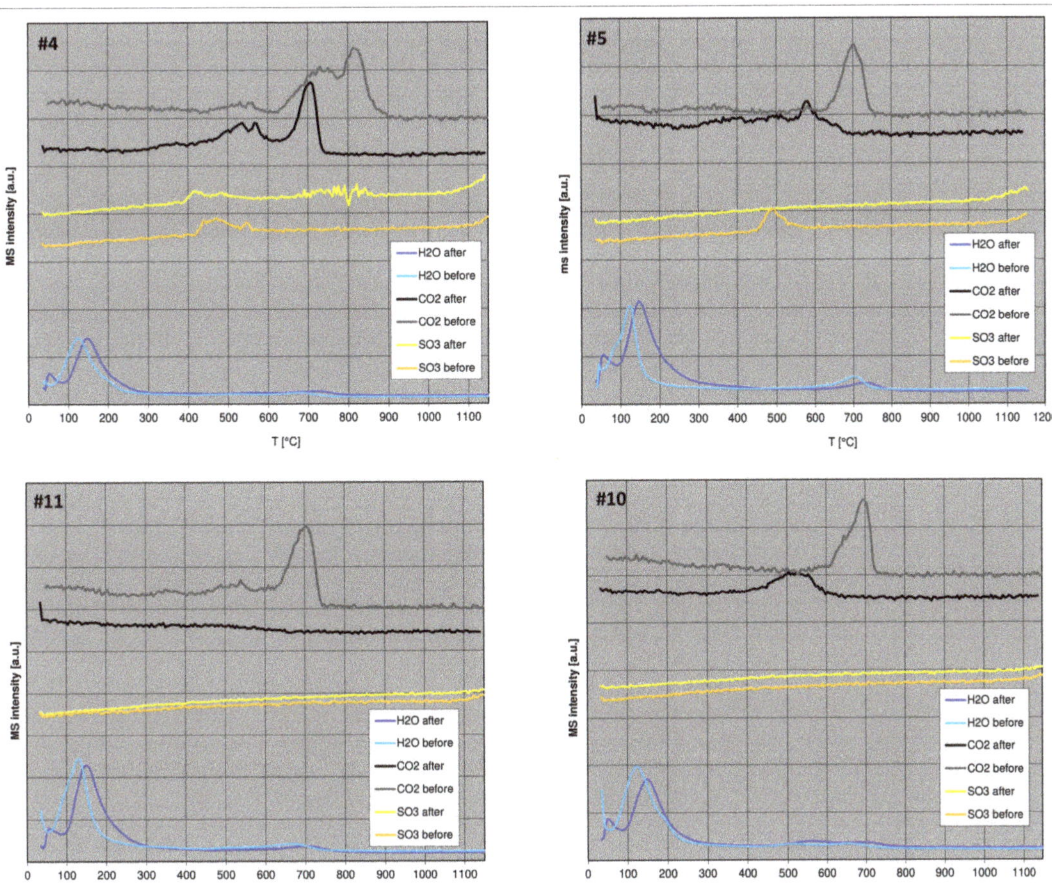

**Figure 7.** STA-MS curves ($H_2O$, $CO_2$, $SO_3$) of four selected blocks ("before" = reference; "after" = contact sample).

Carbonate dissolution and precipitation was found in different medium to large scale deposition tests. Elemental combustion analysis of all samples indicates that different carbonate redistribution reactions occurred. Most samples had a $C_{inorg}$ content of <0.2 mass% and did not show any carbonate redistribution. However, some specific reactions could be identified: A few samples showed some carbonate enrichment in the center of the blocks with larger values compared to the reference (e.g., #1, #14). This indicates that carbonate was either transported into the barrier by the saturating fluid or that carbonate was dissolved in other blocks and transported vertically. Significant carbonate dissolution was observed for both blocks made out of Ibeco Seal (#11, #23; Figure 8). Only the carbonates of the contact sample of #23 seemed to be preserved. Similarly, the blocks made out of Asha NW (#9, #19) showed much lower C values in the blocks compared with the reference which points towards carbonate dissolution. Only the sample taken 1 cm from to the heater showed the same values as the precursor. The largest carbonate content was found in the sample Deponit CAN (about 0.9 mass% $C_{inorg}$). Both blocks were not in the hottest part but showed different reactions. In block #26 carbonate was dissolved near the heater whereas it precipitated near the heater in block #4.

The content of inorganic and organic carbon was determined by combustion (LECO) with and without acid sample pre-treatment. An increase of the C content in the acid treated material would be interpreted as increased organic C because siderite does not fully dissolve in the acid. In the ABM1 and ABM2 tests molykote® was used as lubricant which provided an additional source of organic C and hence complicated the interpretation of elemental C data. The ABM5 was setup without lubricant (it was used during the compaction of the blocks but was removed from the blocks prior to the experiment, by physically removing the outermost 2–3 mm of bentonite from the blocks inner surface by abrasion), which allowed to reveal the formation of siderite at the contact. Some samples showed a decrease of the calcite STA peak (700–800 °C) and an increase of the STA peak between 500 and 600 °C. By elemental C analysis, an increase of the $C_{org}$ content was found for these samples. Both methods hence indicate the formation of siderite at the contact. This reaction, however, was not observed in all blocks. In some, no changes of the carbonates were observed, and in others, dissolution was observed without the precipitation of siderite (e.g., #11). Calcigel showed strong lowering of XRD intensities of dolomite as well as calcite (less clear), along with the formation of siderite (#10, #27).

Siderite formation was observed before. According to Martin et al. [26], Romaine et al. [27], and Mendili et al. [28], magnetite is coating the iron surface and siderite forms on top of the magnetite. They used COX clay which contains carbonates. Moreover, Savoye et al. [29] found siderite formation if carbonate is present, which, however, is not the case in all bentonites. Saheb et al. [30] found Fe-hydroxo-carbonate as corrosion product, which indicates that different Fe-carbonate phases can be present.

The most significant changes detected by XRF, STA, and elemental C are summarized in Table 2. Notably, IR-spectrometry was also used—as in ABM2—but no specific changes except for carbonate dissolution/precipitation could be found. In ABM2, a band at 3740 cm$^{-1}$ pointed towards the formation of free silica. This band could not be observed in any of the ABM5 contact samples.

**Table 2.** Summary of measured differences of the contact sample (0.1 cm) of each block compared with the reference material. Negative values indicate dissolution and positive values correspond to increased concentrations. No contact samples were available of #6, #24, and #27. Bold numbers represent most significant changes, "-" = decrease, "?" = change possible but not unambiguous, "(ox.)" = oxidation (present as sulphate afterwards), "diss" = dissolution.

| | | XRF | XRF | STA | STA | STA | C/S Analyzer | | |
|---|---|---|---|---|---|---|---|---|---|
| | | $\Delta$MgO | $\Delta$Fe$_2$O$_3$ | CaMgCarb | Siderite | Pyrite | $\Delta$C$_{org}$ | $\Delta$C$_{inorg}$ | S |
| 30 | MX80 | −0.4 | **6.3** | | | -, ? | **0.3** | −0.1 | 0.0 |
| 29 | MX80 | −0.1 | 2.6 | | | -, ? | 0.0 | 0.0 | 0.0 |
| 28 | Asha 505 | 0.2 | 1.5 | | | | 0.1 | 0.0 | **0.2** |
| 27 | Calcigel | | | | | | | | |
| 26 | Deponit CAN | −0.2 | 2.9 | - | + | -, (ox.) | 0.1 | **−0.3** | 0.0 |
| 25 | Febex | 0.0 | 1.0 | | | | 0.0 | −0.1 | 0.0 |
| 24 | GMZ | | | | | | | | |

Table 2. *Cont.*

| | | XRF | XRF | STA | STA | STA | C/S Analyzer | | |
|---|---|---|---|---|---|---|---|---|---|
| | | $\Delta$MgO | $\Delta Fe_2O_3$ | CaMgCarb | Siderite | Pyrite | $\Delta C_{org}$ | $\Delta C_{inorg}$ | S |
| 23 | Ibeco SEAL | 1.0 | 0.7 | - | | | −0.1 | −0.1 | **−0.2** |
| 22 | Ikosorb | −0.5 | 2.1 | | | | 0.1 | −0.2 | −0.1 |
| 21 | Kunigel V1 | 0.6 | 0.9 | | +, ? | -, (ox.) | 0.1 | 0.0 | 0.0 |
| 20 | MX80 | 0.1 | 2.3 | | | | 0.0 | 0.0 | −0.1 |
| 19 | Asha NW BFL-L | 0.2 | 2.7 | - | + | | 0.1 | **−0.4** | 0.0 |
| 18 | Rokle | 0.1 | 1.3 | | | | 0.0 | −0.1 | 0.0 |
| 17 | Saponite | −0.9 | 2.6 | - | | | −0.1 | −0.1 | 0.0 |
| 16 | Asha 505 | 0.7 | **4.6** | | | | 0.1 | 0.1 | 0.1 |
| 15 | MX80 | 0.2 | 2.1 | | | -, ? | 0.0 | 0.0 | −0.1 |
| 14 | Rokle | 0.2 | 0.0 | | | | 0.0 | 0.0 | 0.0 |
| 13 | Febex | 0.3 | 1.4 | | | | 0.1 | −0.1 | 0.0 |
| 12 | Saponite | −0.8 | 2.4 | - | + | | 0.1 | −0.2 | 0.0 |
| 11 | Ibeco SEAL | 1.1 | 3.2 | - | | | 0.0 | **−0.6** | **−0.2** |
| 10 | Calcigel | 0.3 | 2.5 | - | + | | **0.2** | −0.2 | 0.0 |
| 9 | Asha NW BFL-L | 0.2 | 1.5 | - | + | | 0.1 | **−0.3** | 0.0 |
| 8 | MX80 | 0.3 | 1.4 | | | | 0.0 | 0.0 | −0.1 |
| 7 | Ikosorb | 0.0 | 1.7 | | | | 0.0 | −0.1 | 0.1 |
| 6 | GMZ | | | | | | | | |
| 5 | Kunigel V1 | 0.4 | 3.7 | - | + | -, (ox.) | 0.1 | 0.0 | 0.0 |
| 4 | Deponit CAN | 0.0 | 1.4 | - | + | no ox. | **0.4** | 0.0 | −0.1 |
| 3 | Asha NW BFL-L | −0.2 | 1.6 | -, (diss) | | | 0.1 | **−0.5** | 0.0 |
| 2 | MX80 | 0.6 | 2.6 | | | -, ? | 0.1 | 0.0 | 0.0 |
| 1 | MX80 | 0.2 | 3.4 | | | -, ? | 0.0 | 0.0 | **0.3** |

*3.4. SEM*

Some blocks which showed the most significant changes (see Table 2) were investigated by scanning electron microscopy (SEM), focusing on the interface.

The lowermost block #1 showed the highest increase of S and appreciable increase of iron. The surface was dark green with some approximately 1 mm² large black spots. Separated particles of the black spots were ferromagnetic, pointing to the presence of magnetite, which is a common corrosion product near the Fe heater. EDX mapping proved the accumulation of Fe at the contact as well as gypsum precipitations (Figure 8a,b).

Block #4 showed a moderate Fe increase. Accordingly, no significant differences were found when mapping the edge of the surface and the bentonite block (Figure 8c). The surface of block #4 showed the largest increase of organic material (LECO), which according to STA could at least partly be explained by siderite. By SEM, however, no newly formed carbonates were found, only a few gypsum crystals at the contact. Instead, a few hyphae were found, which are supposed to result from organic contamination.

The contact of block #10 also showed a significant increase of organic carbon, which based on STA could unambiguously be explained by siderite formation. On the surface, however, traces of microbiological activity were found consisting of hyphae pointing towards the presence of fungi. The concentration is too low to explain the increase of elemental organic-C and the fungi are not believed to have persisted throughout the heat treatment. They most probably formed during storage, although their metabolism could not be characterized further. This has to be systematically investigated in future. The $Fe_2O_3$ increase at the contact was moderate (+2.5 mass%), but SEM-EDX mapping revealed an undulating seem (50–300 μm) in which Fe was enriched.

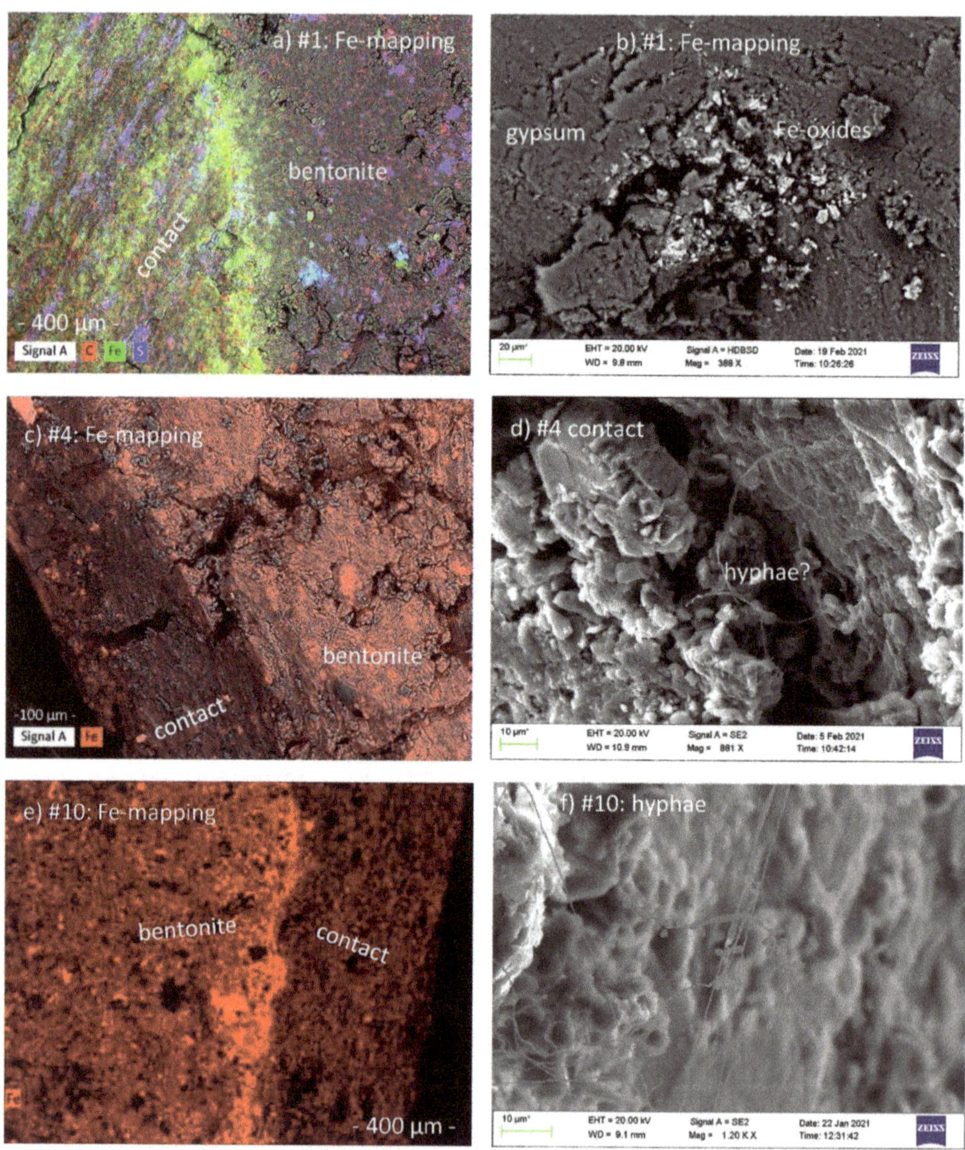



**Figure 8.** SEM images of the surface of selected samples. (**a,b**) block 1; (**c,d**) block 4; (**e,f**) block 10.

Block #16 also showed a significant Fe increase. Figure 9a,b show views on the surface. EDX mapping proved the presence of a homogenous Fe coating (green) with some gypsum spots on top of it.

In block #30 the most significant increase of Fe was found (+6.3 mass%). On the surface, massive Fe coatings were found, which proved to be ferromagnetic. By SEM, however, no small cubic crystals typical of magnetite, but rather a network of fine needles about 500 nm in length was found. Other areas of the surface were coated by gypsum needles (Figure 9e) and isometric gypsum particles (Figure 9f).

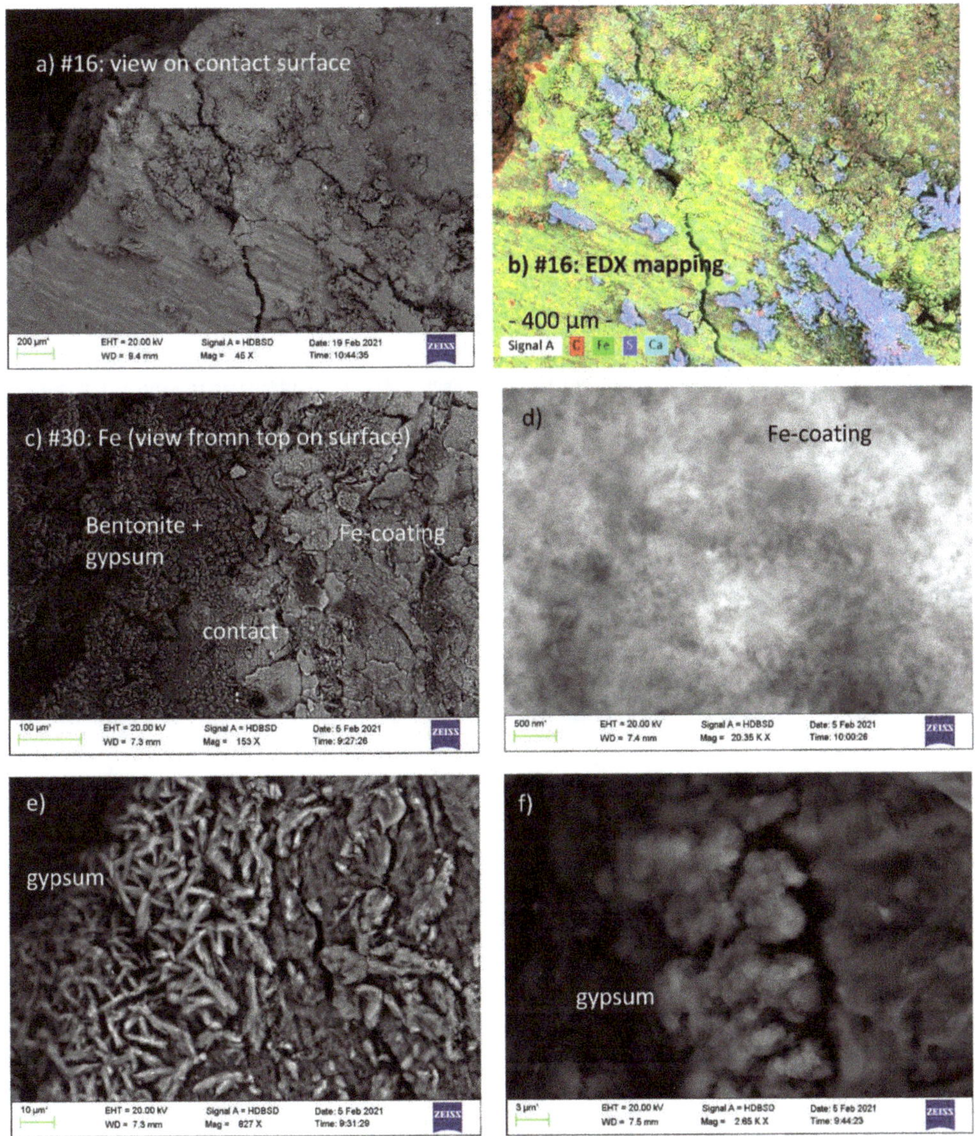

**Figure 9.** SEM images of the surface of selected samples. (**a,b**) block #16, (**c–f**) block #30.

*3.5. Effect of Temperature*

Within an up- or large-scale experiment, locally different conditions may exist, which could lead to different reactions. In the case of the ABM tests, the different materials may have an additional effect on the heterogeneity. The temperature profile (Figure 10) depends on the location and power of the heater and on the thermal conductivity of the adjacent bentonite. The water content (Figure 10) depends on the water uptake of the clays, but also on the drainage system. The bentonite blocks were supposed to be water saturated before heating to the maximum temperature. In the ABM5-test, sand was around the bentonite blocks in which the water could evenly distribute outside the barrier. The hydraulic conditions probably changed with increasing the temperature because of the relatively lower pressure in the sand filter. Moreover, no effect of the fracture in the deposition hole at a depth of about 0.8 m on the water saturation was found. The water content was higher at the bottom probably because the water gathered at the bottom of the deposition hole. The water content and the resulting degree of saturation were therefore larger in the lower 30% of the test, which also affected the dry density (which was lower in the lowermost part). In order to investigate if these locally different conditions could have affected the results, selected chemical differences were plotted depending on depths (Figure 10). No relation of MgO with burial depths could be identified (not shown). The same holds true for inorganic and organic C and total S. The Cl content was on average higher in the lowermost 30% (blocks #1–#10) of the package, which can be explained by the higher water content (and degree of saturation). Most of the Cl is assumed to result from the external saturating fluid. No relation was found between the Fe increase at the contact and the depths. Only block #30 showed a much larger Fe increase, which, however, may have resulted from the special conditions at the top (more oxygen, concrete fluids), but this could not be proven. Interestingly, the Fe increase in 1 cm distance from the heater was more pronounced in the upper blocks, which cannot be explained yet.

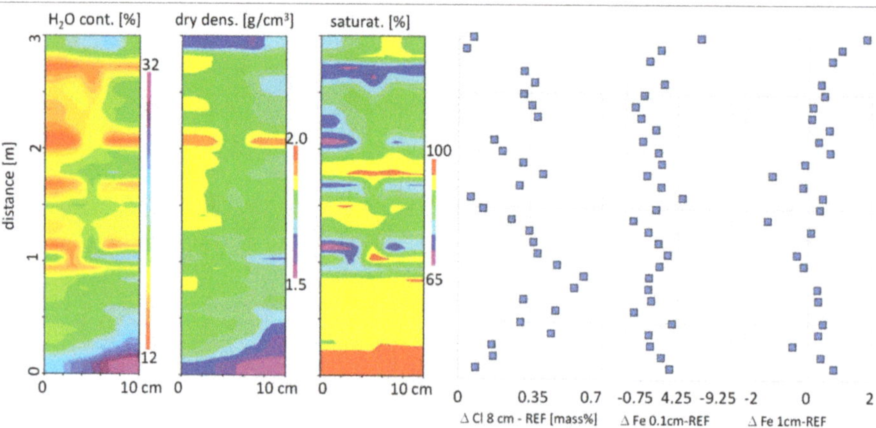


**Figure 10.** Water content, dry density, and degree of saturation measured before termination of the ABM5-test compared with the changes of the Cl- (8 cm distance to the heater) and Fe-content (at 0.1 and 1 cm distance from the heater). Additional information about the T-profile is provided in Figure 1.

### 4. Summary and Conclusions

Results of XRD, XRF, elemental C and S, and STA were not always concise, which results from the fact that many mineralogical changes were close to the detection limit of the methods. Some unambiguous results, however, were found.

Cl concentrations did not depend on material properties or temperature. The Cl content was assumed to result from effects such as inflowing water. On average, a higher Cl increase was found in the lower part of the test, but at the very bottom, the highest water content and lowest Cl increase were found. The reason for different Cl increases could, therefore, not be identified and are supposed to be related to the hydraulic conditions outside and inside the blocks. An additional effect of temperature cannot be excluded.

In blocks #1–#19, the $Fe_2O_3$ increase was restricted to the contact. Blocks #20–#30 all showed a slight to significant increase of the Fe content, even in the 1 cm sample, which indicated both increased corrosion and migration of Fe into the blocks in the upper parts of the test, which can hardly be explained because of the lower water content and hence lower degree of saturation in this zone.

Mg enrichment at the heater, as observed in most of the previous bentonite field experiments, was comparably low. Interestingly the Mg rich 'Saponite' even lost some of its Mg content, also indicating some mobility of Mg within these compacted clays.

Carbonate dissolution and precipitation (redistribution) showed different patterns, which does not allow to derive a general rule. Different profiles were observed for the same material and temperature. Therefore, the differences of the profiles cannot be explained by either material properties (type of carbonate present) or by the different temperatures. Carbonate redistribution is likely more affected by local differences, possibly of porosity and water content. Because of the absence of any lubricant, it was possible to observe a recrystallization of carbonate in some contact samples, where siderite formed at the expense of Ca(Mg)carbonates (proved by STA and LECO and sometimes XRD).

The gypsum redistribution resembled that which was found in the LOT test (increase in the center of the blocks), but this effect was much less pronounced in the ABM5 test. The most significant gypsum increase was found in the upper and lowermost blocks. The identified gypsum may possibly have been anhydrite in the experiment that may have hydrated during storage, as there are indications of moisture in the samples during the storage.

By XRD, no additional interstratification of smectites for any block in comparison with the reference materials was indicated. The neoformation of trioctahedral phases could be indicated only in block #7 (Ikosorb). In the only block with trioctahedral smectites as starting material, #17 ('Saponite'), no evidence for the presence of (additional) dioctahedral phases was observed.

**Supplementary Materials:** The following are available online at https://www.mdpi.com/article/10.3390/min11070669/s1. Table S1: Compilation of geochemical data of all samples taken from the blocks.

**Author Contributions:** STA- and chemical analysis, S.K.; interpretation, S.K. and P.S.; writing, S.K., R.D., K.U., D.S., and P.S.; CEC- and XRD-analysis, R.D. and K.U.; rietveld refinement, K.U.; conduction of test and sampling, D.S.; managing test and sampling, P.S. All authors have read and agreed to the published version of the manuscript.

**Funding:** The work was funded by the EURAD project (work package "HiTec").

**Data Availability Statement:** Not applicable.

**Acknowledgments:** The authors are grateful to Niko Götze, Frank Korte, Natascha Schleuning, Melanie Hein, and Andre Marx for their great analytical work.

**Conflicts of Interest:** The authors declare no conflict of interest.

# References

1. Dohrmann, R.; Kaufhold, S.; Lundqvist, B. The role of clays for safe storage of nuclear waste. In *Handbook of Clay Science, Techniques and Applications*; Bergaya, F., Lagaly, G., Eds.; Elsevier: Amsterdam, The Netherlands, 2013; Volume 5B, pp. 677–710. ISSN 1572-4352.
2. Sellin, P.; Leupin, O. The use of clay as an engineered barrier in radioactive waste management—A review. *Clays Clay Miner.* **2014**, *61*, 477–498. [CrossRef]
3. Kaufhold, S. (Ed.) Bentonites—From mine to application. In *Geologisches Jahrbuch B 107*; Schweizerbart: Stuttgart, Germany, 2020; ISBN 978-3-510-96859-6.
4. Eisenhour, D.D.; Brown, R.K. Bentonite and its impact on modern life. *Elements* **2009**, *5*, 83–88. [CrossRef]
5. Kaufhold, S.; Dohrmann, R. Distinguishing between more and less suitable bentonites for storage of high-level radioactive waste. *Clay Miner.* **2016**, *51*, 289–302. [CrossRef]
6. Eng, A.; Nilsson, U.; Svensson, D.; Äspö Hard Rock Laboratory. *Alternative Buffer Material*; Installation Report SKB IPR-07-15; Svensk Kärnbränslehantering AB: Stockholm, Sweden, 2007. Available online: http://www.skb.com/publication/1633130/ipr-07-15.pdf (accessed on 1 June 2021).
7. Muurinen, A. *Studies on the Chemical Conditions and Microstructure in Package 1 of Alternative Buffer Materials Project (ABM) in Äspö*; Posiva WR 2010–11; Posiva Oy: Olkiluoto, Finland, 2010.

8. Kumpulainen, S.; Kiviranta, L. *Mineralogical, Chemical and Physical Study of Potential Buffer and Backfill Materials from ABM Test Package 1*; Posiva WR 2011–41; Posiva Oy: Olkiluoto, Finland, 2011. Available online: https://inis.iaea.org/collection/NCLCollectionStore/_Public/43/068/43068661.pdf (accessed on 1 June 2021).

9. Svensson, D.; Dueck, A.; Nilsson, U.; Sandén, T.; Lydmark, S.; Jägerwall, S.; Pedersen, K.; Hansen, S. *Alternative Buffer Material. Status of Ongoing Laboratory Investigation of Reference Material and Test Package 1. SKB TR-11-06*; Svensk Kärnbränslehantering AB: Stockholm, Sweden, 2011.

10. Svensson, P.D.; Hansen, S. Redox chemistry in two iron-bentonite field experiments at Äspö hard rock laboratory, Sweden: An XRD and Fe K-edge XANES study. *Clays Clay Miner.* **2013**, *61*, 566–579. [CrossRef]

11. Svensson, P.D. The Bentonite Barrier. Swelling Properties, Redox Chemistry and Mineral Evolution. Ph.D. Thesis, Faculty of Engineering, Lund University, Lund, Sweden, 2015. Available online: https://portal.research.lu.se/ws/files/5584182/5045914.pdf (accessed on 1 June 2021).

12. Kaufhold, S.; Dohrmann, R.; Sandén, T.; Sellin, P.; Svensson, D. Mineralogical investigations of the alternative buffer material test—I. Alteration of bentonites. *Clay Miner.* **2013**, *48*, 199–213. [CrossRef]

13. Kaufhold, S.; Dohrmann, R.; Götze, N.; Svensson, D. Characterisation of the second parcel of the alternative buffer material (ABM) experiment—I mineralogical reactions. *Clays Clay Miner.* **2017**, *65*, 27–41. [CrossRef]

14. Kumpulainen, S.; Kiviranta, L.; Korkeakoski, P. Long-term effects of an iron heater and Äspö groundwater on smectite clays: Chemical and hydromechanical results from the in situ alternative buffer material (ABM) test package 2. *Clay Miner.* **2016**, *51*, 129–144. [CrossRef]

15. Dohrmann, R.; Kaufhold, S. Characterisation of the second package of the alternative buffer material (ABM) experiment—II Exchangeable cation population rearrangement. *Clays Clay Miner.* **2017**, *65*, 104–121. [CrossRef]

16. Sandén, T.; Nilsson, U.; Andersson, L.; Svensson, D. ABM45 Experiment at Äspö Hard Rock Laboratory. Installation Report. SKB Rep. P-18-20. 2018. Available online: https://www.skb.com/publication/2491709/P-18-20.pdf (accessed on 1 June 2021).

17. Kaufhold, S.; Dohrmann, R. Mineralogical and Geochemical Alteration of the MX80 Bentonite from the LOT Experiment—Characterization of the A2 Parcel. Append. 7 SKB-TR-09-29. 2009. Available online: https://www.semanticscholar.org/paper/Mineralogical-and-geochemical-alteration-of-the-the/9005a2659c7419d96b1c81755b44a9a8f7067233 (accessed on 1 June 2021).

18. Olsson, S.; Karnland, O. Mineralogical and chemical characteristics of the bentonite in the A2 test parcel of the LOT field experiments at Äspö HRL, Sweden. *Phys. Chem. Earth* **2011**, *36*, 1545–1553. [CrossRef]

19. Dohrmann, R.; Olsson, S.; Kaufhold, S.; Sellin, P. Mineralogical investigations of the first package of the alternative buffer material test—II. Exchangeable cation population rearrangement. *Clay Miner.* **2013**, *48*, 215–233. [CrossRef]

20. Rousset, D.; Mosser-Ruck, R.; Cathelineau, M.; Villiéras, F.; Pelletier, M. Characterization of the LOT A2, 15 Test Parcel. Appendix 6 in SKB-TR-09-29. 2009. Available online: https://www.skb.com/publications/ (accessed on 1 June 2021).

21. Plötze, M.; Kahr, G.; Dohrmann, R.; Weber, H. Hydro-mechanical, geochemical and mineralogical characteristics of the bentonite buffer in a heater experiment. The HE-B project at the Mont Terri rock laboratory. *Phys. Chem. Earth* **2007**, *32*, 730–740. [CrossRef]

22. Fernandez, A.M.; Kaufhold, S.; Sánchez-Ledesmaa, D.M.; Reya, J.J.; Melóna, A.; Robredoa, L.M.; Fernándeza, S.; Labajoa, M.A.; Claveroa, M.A. Evolution of the THC conditions in the FEBEX in situ test after 18 years of experiment: Smectite crystallochemical modifications after interactions of the bentonite with a C-steel heater at 100 °C. *Appl. Geochem.* **2018**, *98*, 152–171. [CrossRef]

23. Kaufhold, S.; Dohrmann, R.; Degtjarev, A.; Koeniger, P.; Post, V. Mg and silica release in short-term dissolution tests in bentonites. *Appl. Clay Sci.* **2019**, *172*, 106–114. [CrossRef]

24. Kaufhold, S.; Dohrmann, R.; Ufer, K.; Kober, F. Interactions of bentonite with metal and concrete from the FEBEX experiment—mineralogical and geochemical investigations of selected sampling sites. *Clay Miner.* **2018**, *53*, 745–763. [CrossRef]

25. Karnland, O.; Olsson, S.; Nilsson, U. Experimentally determined swelling pressures and geochemical interactions of compacted Wyoming bentonite with highly alkaline solutions. *Phys. Chem. Earth* **2007**, *32*, 275–286. [CrossRef]

26. Martin, F.A.; Bataillon, C.; Schlegel, M.L. Corrosion of iron and low alloyed steel within a water saturated brick of clay under anaerobic deep geological disposal conditions: An integrated experiment. *J. Nucl. Mater.* **2008**, *379*, 80–90. [CrossRef]

27. Romaine, C.A.; Sabot, R.; Jeannin, M.; Necib, S.; Refait, P.H. Electrochemical synthesis and characterization of corrosion productson carbon steel under argillite layers in carbonated media at 80 °C. *Electrochim. Acta* **2013**, *114*, 152–158. [CrossRef]

28. El Mendili, Y.; Abdelouas, A.; Ait Chaou, A.; Bardeau, J.-F.; Schlegel, M.L. Carbon steel corrosion in clay-rich environment. *Corros. Sci.* **2014**, *88*, 56–65. [CrossRef]

29. Savoye, S.; Legrand, L.; Sagon, G.; Lecomte, S.; Chausse, A.; Messina, R.; Toulhoat, P. Experimental investigations on iron corrosion products formed in bicarbonate/carbonate containing solutions at 90 °C. *Corros. Sci.* **2001**, *43*, 2049–2064. [CrossRef]

30. Saheb, M.; Neff, D.; Demory, J.; Foy, E.; Dillmann, P. Characterisation of corrosion layers formed on ferrous archaeological artefacts buried in anoxic media. *Corros. Eng. Sci. Technol.* **2013**, *45*, 381–387. [CrossRef]

 *minerals*

*Article*

# Characterization of Bentonites from the *In Situ* ABM5 Heater Experiment at Äspö Hard Rock Laboratory, Sweden

Ana María Fernández [1,*], José F. Marco [2], Paula Nieto [1], Fco. Javier León [1], Luz María Robredo [1], María Ángeles Clavero [1], Ana Isabel Cardona [1], Sergio Fernández [1], Daniel Svensson [3] and Patrik Sellin [3]

[1] CIEMAT, Centro de Investigaciones Energéticas, Medioambientales y Tecnológicas, Departamento de Medio Ambiente, 28040 Madrid, Spain; paul.nch@gmail.com (P.N.); fcojavier.leon@ciemat.es (F.J.L.); l.robredo@ciemat.es (L.M.R.); ma.clavero@ciemat.es (M.Á.C.); anaisabel.cardona@ciemat.es (A.I.C.); sergio.fernandez@ciemat.es (S.F.)

[2] CSIC, Consejo Superior de Investigaciones Científicas, Instituto de Química Física "Rocasolano", 28006 Madrid, Spain; jfmarco@iqfr.csic.es

[3] Department of Research and Safety Assessment, SKB—Svensk Kärnbränslehantering, SE-57229 Oskarshamn, Sweden; daniel.svensson@skb.se (D.S.); patrik.sellin@skb.se (P.S.)

[*] Correspondence: anamaria.fernandez@ciemat.es

**Citation:** Fernández, A.M.; Marco, J.F.; Nieto, P.; León, F.J.; Robredo, L.M.; Clavero, M.Á.; Cardona, A.I.; Fernández, S.; Svensson, D.; Sellin, P. Characterization of Bentonites from the *In Situ* ABM5 Heater Experiment at Äspö Hard Rock Laboratory, Sweden. *Minerals* **2022**, *12*, 471. https://doi.org/10.3390/min12040471

Academic Editor: Huaming Yang

Received: 28 February 2022
Accepted: 8 April 2022
Published: 12 April 2022

**Publisher's Note:** MDPI stays neutral with regard to jurisdictional claims in published maps and institutional affiliations.

**Abstract:** The Alternative Buffer Material ABM5 experiment is an *in situ* medium-scale experiment performed at Äspö Hard Rock Laboratory (HRL) conducted by SKB in Sweden with the aim of analysing the long-term stability of bentonites used as an engineering barrier for a high-level radioactive waste repository (HLWR). In this work, four different ring-shaped Ca- and Na-bentonite blocks, which were piled around a carbon steel cylindrical heater, subjected to a maximum temperature of 250 °C and hydrated with saline Na-Ca-Cl Äspö groundwater (0.91 ionic strength), were characterized after dismantling. This work allowed us to identify the main geochemical processes involved, as well as the modifications in the physico-chemical properties and pore water composition after 4.4 years of treatment. No significant modifications in mineralogy were observed in samples close to the heater contact, except an increase in Fe content due to C-steel corrosion, carbonate dissolution/precipitation (mainly calcite and siderite) and Mg increase. No magnetite and a low amount of Fe(II) inside the clay mineral structure were detected. No modifications were observed in the smectite structure, except a slight increase in total and tetrahedral charge. A decrease in external surface area and cation exchange capacity (CEC) was found in all samples, with lower values being detected at the heater contact. As a consequence of the diffusion of the infiltrating groundwater, a modification of the composition at clay mineral exchange sites occurred. Ca-bentonites increased their Na content at exchange sites, whereas Na-bentonite increased their Ca content. Exchangeable Mg content decreased in all bentonites, except in MX-80 located at the bottom part of the package. A salinity gradient is observed through the bentonite blocks from the granite to the heater contact due to anions are controlled by diffusion and anion exclusion. The pore water chemistry of bentonites evolved as a function of the diffusion transport of the groundwater, the chemical equilibrium of cations at exchange sites and mineral dissolution/precipitation processes. These reactions are in turn dependent on temperature and water vapor fluxes.

**Keywords:** bentonites; smectites; pore water chemistry; mineralogy; cation exchange; ABM experiment; large-scale tests

## 1. Introduction

Bentonites are essential in the long-term safety of the multi-barrier system for the disposal of nuclear wastes. The efficiency of the bentonite engineered barrier system (EBS) is based on its confinement properties: swelling capacity, low permeability, low diffusivity and high radionuclide retention. Consequently, it is important to have confidence and demonstrate the preservation of these properties under real repository conditions over hundreds of thousands of years [1].

Predicting long-term bentonite barrier behaviour is usually undertaken using results from experiments operated in underground research laboratories (URL) in real conditions, at a real scale and for long test times: Meuse/Haute-Marne in France, HADES in Belgium, Äspö in Sweden, Mont Terri and Grimsel in Switzerland. Different *in situ* large-scale EBS experiments have been accomplished since 1989 by using both compacted bentonite (e.g., prototype, FEBEX, TBT and LOT in situ tests) and high-density bentonite pellets (e.g., FE, RESEAL and EB experiments) (see [2] for references). The aim was to analyse the manufacturing, handling, properties and long-term performance of the bentonite materials (e.g., [3,4] and references therein).

The Alternative Buffer Material (ABM) experiment is a medium-scale field experiment performed at Äspö Hard Rock Laboratory (HRL) in Sweden with the aim of analysing the long-term stability of different bentonites under similar conditions of the current Swedish concept for a high-level radioactive waste repository (HLWR), and under adverse conditions regarding temperature. The test, conducted by SKB (Swedish Nuclear Fuel and Waste Management Company), is based on the KBS-3 (kärnbränslesäkerhet, nuclear fuel safety) concept, in which a repository is placed at approximately 500 m depth in crystalline rock, and a buffer of compacted clay surrounds corrosion-resistant copper canisters containing the waste, in order to minimize water flow and radionuclide transport to the granite host rock [5].

The ABM experiment includes six medium-scale test packages, each one consisting of a central carbon steel tube with heaters, and a buffer of compacted clay artificially hydrated with natural Äspo granitic groundwater for rapid water saturation. Eleven different clays were chosen for the buffers to examine the effects of smectite content, interlayer cations and overall iron content. In addition, bentonite pellets with different proportions of quartz are being tested in some tests. The main purposes of the project were to characterize and compare different bentonite qualities and to identify any differences in behaviour or long-term stability of hydro-mechanical properties, mineralogy and chemical composition after groundwater saturation, heating and interaction with corroding metals (iron–bentonite interactions) [5].

The buffer in package 1 (2006–2009) and package 2 (2006–2013) was subjected to artificial wetting and heating for 28 months and 6.5 years, respectively; the maximum temperatures were 130 °C during the last year for package 1 (ABM1) and 141 °C after the first 2.5 years for package 2 (ABM2). The ABM45 project started in 2012, including three new packages (4, 5 and 6). The ABM5 experiment ran from 2012 to 2017. ABM5 represents an adverse scenario of an HLWR repository, since it is one of the hottest bentonite tests (up to 250 °C) conducted in an underground research laboratory to date (URL). ABM3 (2006–), ABM4 (2012– ) and ABM6 (2012– ) are still running and are expected to be excavated in 2024.

After the experiments, the buffer packages ABM1, ABM2 and ABM5 were retrieved and analysed, with the studies being focused on hydromechanical properties [5], geochemical and mineralogical alteration [5–9], cation exchange rearrangements [10,11], and redox chemistry evolution [12–14]. However, few studies have been performed on the pore water chemistry in these systems [15].

The aim of this work is to analyse the geochemical processes observed in some bentonite samples obtained after dismantling the ABM5 *in situ* experiment, with special emphasis on pore water chemistry studies. The dismantling of this test allowed us to quantify the alteration of the bentonites properties due to different types of perturbations: (a) interactions with artificial granitic groundwater (saturation phase), (b) heat (boiling and desaturation phase due to heating), and (c) interactions with the allochthonous engineered materials (metals, concrete, organics, etc.).

## 2. Materials and Methods

### 2.1. ABM5 Experiment

The project Alternative Buffer Materials 45 (ABM45) was a field experiment consisting of three packages: 4, 5 and 6, each one containing 30 ring-shaped bentonite blocks piled around a cylindrical tube made of carbon steel P235TR1 (Table 1, Figure 1). An electrical heater of 1000 W was placed inside the tube as a heat source. The packages were installed in boreholes (30 cm diameter and 3 m depth) drilled in the tunnel named TASD at ca. 420 m depth in the Äspö Hard Rock Laboratory [16]. The granitic host rock consisted of Äspö diorite and greenstone.

**Table 1.** Chemical composition in wt.% of carbon steel (steel grade P235TR1): EN 10216-1-2014.

| Element | Fe | C | Si | Mn | P | S | Cr | Mo | Ni | Cu | Nb | Ti | V |
|---|---|---|---|---|---|---|---|---|---|---|---|---|---|
| maximum | ball | 0.16 | 0.35 | 1.20 | 0.025 | 0.02 | 0.30 | 0.08 | 0.30 | 0.30 | 0.01 | 0.04 | 0.02 |

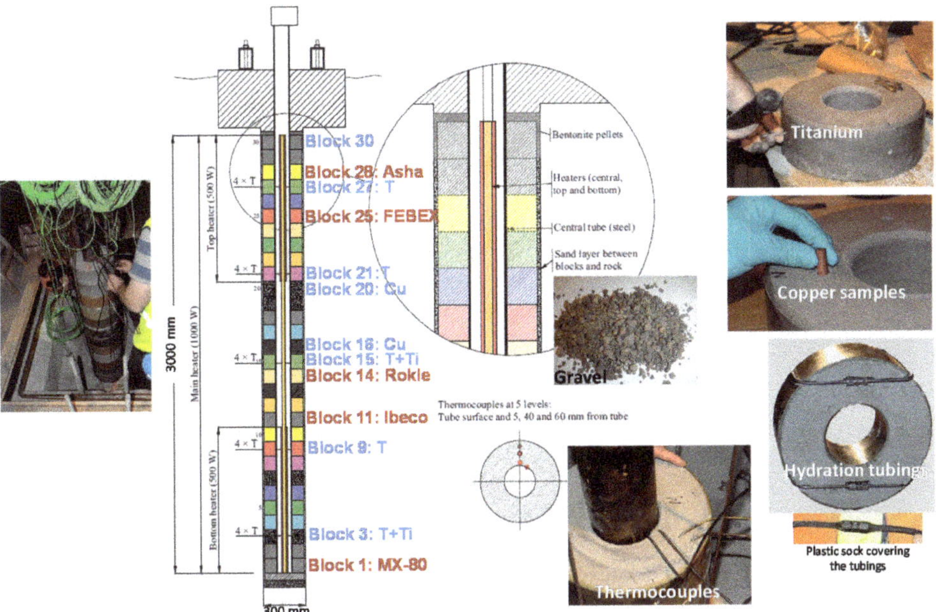


**Figure 1.** Schematic design of the ABM5 test package (after [16], courtesy of SKB). In red: samples analysed in this study. In blue: block containing thermocouples and Cu and Ti coupons. Colours represent different types of bentonites.

The package 5 (ABM5 experiment) was deposited in the borehole named KD0098G01, where there was a groundwater inflow of around 8.5 L/d, coming from a fracture located −0.8 m. However, an artificial water saturation system was used for a rapid saturation of the bentonite. The heating test duration was 4.4 years, starting on 15 November 2012 and finishing on 10 April 2017. The bentonite was heated to 50 °C for the first three and half years, to support saturation and prevent water boiling, and then in 2016 the temperature was increased stepwise to 150 °C and up to 250 °C at the bentonite/heater interface for approximately the last six months of exposure (Figure 2). Estimated maximum temperatures at the heater contact of 240–250 °C were reached between blocks 22 and 8, with decreasing temperatures both at bottom and top part of the bentonite column (<188–156 °C, blocks 3 and 27), and as a function of the granite contact (Figure 2). In the hottest blocks, temperatures of more than 150 °C were reached at a distance of 6 cm from the heater, with the thickness of the bentonite block being 10 cm (see also [8,9]).

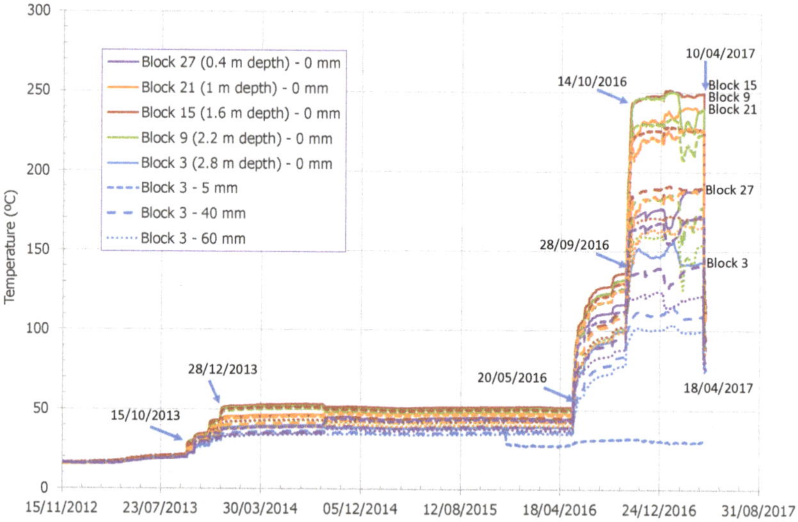

**Figure 2.** Temperatures profiles as a function of the heater contact (0, 5, 40 and 60 mm) measured by thermocouples inside bentonite blocks n° 3, 9 (Asha NW BFL-L), 15 (MX-80), 21 (Kunigel) and 27 (Calcigel).

The bentonites selected for this experiment were of 12 types [16]: MX-80 2012 (Blocks 30, 29, 20, 15, 8, 2, 1), Asha 505 2011 (Blocks 28,16), Asha NW BFL-L (Blocks 19, 9, 3), Deponit CA-N (26, 4), Febex 2012 (Blocks 25, 13), Ikosorb 2011 (Blocks 22, 7), Rokle 2012 (Blocks 18, 14), Kunigel V1 (Blocks 21, 5), GMZ 2011 (Blocks 24, 6), Saponite 2012 (Blocks 17, 12), Calcigel 2012 (Blocks 27, 10) and Ibeco SEAL M-90, old batch (Blocks 23, 11). At least two blocks of each bentonite were placed in the test, except for MX-80 and Asha NW BFL-L, for which 5 and 3 blocks were set, respectively. Therefore, 28 bentonite blocks as compacted rings penetrated by the heater were placed inside the test borehole. Additional MX-80 blocks (#1 and #30) were positioned at the top and bottom to secure a tight sealing. After installation of the test package, concrete plugs were casted above the test borehole in order to prevent the bentonite from swelling upwards.

The ring-shaped blocks (~27.73 cm outer diameter, ~11.0 cm inner diameter and 10 cm height) were compacted uniaxially in a special mould with a pressure between 50 and 100 MPa. A thin layer of molybdenum sulfide containing grease was applied on all steel surfaces in contact with the bentonite powder in order to lubricate to decrease friction. The grease was removed mechanically from the blocks surfaces prior to the bentonites be installed in the test [16].

The outermost gap between bentonite blocks and rock was filled with gravel (2–80 mm), in which 6 mm-diameter titanium tubes with small holes, covered by a plastic sleeve, were placed for a rapid saturation of the bentonite using an artificial groundwater (Table 2). The initial water content of the compacted bentonite was that of "as-received" air dried material. The dry density of the blocks installed in 2013 is listed in Table 3.

**Table 2.** Main composition of the Äspö Groundwater from water supply borehole KA2598A [5].

| Ion | Water-Type | I (M) | Na | K | Ca | Mg | $HCO_3^-$ | $Cl^-$ | $SO_4^{2-}$ | $Br^-$ | $F^-$ | Si | pH |
|---|---|---|---|---|---|---|---|---|---|---|---|---|---|
| mg/L | Ca-Na-Cl | 0.91 | 2470 | 12.4 | 2560 | 64.8 | 51.7 | 8580 | 483 | 59 | 1.5 | 6.3 | 7.33 |

**Table 3.** Average physical properties from bentonite blocks analysed at CIEMAT prior to (initial) and after 4.4 years of test (final).

| Sample | Depth (m) | Max. Temp. (°C) [4] | Bentonite Mass (g) | Water Content w.c. (%) | | Dry Density (g/cm³) | | Grain Density (g/cm³) [2] | Porosity (%) | | Degree of Saturation (%) | | Water Content at Saturation (%) |
|---|---|---|---|---|---|---|---|---|---|---|---|---|---|
| | | | Initial [1] | Initial [1] | Final [3] | Initial [1] | Final [3] | | Initial | Final | Initial | Final | Final |
| Asha Block 28 | 0.3 | 187.0 | 10,820 | 13.1 | 30.1 | 1.84 | 1.49 | 2.869 | 36 | 48 | 67.3 | 93.0 | 32.4 |
| FEBEX Block 25 | 0.6 | 216.5 | 10,640 | 14.3 | 27.9 | 1.80 | 1.51 | 2.735 | 34 | 45 | 75.2 | 94.1 | 29.7 |
| Rokle Block 14 | 1.7 | 252.6 | 10,800 | 17.2 | 29.0 | 1.80 | 1.60 | 2.940 | 39 | 46 | 80.3 | 102.2 | 28.4 |
| IBECO Block 11 | 2.0 | 255.1 | 10,740 | 14.7 | 32.0 | 1.86 | 1.50 | 2.753 | 33 | 45 | 83.6 | 106.0 | 30.2 |
| MX-80 Block 1 | 3.0 | 155.0 | 11,040 | 10.6 | 31.1 | 1.93 | 1.49 | 2.735 | 29 | 46 | 69.9 | 101.7 | 30.6 |

[1] [16]; [2] [5]; [3] Average values; [4] Estimated from thermocouples measurements.

The ABM45 experiments were sparsely instrumented (Figure 1). Twenty thermocouples type T (Chromel-alumel with a shield of cupronickel, 4.5 mm diameter) were installed in blocks numbers 3, 9, 15, 21 and 27. Four thermocouples were installed at each level in pre-drilled holes in the bentonite rings, one on the steel surface and three in the buffer at a radial distance from the heater (5, 40 and 60 mm). In addition, 8 copper specimens (10 mm diameter, 25 mm height, Cu-OFP: oxygen-free phosphorous doped) and 8 titanium specimens (tube of 6 mm outer and 4 mm inner diameters) were positioned. The specimens were installed on the surface in pre-drilled holes followed by a small bentonite cylinder (Figure 1). The copper specimens were positioned in blocks 16 and 20 at 3 cm distance from the heater in four directions (0°, 90°, 180° and 270°). In the case of titanium specimens, four of them were installed in each of the two chosen blocks (#3 and #15), and inside holes were drilled at mid-height of the block periphery in four directions (0°, 90°, 180° and 270°).

The dismantling operation of the ABM5 experiment began in June 2017. Some blocks looked rather intact, while others were highly fractured and very fragile due to the effect of water loss caused by the high temperature. Different bentonite samples were retrieved and analysed for performing different investigations (e.g., [8,9,17]). Most of the bentonite samples were preserved immediately inside vacuum-sealed aluminium-foil bags for avoiding water loss and oxidation due to their exposure to the air-atmosphere.

*2.2. Materials*

For this study, four slices from bentonite samples collected after dismantling were sent to CIEMAT in June 2019 (Figure 1): MX-80 from block position 1 (3 m depth) [18,19], Ibeco from block position 11 (2 m depth) [20], Rokle from block 14 (1.7 m depth) [21,22], FEBEX from block position 25 (0.6 m depth) [23,24], and Asha 505 from block position 28 (0.3 m depth) [25]. These bentonites correspond to mainly Na- (MX-80 and Asha 505) and Ca-Mg-smectites (Rokle, FEBEX and Ibeco) [18]. As additional main differences: (a) Rokle and Asha 505 bentonites have a high content of secondary iron oxides, (b) Ibeco and MX-80 contain pyrite, and (c) Ibeco contains a higher amount of carbonates (total carbon of 1.2 wt.%) than MX-80 and FEBEX bentonites (0.3 wt.% and 0.1%wt. of total C, respectively).

In the laboratory, the received portion of each ring-shaped bentonite block was sampled along one radius. The samples were sliced in several fragments with a knife from the granite contact towards the heater contact. Several subsamples were used for different analyses (Figure 3).

**Figure 3.** Sampling of the bentonites at laboratory for performing different analyses.

### 2.3. Analytical Methods

The objective of the laboratory tests was to undertake a comprehensive analysis of the blocks in order to determine the physical, physico-chemical, mineralogical and geochemical characteristics of the buffer for the assessment of the properties of the bentonite material after the heating and hydration process over the 4.4 years of the experiment.

#### 2.3.1. Physical Properties

The gravimetric water content, w.c., expressed as a percentage, is defined as the ratio between the weight of water lost after heating the sample at 110 °C for 48 h in an oven and the weight of dry solid.

Dry density, $\rho_d$, is defined as the ratio between the weight of the dry sample and its volume occupied prior to drying. The volume of the solid samples was determined by means of the mercury displacement method (UNE Standard 7045).

Total physical porosity or total porosity was calculated by means of the relationship: $n = \left(1 - \left(\rho_{bulk,dry}/\gamma_s\right)\right) \times 100\,[vol\%]$, where $\rho_{bulk,dry}$ is the bulk dry density and $\gamma_s$ is the grain density or specific gravity of the solid sample.

### 2.3.2. Mineralogical Analysis

XRD diffraction patterns were obtained from random powders and oriented aggregates by using a Philips X'Pert –PRO MPD diffractometer, equipped with a fixed divergence slit (0.1245° size), Scientific X´celerator detector and an anticathode Cu-$K_\alpha$ at 45 kV and 40 mA. The samples were analysed from 2° to 70° 2θ with a step size of 0.017° 2θ. The scan rate per step for the powder samples was 50 s. HighScore program v.5 was used for mineral identification and quantitative analyses, with the Power Diffraction File database from the International Center for Diffraction Data (ICDD).

The fine fraction of less than 2 μm was obtained by suspension and sedimentation in deionised water by using the modified Jackson treatment [26,27]. The final clay suspension was ultrasonically dispersed using 1 g in 5 mL of deionized water. Oriented mounts (air dried, ethylene glycol solvated and 550 °C heated) were analysed by using a Bruker D8 Advance diffractometer with an anticathode Cu-Kα at 40 kV and 30 mA, equipped with a fixed divergence slit (0.15° size). The samples were investigated from 2° to 35° 2θ with a step size of 0.02° 2θ and a scan rate of 2 s per step.

Fourier Transform Infrared (FTIR) spectra were obtained in the middle-IR region (4000–400 cm$^{-1}$) by using a Nicolet iS50 with a DTGS KBr detector (resolution 4 cm$^{-1}$, 32 scans) on KBr-pressed discs in transmission technique. The atmosphere was continuously purged from water and atmospheric $CO_2$. Two milligrams of powdered air-dried sample were dispersed in 200 mg of KBr and pressed into a clear disc.

Scanning electron microscopy (SEM). A JEOL JSM-820 SEM microscope coupled with a X Oxford ISIS Link energy dispersive X-ray energy spectrometer (XEDS). Prior to the analysis, the samples were dried at 40–60 °C in an oven overnight, and then subjected to gold metallization by applying $5 \times 10^{-2}$ Torr vacuum and a gold coating of 300 to 400 Å thickness, using a BALZERS SCD 004 sputter coater. The composition of dioctahedral smectites and illites was calculated using the structural formula method according to [27,28], on the basis of 11 oxygen atoms equivalent per half unit cell (e.phuc$^{-1}$), tetrahedral occupation by 4.0 cations ($^{IV}$Si + $^{VI}$Al), octahedral occupation by 2 cations, and complete oxygen/hydroxyl framework of $O_{10}(OH)_2$. For chlorites [ … $O_{10}(OH)_2$-1$H_2O$] and kaolinites [ … $O_5(OH)_4$-1$H_2O$], 14 and 7 oxygen atoms e.phuc$^{-1}$ were used, respectively.

XPS. X-ray photoelectron spectroscopy (XPS) data were recorded using a Phoibos-150 electron analyser (SPECS) under a pressure lower than $2 \times 10^{-9}$ mbar using Al Kα radiation. The wide scan spectra and the narrow (high resolution) spectra were recorded using a constant pass energy of 100 and 20 eV, respectively. The binding energy (BE) scale was referenced to the BE of the main C-C contribution (284.6 eV) of the C 1s spectrum corresponding to the adventitious contamination layer. All of the spectra were computer-fitted using pseudo-Voight line profiles and the CASAXPS software. Relative atomic concentrations were calculated by peak integration after background subtraction using the Shirley method and the atomic sensitivity factors tabulated by Wagner [29].

Mössbauer spectroscopy. $^{57}$Fe Mössbauer spectroscopy data were recorded at room temperature (300 K) in transmission mode using a conventional constant acceleration spectrometer and a $^{57}$Co (Rh) source. Absorbers were prepared to have an effective thickness of about 5–10 mg of natural iron per square centimetre. The velocity scale was calibrated using a 6 μm-thick natural iron foil. All the spectra were computer-fitted using Lorentzian lines and the isomer shifts were referred to the centroid of the α-Fe sextet at room temperature.

### 2.3.3. Geochemical Analysis of Solid Samples

#### X-ray Fluorescence

The major elements ($SiO_2$, $Al_2O_3$, $Fe_2O_3$, MnO, MgO, CaO, $Na_2O$, $K_2O$, $TiO_2$, $P_2O_5$ and $SO_3$ in %) were analysed by using an Axios X-ray Fluorescence (XRF) spectrometer from Panalytical equipped with a rhodium X-ray tube (stimulation power: 1 KW). The dissolution or decomposition of the samples into a homogeneous glass was obtained by fusion, which consisted of heating a mixture of the sample (0.8 g) and a flux of 7.2 g of $Li_2B_4O_7$ at high temperatures (800 to 1000 °C). The end-product after cooling was a one-phase glass. The whole material was melted stepwise in a platinum crucible at a smelting apparatus. After this procedure, the melt was transferred into a platinum jacket and cooled. The loss of ignition was determined separately by oven-drying the dried sample at 1025 °C for 3 h.

*Chemical Total Carbon and Total Sulfur*

Total carbon, total sulfur and total inorganic carbon were determined on 0.2 g of powdered solid samples by means of a LECO CS-244 analyser by combustion. The total inorganic carbon (TIC) content was obtained with a TOC-V$_{CSH}$ analyser (SHIMADZU, Shimadzu Scientific Instruments, Kyoto, Japan) equipped with an SSM-5000A module. The presence of organic carbon was evaluated by the difference between TC and TIC.

*Chemical Fe(II)-Fe$_{total}$ in the Samples*

Fe(III) and Fe(II) were leached from the samples by using a modification of the procedure described in [30] and analysed by the 1,10 phenanthroline spectrophotometric method. To a finely ground sample of 0.1–0.2 g, 1 g of $NH_4HF_2$ and 10 mL of 1:1 $H_2SO_4$ were added in 40 mL high-density polyethylene (HDPE) tubes. The tubes were closed and heated to 90 °C for 60 min in a shaking water bath. After this extraction process, the tubes were allowed to cool to room temperature. The contents were transferred to 50–100 mL volumetric flasks and made up to volume with MilliQ water. The phenanthroline method was used for the determination of Fe(II).

### 2.3.4. Physico-Chemical Characterization

*Cation Exchange Capacity*

Total CEC was measured with 0.01 M copper triethylenetetramine, Cu-Trien or $[Cu(Trien)]^{2+}$, solution [31]. In this process, 200 mg of an air-dried clay sample was weighed in 60 mL centrifuge tubes. Then, 25 mL of deionized water was added, and the suspension was dispersed by ultrasonic treatment for 5 min. Then, 10 mL of 0.01 M $[Cu(Trien)]^{2+}$ was added and allowed to react by end-over-end shaking for 1 h. The suspensions were centrifuged at a constant rotation speed of 11,000 rpm for 30 min. Then, 3 mL of the clear blue solution (filtered through 0.2 μm pore size syringe filter) was filled into 1 cm optical glass cuvettes and the absorbance of the solution was measured at 578 nm wavelength by using a Orion Aquamate 8000 spectrophotometer. The analyses were performed in duplicate, with the standard deviation of the measurement being ±2 meq/100 g.

*Cation Exchange Population*

Cation exchange population was determined by using Cs as the index displacing cation [32]. Solid air-dried samples were equilibrated at a 1:4 solid to liquid ratio (0.25 kg/L) with 0.5 M $CsNO_3$ at pH 8.2 inside a JACOMEX glove box (<1 ppm $O_2$). After phase separation by centrifugation (11,000 rpm, 30 min., outside the glove box), the supernatant solutions were filtered through 0.2 μm pore size syringe filter (inside the anoxic glove box), and the concentration of the major and trace cations was analysed. Sodium and calcium contents were corrected with respect to soluble salts from pore water.

*BET External and Total Surface Area*

Nitrogen adsorption/desorption isotherms were obtained by using a Micromeritics ASAP 2020 V3.02 H sorptometer. Around 0.5 g of the total sample were dried at 90 °C for at least 24 h before the tests. Prior to the nitrogen adsorption, the samples were outgassed by heating at 90 °C for 18 h using a mixture of helium and nitrogen under a residual vacuum between 6 and 10 μmHg. External surface areas, corresponding to both the external faces and the edges of the smectite particles, since in the adsorption of non-polar molecules, the layered structure remains closed, were calculated using the standard N2-BET method, using a series of data points over the P/P0 range of 0.02 to 0.25 on the nitrogen adsorption isotherm [33].

In contrast to nitrogen, which is only adsorbed at the external surface of the stacks of the smectite layers, water molecules can be adsorbed onto the whole surface, including both the internal (interlayer) and external surfaces of the clay minerals. Thus, in order to obtain the total specific surface area (SA), water vapour gravimetric adsorption measurements were performed by Keeling´s hygroscopic method [34] by storing the samples at a constant 75% relative humidity atmosphere in a chamber over-saturated in NaCl solution for 1 month. Prior to the tests, the samples were dried at 110 °C for 24 h. The weight changes of the samples as a result of the adsorbed amount of water were measured and related to the total surface area, SA.

*Soluble Salts by Aqueous Leaching*

Aqueous extract solutions were used for analysing the soluble salts and ion inventories. The crushed subsamples were placed in contact with deionised and degassed water at a 1:4 solid to liquid ratio, shaken end-over-end and allowed to react for one day under anoxic conditions inside an anoxic glove box. After phase separation by centrifugation (30 min at 11,000 rpm), the supernatant solutions were filtered through a 0.2 μm pore-size syringe filter (inside the anoxic glove box) and analysed. Aqueous leaching conditions were selected to suppress mineral oxidation, by working in an oxygen-free atmosphere, and to avoid large modifications of cation concentration at exchange sites by carbonate dissolution, which affects the cation distribution in the aqueous extract solution. All the tests were performed for one day, since a carbonate equilibrium in bentonites usually requires more than six days (e.g., [24,35]).

### 2.3.5. Pore Water Chemistry

The pore water of the bentonite samples was obtained by means of the squeezing technique at high pressures [36–38]. Squeezing is analogous to the natural process of consolidation, caused by the deposition of material in geological time, but at a greatly accelerated rate. The squeezing process involves the expulsion of interstitial fluid from the saturated clayey material being compressed [39].

At CIEMAT, the squeezing rig is similar to that developed by [39,40]. The squeezer was designed to allow a one-dimensional compression of the sample by means of an automatic hydraulic ram operating downwards, the squeezed water being expelled from the top and bottom of the cell into a vacuum vial (Figure 4). The compaction chamber is made of type AISI 329 stainless steel (due to its high tensile strength and resistance to corrosion) with an internal diameter of 70 mm. The compaction chamber is 250 mm high with 20 mm wall thickness and allows pressures up to 100 MPa.

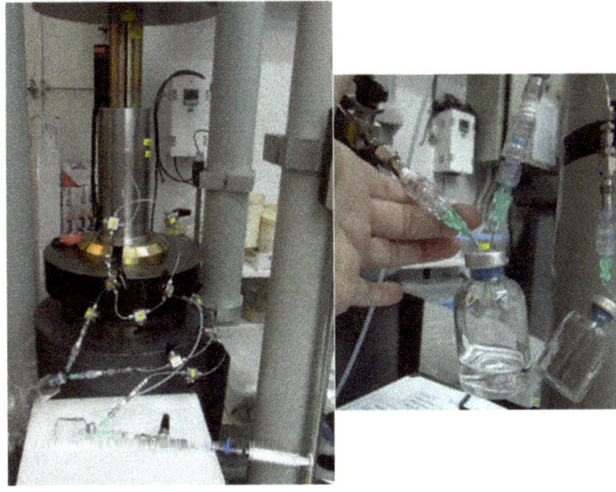

**Figure 4.** Squeezing technique for collecting the pore water from bentonite blocks.

The filtration system allows the extraction of interstitial water by drainage at the top and at the bottom of the sample. This system comprises a 0.5 μm stainless steel AISI 316L porous disk (Cr 17.36%, Ni 11.4%, Mo 2.15%, Si 0.94%, Mn 0.17%, C 0.027%, S 0.011%, P 0.022%, Fe 66.92%) in contact with the sample. The liquid is collected through stainless steel tubes ($^1/_{16}$ inch) in a vacuum vial sealed by a septum. The whole system remains under ambient conditions (room temperature of about 22–25 °C). However, a sampling circuit was designed for collecting the water at anoxic conditions (Figure 4). At the beginning of the test, the squeezing cell and all the sampling tubing and vials were closed to ambient conditions. Then, several Ar-flushing and vacuum cycles were performed, keeping the system under anoxic conditions.

The bentonite samples for squeezing were prepared using a knife to remove the outer part in order to discard possible contaminated material. The sample was cut into a cylindrical shape, weighed and placed into the body of the cell. A small stress of 1 to 5 MPa was initially applied to remove most of the atmospheric gas from the cell and allow the sample to bed in. After applying two additional Ar-flushing and vacuum cycles to the sampling circuit, the stress was progressively increased up to the selected pressure, rather than in a single step. This avoids overconsolidation or collapse of the clay-pore system. When the maximum amount of squeezed water was obtained for a given pressure, the vial was removed, keeping the sample away from any contact with the atmosphere. The water sample collected was weighed and immediately analysed. The bentonite mass was also weighed, and the final water content and dry density were determined. During the squeezing test, the evolution of the pressure, axial strain and changes in the length of the sample due to consolidation were recorded over time by using a data acquisition system. The chemical analysis of the water samples was performed with the methods described in Section 2.3.6.

### 2.3.6. Water Chemical Analyses

The water samples were filtered through 0.2 μm syringe filters, except those for pH and electrical conductivity (EC) measurements. The pH was measured by means of an ORION 720A pH-meter equipped with a Metrohm 6.0224.100 combined pH micro-electrode. The total alkalinity of the water samples was determined by using a Metrohm 888 Titrando equipment (5 mL burette and a 6.0224.100 Metrohm combined pH micro-electrode), with a specific dynamic equivalence point titration (DET) method. The major and trace cations, including silica, were analysed by means of inductively coupled plasma optical emission spectrometry (ICP-OES) in an Agilent 5900 synchronous vertical dual view (SVDV) spectrometer. Sodium and potassium were determined by atomic absorption spectrometry in an Agilent AA 240 FS spectrometer. Anions were analysed by ion chromatography by using a Dionex ICS-2000 equipment.

## 3. Results

### 3.1. Physical Properties

The characteristics of the bentonite blocks at initial conditions before the *in situ* experiment are given in Tables 3 and 4. The initial water content is between 10.6 and 17.2%, the dry density is between 1.80 and 1.93 g/cm$^3$ and the degree of saturation is between 67 and 84%. After 4.4 years of artificial hydration and heating, the samples increased their water content up to ~30 ± 2%, showing lower values closer to the heater contact. The dry density decreased up to ~1.52 ± 0.05 g/cm$^3$, with a tendency to increase closer to the heater contact (Table 4). It is interesting to note that although complete saturation was achieved, the degree of saturation increased from the top to the bottom of the bentonite package (from 93 to 100% of saturation), as described in [8], probably due to gravity, the highest temperatures in the central part, and the steam–bentonite interactions at the top of the bentonite package (Figure 2).

**Table 4.** Physical properties of bentonite blocks analysed (see locations in Figure 3).

| Subsample | Distance to Heater (cm) | Water Content w.c. (%) | Grain Density (g/cm$^3$) [1] | Dry Density (g/cm$^3$) | Porosity (%) | Degree of Saturation (%) |
|---|---|---|---|---|---|---|
| Asha Block 28 granite (G) | 8.33 | 30.3 | 2.869 | 1.49 | 0.48 | 93 |
| Asha Block 28 central (M) | 5.00 | 29.9 | 2.869 | 1.49 | 0.48 | 92 |
| FEBEX Block 25 granite (G) | 8.33 | 28.1 | 2.735 | 1.51 | 0.45 | 94 |
| FEBEX Block 25 central (M) | 5.00 | 27.8 | 2.735 | 1.51 | 0.45 | 94 |
| Rokle Block 14 granite (G) | 8.33 | 29.9 | 2.940 | 1.63 | 0.45 | 109 |
| Rokle Block 14 central (M) | 5.00 | 28.2 | 2.940 | 1.58 | 0.46 | 96 |
| IBECO Block 11 granite (G) | 8.33 | 33.1 | 2.753 | 1.44 | 0.48 | 100 |
| IBECO Block 11 heater (H) | 1.67 | 30.9 | 2.753 | 1.56 | 0.43 | 112 |
| MX-80 Block 1 granite (G) | 8.33 | 32.5 | 2.735 | 1.46 | 0.47 | 102 |
| MX-80 Block 1 heater (H) | 1.67 | 29.8 | 2.735 | 1.52 | 0.45 | 101 |

[1] [5].

### 3.2. Mineralogy

#### 3.2.1. XRD Analysis

The random powder patterns of the samples analysed show the strongest (001) reflection of smectite located at ~14–15Å indicative of a two-layer hydrate Ca-Mg-Montmorillonite (Figure S1 from Supplementary Data, Figure 5). This is due to the saline Na-Ca-Cl-groundwater/bentonite interactions which provoked cation exchange reactions. Na-bentonites (MX-80 and Asha 505) have changed the character of the smectite particles acquiring a Ca-character, shifting d(001) values from 12 Å to 15 Å. In turn, the initial Ca-bentonites (FEBEX, IBECO, Rokle) reveal two populations of water-hydrated montmorillonite clay particles, increasing the one-layer hydrate Na-montmorillonite component (~12 Å). None of the bentonites showed significant indication of any smectite loss or transformation.

d(006) XRD reflexion was centred at 1.49-150 Å both in reference and retrieved samples, with no changes being observed even in the samples located at the heater contact (Figure 5). This indicates that the presence of neoformed trioctahedral smectites (saponite), appearing at 60–61° 2θ, can be ruled out after the experiment, in contrast to former ABM experiments [5,6,41] or in other *in situ* experiments, such as the FEBEX *in situ* test [42].

XRD analyses of oriented aggregates were performed to verify possible changes in clay mineral particles, increasing Ilt/Sme interstratified and/or mixed-layered clay minerals after the high-temperature heating experiment. However, no significant differences are observed in the oriented aggregates (Figure 6, Figure S2 from Supplementary Material). The dioctahedral smectites preserve their expandability (i.e., no illitization), according to the (001) reflexion around 17 Å in the EG XRD pattern for all samples, which indicates complete expansion of the interlayer sites. Kaolinite was present in the reference sample of Asha 505 bentonite, and it is not a neoformed clay mineral.

Regarding other accessory mineral phases (Figure S1 from Supplementary Material and Figure 5), Asha 505 bentonite block 28 showed the highest mineralogical differences in the sample at contact with the heater interface, in which gypsum, Na-clinoptilolite and an increase in calcite and Fe-oxides (goethite, hematite) were observed. FEBEX bentonite block 25 shows an increase in calcite, feldspars and the presence of gothite (Figure S1). Rokle bentonite block 14 increased the amount of goethite, hematite, calcite, dolomite and the presence of siderite. IBECO bentonite block 11 showed similar patterns to the original bentonite sample, except for the absence of calcite. MX-80 bentonite block 1 shows increased amounts of calcite, gypsum, goethite and hematite and the presence of monohydrocalcite, siderite and pyrite.

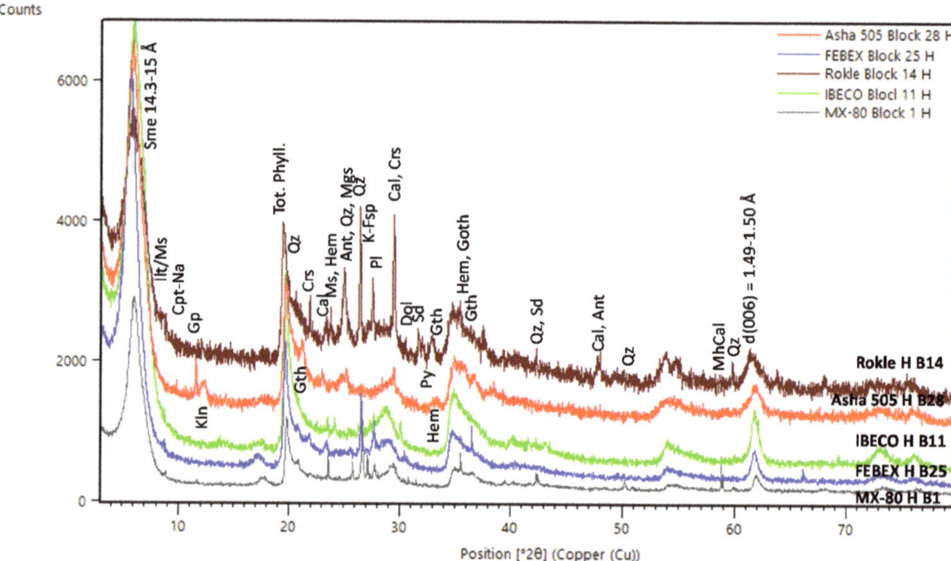

**Figure 5.** XRD patterns of bulk samples from ABM5 experiment. Symbols according to [43]. Sme: smectite, Ilt/Ms: illite/muscovite, Cpt-Na: Clinoptilolite, Gp: Gypsum, Tot. Phyll: total phyllosilicates, Crs: cristobalite, Cal: calcite, MhCal: monohydrocalcite, Qz: quartz, K-Fsp: potassium feldspar, Pl: Plagioclase, Dol: dolomite, Gth: goethite, Hem: hematite.

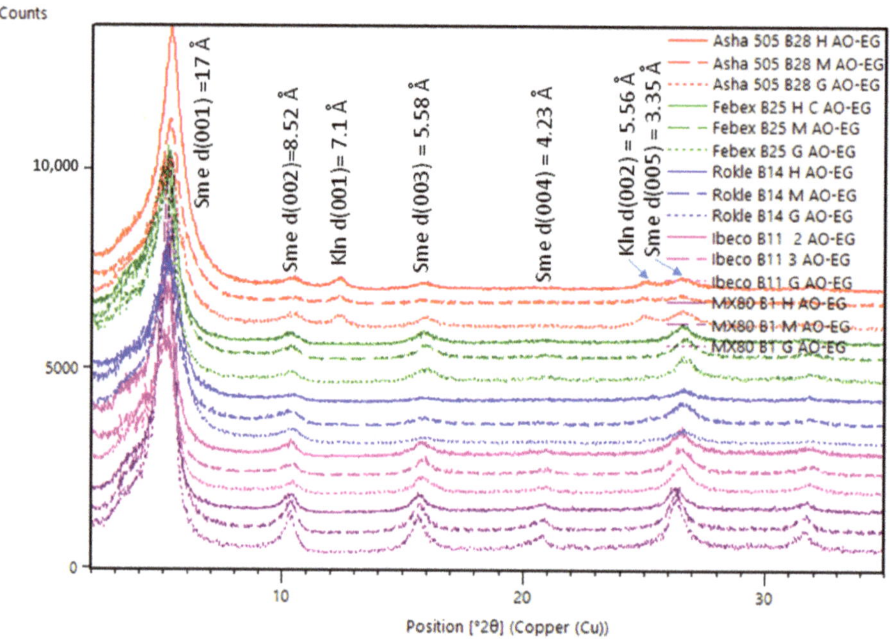

**Figure 6.** XRD patterns of oriented aggregate samples (after ethylene glycol treatment).

### 3.2.2. FTIR Analysis

The bentonite samples were also analysed by FTIR for acquiring information on possible changes in the smectite structure, chemical composition and surface properties due to chemical modifications, as well as to investigate mineral neoformations.

The spectra from the bulk samples of different bentonite blocks are shown in Figures 7 and 8. All spectra include the main typical dioctahedral smectite bands (Table S2 from Supplementary Material). The spectra show a band at 3627 cm$^{-1}$ which corresponds to the typical OH stretching region of structural hydroxyl groups for dioctahedral smectites with Al-rich octahedral sheets. These are inner hydroxyl groups lying between the tetrahedral and octahedral sheets. The broad band near 3426 cm$^{-1}$ is due to stretching H-O-H vibrations of adsorbed water, while the band at 1642 cm$^{-1}$ corresponds to the OH deformation or bending adsorption of water. However, additional bands around 3697 cm$^{-1}$ indicate the presence of kaolinite in the Asha 505 and Rokle samples, initially present in the raw samples. If the Si-O absorptions and OH bending bands in the 1300–400 cm$^{-1}$ range are examined, only one broad, complex Si-O stretching vibration band at around 1030 cm$^{-1}$ is seen, which is typical of dioctahedral montmorillonite. In this range, the occupancy of the octahedral sheet can be distinguished due to each cation strongly influencing the position of the OH bending bands, which arise from vibrations of the inner and surface OH groups. In all samples, the presence of a peak at 915 cm$^{-1}$ ($\delta$AlAlOH) is observed, which is typical of dioctahedral smectites. All samples from bentonites Febex, Ibeco and MX-80 show a band at ~840 cm$^{-1}$ ($\delta$AlMgOH), indicating a partial substitution of octahedral Al by Mg. Asha 505, Rokle and MX-80 samples reflect an additional partial substitution of aluminium by iron ($\delta$AlFeOH), with the band at 874-885 cm$^{-1}$. However, the Fe substitution is much higher in Asha 505 samples due to the decrease in the $\delta$AlMgOH peak and the increase in the $\delta$AlFeOH peak. The adsorption band at 622 cm$^{-1}$ can be attributed to a R-O-Si vibrations (R = Al, Fe, Mg) in smectites and indicates a perpendicular vibration of the octahedral cations and their connection to the tetrahedral sheet. The bands at 520 cm$^{-1}$ and 466 cm$^{-1}$ correspond to Si-O-Al vibration of aluminium in the tetrahedral sheet and Si-O-Si bending vibrations, respectively. All samples show the weak band at 798 cm$^{-1}$ caused by the Si-O stretching of quartz. Calcite (~1426 cm$^{-1}$) is observed in Asha 505, Rokle, Ibeco and MX-80 samples. However, this peak disappears in all the analysed retrieved samples from the Ibeco bentonite block 11, as seen in XDR patterns, indicating a high mineral dissolution process in this part of the bentonite package. No other alterations are observed between the initial and retrieved samples from the bentonite blocks analysed.

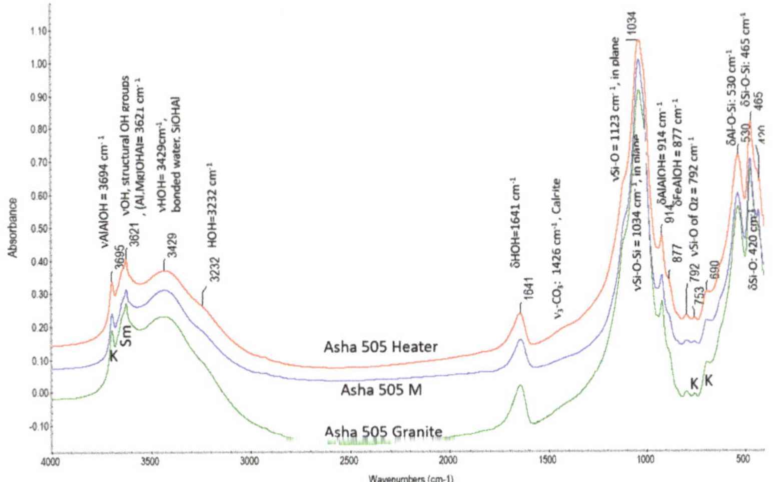


**Figure 7.** *Cont.*

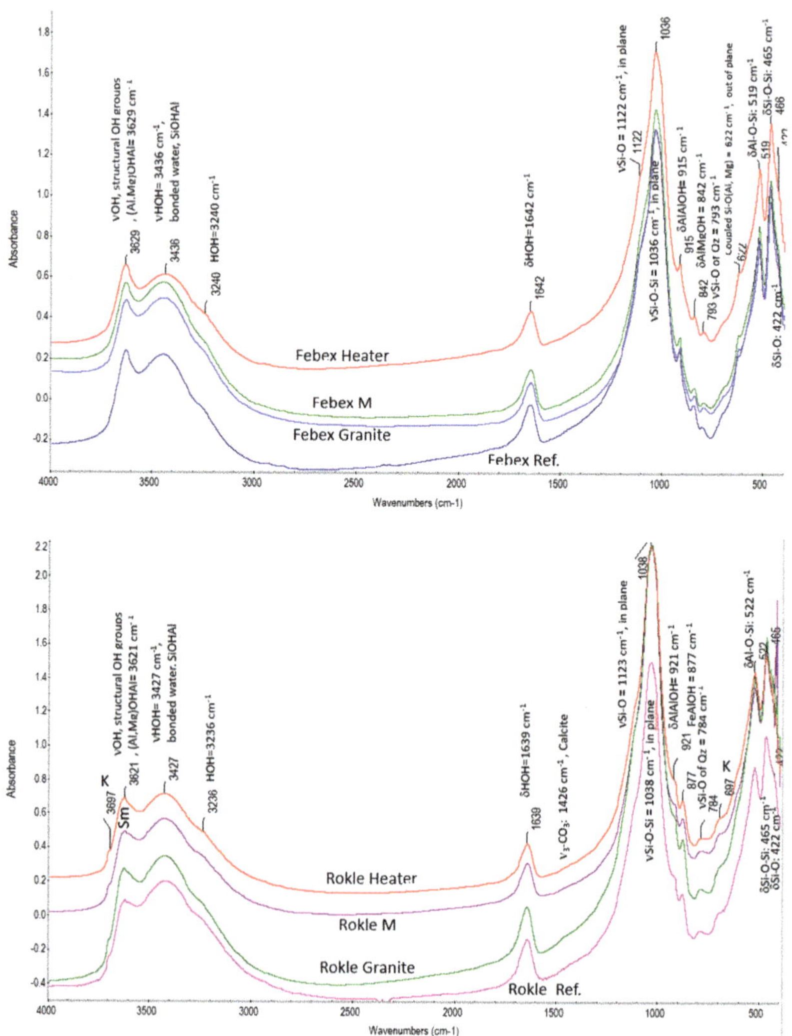

**Figure 7.** FTIR spectra from samples taken from the ABM5 experiment: Asha 505 block 28 Febex block 25 and Rokle Block 14.

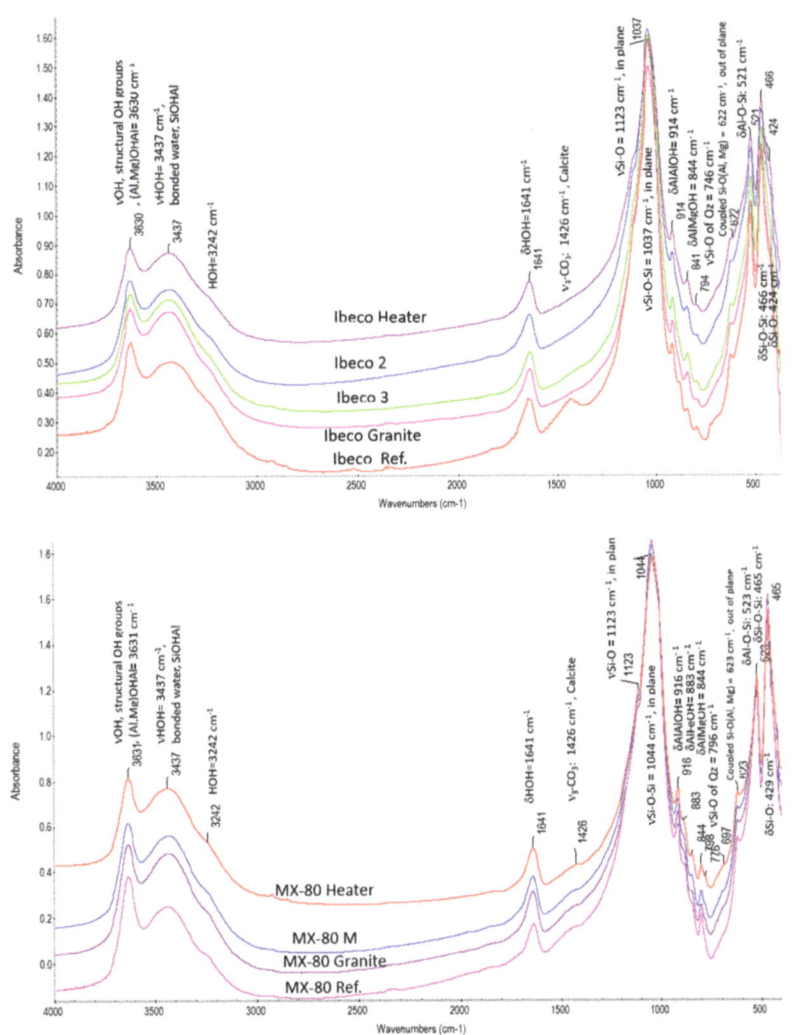

**Figure 8.** FTIR spectra from samples taken from the ABM5 experiment: Ibeco block 11 and MX-80 Block 1.

### 3.2.3. SEM Analysis

Bentonite samples in contact with the heater were analysed by SEM. All of them showed Fe-hydro/oxides indicating the presence of corrosion products near the heater. Barite was found in FEBEX and Rokle samples and pyrite ($FeS_2$) and sphalerite ($ZnS$) in MX-80 samples. All the samples contained organic carbon in some parts (Asha 505, FEBEX, Ibeco). Microorganisms of fungi type (hyphae form) found in FEBEX, Rokle, and Ibeco samples (see OM in Figure 9) may have formed after dismantling during storage due to the high water content of the samples.

**Figure 9.** SEM photomicrographs from different bentonites in direct contact with the heater (Sme: esmectite, Ilt/Ms: illite/muscovite, Qz: quartz, ZnS: sphalerite, Fe-ox: Fe-oxyhydroxides, OM: organic matter (hyphae), Py: pyrite, KFds: potassium feldspars).

The microstructure and composition of the clay mineral particles from samples at contact with the heater contact were analysed by SEM-EDX. The SEM photomicrographs and the crystallochemical formula of the smectite clay minerals analysed are shown in Figure 9 and Table 5. The structure of the clay mineral particles remains intact, showing a porous honeycomb microstructure, following similar patterns observed in original samples prior to the hydration-heating treatment (see Sme particles in Figure 9). The clay mineral particles analysed belong to the same chemical composition domain of the smectite clay mineral from the reference samples. However, some particles deviate from the reference crystallochemical structure, the most significant changes being related to changes in the cation composition in tetrahedral sheets, decreasing tetrahedral silicon, increasing tetrahedral charge and total charge of the smectite clay particles. This is true for all samples except for the Rokle bentonite samples, in which the excess of charge is located in tetrahedral sheets originally, and after treatment, there is an increase in the octahedral charge, losing octahedral Al. Fe and Mg content increased in the octahedral sheets, but variations are not significant.

**Table 5.** Structural formula of clay particles analysed in reference samples and close to the heater contact (h.u.c: half unit cell, p.f.u: per full unit cell, $\tau$ charge: tetrahedral charge, O. charge: octahedral charge).

| Sample | Structural Formula | Layer Charge (eq/h.u.c.) | $\Sigma_{tet}$ | $\Sigma_{oct}$ | $\tau$ Charge (%) | O. Charge (%) | Weight (g/mol) p.f.u. |
|---|---|---|---|---|---|---|---|
| Asha 505 Ref. | $(Si_{3.81}Al_{0.19})^{IV}\ (Al_{1.28}Fe^{3+}_{0.48}Mg_{0.24})^{VI}$ $O_{10}(OH)_2\ (Ca_{0.12}Na_{0.21})_{0.32}$ | 0.43 | 4.0 | 2.00 | 44 | 56 | 764.87 |
| Asha 505 Block 28 | $(Si_{3.81}Al_{0.19})^{IV}\ (Al_{1.10}Fe^{3+}_{0.74}Mg_{0.16}Ti^{4+}_{0.01})^{VI}$ $O_{10}(OH)_2\ (Ca_{0.12}Na_{0.05}K_{0.05})_{0.33}$ | 0.33 | 4.0 | 2.01 | 59 | 41 | 778.21 |
| | $(Si_{3.84}Al_{0.16})^{IV}\ (Al_{1.43}Fe^{3+}_{0.32}Mg_{0.23})^{VI}$ $O_{10}(OH)_2\ (Ca_{0.17}Na_{0.10}K_{0.02})_{0.29}$ | 0.46 | 4.0 | 1.98 | 36 | 64 | 756.05 |
| | $(Si_{3.78}Al_{0.22})^{IV}\ (Al_{1.17}Fe^{3+}_{0.60}Mg_{0.19}Ti^{4+}_{0.01})^{VI}$ $O_{10}(OH)_2\ (Ca_{0.17}Na_{0.10}K_{0.02})_{0.29}$ | 0.48 | 4.0 | 1.97 | 46 | 54 | 772.63 |
| | $(Si_{3.62}Al_{0.68})^{IV}\ (Al_{0.81}Fe^{3+}_{0.98}Mg_{0.24}Ti^{4+}_{0.004})^{VI}$ $O_{10}(OH)_2\ (Ca_{0.22}Na_{0.05}K_{0.03})_{0.29}$ | 0.51 | 4.0 | 2.04 | 75 | 25 | 798.99 |
| | $(Si_{3.83}Al_{0.17})^{IV}\ (Al_{1.38}Fe^{3+}_{0.35}Mg_{0.27}Ti^{4+}_{0.03})^{VI}$ $O_{10}(OH)_2\ (Ca_{0.13}Na_{0.08}K_{0.001})_{0.21}$ | 0.34 | 4.0 | 2.02 | 50 | 50 | 755.41 |
| FEBEX Ref. | $(Si_{3.96}Al_{0.04})^{IV}\ (Al_{1.49}Fe^{3+}_{0.13}Mg_{0.38})^{VI}$ $O_{10}(OH)_2\ (Mg_{0.10}Na_{0.20}K_{0.03})_{0.32}$ | 0.42 | 4.0 | 2.00 | 10 | 90 | 742.12 |
| FEBEX Block 25 | $(Si_{3.96}Al_{0.04})^{IV}\ (Al_{1.44}Fe^{3+}_{0.12}Mg_{0.43}Ti^{4+}_{0.01})^{VI}$ $O_{10}(OH)_2\ (Ca_{0.15}Na_{0.13}K_{0.05})_{0.32}$ | 0.47 | 4.0 | 2.00 | 9 | 91 | 746.71 |
| | $(Si_{3.72}Al_{0.28})^{IV}\ (Al_{1.32}Fe^{3+}_{0.39}Mg_{0.26}Ti^{4+}_{0.05})^{VI}$ $O_{10}(OH)_2\ (Ca_{0.15}Na_{0.11}K_{0.03})_{0.29}$ | 0.44 | 4.0 | 2.02 | 63 | 37 | 763.51 |
| Rokle Ref. | $(Si_{3.75}Al_{0.25})^{IV}(Al_{1.19}Fe^{3+}_{0.67}Mg_{0.15}Ti^{4+}_{0.02}Mn^{2+}_{0.01})^{VI}$ $O_{10}(OH)_2\ (Ca_{0.03}Na_{0.14}K_{0.07})_{0.24}$ | 0.27 | 4.0 | 2.04 | 93 | 7 | 776.00 |
| Rokle Block 14 | $(Si_{3.89}Al_{0.11})^{IV}\ (Al_{0.98}Fe^{3+}_{0.65}Mg_{0.15}Ti^{4+}_{0.08})^{VI}$ $O_{10}(OH)_2\ (Ca_{0.14}Na_{0.04}K_{0.08})_{0.26}$ | 0.40 | 4.0 | 1.96 | 27 | 73 | 777.36 |
| | $(Si_{3.90}Al_{0.10})^{IV}\ (Al_{0.98}Fe^{3+}_{0.60}Mg_{0.34}Ti^{4+}_{0.06})^{VI}$ $O_{10}(OH)_2\ (Ca_{0.17}Na_{0.03}K_{0.10})_{0.30}$ | 0.47 | 4.0 | 1.97 | 21 | 79 | 776.70 |
| IBECO Ref. | $(Si_{3.96}Al_{0.04})^{IV}\ (Al_{1.48}Fe^{3+}_{0.20}Mg_{0.37}Ti^{4+}_{0.01})^{VI}$ $O_{10}(OH)_2\ (Ca_{0.04}Na_{0.09}K_{0.02})_{0.15}$ | 0.20 | 4.0 | 2.07 | 18 | 82 | 743.37 |
| IBECO Block 11 | $(Si_{4.0})^{IV}\ (Al_{1.44}Fe^{3+}_{0.12}Mg_{0.39}Ti^{4+}_{0.04})^{VI}$ $O_{10}(OH)_2\ (Ca_{0.08}Na_{0.18}K_{0.05})_{0.32}$ | 0.40 | 4.0 | 1.99 | 1 | 99 | 745.27 |
| MX-80 Ref. | $(Si_{4.00})^{IV}\ (Al_{1.51}\ Fe^{3+}_{0.22}Ti_{0.01}Mg_{0.24})^{VI}$ $O_{10}\ (OH)_2\ (Ca_{0.06}Na_{0.15}K_{0.01})$ | 0.29 | 4.00 | 1.98 | 0 | 100 | 744.34 |
| MX-80 Block 1 | $(Si_{3.77}Al_{0.23})^{IV}\ (Al_{1.20}Fe^{3+}_{0.60}Mg_{0.17}Ti^{4+}_{0.02})^{VI}$ $O_{10}(OH)_2\ (Ca_{0.18}Na_{0.04}K_{0.03})_{0.26}$ | 0.44 | 4.0 | 1.98 | 53 | 47 | 772.58 |
| | $(Si_{3.92}Al_{0.08})^{IV}\ (Al_{1.65}Fe^{3+}_{0.14}Mg_{0.26})^{VI}$ $O_{10}(OH)_2\ (Ca_{0.05}Na_{0.06}K_{0.01})_{0.12}$ | 0.17 | 4.0 | 2.06 | 48 | 52 | 738.05 |

### 3.2.4. XPS and Mössbauer Analysis

For analysing the Fe distribution and speciation inside the bentonite block after the *in situ* test, Rokle bentonite was selected due to its high Fe content. Figure S3 from the Supplementary Material shows the wide scan XPS spectra recorded from a representative Rokle retrieved sample. The rest of the samples gave very similar spectra. Table 6 collects the atomic percentages obtained from the quantification of such spectra. The atomic concentrations show only minute variations from sample to sample which are within the error of the experimental determination. We must recall that XPS is a surface sensitive technique and, therefore, the quantitative analysis refers to the composition within the characteristic XPS depth probe, which is around 3–5 nm.

**Table 6.** Surface atomic concentrations calculated from the wide scan XPS spectra (in wt.%).

| Sample | Fe | O | Ti | Ca | Mg | C | Si | Al |
|---|---|---|---|---|---|---|---|---|
| Reference | 3.3 | 52.2 | 0.7 | 1.5 | 3.3 | 15.4 | 20.5 | 3.1 |
| 6G | 4.4 | 56.6 | 0.8 | 1.3 | 2.9 | 11.5 | 19.5 | 3.0 |
| 5M | 4.0 | 54.7 | 0.5 | 1.3 | 3.2 | 14.5 | 15.8 | 6.0 |
| 4H | 3.3 | 53.7 | 0.9 | 1.3 | 3.2 | 14.8 | 18.7 | 4.0 |

In order to gain insight on the surface chemical states of iron and oxygen, high-resolution spectra were recorded in the Fe 2p and O 1s spectral regions. The spectra were computer-fitted using previous models [44] and the results are presented in Figure 10. The Fe 2p spectra are all very similar. They are composed of an intense main spin orbit doublet (BE Fe $2p_{3/2}$ = 712.3 eV; BE Fe $2p_{1/2}$ = 726.5 eV) and secondary structure: two small shake-up satellites at 722.3 eV and 735.5 eV and a multiplet splitting component at 717.6 eV. These binding energies values and spectral features are all compatible with the presence of $Fe^{3+}$. It is difficult, based on these spectra, to ascertain if there is any small $Fe^{2+}$ contribution. As it can be observed in Figure S4 of the Supplementary Information, it is very complicated to separate, in an XPS spectrum, small concentrations of $Fe^{2+}$ from the majority presence of $Fe^{3+}$ and vice versa.

The O 1s spectra showed in all the cases four different contributions located at 529.7 eV, 531.5 eV, 532.7 eV and 534.7 eV, which correspond to metal-O bonds, Al-O bonds, Si-O bonds and organic carbon/adsorbed water, respectively [45–48]. The spectra show only small differences, except for the increase in intensity of the contribution at 532.7 eV over the course of the series. Although this component is mainly due to Si-O bonds, other contributions due to oxygen-containing functional organic groups or chemi-/physi-sorbed water cannot be discarded either.

Although XPS is a useful technique to determine oxidation states, it lacks the specificity of other techniques, such as Mössbauer spectroscopy, to characterize iron compounds. Hence, $^{57}$Fe Mössbauer spectroscopy was employed to study the speciation of the Rokle samples. Figure 11 shows the room temperature Mössbauer spectra recorded from these materials.

All the Mössbauer spectra were fitted using six different components whose relative areas varied from sample to sample. Whilst the reference samples, 6G (at granite contact) and 5M (at middle bentonite block), gave spectra showing the same contributions in different proportions, the sample 4H close to the heater contact gave a quite different spectrum, showing an additional contribution which was not present in the spectra of the other three. Table 7 collects the hyperfine parameters of the different spectral components and Table 8 shows the relative areas of the components obtained from the fit of the spectra.

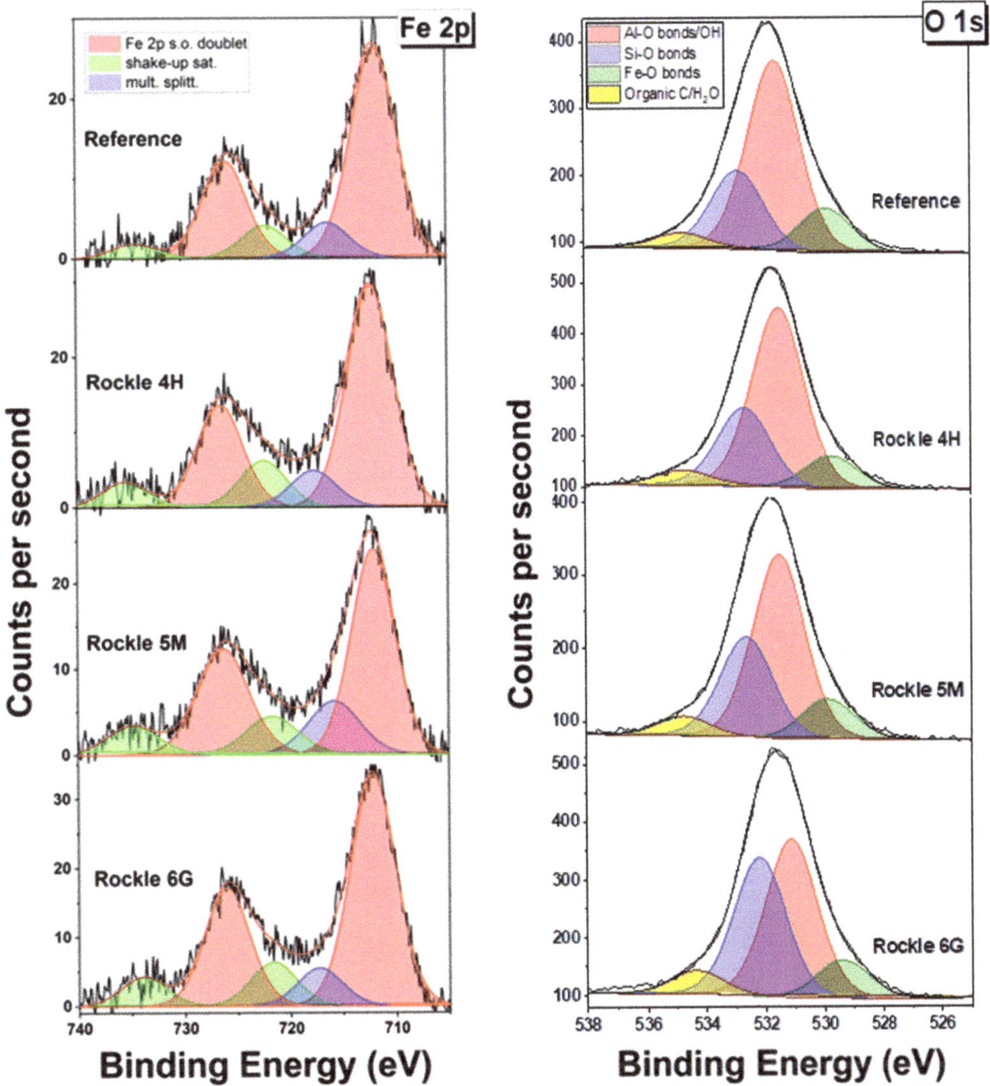



**Figure 10.** XPS narrow scan Fe 2p and O 1s spectra recorded from the Rokle samples: reference, 4H: close to heater interface, 5M: middle, and 6G: close to granite interface; i.e., at 1.67 cm, 5.00 cm and 8.33 cm from the heater contact, respectively.

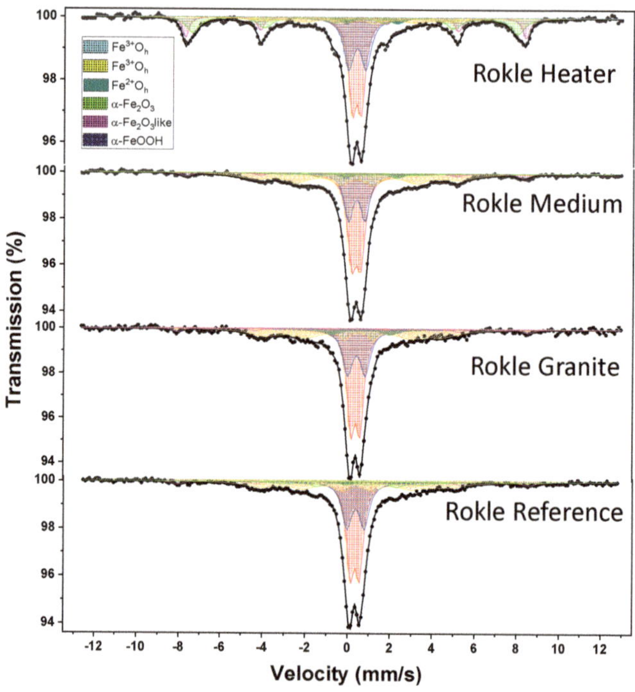


**Figure 11.** Room temperature $^{57}$Fe Mössbauer spectra recorded from the Rokle samples: reference, 4H: close to heater interface, 5M: middle, and 6G: close to granite interface; i.e., at 1.67 cm, 5.00 cm and 8.33 cm from the heater contact, respectively.

**Table 7.** Hyperfine parameters obtained from the fit of the spectra presented in Figure 12.

| Hyperfine Parameter | D1 | D2 | D3 | S1 | S2 | S3 | S4 |
|---|---|---|---|---|---|---|---|
| $\delta$ (mms$^{-1}$) | 0.36 | 0.38 | 1.03 | 0.37 | 0.33 | 0.35 | 0.37 |
| $\Delta/2\varepsilon$ (mms$^{-1}$) | 0.44 | 0.81 | 2.41 | −0.15 | −0.14 | −0.20 | −0.20 |
| H (T) | | | | 27.4 | 20.1 | 51.0 | 47.6 |

$\delta$, isomer shift; $\Delta$, quadrupole splitting, aplies to doublets; $2\varepsilon$, quadrupole shift, apllies to sextets; H, hyperfine magnetic field.

**Table 8.** Relative areas obtained from the fit of the spectra presented in Figure 12.

| Area (%) | Distance to Heater (cm) | D1 Fe³⁺ Oₕ | D2 Fe³⁺ Oₕ | D3 Fe²⁺ | S1 α-FeOOH | S2 α-FeOOH | S3 α-Fe₂O₃ | S4 α-Fe₂O₃ like |
|---|---|---|---|---|---|---|---|---|
| Reference | | 41 | 29 | 3 | 16 | 7 | 5 | |
| 6G | 8.33 | 37 | 24 | 2 | 25 | 7 | 5 | |
| 5M | 5.00 | 37 | 21 | 3 | 29 | 5 | 5 | |
| 4H | 1.67 | 30 | 23 | 4 | 20 | | 12 | 21 |

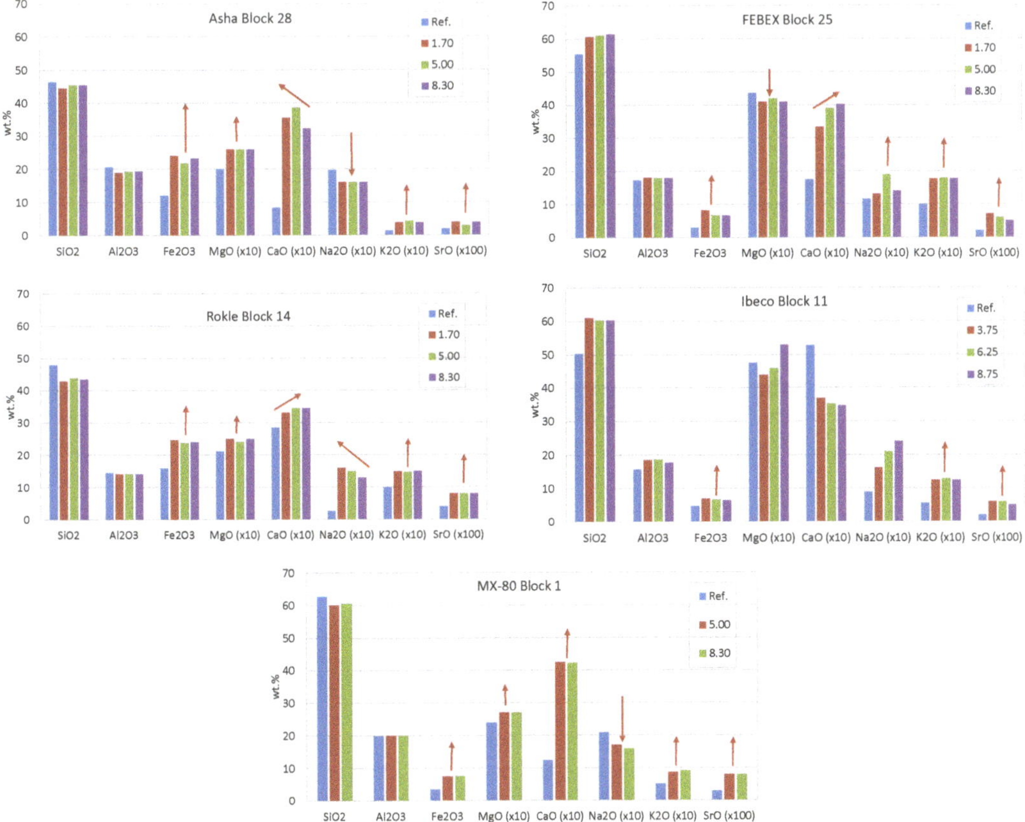

**Figure 12.** Chemical composition of the solid phase (total fraction) prior to and after the dismantling of the ABM5 experiment as a function of the heater distance (1.70 (H), 5.00 (M) and 8.30 (G) cm). Arrows: variations of content (↑: increases or ↓: decreases).

The spectra of the four samples were dominated by an intense central paramagnetic component which was best-fitted to two different quadrupole doublets (D1 and D2). The hyperfine parameters of these doublets (Table 7) are characteristic of high spin $Fe^{3+}$ in octahedral oxygen coordination [49]. Since for a high spin $Fe^{3+}$ ion (having a half filled $3d^5$ spherically symmetric configuration) the electric field gradient depends only on the lattice charge distribution, a larger quadrupole splitting, $\Delta$, implies a larger distortion from the perfectly symmetrical octahedral coordination [50]. Thus, D1 corresponds to a situation where the $Fe^{3+}$ is in a lesser distorted octahedral configuration than that represented by doublet D2. These types of quadrupole doublets are very common in the Mössbauer spectra of bentonites. Sometimes they have been interpreted in terms of cis and trans configurations, associated with the lower and higher $\Delta$, respectively [51–53], although some other interpretations consider only that the different $\Delta$ values result from different geometrical distortions of the coordination polyhedra and/or the existence of different ligands beyond the first coordination shell [54–56]. The highest paramagnetic contribution corresponds to the reference sample while the smallest occurs in the 4H sample (close to heater contact).

Apart from these two $Fe^{3+}$ doublets, all the spectra contain a tiny $Fe^{2+}$ doublet, accounting for 3–4% of the total spectral area. Because of its small intensity and the many overlapping components, its Mössbauer parameters were best determined by recording spectra in a lower range of velocities (Figure S4 from Supplementary Material). As shown in Table 7, these Mössbauer parameters (D3) are compatible with an octahedral high spin $Fe^{2+}$ species [49]. The spectra recorded from the reference samples 6G and 5M contain a broad magnetic component which was fitted to two broad sextets, S1 and S2. The Mössbauer parameters of these sextets are compatible with microcrystalline goethite [57]. The occurrence of two different goethite sextets might be due to goethite fractions having different particle size or Al substitution. These goethite contributions increase in the order reference < 6G < 5M. The spectra also show a low intensity (around 5%) magnetic sextet, S3, with parameters typical of hematite [57,58].

As mentioned above, however, the spectrum recorded from the sample 4H located close to the heater contact is quite different. Although the nature of the central paramagnetic region appears to be the same as in the rest of the samples, it contains a much more intense and defined magnetic component characterized by larger hyperfine magnetic fields. This component was fitted to two different sextets, S3 and S4. Sextet S3 is typical of hematite and S4 probably corresponds to a defective hematite phase ("hematite-like phase") [59]. This sample still contains a noticeable goethite concentration. It is well-known that the dehydration of goethite may result in the formation of hematite [60,61]. Therefore, it seems plausible that this sample, which has been collected close to the heater, contains hematite as a consequence of the loss of OH groups in goethite brought about by the local heating. In addition, these species, or at least a fraction of them, may result from a high-temperature corrosion process occurring at the C-steel heater surface. This sample also contains $Fe^{2+}$. In fact, it is in this sample that it can be observed more clearly due to the lower concentration of goethite whose spectrum overlaps strongly with the $Fe^{2+}$ doublet. This result indicates that heating does not affect the $Fe^{2+}$ species, probably because it is comfortably sited within the clay mineral structure and is thus difficult to oxidize or cause a slight increase in $Fe^{2+}$ within the smectite structure. The presence of magnetite is discarded in all samples analysed.

### 3.3. Geochemistry of the Solid Samples

The geochemical analysis of the main elements in the total fraction of the bentonite material was performed by means of: (a) XRF, (b) combustion (total carbon and sulfur content), and (c) Fe(II)/Fe(III) speciation analysed via $NH_4HF_2$-$H_2SO_4$ leaching tests.

The main changes observed in all the samples are related to iron content, which increased in all samples with respect their reference values (Figure 12 and Tables S3–S4 from Supplementary Material), although other variations are observed. In all of the Ca/Mg-bentonites (Febex and Ibeco), the calcium content slightly decreased, increasing sodium, potassium and strontium contents. However, in all of the Na-bentonites (Asha 505 and MX-80), sodium content decreased, increasing calcium, potassium and strontium content. This seems to indicate a tendency to equilibrate the composition at interlayer sites in all type of bentonites, with a predominance of bivalent cations (see Section 3.4). In addition, those bentonites with an initial higher iron content (Asha 505, Rokle and MX-80) are able to acquire more iron in their structure than those having an initial lower iron content (Figure 12). On the other hand, although the amount of Fe(II) is low with respect to total Fe, Fe(II) content increased in samples close to the heater (Figure 13, Table S5 from Supplementary Material). Variations in the Mg content at exchange sites are less significant in comparison to sodium and calcium, but there is a tendency to increase in the samples with a higher initial Fe content.

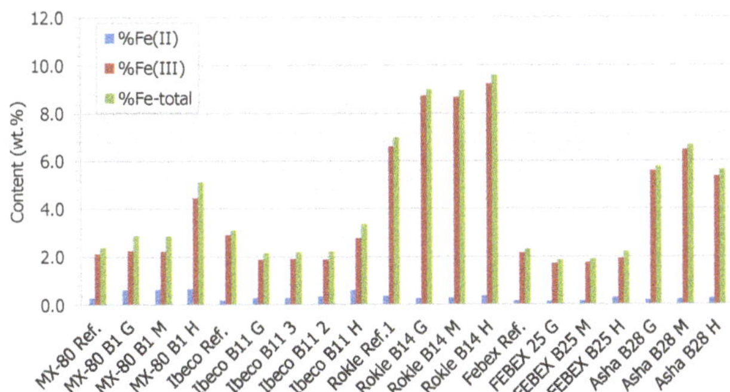


**Figure 13.** Iron content in the solid samples prior to and after the dismantling of the ABM5 experiment: Fe(II), Fe(III) and total Fe.

Regarding carbon and sulfur contents (Table 9), no significant differences are observed, except for the presence of organic matter in the MX-80 bentonite block 1, but the values are lower than in the reference value.

**Table 9.** Total carbon, total sulfur, total surface area (SA) and BET surface area obtained after the dismantling of ABM5.

| Sample | Distance to Heater (cm) | Water Content (%) [1] | $C_{Total}$ (wt.%) | $C_{Inorg.}$ (wt.%) | $C_{Org.}$ (wt.%) | $S_{Total}$ (wt.%) | Total SA ($m^2$/g) | $S_{BET}$ ($m^2$/g) | CEC (meq/100 g) |
|---|---|---|---|---|---|---|---|---|---|
| Asha 505 Ref. [2] | | 13.1 | 0.03 | 0.02 | 0.01 | 0.02 | | | 88.6 |
| Asha B28 G | 8.3 | 16.8 | 0.10 | <0.1 | <0.1 | <0.1 | 585 | 65.41 | 85.8 |
| Asha B28 M | 5.0 | 17.6 | 0.16 | <0.1 | <0.1 | <0.1 | 574 | 55.99 | 83.9 |
| Asha B28 H | 1.7 | 13.4 | 0.19 | <0.1 | <0.1 | <0.1 | 584 | 61.49 | 81.3 |
| FEBEX Ref. | | 14.3 | 0.12 | 0.08 | 0.04 | <0.05 | 628 ± 4 | 59.2 ± 5 | 98.1 |
| Febex B25 G | 8.3 | 13.8 | 0.08 | <0.1 | <0.1 | <0.1 | 640 | 38.01 | 97.6 |
| Febex B25 M | 5.0 | 13.3 | 0.09 | <0.1 | <0.1 | <0.1 | 627 | 45.76 | 96.5 |
| Febex B25 H | 1.7 | 13.2 | 0.05 | <0.1 | <0.1 | <0.1 | 622 | 36.20 | 93.1 |
| Rokle Ref. | | 17.2 | 0.27 | 0.10 | 0.17 | 0.02 | 573 ± 5 | 82.8 ± 0.3 | 73.8 |
| Rokle B14 G | 8.3 | 8.6 | 0.26 | <0.1 | <0.1 | <0.1 | 538 | 66.90 | 73.1 |
| Rokle B14 M | 5.0 | 8.9 | 0.22 | <0.1 | <0.1 | <0.1 | 549 | 64.09 | 73.0 |
| Rokle B14 H | 1.7 | 8.1 | 0.21 | <0.1 | <0.1 | <0.1 | 517 | 58.70 | 70.3 |
| Ibeco Ref. | | 14.7 | 0.79 | 0.62 | 0.17 | 0.23 | 611 ± 2 | 57.4 ± 0.4 | 90.2 |
| Ibeco B11 G | 8.75 | 13.1 | 0.13 | <0.1 | <0.1 | <0.1 | 706 | 49.75 | 97.1 |
| Ibeco B11 3 | 6.25 | 13.1 | 0.13 | <0.1 | <0.1 | <0.1 | 688 | 48.98 | 97.9 |
| Ibeco B11 2 | 3.75 | 13.6 | 0.15 | <0.1 | <0.1 | <0.1 | 660 | 45.04 | 88.6 |
| Ibeco B11 H | 1.25 | 13.6 | 0.10 | <0.1 | <0.1 | <0.1 | 680 | 51.71 | 95.0 |
| MX-80 Ref. | | 10.6 | 0.28 | 0.08 | 0.20 | 0.24 | 481 ± 1 | 29.7 ± 0.2 | 83.6 |
| MX 80 B1 G | 8.3 | 11.6 | 0.27 | 0.16 | 0.11 | 0.16 | 523 | 21.72 | 84.1 |
| MX 80 B1 M | 5.0 | 11.6 | 0.29 | 0.22 | 0.07 | 0.22 | 542 | 18.91 | 87.6 |
| MX 80 B1 H | 1.7 | 11.4 | 0.24 | 0.22 | 0.02 | 0.22 | 528 | 22.48 | 87.8 |

[1] from air-dried samples; [2] [5]; G: granite contact; H: heater contact.

### 3.4. Physico-Chemical Properties

The main physico-chemical properties analysed were external (BET) and total surface (SA) areas, total cation exchange capacity (CEC), cation exchange population and the distribution of soluble ions in aqueous extracts.

Total surface area is an important parameter related to the water absorption and swelling capacity of bentonites. It determines the amount of water needed for hydrating all of the clay particles. The lower the particle size, the higher the total surface area. The external surface area of the stacks of layers was determined by $N_2$-BET measurements. External BET and total surface area values are given in Table 9. No significant changes are observed in total SA, but a decrease in the external BET surface area is observed in all of the retrieved samples, with the values being much lower at the heater contact (Figure S5 and Table S6 from the Supplementary Material).

The $N_2$ adsorption-desorption curves of the samples (reference and retrieved) are shown in Figure S6 from Supplementary Material. The isotherms are Type IV, indicating adsorption on the mesoporous material (IUPAC classification). When the relative pressure $(P/P_0)$ is higher than 0.4, a hysteresis loop appears, indicating that capillary condensation occurs in the mesopores and macropores, with adsorption being dominated by capillary condensation and evaporation. At $P/P_0 < 0.4$, the adsorption and desorption isotherms coincide with each other, and the adsorption is dominated by van der Waals force at this stage. The $N_2$ adsorption–desorption curves differ from all of the reference values and are significantly affected by heating. Regarding the types of hysteresis loops, they can reflect pore morphology in the porous material. The hysteresis loop for all of the samples, except for the MX-80 bentonite, is between the H2 and H3 type, indicating the coexistence of ink bottle-like pores and many slit pores formed by the plate-shaped minerals [62]. For MX-80, the hysteresis loop is close to the H4 type, indicating the existence of slit-like pore produced by similar layered minerals. According to the modification of the hysteresis loop, all retrieved samples seem to increase the ink bottle-like pores.

Cation exchange capacity, CEC, is one of the most important parameters for characterising the bentonite adsorption behaviour. This parameter is equivalent to the total negative surface charge of a clay mineral and reflects the degree of reactivity of the bentonite because CEC is related to the capacity of adsorption and retention of ions and to the swelling capacity of the bentonites. The CEC parameter is linked with the interlayer cations, who nature is also an important issue because the cation composition at exchange sites affects not only the exchange properties but also the plasticity, the swelling capacity and the rheological behaviour. Values of CEC and cation exchange population are observed in Figure 14 and Table S7 from the Supplementary Material. Slight but not significant variations are observed in CEC values with respect to initial values, slightly decreasing towards the heating source, except for the MX-80 bentonite block located at the bottom part. However, there is a complete readjustment in the cation exchange population in all bentonites. Na-bentonites (Asha 505 and MX-80) reduce their sodium content, increasing calcium. In the case of Mg-Ca bentonites (Rokle, Ibeco, FEBEX), sodium content increased, as well as the calcium content, and the higher the sodium increase content, the lower the initial sodium content at exchange sites in the bentonite (e.g., see Rokle with respect to FEBEX bentonite). Magnesium content tends to decrease in all samples, except in MX-80 bentonite, where values increase at the heater contact.

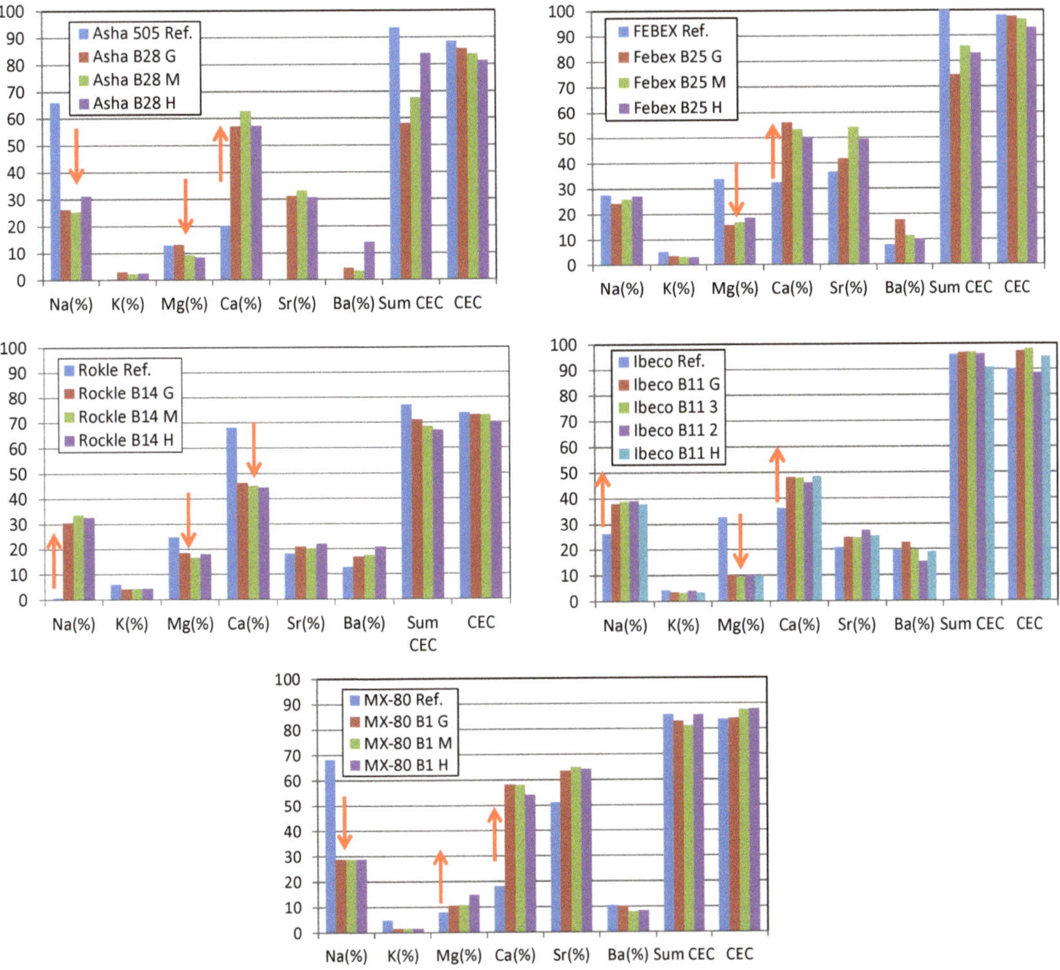

**Figure 14.** Cation exchange population after dismantling of the ABM5 experiment. Sr and Ba percentage is multiplied by 100 and 1000, respectively. Arrows: variations of content (↑: increases or ↓: decreases).

Aqueous extracts were performed for obtaining ion inventories and ion distributions along the bentonite blocks (Tables S8 and S9 from Supplementary Material). It should be taken into account that only $Cl^-$, $F^-$, and $Br^-$ can be considered as conservative ions (only affected by anion exclusion phenomena), whereas the rest of the anions and cations are controlled by mineral dissolution/precipitation processes and ion exchange reactions during the leaching tests.

The soluble salts obtained from the samples taken after the ABM5 experiment are shown in Figure 15. The pH in the aqueous extracts is neutral with values between 7.6 and 7.8. Chloride values increased in all samples with respect to the initial reference values, but the content is lower at the heater contact (**H**) than at the granite contact (**G**), indicating a diffusion process through the compacted bentonite. It is interesting to see also the increased values of fluoride and bromide contents coming from the infiltrating Äspö groundwater (Table 2). Sulfate content depends on the type of bentonite and its initial pore composition, although values increase towards the heater contact in most samples (Figure 16). Alkalinity values range between 0.3 and 0.7 mmol/100 g, expressed as $HCO_3^-$. Therefore, dissolution/precipitation processes of carbonates control the pH of the bentonites' pore water and cation exchange reactions. This is observed by cation content variations (Figure 16), which increased in all samples with respect to the reference samples, although with a tendency to decrease as a function of the heater contact, except for the Na-bentonites. Mg also increased at the heater contact in the Na-bentonites (Figure 16).

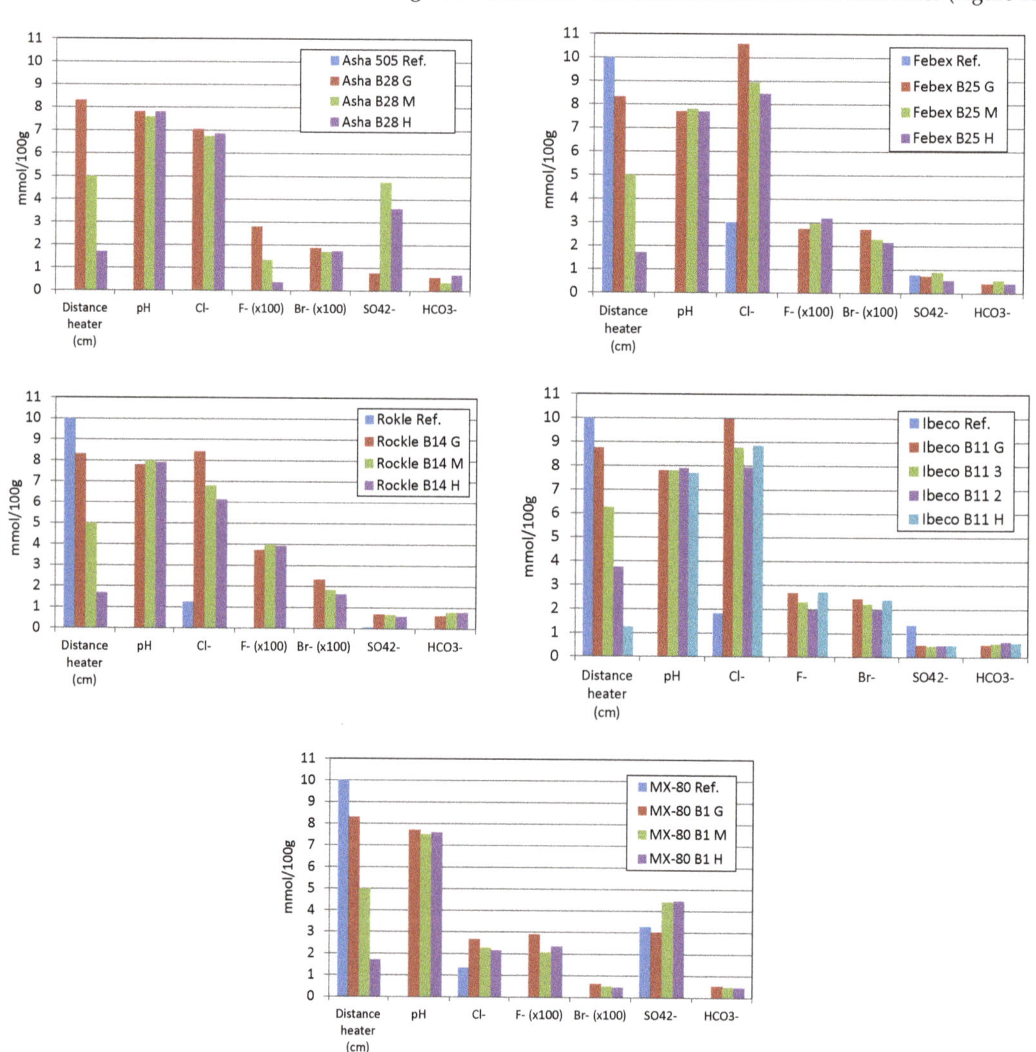

**Figure 15.** Anion distribution as a function of the heating contact obtained from leaching tests (in mmol/100 g). Reference data for chloride and sulfate are given for comparison.

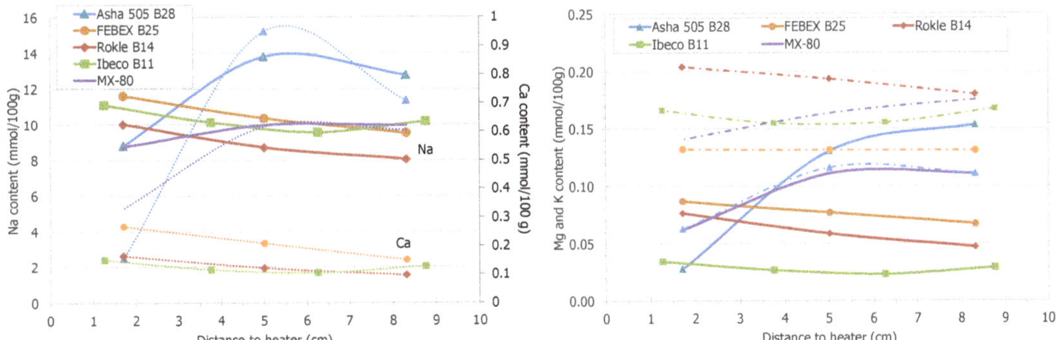

**Figure 16.** Cation distribution as a function of the heating contact obtained from leaching tests (in mmol/100 g). Reference data for chloride and sulfate are given for comparison.

### 3.5. Pore Water Chemistry

Pore water chemistry was obtained by squeezing at a pressure ranging between 20 and 30 MPa (Table 10), depending on the water content and dry density of the samples. The greater the water content of the sample, the lower the squeezing pressure that was needed (otherwise for dry density). The squeezing tests were performed with the pre-requirement of obtaining some amount of water enough for chemical analysis at the lowest squeezing pressure. This is because the greater the squeezing pressure, the greater the ease with which water bound to clay particle surfaces is expelled [38]. A representative pore water sample of homogeneously distributed electrolytes in the bulk aqueous phase requires avoiding the sampling of different portions of the diffuse layer water. Because of the high water content of the samples, pore water could be extracted at low squeezing pressures in a few days (3–7 days). It is worth noting that the whole piece of a bentonite block was used for squeezing tests, so the pore water chemical composition represents an average value from each bentonite block from the granite contact to the heater contact, in spite of there being a small solute gradient between these interfaces, as shown in Figures 15 and 16.

**Table 10.** Characteristics of the squeezing tests performed from bentonites after ABM5 dismantling.

| Core Sample | Initial Mass (g) | Initial Dry Density (g/cm$^3$) | Initial w.c. (%) | Time Elapsed (days) | Squeezing Pressure (MPa) | Pore Fluid Extracted/ Mass Loss (g) | Final Mass (g) | Final Dry Density (g/cm$^3$) | Final w.c. (%) | Efficiency (%) [1] | Efficiency (%) [2] |
|---|---|---|---|---|---|---|---|---|---|---|---|
| Asha B28 | 208.74 | 1.53 | 28.6 | 7 | 20 | 7.3 | 201.44 | 1.62 | 26.4 | 45.3 | 15.7 |
| FEBEX B25 | 369.23 | 1.49 | 26.6 | 7 | 30 | 10.1 | 359.15 | 1.62 | 24.8 | 58.8 | 13.0 |
| Rokle B14 | 292.90 | 1.60 | 31.2 | 5 | 20 | 10.8 | 282.06 | 1.67 | 25.4 | 23.3 | 15.6 |
| IBECO B11 | 187.07 | 1.49 | 36.0 | 3 | 20 | 7.1 | 180.00 | 1.55 | 28.7 | 49.5 | 14.3 |
| MX-80 B1 | 201.08 | 1.51 | 35.8 | 3 | 30 | 5.8 | 195.26 | 1.60 | 25.9 | 17.2 | 11.0 |

[1] Efficiency (%) = (Collected water × 100)/Extracted water; [2] Total Efficiency (%) = (100 × extracted water)/($P_{initial} - P_{dry}$).

The pH of the pore waters is neutral/slightly alkaline, with values ranging between 7.1 and 7.9 (Table 11). The salinity of the pore water is very high due to both the infiltrating marine Äspö groundwater (chloride content of 8580 mg/L (241.9 mmol/L, Table 2), and the high salinity of the initial bentonites pore water (Table S11 from Supplementary Material). After dismantling, ionic strength of pore waters ranges from 0.26 to 0.91 M, the lower values being at the bottom part of the bentonite column (MX-80 bentonite block 1).

**Table 11.** Chemical composition of the pore water extracted from the AMB5 bentonite blocks (n.d.: non determined).

| Sample | Asha<br>Block 28 S | FEBEX<br>Block 25 W | Rokle<br>Block 14 NW | IBECO<br>Block 11 S | MX-80<br>Block 1 IN |
|---|---|---|---|---|---|
| Sq. Pressure (MPa) | 20 | 30 | 20 | 20 | 30 |
| Water content (%) | 28.6 | 26.6 | 31.2 | 36.0 | 35.8 |
| Total weight (g) | 208.74 | 369.23 | 292.90 | 187.07 | 201.08 |
| Water type | Na-Ca-Cl | Na-Ca-Cl | Na-Ca-Cl | Na-Ca-Cl | Na-Cl |
| pH | 7.9 | 7.3 | 7.7 | 7.1 | 7.5 |
| Alkalinity (meq/L) | 0.94 | 0.90 | 0.90 | 0.76 | 1.18 |
| $F^-$ (mg/L) | <1 | <1 | <1 | <1 | <1 |
| $Br^-$ (mg/L) | 152 | 195 | 128 | 112 | 28 |
| $Cl^-$ (mg/L) | 25,000 | 24,000 | 24,000 | 19,000 | 5800 |
| $SO_4^{2-}$ (mg/L) | 1100 | 1100 | 1100 | 1700 | 1800 |
| $NO_3^-$ (mg/L) | 6.1 | <1 | <1 | 3.3 | 1.3 |
| Si (mg/L) | 210 | 7.0 | 3.6 | 200 | 8.5 |
| Al (mg/L) | <0.3 | <0.3 | 0.08 | <0.3 | 0.09 |
| Na (mg/L) | 9620 | 8100 | 9460 | 8040 | 3410 |
| K (mg/L) | 185 | 110 | 150 | 256 | 59 |
| Ca (mg/L) | 6100 | 4450 | 4350 | 3175 | 575 |
| Mg (mg/L) | 450 | 1500 | 560 | 850 | 180 |
| Sr (mg/L) | 49 | 60 | 47 | 30 | 12 |
| Ba (mg/L) | 0.45 | 0.52 | 0.55 | <0.3 | 0.33 |
| Fe (mg/L) | <0.3 | 1.4 | <0.03 | <0.3 | 0.04 |
| Cu (mg/L) | 0.67 | 0.70 | 0.39 | 1.0 | 0.19 |
| B (mg/L) | 4.5 | 0.90 | 0.26 | 3.1 | 0.07 |
| Mn (mg/L) | 1.3 | 2.8 | 0.52 | 8.9 | 0.59 |
| Mo (mg/L) | <0.3 | <0.3 | 0.37 | <0.3 | 0.67 |
| Ni (mg/L) | <0.3 | 1.7 | 0.14 | 3.9 | 2.0 |
| Ti (mg/L) | <0.3 | <0.3 | <0.03 | <0.3 | <0.03 |
| V (mg/L) | <0.3 | <0.3 | <0.03 | <0.3 | <0.03 |
| Zn (mg/L) | 0.41 | <0.3 | 0.06 | 0.51 | 0.23 |
| $S_2O_3^{2-}$ (mg/L) | <1 | <1 | <1 | <1 | <1 |
| Acetate (mg/L) | 51 | <1 | 48 | 21 | 1.8 |
| Formate (mg/L) | 35 | <1 | 2.0 | 741 | 14 |
| TOC (mg/L) | n.d. | 57.1 | 97.1 | n.d | 44.2 |

The chemical composition of all bentonite pore waters is dominated by $Cl^-$, Na and Ca, except for the more dilute pore water from MX-80 block 1, which is Na-Cl water type. The Äspö granitic groundwater (Ca-Na-Cl water type) is very saline with an ionic strength of 0.91 M (Table 2). During infiltration, the bentonite pore waters changes from mainly initial Na-Cl or Na-$SO_4$ water type (see Table S11 from Supplementary Material), towards a Na-Ca-Cl pore water in all bentonite blocks analysed, except for MX-80 block 1, which changes from Na-$SO_4$ to Na-Cl water-type.

Chloride concentration increased in all samples, with values ranging between 25,000 mg/L (705.2 mmol/L) and 5800 mg/L (163.6 mmol/L) from the top to the bottom of the bentonite package, with the lower values being located at the bottom part (MX-80 block 1). It should be taken into account the initial chloride contents of 6600, 2800 and 1100 mg/L for reference FEBEX, IBECO and MX-80 bentonites, respectively (Table S11 from Supplementary Material). The same behaviour occurs for Na and Ca, showing increased values.

Mg concentration in the pore water increased in all samples after the experiment, but variations depend on the initial composition of the bentonite pore water, being higher for the FEBEX bentonite [24,63] (Table S11). Sulfate contents are similar in all the bentonites, with concentrations ranging between 1100 and 1800 mg/L, and the higher values corresponding to Ibeco block 11 and MX-80 block 1 bentonites, which have higher initial reference values (4300 and 9300 mg/L, respectively, Table S11).

Alkalinity values are low, similar to that of the groundwater, with values ranging from 0.76 to 1.18 meq/L, and the highest value belonging to the MX-80 Block 1 pore water. However, these values seem to be much lower than those detected in reference samples (Table S11). It is interesting to note the presence of TOC, acetate, and formate in the pore waters (Table 11).

The geochemical code PHREEQC, version 3.4 [64] and the Thermoddem database [65] were used for equilibrium modelling and saturation indices calculations (Table S10 from Supplementary Material) at 25 °C. The saturation indices indicate that the pore waters from all samples are saturated with respect to calcite, gypsum, celestine, barite, calcite, dolomite, magnesite, barite and quartz (except Rokle) and undersaturated with respect to halite.

## 4. Discussion

From the geochemical point of view, specific long-term requirements for the function of a bentonite to isolate the canisters from water and retard the migration of radionuclides is to maintain a suitable chemical environment for the canister integrity and radionuclide retention over time, buffering possible alteration/deterioration processes within the geological and engineered barrier systems. In this section, a discussion of results regarding the possible alterations of the bentonites after 4.4 years of the experiment is given.

### 4.1. Physical Properties Variations

After 4.4 years of artificial hydration and heating, samples increased their water content by up to ~30 ± 2%, and the dry density decreased by up to ~1.52 ± 0.05 g/cm$^3$ due to bentonite swelling capacity. Therefore, the total physical porosity increased from 29–39% to 45–48% of the compacted bentonite blocks after the infiltration of the saline groundwater (see microstructure of clay minerals Figure 9). Two factors probably influenced the mineralogical/geochemical behaviour in this experiment and the physico-chemical properties of the bentonites. First, due to the bentonite/groundwater infiltration and final salinity of the pore water, the relationship between external (pore water) and internal (interlayer water) must have been readjusted during the experiment due to the decrease in the thickness of the double diffuse layer (DDL), increasing the amount of available water (free water layer) for geochemical reactions. Second, the experiment was performed at a constant temperature of 50 °C during most of the experimental period (3.45 years). The temperature was increased to 150 °C for a period of 6 months and to 250 °C for the final 6 months (Figure 2). These final overheating periods probably provoked an impact on water distribution and final density, as observed after dismantling [8]. Dry density tends to increase towards the heating contact, indicating a collapse or shrinkage process of the clay minerals, which increases the inter-aggregate porosity and, hence, the external free water. Some micro-cracks are observed in samples at heater contact (e.g., Ibeco, MX-80 in Figure 3), and temperature increases the mineral dissolution. Both effects favour the increase in porosity. In addition, water content is higher at the bottom part of the bentonite package, in spite of supposed complete saturation of the bentonites. Indeed, the degree of saturation varies from 93 to 100% from the top to the bottom of the bentonite package probably due to steam–bentonite interactions which affected the bentonites/groundwater interactions. After dismantling, some geochemical modifications were observed depending on the location of the bentonite block inside the package. For example, calcite content in IBECO bentonite samples (Block 11) was drastically decreased, whereas calcite precipitation is observed in the bentonite blocks below and above. The variation in the sum of cation population and CEC values is not the same in samples located at the bottom part (IBECO block 11 and MX-80 Block 1) with respect to those located at the top part (Asha 505 Block 28, FEBEX Block 25 and Rokle block 14).

### 4.2. Mineralogical and Geochemical Alterations

The high thermal conditions of this experiment (up to 250 °C) could have changed the bentonite clay minerals to non-swelling minerals, affecting the performance of the barrier.

However, there is no evidence for illitization, probably due to the low potassium contents both in pore water and exchange sites, despite the high temperatures. No modifications are observed in the clay mineral fraction after ethylene glycol treatment of the samples analysed (Figure 6).

In addition, metallic iron produced by corrosion of the heater could lead to [41]: (a) Fe(III) to Fe(II) reduction inside the clay mineral structure, affecting the layer charge, (b) dissolution/alteration of smectite to other minerals, and (c) cementation processes affecting plasticity, swelling or hydraulic conductivity of the bentonites. Trioctahedral smectites were found in previous experiments, such as the ABM1 and ABM2 experiments [6,12], the FEBEX *in situ* test [35,42], the TBT experiment [12], and the Prototype experiment [66]. However, in this work, no Mg- or Fe-rich trioctahedral smectites were detected despite the decrease in Mg at exchange sites and the increase in iron due to the corrosion of the C-steel heater. This is in agreement with other ABM5 studies [8,9]. Fully reducing conditions are important constraints for the alteration of diocthedral smectites to trioctahedral ones.

The most interesting modifications found in the samples close to the heater were the increase in gypsum and calcite, and the presence of siderite, monohydrocalcite, pyrite and Na-clinoptilolite. These mineral phases decrease the porosity of the bentonite by means of a cementation process, and act as a sink of Fe(II), avoiding the modification of the clay mineral structure (e.g., transformation of montmorillonite to nontronite or saponite). However, not all of these mineral phases were found in all bentonites (Figure 5 and Table S1 from Supplementary Material). Goethite and hematite (due to heat), as the main Fe-oxides, and siderite were the corrosion products detected. Magnetite was not found (even by Mössbauer spectroscopy analysis) in spite of the increase in Fe(II) at the heater contact. Siderite neoformation was also observed in other ABM5 studies [8]. Hematite was not observed in all samples, only in Asha 505 Block 18, Rokle Block 14 and MX-80 Block 1. These bentonites contained higher amounts of goethite (Asha 505, Rokle) or siderite/pyrite (MX-80) in their reference samples. Precipitation of hematite may be due to heating (goethite to hematite transformation by heat), or due to siderite dissolution/pyrite oxidation in the case of MX-80. However, pyrite and siderite are still observed in all retrieved MX-80 samples analysed. Therefore, the presence of goethite or hematite may be related with a heater corrosion process under different oxic/anoxic environmental conditions, with the neoformation of goethite being favoured in presence of oxygen.

Dissolution/precipitation processes of calcite are observed, which seems to be affected by temperature and water vapour fluxes. Indeed, calcite dissolution increases with temperature. It is interesting to note the dissolution of calcite in the IBECO bentonite Block 11, which is not observed in the rest of the bentonite blocks located at the top and the bottom of the bentonite package, where an increase in carbonates is indicated. Monohydrocalcite is neoformed at the heater contact in Asha 505 bentonite. Gypsum content decreased in MX-80 bentonite Block 1 at heater contact but increased in Asha 505 Block 28. Anhydrite has not been found, although saturation indices from pore waters are oversaturated with respect to gypsum and anhydrite.

An increase in free silica can be ruled out, although an increase in tetrahedral charge in some clay minerals particles is observed. No significant variations in cristobalite/trydimite and feldspars mineral phases were found, although the intensity of their reflections increased in some samples.

### 4.3. Redistribution of CEC and Exchangeable Cations

The cation composition at exchange sites for the different bentonites was modified after 4.4 years. Changes in exchangeable cation composition can be explained on the basis of equilibration with Äspo saline (I = 0.91 M) groundwater (Table 2), which is enriched with Ca and Na salts, due to chemical interactions between the buffer blocks and temperature. The variation of the type of exchangeable cation in all bentonites seems to indicate that concentration tends to reach equilibrium with groundwater/pore water and tends to be homogeneous in the whole bentonite column, with predominance of bivalent cations. However, complete equilibration of the cation occupancy was not achieved after 4.4 years of experiment due to the differences observed in the cation content at exchange sites and in the pore water chemistry both from the granite to the heater contact and from the top to the bottom of the column. The calcium bentonites increased their sodium content, and the sodium bentonites increased their calcium content, with exchangeable magnesium being replaced and decreased in all the samples, except in MX-80 block 1 located at the bottom of the column. In spite of the available and increased magnesium content in the pore water, no other additional magnesium-bearing mineral phases, such as saponite, were detected in the bentonites by SEM, XRD or FTIR (Table S1).

These cation exchange variations may involve the decrease in the external surface area noticed in all samples. This decrease may be explained by two mechanisms: (a) Missana et al. [67] showed that the presence of over 80% divalent cations at exchange sites favours the formation of large clay colloid particles, as well as that an increase in clay colloid size is produced when there is an increase in the tetrahedral charge, independently of the main interlayer cation. An increase in the particle size implies a decrease in the external surface area; (b) an increase in the electrolyte concentration affects the colloid properties (size and mobility) of the smectite due to the aggregation of clay particles. In this experiment, an increase in the size of clay particles is expected due to the increase in pore water salinity, divalent cations and the tetrahedral charge observed, favouring the decrease in the external surface area.

The CEC values seems to slightly decrease towards the heater contact from the upper part of the bentonite column towards the middle part (Asha 505 Block 28, FEBEX Block 25 and Rokle Block 14), with variations of −7.3, 5.0 and −3.5 meq/100 g in samples at the heater contact with respect to reference values. However, this decrease is not observed in the two blocks located at the bottom part of the column (IBECO block 11 and MX-80 Block 1), where changes are positive, increasing by about 4.5 meq/100 g.

The decrease in CEC at the heater contact has been observed in other *in situ* experiments [9,11,42,68,69]. The decrease in CEC values may be related to high-temperature conditions, water vapor formation [70], a lower content of smectite or to the decrease in the layer charge of the smectite clay particles. The smectite and total phyllosilicates contents in most of the samples are similar to the reference values, within the experimental error. Thus, other explanations need to be given. Interestingly, the sum of cations at exchange sites is similar to the total CEC value, except in the upper bentonite blocks (Asha 505 Block 28, FEBEX Block 25), where a decrease in exchangeable Mg was detected as an expense of an increase in exchangeable Ca. This observation was observed twice after repeating the determination, and also detected by [9]. The increase in calcium at exchange sites may be indicative of an increase in the calcite content in the retrieved samples, as observed in XRF data (Tables S3 and S4) and in [8], which could lead to its dissolution during the aqueous extraction. However, this should not affect the magnesium content at exchange sites, which decreases. Thus, this Mg decrease is not well understood.

A decrease in the CEC may be caused in this experiment by: (a) the collapse of clay particles, particularly of highly charged smectites [71], (b) the collapse of clay particles due to the high ionic strength of the pore water, reducing the swelling pressure (e.g., [18,72]), (c) extensive drying because of the increased Ca/Mg fixation and/or K fixation at interlayer sites [73]. All of these factors—the increase in layer charge, the increase in tetrahedral charge (Table 5), high salinity and drying—are implicated in this package system. These factors are controlled by temperature and water saturation. For this reason, the variations observed in the CEC parameter, decreasing at the top and increasing at the bottom of the bentonite column, seem to reflect that CEC is, in turn, affected by the temperature and water vapor fluxes.

*4.4. Pore Water Chemistry Variations*

Intrinsic properties of bentonites, such as CEC, the presence of soluble minerals (e.g., calcite, gypsum, … ) and the low hydraulic conductivity and permeability of compacted material (i.e., solute transport by diffusion) imply a large buffering capacity for many geochemical processes. The chemical state of the buffer is defined by the bentonite composition (clay minerals and accessory minerals) and the pore water composition. Consequently, the chemical stability of the bentonite, that is, the alkalinity and redox buffering capacity, will be firstly controlled by the bentonite–water interactions and the resulting pore water chemistry (pH, redox potential, ionic strength, ionic composition, speciation and complexation). In addition, the knowledge of the porewater chemistry in the clay barrier is essential for performance assessment purposes in a nuclear waste repository, since the porewater composition controls the processes involved in the release and transport of the radionuclides.

In most studies performed with bentonites, it is observed that the bentonite composition controls the pore water chemistry evolution after their interaction with an infiltrating groundwater, and it is basically controlled by ion exchange reactions and dissolution/precipitation of the more soluble trace minerals of the bentonite [74–76]. However, this depends on the salinity of the infiltrating water, as observed in this experiment. In the case of the saline Äspö groundwater/bentonite interaction, the pore water chemistry depends on the diffusion rate of the infiltrating groundwater (which varies with the bentonite dry density and water content, i.e., porosity), and it is basically controlled by ion exchange and mineral dissolution/precipitation reactions.

The initial bentonite pore water was modified in all bentonite blocks after their interaction with the saline Na-Ca-Cl groundwater. However, the pore water composition is not at equilibrium in each bentonite block and in the whole bentonite package, since a diffusion infiltration process of the conservative ions ($Cl^-$, $Br^-$, $F^-$) is still observed towards the heater after 4.4 years of the experiment. Anions diffuse more slowly than cations through the bentonite as a result of anion exclusion. Thus, a clear evolution of the pore water is observed after groundwater infiltration. The bentonite pore water changes from mainly initial Na-Cl or Na-$SO_4$ water-type (see Table S11 from Supplementary Material) towards Na-Ca-Cl pore water in all bentonite blocks analysed, except for MX-80 block 1, which changes from Na-$SO_4$ to Na-Cl water-type. The final ionic strength values of the pore waters (0.82–0.91 M) are similar in the upper bentonite blocks (Asha 505, FEBEX, Rokle) but lower at the bottom part of the package (IBECO and MX-80), with values of 0.69 and 0.26 M, respectively. Water vapor fluxes probably increased the salinity of the pore waters at the top part. However, variations in ion composition of the pore water indicate that pore water is controlled by the equilibration of infiltrating water with main accessory minerals and the exchanger (cation exchange sites and surface complexation sites). In any case, pore water chemistry data (Table 11) and anion inventory (Figure 15) indicate that saturated bentonites (including cation occupancy and pore water composition) are not at equilibrium with the external infiltrating groundwater. At each time, an equilibrium between pore water and exchange sites is established, but pore water chemistry is continuously modified during the diffusion transport of anions, which are affected by anion exclusion.

pH values are neutral in all pore water samples analysed, indicating the buffering capacity of the bentonite via protonation/deprotonation reactions and dissolution/precipitation of mainly calcite. Alkalinity values decreased in all samples and saturation indices from pore waters indicate saturated conditions with respect to calcite.

Organic matter (TOC) and acetate and formate (volatile fatty acids, VFAs) were found in pore water samples from the different bentonites (Table 11). At this interface, factors of temperature and possible hydrogen, available due to corrosion processes, may reflect the reduction of $CO_2$ to produce acetate and then formate, via abiotic or microbially induced reactions. However, the presence of microorganisms was not studied in this experiment and their implication in different reactions at the heater interface is a pure speculation. Another idea is that the hydrolysis of carbon-containing steels, leading to the generation of $H_2$, $CO_2$

and hydrocarbons, may be another possible process for the presence of acetate and formate in the pore water close to the heater, as discussed in [35,42]. Organics were not studied in this experiment, but their evolution should be taken into account in further studies.

*4.5. Iron–Bentonite Interactions*

Metal corrosion was characterised by an increase in the Fe content in the samples and the presence of corrosion products, mainly close to the bentonite/heater interface. Most of the heater corrosion probably occurred in oxic conditions due to the observed corrosion products, goethite and hematite. However, oxygen may rapidly be depleted at the surface of steel by the reactions of Fe-bearing minerals in the bentonites: pyrite, siderite, Fe(II) in octahedral sites of montmorillonite particles, etc. However, this redox buffer capacity is limited due to the low amounts of reduced components, which will depend on the initial composition of each bentonite. In some zones, after oxygen depletion/consumption, the possible anoxic corrosion could produce some amount of Fe(II), which was transported by diffusion through the bentonite from the heater contact [14]. Indeed, an increase in Fe(II) is observed at the heater contact (Figures 11 and 13), with the content being much lower than Fe(III). This Fe(II) precipitated as siderite and pyrite, as shown by XRD in different samples (Figure 5), but not in all bentonites. Therefore, redox conditions must have been locally different. However, it seems that in this package system, oxygen is not completely depleted. Magnetite was not detected, in contrast to former ABM experiments [14]. In addition, variations in sulfur contents are insignificant.

## 5. Conclusions

Geochemical modifications of different bentonites (Asha 505, FEBEX, Rokle, IBECO and MX-80) used as engineered barriers were studied in the ABM5 experiment. In this medium large-scale *in situ* test, bentonite compacted blocks were artificially saturated with saline Äspö groundwater and heated progressively to 150 °C and finally up to 250 °C during the last six months.

The main change observed in all bentonites is the modification of exchangeable cation composition, explained on the basis of equilibration with Äspo saline groundwater enriched with Ca and Na salts. Calcium bentonites (FEBEX, Rokle, IBECO) increased their sodium content, and sodium bentonites (Asha 505, MX-80) increased their calcium content. Exchangeable magnesium decreased in all of the samples, except in MX-80 block 1 located at the bottom of the column. The variation of the type of exchangeable cations in all bentonites seems to indicate that concentration tends to reach equilibrium with groundwater/pore water and to be homogeneous in the whole bentonites column, with predominance of bivalent cations at exchange sites.

Cation exchange variations and/or salinity of the pore waters may explain the decrease in the external surface area values observed in all samples. However, swelling capacity is not affected due to the fact that the total surface area data are not modified.

A decrease in CEC values is observed towards the heating surface, which may be related to the high salinity of the pore water, a modification of the crystal structure of the smectite clay particles, increasing the layer charge, and the drying. These factors are driven by temperature, water vapor fluxes and water saturation.

In spite of the available and increased magnesium content in the pore water, no other additional magnesium-bearing mineral phases and/or trioctahedral smectites were detected, as found in ABM1 and ABM2 experiment. Furthermore, the transformation of montmorillonite to illite is discarded, probably due to the low potassium concentrations both in pore waters and exchange sites.

Fe increased as a function of the distance to the heater contact. Heater corrosion provoked the increase in iron in the bentonite, and goethite, hematite and siderite were found as corrosion products. No magnetite was detected. Although the ratio of ferrous to ferric iron increased in the close vicinity of the C-steel heater, the major Fe content is as Fe(III). No indications of Fe-montmorillonite were detected.

The initial bentonite pore water was modified in all bentonite blocks after their interaction with the saline Na-Ca-Cl groundwater. Pore waters changed from mainly an initial Na-Cl or Na-SO$_4$ water type towards a Na-Ca-Cl pore water in all bentonite blocks analysed, except for MX-80 block 1, which changes from Na-SO$_4$ to Na-Cl water-type. Water vapor fluxes probably increased the salinity of the pore waters at the top part of the package, since the final ionic strength (0.82–0.91 M) was similar in the upper bentonite blocks (Asha 505, FEBEX, Rokle) but lower in the bottom part of the package (IBECO and MX-80), with values of 0.69 and 0.26 M, respectively. The pore water chemistry of bentonites evolved as a function of the diffusion transport of the saline infiltrating groundwater (anions being affected by anion exclusion), the chemical equilibrium of cations at exchange sites and mineral dissolution/precipitation processes. These reactions are in turn dependent on temperature and water vapor fluxes. All bentonites preserved their hydro-geochemical properties, after being subjected to saline groundwater infiltration, heating and interaction with corroding metals during the 4.4 years of the experiment.

**Supplementary Materials:** The following are available online at https://www.mdpi.com/article/10.3390/min12040471/s1, Table S1. Mineral phases detected in the bentonite samples (by XRD, FTIR and SEM techniques); Table S2. Positions and assignments of vibrational bands of dioctahedral smectites, kaolinite and illite; Table S3. Chemical composition of the solid phase (total fraction) for different samples obtained after the dismantling of the ABM5 experiment: Asha Block 28, Febex Block 25 and Rokle Block 11; Table S4. Chemical composition of the solid phase (total fraction) for different samples obtained after the dismantling of the ABM5 experiment: Ibeco Block 11, MX-80 Block 1; Table S5. Fe(II), Fe(II) and total Fe contents obtained after the dismantling of theABM5 experiment; Table S6. Parameters deduced from the BET and t-plot treatment on the adsorption of N2 at 77 K from samples obtained after the dismantling of the ABM5 experiment, Table S7. Total cation exchange capacity (CEC) and cation exchange population prior to and after the dismantling of the ABM5 experiment (in meq/100 g); Table S8. Soluble salts from aqueous leaching tests a 1:4 solid to liquid ration, in mg/L; Table S9. Ion inventory obtained from aqueous leaching tests, in mol/100 g; Table S10. Calculated parameters and saturation indexes of the pore waters; Table S11. Chemical composition of the pore waters obtained by squeezing at 25 MPa for water vapour saturated FEBEX, IBECO RW C16, and MX-80 bentonites at initial conditions; Figure S1. XRD patterns of total fraction samples from ABM5 experiment; Figure S2. XRD patterns of oriented aggregate samples from ABM5 experiment (normal and after ethylene glycol and heating at 550 °C treatments); Figure S3. Wide scan XPS spectrum recorded from the sample Rokle 4H at heater contact, Figure S4. Fe 2p XPS spectra recorded from samples containing different concentrations of Fe(III) and Fe(II) standard compounds; Figure S5. Room temperature Mössbauer spectra recorded in a narrow range of velocities for Rokle samples: reference, 4H: close to the heater interface, 5M: middle, and 6G: close to granite interface, i.e., at 1.67 cm, 5.00 cm and 8.33 cm from heater contact, respectively, Figure S6. N2 adsorption/desorption isotherms from reference and retrieved ABM5 samples.

**Author Contributions:** Conceptualization, A.M.F., J.F.M.; D.S. and P.S.; methodology, A.M.F., J.F.M. and D.S.; validation, A.M.F., J.F.M. and D.S.; formal analysis, A.M.F., J.F.M., D.S., P.N., F.J.L., L.M.R., M.Á.C., A.I.C. and S.F.; investigation, A.M.F.; P.N. and F.J.L., resources, CIEMAT, CSIC; data curation, A.M.F., J.F.M., D.S., P.N., F.J.L., L.M.R. and M.Á.C.; writing—original draft preparation, A.M.F.; writing—review and editing, A.M.F., J.F.M., D.S. and P.S. All authors have read and agreed to the published version of the manuscript.

**Funding:** This work was funded by the EURAD-Concord European Commission Project and CIEMAT. Financial support for grant RTI2018-095303-B-C51 funded by MCIN/AEI/10.13039/501100011033 and "ERDF A way of making Europe" is gratefully acknowledged.

**Data Availability Statement:** Not applicable.

**Acknowledgments:** We acknowledge to UCM (Madrid, Spain) and the Chemical Department from CIEMAT for performing the XRF, XRD, SEM and chemical analyses.

**Conflicts of Interest:** The authors declare no conflict of interest.

## References


1. Posiva SKB. Safety functions, performance targets and technical design requirements for a KBS-3V repository. Conclusions and recommendations from a joint SKB and Posiva working group. In *Posiva SKB Report 01*; Svensk Kärnbränslehantering AB: Stockholm, Sweden, 2017; p. 116.
2. Villar, M.V.; Armand, G.; Conil, N.; de Lesquen, C.; Herold, P.; Simo, E.; Mayor, J.C.; Dizier, A.; Li, X.; Chen, G.; et al. D7.1 HITEC. Initial State-of-the-Art on THM Behaviour of (i) Buffer Clay Materials and of (ii) Host Clay Materials. 2020. Deliverable D7.1 HITEC. EURAD Project, Horizon 2020 No 847593. 214p. Available online: https://www.ejp-eurad.eu/sites/default/files/2021-0 4/EURAD%20-%20D7.1%20Initial%20SotA%20HITEC.pdf (accessed on 7 April 2022).
3. Kaufhold, S.; Dohrmann, R. Distinguishing between more and less suitable bentonites for storage of high-level radioactive waste. *Clay Miner.* **2016**, *51*, 289–302. [CrossRef]
4. Dohrmann, R.; Kaufhold, S.; Lundqvist, B. The role of clays for safe storage of nuclear waste. In *Handbook of Clay Science*; Bergaya, F., Lagaly, G., Eds.; Elsevier: Amsterdam, The Netherlands, 2013; Volume 5B, pp. 677–710.
5. Svensson, D.; Sandén, T.; Olsson, S.; Dueck, A.; Eriksson, S.; Jägerwall, S.; Hansen, S. Alternative buffer material Status of the ongoing laboratory investigation of reference materials and test package 1. In *SKB TR-11-06*; Svensk Kärnbränslehantering AB: Stockholm, Sweden, 2011; p. 146.
6. Kaufhold, S.; Dohrmann, R.; Sandén, T.; Sellin, P.; Svensson, D. Mineralogical investigations of the first package of the alternative buffer material test—I. Alteration of bentonites. *Clay Miner.* **2013**, *48*, 199–213. [CrossRef]
7. Kaufhold, S.; Dohrmann, R.; Götze, N.; Svensson, D. Characterization of the second parcel of the Alternative Buffer Material (ABM) Experiment—I mineralogical reactions. *Clays Clay Miner.* **2017**, *65*, 27–41. [CrossRef]
8. Kaufhold, S.; Dohrmann, R.; Ufer, K.; Svensson, D.; Sellin, P. Mineralogical analysis of bentonite from the ABM5 heater experiment at Äspö hard rock laboratory, Sweden. *Minerals* **2021**, *11*, 669. [CrossRef]
9. Sudheer, K.R.; Podlech, C.; Grathoff, G.; Warr, L.N.; Svensson, D. Thermally induced bentonite alterations in the SKB ABM5 hot bentonite experiment. *Minerals* **2021**, *11*, 1017. [CrossRef]
10. Dohrmann, R.; Olsson, S.; Kaufhold, S.; Sellin, P. Mineralogical investigations of the first package of the alternative buffer material test—II. Exchangeable cation population rearrangement. *Clay Miner.* **2013**, *48*, 215–233. [CrossRef]
11. Dohrmann, R.; Kaufhold, S. Characterisation of the second package of the alternative buffer material (ABM) experiment–II Exchangeable cation population rearrangement. *Clays Clay Miner.* **2017**, *65*, 104–121. [CrossRef]
12. Svensson, P.D.; Hansen, S. Redox chemistry in two iron-bentonite field experiments at Äspö hard rock laboratory, Sweden: An XRD and Fe k-edge XANES study. *Clays Clay Miner.* **2013**, *61*, 566–579. [CrossRef]
13. Wersin, P.; Jenni, A.; Mäder, U.K. Interaction of corroding iron with bentonite in the ABM1 experiment at Äspö, Sweden: A microscopic approach. *Clays Clay Miner.* **2015**, *63*, 51–68. [CrossRef]
14. Wersin, P.; Hadi, J.; Jenni, A.; Svensson, D.; Grenèche, J.-M.; Sellin, P.; Leupin, O.X. Interaction of Corroding Iron with Eight Bentonites in the Alternative Buffer Materials Field Experiment (ABM2). *Minerals* **2021**, *11*, 907. [CrossRef]
15. Muurinen, A. *Studies on the Chemical Conditions and Microstructure in Package 1 of Alternative Buffer Materials Project (ABM) in Äspö. Posiva Working Report*; Posiva: Eurajoki, Finland, 2010; p. 39.
16. Sandén, T.; Nilsson, U.; Svensson, D. *ABM45 Experiment at Äspö Hard Rock Laboratory*; Installation Report; SKB Report P-18-20; Svensk Kärnbränslehantering AB: Stockholm, Sweden, 2018; p. 49.
17. Gordon, A.; Pahverk, H.; Börjesson, E.; Johanss, A.J. *Examination of Copper Corrosion Specimens from ABM45, Package 5*; Technical Report TR-18-17; Svensk Kärnbränslehantering AB: Stockholm, Sweden, 2018; p. 27.
18. Karnland, O.; Olsson, S.; Nilsson, U. *Mineralogy and Sealing Properties of Various Bentonites and Smectite-Rich Clay Materials*; SKB TR-06-30; Svensk Kärnbränslehantering AB: Stockholm, Sweden, 2006.
19. Slaughter, M.; Earley, J.W. *Mineralogy and Geological Significance of the Mowry Bentonites, Wyoming*; Special Paper N° 87; Geological Society of America: New York, NY, USA, 1965.
20. Christidis, G.; Scott, P.W.; Marcopoulos, T. Origin of the bentonite deposits of Eastern Milos, Aegean, Greece: Geological, Mineralogical and Geochemical Evidence. *Clays Clay Miner.* **1995**, *43*, 63–77. [CrossRef]
21. Konta, J. Textural variation and composition of bentonite derived from basaltic ash. *Clays Clay Miner.* **1986**, *34*, 257–265. [CrossRef]
22. Přikryl, R.; Woller, F. Going underground: A new market for Czech bentonite in nuclear waste disposal. *Ind. Miner.* **2002**, *415*, 72–77.
23. Huertas, F.; Farina, P.; Farias, J.; Garcia-Sineriz, J.L.; Villar, A.M.; Fernandez, A.M.; Martin, P.L.; Elorza, F.J.; Gens, A.; Sanchez, M.; et al. *FEBEX Project. Full-Scale Engineered Barriers Experiment for a Deep Geological Repository for High Level Radioactive Waste in Crystalline Host Rock*; Updated Final Report 1994-2004; Technical Publication ENRESA 5-0/2006; ENRESA: Madrid, Spain, 2006; p. 590.
24. Fernandez, A.M.; Baeyens, B.; Bradbury, M.; Rivas, P. Analysis of the pore water chemical composition of a Spanish compacted bentonite used in an engineered barrier. *Phys. Chem. Earth* **2004**, *29*, 105–118. [CrossRef]
25. Shah, N.R. Indian bentonite: Focus on the Kutch region. *Ind. Miner.* **1997**, *359*, 43–47.
26. Jackson, M.L. *Soil Chemical Analysis: Advanced Course*; Revised Second Edition; Parallel Press University of Wisconsin Madison Libraries: Madison, WI, USA, 2005; p. 930.
27. Moore, D.; Reynolds, R. *X-ray Diffraction and the Identification and Analysis of Clay Minerals*; Oxford University Press: New York, USA, 1989; p. 332.

28. Newman, A.C.D. Chemistry of Clays and Clay Minerals. Mineral Society. Monograph N° 6. *Longman Sci. Tech.* **1987**, *6*, 480.

29. Wagner, C.D.; Davis, L.E.; Zeller, M.V.; Taylor, J.A.; Raymond, R.M.; Gale, L.H. Empirical atomic sensitivity factors for quantitative analysis by electron spectroscopy for chemical analysis. *Surf. Interface Anal.* **1981**, *3*, 211–225. [CrossRef]

30. Tarafder, P.K.; Thakur, R. An optimised 1,10-Phenathroline method for the determination of ferrous and ferric oxides in silicate rocks, soils and minerals. *Geostand. Geoanal. Res.* **2012**, *37*, 155–168. [CrossRef]

31. Amman, L.; Bergaya, F.; Lagaly, G. Determination of the cation exchange capacity of clays with copper complexes revisited. *Clay Miner.* **2005**, *40*, 441–453. [CrossRef]

32. Sawhney, B.L. Potassium and caesium ion selectivity in relation to clay mineral microstructure. *Clays Clay Miner.* **1970**, *18*, 47–52. [CrossRef]

33. Gregg, S.J.; Sing, K.S.W. *Adsorption, Surface area and Porosity*; Academic Press: London, UK, 1982; p. 303.

34. Keeling, P.S.; Kirby, E.C.; Robertson, R.H.S. Moisture adsorption and specific surface area. *J. Brit. Ceram. Soc.* **1980**, *79*, 36–40.

35. Fernández, A.M.; Sánchez-Ledesma, D.M.; Melón, A.; Robredo, L.M.; Rey, J.J.; Labajo, M.; Clavero, M.A.; Fernández, S.; González, A.E. *Thermo-Hydro-Chemical (THC) Behaviour of a Spanish Bentonite after Dismantling Heater#1 and Heater#2 of the FEBEX In Situ Test at the Grimsel Test Site*; Nagra Working Reports. NAB 16-25; Nagra: Wettingen, Switzerland, 2018; p. 583.

36. Fernández, A.M.; Bath, A.; Waber, H.N.; Oyama, T. Annex 2: Water sampling by squeezing drillcores. In *Geochemistry of Water in the Opalinus Clay Formation at the Mont Terri Rock Laboratory*; Pearson, F.J., Arcos, D., Bath, A., Boisson, J.Y., Fernández, A.M., Gäbler, H.-E., Gaucher, E., Gautschi, A., Griffault, L., Hernán, P., et al., Eds.; Geology Series No. 5; Federal Office for Water and Geology (FOWG): Bern, Switzerland, 2003; pp. 171–199.

37. Fernández, A.M. Caracterización y Modelización del Agua Intersticial en Materiales Arcillosos: Estudio de la Bentonita de Cortijo de Archidona. Ph.D. Thesis, CIEMAT, Madrid, Spain, 2004; p. 505.

38. Fernández, A.M.; Sánchez-Ledesma, D.M.; Tournassat, C.; Melón, A.; Gaucher, E.C.; Astudillo, J.; Vinsot, A. Applying the squeezing technique to highly consolidated clayrocks for pore water characterisation: Lessons learned from experiments at the mont terri rock laboratory. *Appl. Geochem.* **2014**, *49*, 2–21. [CrossRef]

39. Entwisle, D.C.; Reeder, S. New apparatus for pore fluid extraction from mudrocks for geochemical analysis. In *Geochemistry of Clay-Pore Fluid Interactions*; The Mineralogical Society Series; Chapman and Hall: London, UK, 1993; pp. 365–388.

40. Peters, C.A.; Yang, Y.C.; Higgins, J.D.; Burger, P.A. A preliminary study of the chemistry of pore water extracted from tuff by one-dimensional compression. In *Water-Rock Interaction*; Kharaka, Y.K., Maes, A.S., Eds.; A.A.Balkema: Rotterdam, The Netherlands, 1992; pp. 741–745.

41. Svensson, D. The Bentonite Barrier. Swelling Properties, Redox Chemistry and Mineral Evolution. Ph.D. Thesis, Lund University, Lund, Sweden, 2015.

42. Fernández, A.M.; Kaufhold, S.; Sánchez-Ledesma, D.M.; Rey, J.J.; Melón, A.; Robredo, L.M.; Fernández, S.; Labajo, M.A.; Clavero, M.A. Evolution of the THC conditions in the FEBEX *In Situ* test after 18 years of experiment: Smectite crystallochemical modifications after interactions of the bentonite with a C-steel heater at 100 °C. *Appl. Geochem.* **2018**, *98*, 152–171. [CrossRef]

43. War, L. IMA–CNMNC approved mineral symbols. *Mineral. Mag.* **2021**, *85*, 291–320. [CrossRef]

44. Alburquenque, D.; Márquez, P.; Troncoso, L.; Pereira, A.; Celis, F.; Sánchez-Arenillas, M.; Marco, J.F.; Gautier, J.L.; Escrig, J. $LiM_{0.5}Mn_{1.5}O_{4-\delta}$ (M = Co or Fe) spinels with a high oxidation state obtained by ultrasound-assisted thermal decomposition of nitrates. Characterization and physicochemical properties. *J. Solid State Chem.* **2020**, *284*, 121175. [CrossRef]

45. López, G.P.; Castuer, D.G.; Ratner, B. XPS 0 1s Binding Energies for Polymers Containing Hydroxyl, Ether, Ketone and Ester Groups. *Surf. Interf. Anal.* **1991**, *17*, 267–272. [CrossRef]

46. Marco, J.F.; Gancedo, J.R.; Ortiz, J.; Gautier, J.L. Appl. Characterization of the spinel-related oxides $Ni_xCo_{3-x}O_4$ (x=0.3,1.3,1.8) prepared by spray pyrolysis at 350 °C. *Surf. Sci.* **2004**, *227*, 175–186. [CrossRef]

47. Smirnov, M.Y.; Kalinkin, A.V.; Bukhtiyarov, V.L. X-ray photoelectron spectroscopic study of the interaction of supported metal catalysts with NOx. *J. Struct. Chem.* **2007**, *48*, 1053–1066. [CrossRef]

48. Meskinis, S.; Vasiliauskas, A.; Androlevicious, M.; Peckus, D.; Tamulevicius, S.; Viskontas, K. Diamond Like Carbon Films Containing Si. *Struct. Nonlinear Opt. Prop. Mater.* **2020**, *13*, 1003.

49. Bancroft, G.M. *Mössbauer Spectroscopy: An Introduction for Inorganic Chemists and Geochemists*; McGraw Hill: Maidenhead, UK, 1973.

50. Maddock, A.G. Mössbauer Spectrometry in Mineral Chemistry. In *Chemical Bonding and Spectroscopy in Mineral Chemistry*; Berry, F.J., Vaughan, D.J., Eds.; Springer: Berlin/Heidelberg, Germany, 1985; pp. 141–208.

51. Coey, J.M.D. Clay minerals and their transformations studied with nuclear techniques: The contribution of Mössbauer spectroscopy. *At. Energy Rev.* **1980**, *18*, 73–124.

52. Heller-Kallai, L.; Rozenson, I. The use of Mössbauer spectroscopy of iron in clay mineralogy. *Phys. Chem. Miner.* **1981**, *7*, 223–238. [CrossRef]

53. De Grave, E.; Vandenbruwaene, J.; Elewaut, E. An $^{57}$Fe Mössbauer effect study on glauconites from different locations in Belgium and northern France. *Clay Miner.* **1985**, *20*, 171–179. [CrossRef]

54. Dainyak, L.G.; Drits, V.A. Interpretation of Mössbauer spectra of nontronite, celadonite and glauconite. *Clays Clay Miner.* **1987**, *35*, 363–373. [CrossRef]

55. Dainyak, L.G.; Drits, V.A.; Heifits, L.M. Computer simulation of cation distribution in dioctahedral 2:1 layer silicates using IR-data: Application to Mössbauer spectroscopy of a glauconite sample. *Clays Clay Miner.* **1992**, *40*, 470–479. [CrossRef]

56. Pelayo, M.; Marco, J.F.; Fernández, A.M.; Vergara, L.; Melón, A.M.; del Villar, L.P. Infrared and Mössbauer spectroscopy of Fe-rich smectites from Morrón de Mateo bentonite deposit (Spain). *Clay Miner.* **2018**, *53*, 17–28. [CrossRef]
57. Murad, E.; Johnston, J.H. *Iron Oxides and Oxyhydroxides. Mössbauer Spectroscopy Applied to Inorganic Chemistry*; Gary, J., Ed.; Plenum Press: New York, NY, USA, 1987; Volume 2, Chapter 12; pp. 507–582.
58. Vandenberghe, R.E.; De Grave, E. Application of Mössbauer spectroscopy in earth sciences. In *Mössbauer Spectroscopy. Tutorial Book*; Yoshida, Y., Langouche, G., Eds.; Springer: Berlin/Heidelberg, Germany, 2013; Chapter 3; pp. 91–186.
59. Dang, M.Z.; Rancourt, D.G.; Dutrizac, J.E.; Lamarche, G.; Provencher, R. Interplay of surface conditions, particle size, stoichiometry, cell parameters, and magnetism in synthetic hematite-like materials. *Hyperfine Interact.* **1998**, *117*, 271–319. [CrossRef]
60. Goss, C.J. The kinetics and reaction mechanism of the goethite to hematite transformation. *Mineral. Mag.* **1987**, *51*, 437–451. [CrossRef]
61. Hernández, T.; Sánchez, F.J.; Moroño, A.; Aristu, M.; Marco, J.F. Effect of irradiation on the stability of the corrosion layer produced in Eurofer by contact with lithium ceramics. *J. Nucl. Mater.* **2021**, *545*, 152614. [CrossRef]
62. Sing, K.S.W.; Everett, D.H.; Haul, R.A.W.; Moscou, L.; Pierotti, R.A.; Rouquerol, J.; Siemieniewska, T. Reporting physisorption data for gas/solid systems with Special Reference to the Determination of Surface Area and Porosity. *Pure Appl. Chem.* **1985**, *57*, 603–619. [CrossRef]
63. Fernández, A.M.; Rivas, P. Pore water chemistry of saturated Febex bentonite compacted at different densities. In *Advances in Understanding Engineered Clay Barriers*; Alonso, E.E., Ledesma, A., Eds.; Balkema Publishers: Leiden, The Netherlands, 2005; pp. 505–514.
64. Parkhurst, D.L.; Appelo, C.A.J. Description of input and examples for PHREEQC version 3—A computer program for speciation, batch-reaction, one-dimensional transport, and inverse geochemical calculations: U.S. In *Geological Survey Techniques and Methods*; U.S. Geological Survey: Denver, CO, USA, 2003; Chapter A43; p. 497.
65. Blanc, P. *Thermoddem: Update for the 2017 Version*; Report BRGM/RP-66811-FR; BRGM: Orleans, France, 2017; p. 20.
66. Olsson, S.; Jensen, V.; Johanesson, L.E.; Hansen, E.; Karnland, O.; Kumpulainen, S.; Kirivanta, L.; Svensson, D.; Hansen, S.; Lindén, J. *Prototype Repository-Hydromechanical, Chemical and Mineralogical Characterization of the Buffer and Backfill Material from the Outer Section of the Prototype Repository*; Technical Report TR-13-21; Svensk Kärnbränslehantering AB: Stockholm, Sweden, 2013.
67. Missana, T.; Alonso, U.; Fernández, A.M.; García-Gutiérrez, M. Analysis of the stability behaviour of colloids obtained from different smectite clays. *Appl. Geochem.* **2018**, *92*, 180–187. [CrossRef]
68. Kaufhold, S.; Dohrmann, R.; Ufer, K.; Kober, F. Interactions of bentonite with metal and concrete from the FEBEX experiment-mineralogical and geochemical investigations of selected sampling sites. *Clay Miner.* **2018**, *53*, 745–763. [CrossRef]
69. Dohrmann, R.; Kaufhold, S. Cation exchange and mineral reactions observed in MX 80 buffer samples from the Prototype repository *in situ* experiment in Äspö, Sweden. *Clays Clay Miner.* **2014**, *5*, 357–373. [CrossRef]
70. Heuser, M.; Weber, C.; Stanjek, H.; Chen, H.; Jordan, G.; Schmahl, W.W.; Natzeck, C. The interaction between bentonite and water vapor. I: Examination of physical and chemical properties. *Clays Clay Miner.* **2014**, *62*, 188–202. [CrossRef]
71. Kaufhold, S.; Dohrmann, R. Stability of bentonites in salt solutions: II. Potassium chloride solution. Initial step of illitization? *Appl. Clay Sci.* **2010**, *49*, 98–107. [CrossRef]
72. Herbert, H.-J.; Kasbohm, J.; Sprenger, H.; Fernández, A.M.; Reichelt, C. Swelling pressures of MX-80 bentonite in solutions of different ionic strength. *Phys. Chem. Earth* **2008**, *33*, 327–342. [CrossRef]
73. Kaufhold, S.; Dohrmann, R. Effect of extensive drying on the cation exchange capacity of bentonites. *Clay Miner.* **2010**, *45*, 441–448. [CrossRef]
74. Fernández, A.M.; Cuevas, J.; Rivas, P. Pore water chemistry of the FEBEX bentonite. *Mat. Res. Soc. Symp. Proc.* **2001**, *663*, 573–588. [CrossRef]
75. Wersin, P. Geochemical modelling of bentonite pore water in high-level waste repositorys. *J. Contam. Hydrol.* **2003**, *61*, 405–422. [CrossRef]
76. Bradbury, M.H.; Berner, U.; Curti, E.; Hummel, W.; Kosakowski, G.; Thoenen, T. *The Long Term Geochemical Evolution of the Nearfield of the HLW Repository*; Nagra Technical Report NTB 12-01; NAGRA: Wettingen, Switzerland, 2014.

 *minerals*

*Article*

# Alteration of Bentonite Reacted with Cementitious Materials for 5 and 10 years in the Mont Terri Rock Laboratory (CI Experiment)

Shingo Yokoyama [1,*], Misato Shimbashi [1], Daisuke Minato [1], Yasutaka Watanabe [1], Andreas Jenni [2] and Urs Mäder [2]

[1] Nuclear Fuel Cycle Backend Research Center, Central Research Institute of Electric Power Industry, 1646 Abiko, Chiba 270-1194, Japan; s-misato@criepi.denken.or.jp (M.S.); d-minato@criepi.denken.or.jp (D.M.); yasutaka@criepi.denken.or.jp (Y.W.)
[2] RWI, Institute of Geological Sciences, University of Bern, Baltzerstrasse 3, CH-3012 Bern, Switzerland; andreas.jenni@geo.unibe.ch (A.J.); urs.maeder@geo.unibe.ch (U.M.)
* Correspondence: shingo@criepi.denken.or.jp; Tel.: +81-70-6568-9742

**Citation:** Yokoyama, S.; Shimbashi, M.; Minato, D.; Watanabe, Y.; Jenni, A.; Mäder, U. Alteration of Bentonite Reacted with Cementitious Materials for 5 and 10 years in the Mont Terri Rock Laboratory (CI Experiment). *Minerals* **2021**, *11*, 251. https://doi.org/10.3390/min11030251

Academic Editors: Ana María Fernández, Stephan Kaufhold, Markus Olin, Lian-Ge Zheng, Paul Wersin and James Wilson

Received: 4 January 2021
Accepted: 22 February 2021
Published: 28 February 2021

**Publisher's Note:** MDPI stays neutral with regard to jurisdictional claims in published maps and institutional affiliations.

**Abstract:** The cement–clay interaction (CI) experiment was carried out at the Mont Terri rock laboratory to complement the current knowledge on the influence that cementitious materials have on Opalinus Clay (OPA) and bentonite (MX). Drill cores including the interface of OPA, concrete (LAC = low-alkali binder, and OPC = ordinary Portland cement), and MX, which interacted for 4.9 and 10 years, were successfully retrieved after drilling, and detailed analyses were performed to evaluate potential mineralogical changes. The saturated compacted bentonites in core samples were divided into ten slices, profiling bentonite in the direction towards the interface, to evaluate the extent and spatial variation of the mineralogical alteration of bentonite. Regarding the mineral compositions of bentonite, cristobalite was dissolved within a range of 10 mm from the interface in both LAC-MX and OPC-MX, while calcite precipitated near the interface for OPC-MX. In LAC-MX and OPC-MX, secondary products containing Mg (e.g., M-S-H) also precipitated within 20 mm of the interface. These alterations of bentonite developed during the first 4.9 years, with very limited progress observed for the subsequent 5 years. Detectable changes in the mineralogical nature of montmorillonite (i.e., the formation of illite or beidellite, increase in layer charge) did not occur during the 10 years of interaction.

**Keywords:** cement–clay interaction; bentonite; cementitious materials; alteration; alkaline conditions

## 1. Introduction

Bentonite is used in radioactive waste disposal facilities as an engineered barrier owing to its suitable physicochemical properties, such as swelling properties [1], low hydraulic permeability under saturation conditions [1] and cation exchange capacity for radionuclide sorption [2]. Cementitious materials that are used as part of the engineered barrier degrade and generate alkaline pore water, which interacts with natural clays or bentonite. Although bentonite should inhibit the migration of radionuclides, its physicochemical properties change with mineralogical alterations when under alkaline conditions for a long period of time. Therefore, the evaluation of the alteration behavior of bentonite during its interaction with cementitious materials is important for performance assessments and safety analysis for radioactive waste disposal.

The mineralogical alteration of bentonite under alkaline conditions has been reviewed in previous studies [3–5]. Based on these reviews, the mineralogical alteration of bentonite has been attributed to the dissolution of the minerals present in bentonite and the subsequent precipitation of secondary phases. The dissolution of the major minerals of bentonite has been extensively investigated; previous studies have summarized the dissolution rates of these minerals [6,7], as well as the potential secondary phases in various chemical conditions [3,5]. Savage et al. (2007) suggested that calcite, dolomite, chalcedony, C(A)SH at variable Ca/Si ratios, K-feldspar, illite, phillipsite, analcime, clinoptilolite, and heulandite are the most likely secondary phases to form in a low-temperature cement–bentonite systems [5].

In addition to the above mineralogical alterations (i.e., dissolution and precipitation), the mineralogical nature of montmorillonite, which is the main component of bentonite, changes as it reacts with alkaline solutions. The illitization of montmorillonite is one of the major alteration phenomena that occur when montmorillonite reacts with a KOH solution [8–13]. Beidellitization and a change in the layer charge of montmorillonite also occur when reacting with alkaline solutions [8,9]. These mineralogical alterations of montmorillonite affect its swelling properties. Consequently, there are changes to be expected in the physicochemical properties of bentonite. Therefore, the progress of these changes in the mineralogical nature of montmorillonite with time, is also important for understanding the alteration behavior of bentonite under alkaline conditions.

In radioactive waste disposal facilities, a highly compacted form of bentonite is used to obtain low hydraulic permeability. Many studies on the mineralogical alteration of bentonite, as described above, have been carried out based on a high liquid–solid ratio used in powder samples. In contrast, there are a number of uncertainties that must be considered as the mineralogical alteration of compacted bentonite materials under alkaline conditions has not been extensively investigated [14–19]. Long-term alteration of bentonite materials in radioactive waste disposal is predicted by numerical analysis, such as in reactive transport modelling. Validation of the numerical results requires experimental results that take into account specific material properties, such as compacted bentonite and the type of cementitious materials. Such experimental results will also be useful for examining unknown parameters in the numerical analysis, such as reactive surface area of minerals that undergo dissolution. Furthermore, because the performance assessments and safety analysis of radioactive waste disposal targets long-term processes, it is valuable that the results of these experiments, such as those produced by Alonso et al. (2017) [20] and Fernández et al. (2017) [21], are produced under the assumed environmental conditions (e.g., temperature) for as long as possible.

The cement–clay interaction (CI) experiment is carried out at the Mont Terri rock laboratory [22]. The aim of the CI experiment is to reduce uncertainty in process-related cement–clay interactions, such as mineralogical alteration, generation of hyperalkaline pore water chemistry, and changes in the porosity of each material over relatively long-term periods. In the CI experiment, the compacted bentonite and cementitious materials were emplaced in Opalinus Clay (OPA), which is a claystone formation of mid-Jurassic age, investigated at the Mont Terri rock laboratory, and the host rock was foreseen for a deep geological repository envisaged in Switzerland [23]. The field experiment was repeatedly sampled, including the interfaces of each material: OPA, bentonite, and cementitious materials. Interface regions were analyzed to evaluate the mineralogical alteration as a function of space and time [22]. In this study, we characterized bentonite that reacted with cementitious materials for 4.9 and 10 years to specifically estimate the extent of the mineralogical alteration of bentonite.

## 2. Overview of CI Experiment

Previous studies have described the detailed setup of the CI experiment [22,24]. The concrete and bentonite were installed in two vertical boreholes (386 mm in diameter, up to 9 m in length) in the OPA. Three different cements, i.e., ordinary Portland cement (OPC), ESDRED cement, and low alkali cement (LAC), were used to prepare the concrete with common aggregate content and grain size distributions (sand and gravel up to 16 mm). These concretes were set in each part, as shown in Figure 1. Previous studies have also summarized the details of the materials [22,24]. In particular, the pH of OPC and LAC pore water, which influences the chemical interactions between concretes and bentonite, was reported to be equal at 13–13.3 and 12.2–12.4, respectively [22]. MX-80 bentonite (MX), also called Wyoming bentonite, was emplaced as a mixture of highly compacted pellets and fines, covered with a prefabricated concrete lid with a water saturation system and sealed against the overlying concrete section. In other words, MX was installed by sandwiching the top and bottom surfaces with different types of concrete (Figure 1). MX was saturated by artificial pore water (APW) using a water saturation system. APW mainly contained Na, Ca, Mg, Cl, SO$_4$, K, and TIC (Table 1) and provides an approximate of the pore water composition of the OPA at the experimental location in the Mont Terri rock laboratory. The bentonite sections were fully saturated after approximately 1 year (monitoring water uptake and swelling pressure).

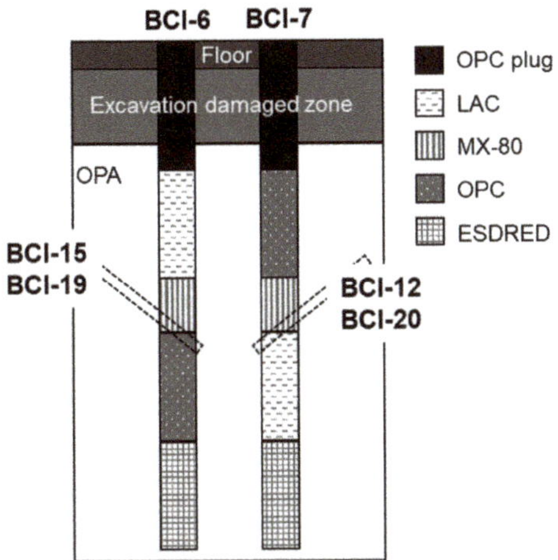


**Figure 1.** Schematic diagram of the settings for the CI experiment [22]. Dotted rectangles indicate the sampling location for each core sample. Samples BCI-12 and BCI-15 were sampled after 4.9 years, samples BCI-19 and BCI-20 10 years after emplacement of the materials. OPC: ordinary Portland cement; LAC: low-alkali binder; MX-80: bentonite.

**Table 1.** Composition of the artificial pore water (APW).

|  | Na$^+$ | K$^+$ | Mg$^{2+}$ | Ca$^{2+}$ | SO$_4^{2-}$ | Cl$^-$ | TIC * | pH ** |
|---|---|---|---|---|---|---|---|---|
| Concentration (mmol/L) | 204.8 | 2.33 | 14.33 | 12.65 | 11.76 | 236.9 | 0.359 | 9.6 |

* Toral inorganic carbon. ** pH was calculated from the charge balance of each cation and anion concentration using the Geochemist's Workbench (GWB) software package.

The MX employed in the scope of the CI experiment was the same granular bentonite as used in a large-scale emplacement test in the scope of the ESDRED project [25]. The bentonite consisted of small pellets (8–15 mm) with high density (ca. 2000 kg/m$^3$) and low water content (4 wt%), and variable amounts of fine fractions. The CI experiment performed a full-scale emplacement test for bentonite (92 kg), with a relatively low fine fraction (about 20%) which was partially removed by sieving, with an average moisture content of 6.1 wt%. It achieved a bulk density (moist) of 1548 kg/m$^3$ and developed a swelling pressure of >3.3 MPa. A similar mixture was emplaced in the field test, but the emplacement density is not exactly known, approximately 1450 kg/m$^3$. Swelling pressure was monitored at the top of the bentonite sections, and reached 0.6–1.3 MPa after 4 months of saturation, 1.3–1.6 MPa (BCI-6) and 1.4–1.9 MPa (BCI-7) after 12 months. Pressures of 2.2–2.5 MPa (BCI-6) and 2.6 MPa (BCI-7) were reached after 4 years, and remained at similar values 10 years after emplacement.

## 3. Samples and Methods

### 3.1. Samples

A drilling campaign to collect the core samples, including the interface of bentonite and concrete, was carried out 4.9 and 10 years after the installation of each material. Core samples were taken from the bottom interface of the bentonite to avoid any complications from the water saturation system equipped at the top of the bentonite. (Figure 1). The core samples used in this study were BCI-12, BCI-15, BCI-19, and BCI-20. BCI-12 and BCI-20 included the interface of MX and LAC which reacted for 4.9 and 10 years, respectively. BCI-15 and BCI-19 included the interface of MX and OPC which reacted for 4.9 and 10 years, respectively. Figure 2 shows photos and X-ray CT images of BCI-15 and a schematic diagram of sample separation. To evaluate the mineralogical alteration in bentonite, it was divided into ten sections in the range of 50 mm from the interface. The sample ID was numbered sequentially from the interface.

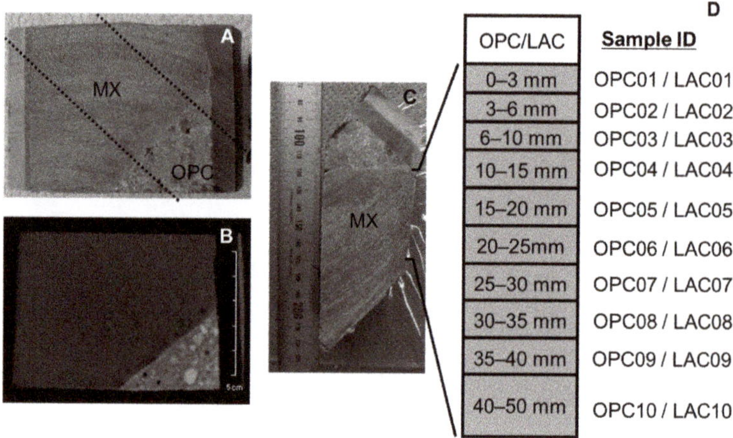


**Figure 2.** Core sample of BCI-15. (**A**) Photo of core sample, (**B**) X-ray CT image of core sample, (**C**) photo of core sample after cutting along the dotted line in (**A**), and (**D**) division thickness of bentonite and sample ID.

### 3.2. Sample Characterization

For mineral identification, the X-ray powder diffraction (XRD) patterns of the bentonite were obtained using a randomly oriented sample by XRD with CuK$\alpha$ radiation (RIGAKU RINT 2500, Tokyo, Japan). Samples divided as shown in Figure 2 were dried under vacuum conditions and manually ground using an agate mortar. The obtained powders were loosely compacted in an aluminum folder for XRD measurements to avoid particle orientation. Each sample was scanned at 40 kV, 30 mA, 0.01° 2$\theta$ steps and a scan range of 2.5–65° 2$\theta$. The morphologies and major elemental distribution of the bentonite samples were investigated by scanning electron microscopy (FE-SEM: JEOL JSM-7001F, Tokyo, Japan) and electron probe microanalyzer (EPMA: SHIMADZU EPMA1610, Kyoto, Japan; JEOL JXA-8100, Tokyo, Japan), respectively. For FE-SEM imaging, the vacuum-dried bentonite granules were split, and the fracture surface was coated with osmium and observed. EPMA samples, containing the interface of concrete and bentonite, were prepared by encapsulating vacuum-dried samples in resin and polishing the analysis surface.

The extracted cations of bentonite were evaluated as follows. The dry sample (300–500 mg) was dispersed in 10 mL of 1 M $NH_4Cl$ solution for 1 day and was solid–liquid separated by centrifugation. This process was repeated five times. All collected solutions were used to measure the concentrations of Na, K, Ca, and Mg by inductively coupled plasma-atomic emission spectrometry (ICP-AES: SHIMADZU ICPS-7510, Kyoto, Japan). This method may indicate a Ca concentration that is overly high compared with the exchangeable Ca content because of the possible dissolution of Ca carbonate contained in the sample, as discussed in a previous study [26]. Therefore, the term "extracted cation" was used in the present study.

The montmorillonite content of bentonite was estimated using the methylene blue absorption test. The methylene blue absorbed on bentonite was measured based on the Japanese Industrial Standard (JIS Z 2451) [27]. The montmorillonite content was calculated by dividing the amount of methylene blue absorbed on bentonite by that of pure montmorillonite, which was assumed to be 140 mmol/100 g [28]. According to JIS Z 2451, the standard deviation of the adsorbed amount is 4–6 mmol/100 g [27]. Therefore, the error in calculated montmorillonite content was $\pm 3$–4%.

The mineralogical alteration of montmorillonite in bentonite under alkaline conditions was examined by detecting illitization, beidellitization, and an increase in the layer charge [8–13]. To evaluate these alterations, the clay particle fraction of bentonite (0.2–2 μm) was collected by centrifugation. For oriented clay sample preparation, dispersions of the collected clay samples in deionized water were spread onto glass slides and air-dried. To evaluate illitization, the oriented clay sample, after saturation with ethylene glycol (EG), was analyzed by XRD. The percentage of expandable layers (smectite) of the sample was determined using the $\Delta 2\theta_1$–$\Delta 2\theta_2$ diagram [29]. To evaluate beidellitization, a Greene–Kelly (GK) test was carried out for all clay samples. Each sample (30–50 mg) was immersed three times in 10 mL of 1 M LiCl solution for 1 day. The samples were subsequently washed five times with 80% alcohol until the samples were chloride free ($AgNO_3$ test). The Li-saturated samples were spread with water onto silica slide glass. The Li-saturated samples were air-dried and then heated at 300 °C for 24 h in a muffle furnace. Afterwards, the samples were saturated with glycerol (GLY) vapor at 110 °C for 1 day before XRD measurements. The mean layer charge of the clay samples was analyzed using dodecylammonium chloride and octadecylammonium chloride according to the alkylammonium ion exchange method presented in previous studies [30–32]. The clay samples (15 mg) were dispersed in 5 mL of 0.1 M dodecylammonium chloride solution and 0.05 M octadecylammonium chloride solution and heated at 65 °C for 2 days. The suspensions were solid–liquid separated by a centrifuge and then treated once again with fresh alkylammonium chloride solution for one day. The samples were then washed with alcohol until the samples were chloride free. The washed clay sample was spread onto glass slides and dried at 65 °C in an oven. The dried samples were analyzed by XRD. The mean layer charge of the clay sample was calculated using the formula proposed by Olis et al. (1990) [31].

## 4. Results and Discussion

### 4.1. Mineral Compositions and Element Distributions

Figure 3 shows the XRD patterns of the bentonite samples. Although no XRD analysis of the initial MX used for the CI experiment was conducted, Karnland (2010) summarized the mineral composition of MX-80 for the commercial batches of each year, with the mean composition determined as montmorillonite (81.4%), illite (0.8%), calcite (0.2%), cristobalite (0.9%), gypsum (0.9%), muscovite (3.4%), plagioclase (3.5%), pyrite (0.3%), quartz (3.0%), and tridymite (3.8%) [2]. The MX used in the present study mainly comprised montmorillonite, quartz, cristobalite, and feldspar and contained extremely small amounts of mica minerals, gypsum, and calcite. In the XRD patterns recorded to evaluate mineral composition, each peak attributed to feldspar occurred randomly regardless of the distance from the interface, although sufficient attention was paid to sample homogenization and orientation randomization (see above). Therefore, all samples were concluded to contain feldspar, but the exact variation of feldspar amount could not be assessed. In contrast, the peak intensity of cristobalite was weak near the interface (within 10 mm) for both LAC-MX and OPC-MX (Figure A1). Figure 4A,B show SEM images of cristobalite in BCI-15. These cristobalites were observed at a location more than 60 mm from the interface, but they were hardly observed near the interface. From these results, as cristobalite was characterized as fine particles with a diameter of several hundred nm, cristobalite had a high reactivity and dissolved by reaction with cementitious fluids near the interface. Comparing between the 4.9- and 10-year samples, the range in which cristobalite was dissolved was slightly more advanced in LAC-MX after 10 years but was almost the same in OPC-MX. This indicates that the advance of the dissolution front for cristobalite occurred during the first 4.9 years of interaction, after which there was a strong decrease in its propagation rate. A small peak of gypsum was observed far from the interface, disappearing at a distance of ~20 mm from the interface. As Karnland (2010) reported that gypsum is initially contained in MX-80 [2], this behavior was ascribed to the dissolution of gypsum near the interface.

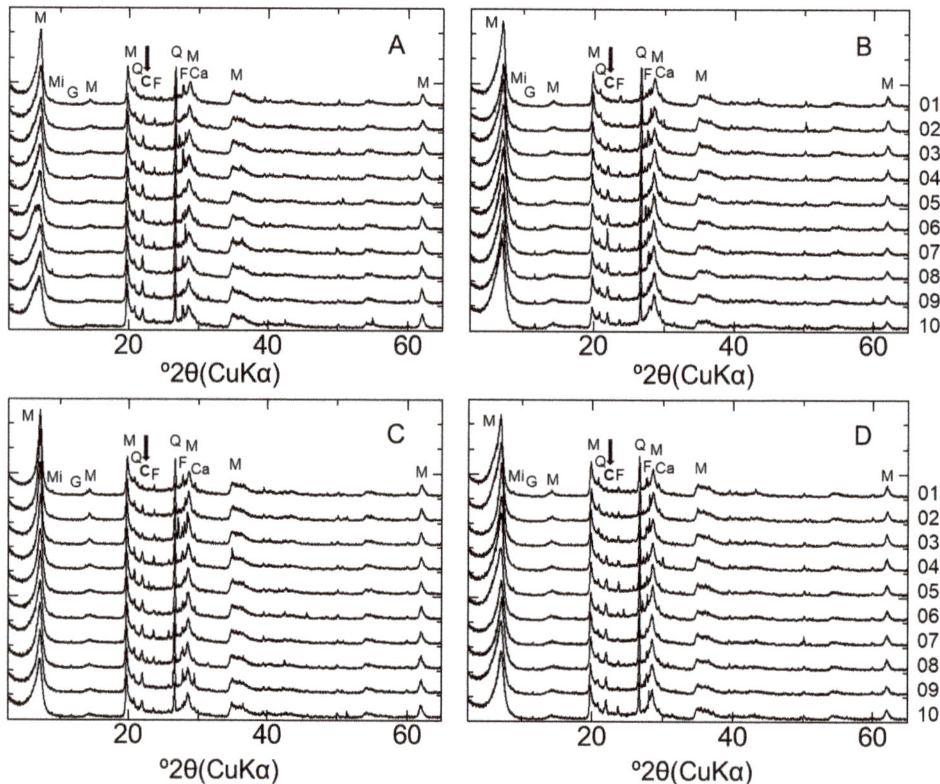

**Figure 3.** XRD patterns of bentonite in LAC-MX and OPC-MX. (**A**) LAC-MX_4.9 y, (**B**) LAC-MX_10 y, (**C**) OPC-MX_4.9 y, and (**D**) OPC-MX_10 y. M: montmorillonite; Q: quartz; C: cristobalite; F: feldspar; Mi: mica mineral; G: gypsum; Ca: calcite. Arrow pointing to cristobalite peak.

Figure 5 shows the EPMA-determined elemental distributions of 10-year core samples. For OPC-MX (BCI-19), a high Ca concentration was detected near the interface, as was also observed in the SEM and EPMA analyses of BCI-15 (Figures 4C and A4). Figure 4C shows an SEM image of calcite grains near the interface (~1 mm). In SEM images, calcite was hardly observed far from the interface because of its extremely small content. The results of SEM and EPMA analyses suggest that calcite precipitated only near the interface because of the interaction between MX and OPC.

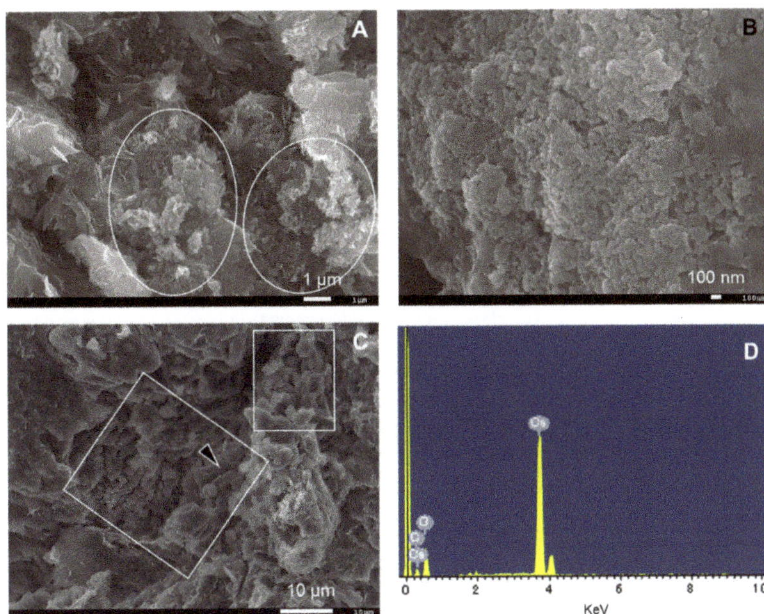

**Figure 4.** SEM images and EDS analysis of bentonite in BCI-15. (**A**) Montmorillonite and cristobalite (in white circle), (**B**) fine particles of cristobalite, (**C**) calcite (in white square), and (**D**) EDS spectrum of calcite at point indicated by the arrow in (**C**).

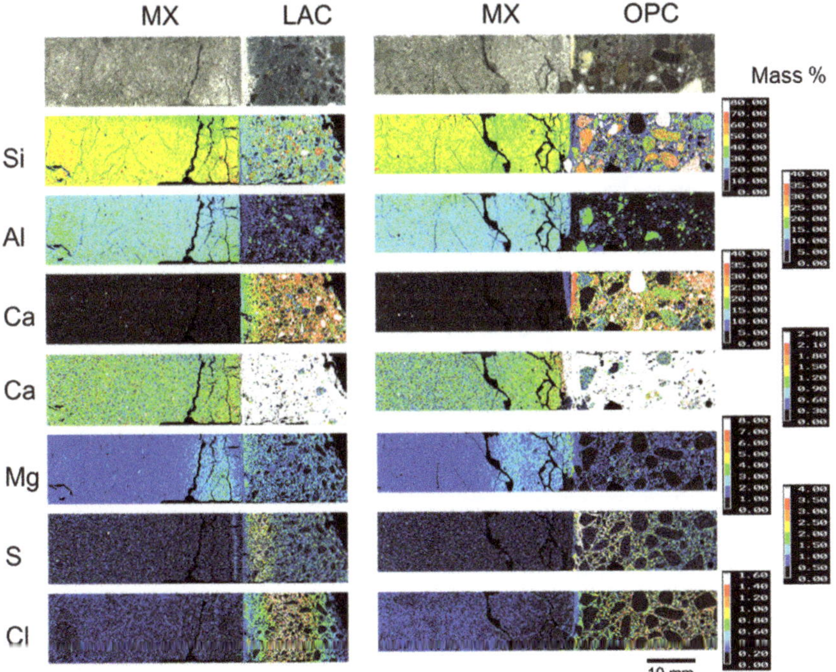

**Figure 5.** Elemental distribution of the 10-year core sample (BCI-19 and BCI-20).

In Figure 5, a notable elemental re-distribution in the bentonite component is the high Mg concentration near the interface in both LAC-MX and OPC-MX. This high Mg concentration was also observed in the 4.9-year core sample (Figures A2 and A3). This suggests that secondary products containing Mg were precipitated near the interface. Mäder et al. (2017) and Bernard et al. (2020) reported the possible formation of magnesium-silicate-hydrate (M-S-H), or hydrotalcite-like phases, due to the interaction between OPA and cementitious materials in the CI experiment [24,33]. The Mg-containing secondary products observed herein were also presumed to be M-S-H, hydrotalcite-like, or similar phases, as OPA and MX are similar clayey materials. In the concrete section, the distribution of the high concentrations occurred from the interface in the order of S and Cl in both LAC and OPC (Figure 5). In other words, Cl penetrated the concrete section more deeply than S did. The depth of S and Cl penetration into concrete was deeper in OPC than in LAC. The enrichment zones of these elements were distributed parallel to the interface between each concrete and MX. Therefore, these elements were concluded to penetrate concrete through MX. As the APW used to saturate MX and the pore water of OPA [34] contained substantial amounts of S and Cl, these elements penetrated the concrete along with the saturation of MX and precipitated. Although the precipitates were not analyzed, they presumably contained new products such as monosulfate and/or ettringite and Friedel's salt.

### 4.2. Extracted Cations and Montmorillonite Contents in Bentonite

Figure 6 shows the compositions of the extracted cations in bentonite. As MX was saturated by APW and the pore water of OPA containing various ions, the extracted cation composition in Figure 6 represents that after the reaction of MX and these solutions. Karnland (2010) summarized the exchangeable cations in MX-80 for batches of each year [2]. From this report, the mean exchangeable Na, Ca, K and Mg contents of MX-80 were obtained as 55, 13, 1 and 5 meq/100 g, respectively. The extracted Na, K, and Mg contents of MX far from the interface were similar to the above values, whereas the extracted Ca content exceeded the exchangeable Ca content reported by Karnland (2010) [2]. This difference was due to the dissolution of the small amounts of calcite contained in MX during extraction by 1 M $NH_4Cl$ solution, as described above.

Here, the amounts of extracted Ca and Mg near the interface (0–15 mm) were higher than those far from the interface. The amounts of extracted Na and K were almost constant, independent of the distance from the interface. The ion exchange reaction of bentonite results in an increased content of a certain cation while decreasing the contents of other cations. Therefore, this result suggests that the ion exchange reaction is not the main cause of the increase in Ca and Mg contents near the interface. Furthermore, the total amount of extracted cations increases near the interface. If the extracted cation is electrostatically adsorbed, the negative charge of bentonite, especially montmorillonite, should have increased near the interface. However, no change was observed in the mean layer charge of montmorillonite, as described below. Based on these observations, we suggest that the increase in the amount of extracted Ca and Mg is due to the dissolution of the mentioned secondary phases (i.e., calcite and Mg-binding phases) during the extraction treatment. In the comparison between the 4.9- and 10-year samples, both Ca and Mg in OPC-MX and LAC-MX showed almost the same value and distance from the interface with no change in the increasing total extracted cations. These results indicate that the precipitation of secondary phases occurred mainly during the 4.9-year interaction and had little progress over the next five years.

Figure 7 shows the content of montmorillonite in bentonite. The montmorillonite content of MX was approximately 78~80%. This value is almost identical to the montmorillonite content obtained by Karnland (2010) [2] when the reported variation in each year is considered. In LAC-MX, no change was observed in the 4.9- and 10-year samples, such that we consider that there was no montmorillonite dissolution. In OPC-MX, the montmorillonite content appears to show a slight decrease near the interface, but it was not a significant difference considering the error of the methylene blue adsorption experiment.

These results indicate that the dissolution of montmorillonite did not significantly progress in the interaction between LAC-MX and OPC-MX for 10 years.

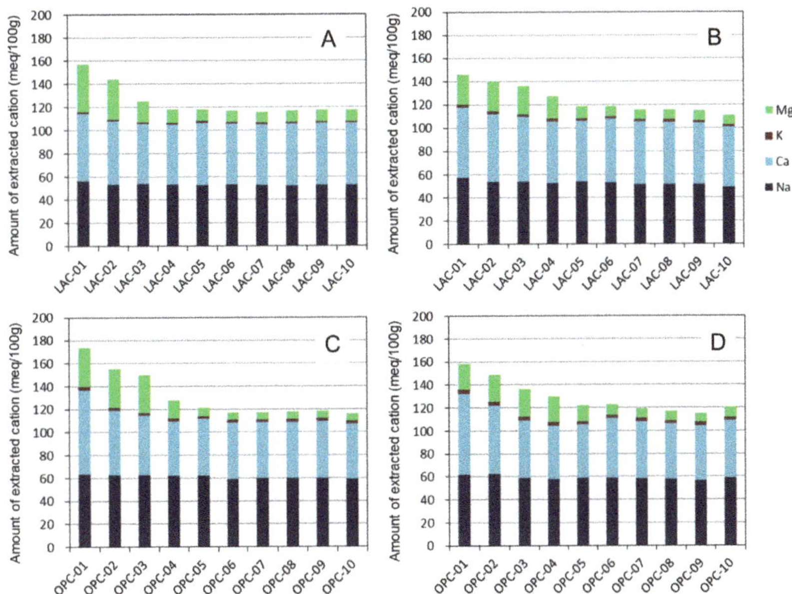

**Figure 6.** Extracted cation compositions of bentonite by a 1 M NH$_4$Cl solution. (**A**) LAC-MX_4.9 y, (**B**) LAC-MX_10 y, (**C**) OPC-MX_4.9 y, and (**D**) OPC-MX_10 y.

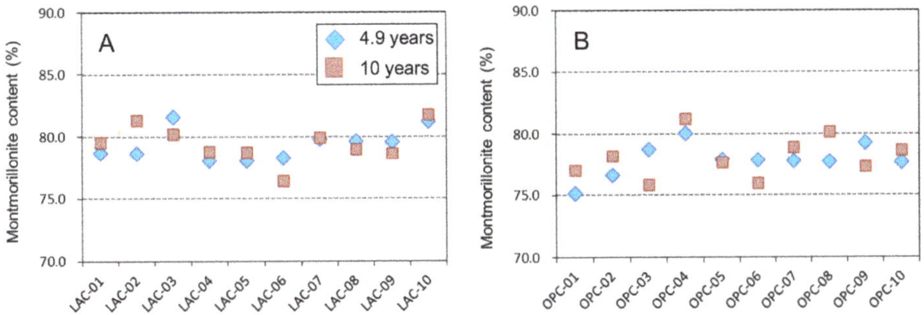

**Figure 7.** Montmorillonite content of bentonite. (**A**) LAC-MX and (**B**) OPC-MX for 4.9 (blue diamonds) and 10 (brown squares) years.

The dissolution rate of montmorillonite under alkaline conditions depends on the pH, temperature, and degree of undersaturation ($\Delta$Gr) [35–37]. Furthermore, for compacted bentonite, the reactive surface area for dissolution, i.e., the edge surface area [38], would be reduced as the bentonite is compacted and the component minerals are coated with each other. To evaluate the dissolution rate of montmorillonite in a realistic compacted bentonite–concrete interaction, it is important to analyze core samples that have reacted for an extended period to detect dissolution and obtain data on the above influencing factors

### 4.3. Mineralogical Characteristics of Montmorillonite

The illite % of the illite/smectite interstratified mineral (I/S) was determined by XRD analysis of the sample treated with EG. If smectite-rich R0 I/S forms from smectite, the peaks of $d_{002}$ and $d_{003}$ of the sample shift to low and high angles, respectively [39]. Figure 8 shows the XRD patterns of the clay fraction samples collected from each bentonite sample after EG treatment. No peak shifts in $d_{002}$ and $d_{003}$ were observed in any of the samples (Figure A5). The illite% was evaluated using the $\Delta 2\theta_1$–$\Delta 2\theta_2$ diagram [29]; consequently, the formation of an illite layer was not observed. Therefore, illitization of montmorillonite did not occur during the 10-year interaction in both LAC-MX and OPC-MX.

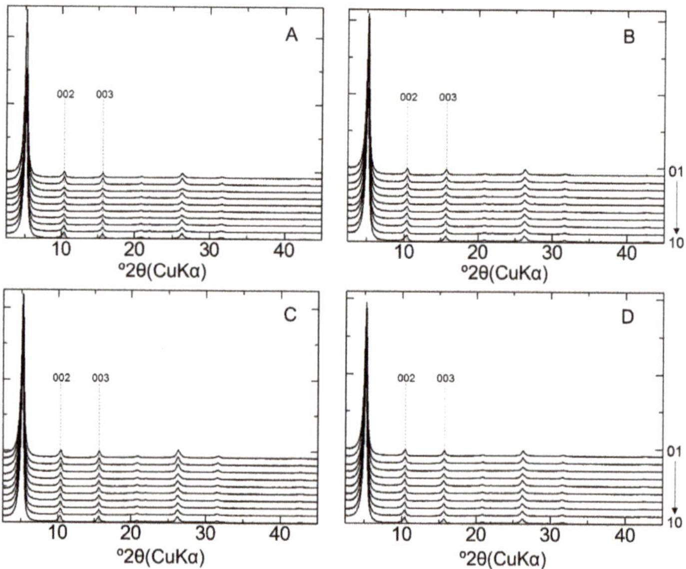

**Figure 8.** XRD patterns of montmorillonite after ethylene glycol (EG) treatment. (**A**) LAC-MX_4.9 y, (**B**) LAC-MX_10 y, (**C**) OPC-MX_4.9 y, and (**D**) OPC-MX_10 y. Sample IDs are arranged 1–10 from the top to bottom of each figure.

Montmorillonite and beidellite have mainly negative charges (layer charge) on octahedral and tetrahedral sheets, respectively. These clay minerals can be identified by the GK test, featuring characteristic XRD peaks at 0.95 nm (montmorillonite) and 1.77 nm (beidellite) [40]. Figure 9 shows the XRD patterns of the clay fraction samples after the GK test. Although some patterns appear to show a weak broad shoulder around 3~7° 2θ, this remains unexplained at present. On the other hand, all samples showed a peak near 0.95 nm, which indicated that almost no beidellite was formed. The mean layer charge of montmorillonite was evaluated from the XRD patterns of the alkylammonium treated samples (Table 2). From these results, the mean layer charge shows almost the same value, where no change was observed during the 10-year interaction in both LAC-MX and OPC-MX. From the above evaluation, we clarified that illitization and changes in layer charge characteristics (i.e., position and the mean layer charge) of montmorillonite did not occur during the 10-year interaction. The change in the mineralogical nature of montmorillonite, as described above, and the change in the exchangeable cation compositions affect the swelling property of bentonite. As these changes were not observed in this study, there would be no change in the swelling property of MX due to that effect for 10 years.

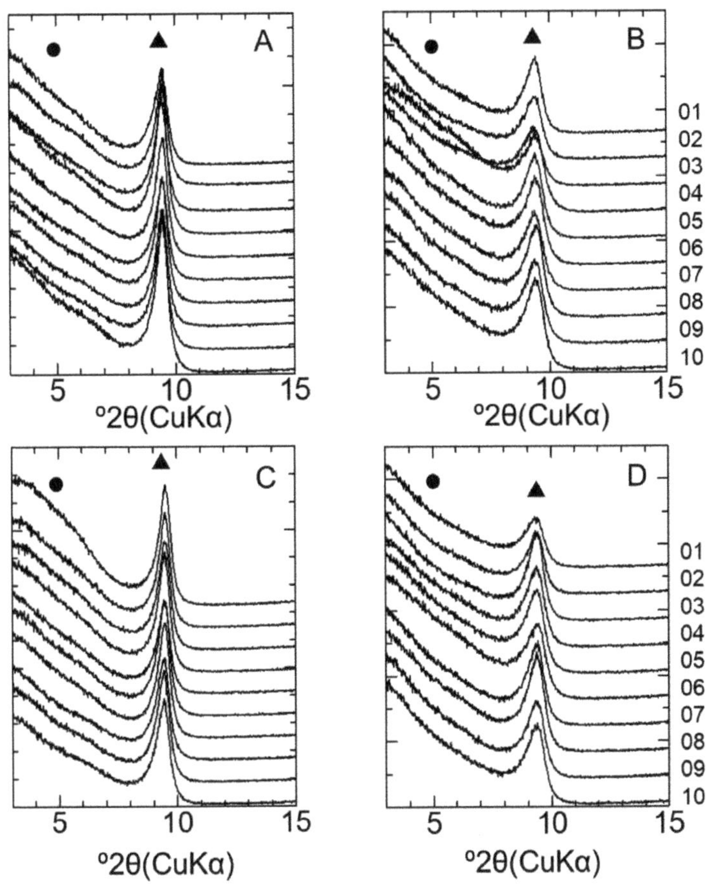

**Figure 9.** XRD patterns of montmorillonite after the GK test. (**A**) LAC-MX_4.9 y, (**B**) LAC-MX_10 y, (**C**) OPC-MX_4.9 y, and (**D**) OPC-MX_10 y. Sample IDs are arranged 1–10 from the top to bottom of each figure. Triangles (▲) and circles (●) indicate the ideal peak positions of montmorillonite and beidellite, respectively.

**Table 2.** Mean layer charge * of montmorillonite in each compacted bentonite.

| Sample ID | LAC-MX | | OPC-MX | |
|---|---|---|---|---|
| | 4.9 y | 10 y | 4.9 y | 10 y |
| 01 | 0.28 | 0.29 | 0.28 | 0.29 |
| 02 | 0.28 | 0.29 | 0.28 | 0.29 |
| 03 | 0.28 | 0.29 | 0.28 | 0.29 |
| 04 | 0.28 | 0.29 | 0.28 | 0.28 |
| 05 | 0.27 | 0.28 | 0.27 | 0.28 |
| 06 | 0.28 | 0.28 | 0.27 | 0.28 |
| 07 | 0.28 | 0.29 | 0.27 | 0.28 |
| 08 | 0.27 | 0.28 | 0.27 | 0.28 |
| 09 | 0.27 | 0.28 | 0.27 | 0.28 |
| 10 | 0.27 | 0.28 | 0.27 | 0.28 |

* The mean layer charge is the average of the values evaluated using dodecylammonium chloride and octadecy-lammonium chloride solutions.

*4.4. Bentonite Alteration during 10 Years of Interaction*

In the CI experiment, the main alterations of bentonite due to interaction with cementitious materials were the dissolution of cristobalite and the precipitation of secondary phases containing Ca and Mg. The initial pH of MX-80 slurry and the pore water of compacted samples has been previously reported to be approximately 8~9 [41,42]. The direct measurement of pH in the alteration zone is not possible, but the observed dissolution of cristobalite and precipitation of secondary products mentioned imply a high-pH perturbation extending some distance into the bentonite. Particularly, all secondary Mg-phases showed a strongly reduced solubility with increasing pH, and therefore, zones of Mg precipitation outline a high-pH front. The complexity of such mineral reaction fronts is best clarified by using reactive-transport modeling.

These alterations mainly progressed in the first 4.9 years with little progress during the next five years. Although the cause of the change in expansion rate of alteration zone is not yet clear, the following two possibilities can be considered. The first is the possibility that alteration occurred in a relatively short time after each material was installed in the vertical boreholes. Cementitious materials had excess water and pore water when the hydration reaction was in progress after casting. In contrast, bentonite was not completely saturated with water immediately after installation. Therefore, this would suggest that the highly alkaline solution in the cementitious materials was easily transferred to the bentonite by suction. Due to this rapid migration of highly alkaline solutions, the dissolution of highly reactive minerals (e.g., cristobalite with a fine particle size) and the precipitation of secondary products would proceed in a relatively short period of time. The permeation of the solution into the unsaturated bentonite causes the swelling of montmorillonite near the interface. The swelling of montmorillonite improves the water tightness and consequently inhibits further penetration of the solution. In addition, the supply of APW and the pore water of OPA could also have contributed to a reduction in suction. Therefore, it would suggest that the distance of the observed alteration area, i.e., the range of the initial alkaline solution penetration, was limited to approximately 15 mm or less.

The second possibility is the effect of clogging by secondary products. The precipitation of secondary products, calcite, and Mg-binding phases may reduce the porosity of bentonite, resulting in limited mass transfer. Similar to this manner, the precipitation of S- and Cl-binding phases and the progressing hydration in cementitious materials may also contribute to the reduction in porosity. As a result, this would inhibit the expansion rate of the alteration zones. This second possibility, especially the effect that a decrease in the porosity has on the expansion rate of the alteration zones, will be also discussed in future studies using reactive-transport models. The finding that the observed expansion rate of the alteration zones occurs in a non-linear manner is important for predictions of long-term bentonite–cement interactions. For predictions of long-term interactions, one must not only elucidate the mechanism of the observed alteration behavior (the two possibilities mentioned above), but also obtain data for the slow alteration rates (e.g., dissolution of montmorillonite). This will be revealed in future analyses of long-term interacting core samples from the CI experiment (e.g., 20 years).

## 5. Conclusions

To understand the alteration behavior of bentonite caused by the interaction with cementitious materials, we analyzed core samples collected from the CI experiment conducted at the Mont Terri rock laboratory. Cristobalite dissolved within the range of 10 mm from the interface and calcite precipitated near the interface. Secondary products containing Mg (e.g., M-S-H or hydrotalcite-like phases) probably also precipitated in the range of 20 mm from the interface. In the extracted cation composition, the amounts of Ca and Mg increased near the interface due to the dissolution of these secondary products during the extraction process. These alterations of bentonite mainly progressed during the first 4.9 years (possibly within a much shorter time), with little progress occurring during the next five years. It must be noted that the mineralogical characteristics of montmorillonite in bentonite did not change during the 10-year interaction; consequently, the swelling property of MX should not have changed due to this effect for 10 years. Although the mechanism of the observed alteration behavior—as mentioned above—is not yet entirely clear, the finding that the observed expansion rate of the alteration zones is highly non-linear is important for predictions of long-term bentonite–cement interactions.

The CI experiment represents a long-term experiment at a relatively large scale, evaluating the interaction of compacted bentonite with concrete, taking into account relevant materials expected in radioactive waste repositories. We believe that the results of the detailed analysis in this paper can be used to advance numerical analyses of long-term interactions by reactive transport modelling.

**Author Contributions:** Conceptualization, S.Y.; Investigation, S.Y., M.S., D.M., and Y.W.; Project administration, S.Y.; Resources, A.J. and U.M.; Writing—original draft, S.Y. and U.M.; Writing—review & editing, M.S., D.M., Y.W., and A.J. All authors have read and agreed to the published version of the manuscript.

**Funding:** The CI Project is funded by ANDRA (France), CRIEPI (Japan), IRSN (France), FANC (Belgium), Nagra (Switzerland), NWMO (Canada), Obayashi (Japan), RWM (UK) and SCK-CEN (Belgium). It financed all drilling campaigns, a part of sample preparation, and some earlier analytical costs.

**Data Availability Statement:** Data is contained within this article.

**Acknowledgments:** Discussions with Lukas Martin of Nagra are greatly appreciated. The authors like to acknowledge the support by the funding organizations of the CI Project and the Mont Terri Consortium, specifically the scientific and technical team operating the Mont Terri rock laboratory (Swisstopo) for field support.

**Conflicts of Interest:** The authors declare no conflict of interest.

## Appendix A

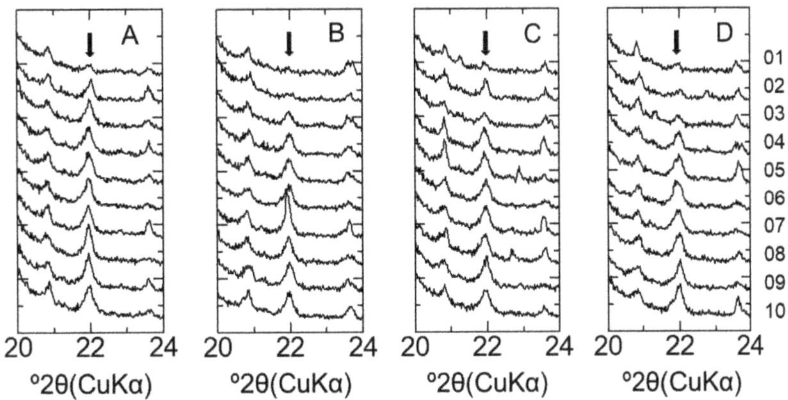


**Figure A1.** XRD patterns of bentonite in LAC-MX and OPC-MX. (**A**) LAC-MX_4.9 y, (**B**) LAC-MX_10 y, (**C**) OPC-MX_4.9 y, and (**D**) OPC-MX_10 y. Arrows indicate the peak position of cristobalite.

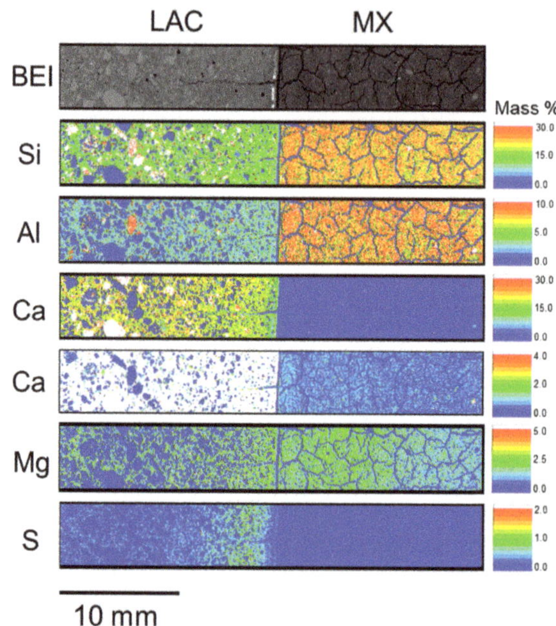



**Figure A2.** Elemental distribution of LAC-MX_4.9 y (BCI-12).

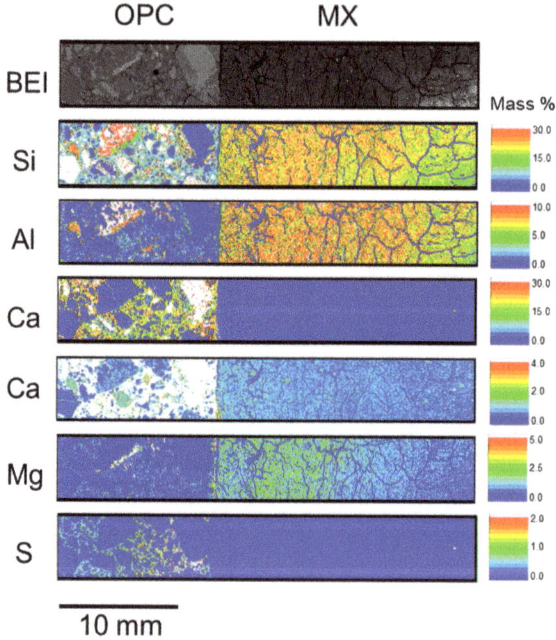



**Figure A3.** Elemental distribution of OPC-MX_4.9 y (BCI-15).

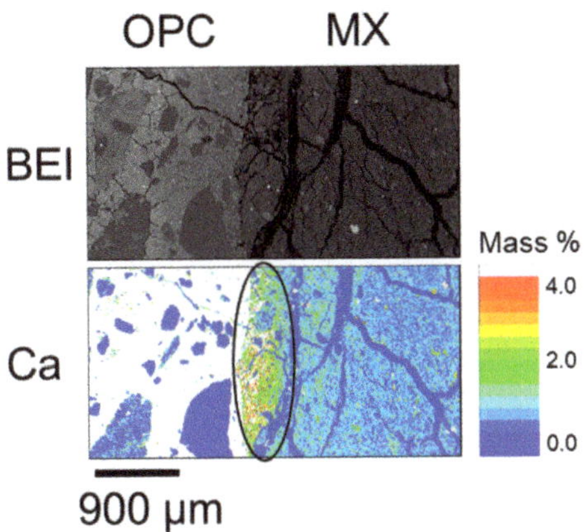



**Figure A4.** Calcium distribution of OPC-MX_4.9 y (BCI-15).

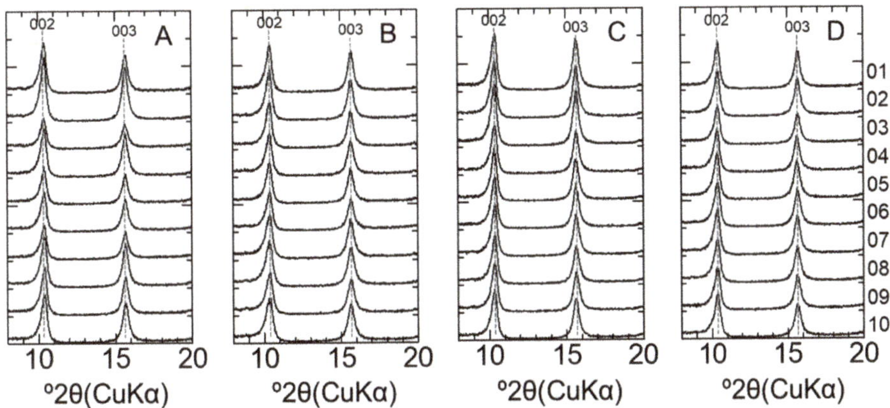



**Figure A5.** XRD patterns of montmorillonite after EG treatment. (**A**) LAC-MX_4.9 y, (**B**) LAC-MX_10 y, (**C**) OPC-MX_4.9 y, and (**D**) OPC-MX_10 y. This figure shows an enlargement of 8–20° 2θ in Figure 8.

## References


1. Karnland, O.; Olsson, S.; Nilsson, U. *Mineralogy and Sealing Properties of Various Bentonites and Smectite-Rich Clay Materials*; SKB Technical Report TR-06-30; SKB: Stockholm, Sweden, 2006.
2. Karnland, O. *Chemical and Mineralogical Characterization of the Bentonite Buffer for the Acceptance Control Procedure in a KBS-3 Repository*; SKB Technical Report TR-10-60; SKB: Stockholm, Sweden, 2010.
3. Oda, C.; Honda, A.; Savage, D. *An Analysis of Cement-Bentonite Interaction and Evolution of Pore Water Chemistry*; NUMO-TR-04-05; NUMO: Tokyo, Japan, 2004; pp. 74–79.
4. Gaucher, E.C.; Blanc, P. Cement/clay interactions—A review: Experiments, natural analogues, and modeling. *Waste Manag.* **2006**, *26*, 778–788. [CrossRef] [PubMed]
5. Savage, D.; Walker, C.; Arthur, R.; Rochelle, C.; Oda, C.; Takase, H. Alteration of bentonite by hyperalkaline fluids: A review of the role of secondary minerals. *Phys. Chem. Earth* **2007**, *32*, 287–297. [CrossRef]
6. Bandstra, J.Z.; Buss, H.L.; Campen, R.K.; Liermann, L.J.; Moore, J.; Hausrath, E.M.; Navarre-Sitchler, A.K.; Jang, J.H.; Brantley, S.L. Compilation of mineral dissolution rates. In *Kinetics of Water-Rock Interaction*; Brantley, S.L., Kubicki, J.D., White, A.F., Eds.; Springer: New York, NY, USA, 2008; pp. 737–823, (Appendix).
7. Cama, J.; Ganor, J. Dissolution kinetics of clay minerals. In *Natural and Engineered Clay Barriers*; Tournassat, C., Steefel, C.I., Bourg, I.C., Bergaya, F., Eds.; Elsevier: Amsterdam, The Netherlands, 2015; Volume 6, pp. 101–153.
8. Inoue, A. Potassium fixation by clay minerals during hydrothermal treatment. *Clays Clay Miner.* **1983**, *31*, 81–91. [CrossRef]

9.  Eberl, D.D.; Środoń, J.; Northrop, H.R. Potassium fixation in smectite by wetting and drying. In *Geochemical Processes at Mineral Surfaces*; Davis, J.A., Hayes, K.F., Eds.; ACS: Washington, DC, USA, 1986; pp. 296–326.
10. Eberl, D.D.; Velde, B.; McCormick, T. Synthesis of illite-smectite from smectite at earth surface temperatures and high pH. *Clay Miner.* **1993**, *28*, 49–60. [CrossRef]
11. Bauer, A.; Velde, B. Smectite transformation in high molar KOH solutions. *Clay Miner.* **1999**, *34*, 259–273. [CrossRef]
12. Drief, A.; Martinez-Ruiz, F.; Nieto, F.; Sanchez, N.V. Transmission electron microscopy evidence for experimental illitization of smectite in K-enriched seawater solution at 50 °C and basic pH. *Clays Clay Miner.* **2002**, *50*, 746–756. [CrossRef]
13. Fernández, R.; Ruiz, A.I.; Cuevas, J. The role of smectite composition on the hyperalkaline alteration of bentonite. *Appl. Clay Sci.* **2014**, *95*, 83–94. [CrossRef]
14. Nakayama, S.; Sakamoto, Y.; Yamaguchi, T.; Akai, M.; Tanaka, T.; Sato, T.; Iida, Y. Dissolution of montmorillonite in compacted bentonite by highly alkaline aqueous solutions and diffusivity of hydroxide ions. *Appl. Clay Sci.* **2004**, *27*, 53–65. [CrossRef]
15. Yamaguchi, T.; Sakamoto, Y.; Akai, M.; Takazawa, M.; Iida, Y.; Tanaka, T.; Nakayama, S. Experimental and modeling study on long-term alteration of compacted bentonite with alkaline groundwater. *Phys. Chem. Earth* **2007**, *32*, 298–310. [CrossRef]
16. Yamaguchi, T.; Sawaguchi, T.; Tsukada, M.; Kadowaki, M.; Tanaka, T. Changes in hydraulic conductivity of sand-bentonite mixtures accompanied by alkaline alteration. *Clay Miner.* **2013**, *48*, 403–410. [CrossRef]
17. Fernández, R.; Ruiz, A.I.; Cuevas, J. Formation of C-A-S-H phases from the interaction between concrete or cement and bentonite. *Clay Miner.* **2016**, *51*, 223–235. [CrossRef]
18. González-Santamaría, D.E.; Fernández, R.; Ruiz, A.I.; Ortega, A.; Cuevas, J. Bentonite/CEM-II cement mortar INTERFACE EXPERIMENTS: A proxy to in situ deep geological repository engineered barrier system surface reactivity. *Appl. Geochem.* **2020**, *117*, 104599. [CrossRef]
19. González-Santamaría, D.E.; Fernández, R.; Ruiz, A.I.; Ortega, A.; Cuevas, J. High-pH/low pH ordinary Portland cement mortars impacts on compacted bentonite surfaces: Application to clay barriers performance. *Appl. Clay Sci.* **2020**, *193*, 105672. [CrossRef]
20. Alonso, M.C.; Calvo, J.L.G.; Cuevas, J.; Turrero, M.J.; Fernández, R.; Torres, E.; Ruiz, A.I. Interaction processes at the concrete-bentonite interface after 13 years of FEBEX-Plug operation. Part I: Concrete alteration. *Phys. Chem. Earth* **2017**, *99*, 38–48. [CrossRef]
21. Fernández, R.; Torres, E.; Ruiz, A.I.; Cuevas, J.; Alonso, M.C.; Calvo, J.L.G.; Rodríguez, E.; Turrero, M.J. Interaction processes at the concrete-bentonite interface after 13 years of FEBEX-Plug operation. Part II: Bentonite contact. *Phys. Chem. Earth* **2017**, *99*, 39–63. [CrossRef]
22. Jenni, A.; Mäder, U.; Lerouge, C.; Gaboreau, S.; Schwyn, B. In situ interaction between different concretes and Opalinus Clay. *Phys. Chem. Earth Parts A/B/C* **2014**, *70*, 71–83. [CrossRef]
23. Bossart, P.; Bernier, F.; Birkholzer, J.; Bruggeman, C.; Connolly, P.; Dewonck, S.; Fukaya, M.; Herfort, M.; Jensen, M.; Matray, J.M.; et al. Mont Terri rock laboratory, 20 years of research: Introduction, site characteristics and overview of experiments. *Swiss J. Geosci.* **2017**, *110*, 3–22. [CrossRef]
24. Mäder, U.; Jenni, A.; Lerouge, C.; Gaboreau, S.; Miyoshi, S.; Kimura, Y.; Cloet, V.; Fukaya, M.; Claret, F.; Otake, T.; et al. 5-year chemico-physical evolution of concrete-claystone interfaces. *Swiss J. Geosci.* **2017**, *110*, 307–327. [CrossRef]
25. Plötze, M.; Weber, H.P. ESDRED: Emplacement tests with granular bentonite MX-80: Laboratory results from ETH Zürich. *Nagra Arbeitsbericht* **2007**, *7*, 1–10.
26. Dohrmann, R. Cation exchange capacity methodology I: An efficient model for the detection of incorrect cation exchange capacity and exchangeable cation results. *Appl. Clay Sci.* **2006**, *34*, 31–37. [CrossRef]
27. Japanese Industrial Standard Committee. *Test Method for Methylene Blue Adsorption on Bentonite and Acid Clay*; JISC: Tokyo, Japan, 2019; p. 2451.
28. White, D.; Michel, G.P. A proposed method for the determination of small amounts of smectites in clay mineral mixtures. *Proc. Br. Ceram. Soc.* **1979**, *28*, 137–145.
29. Watanabe, T. The structural model of illite/smectite interstratified mineral and the diagram for its identification. *Clay Sci.* **1988**, *7*, 97–114.
30. Lagaly, G.; Weiss, A. Determination of the layer charge in mica-type layer silicates. In Proceedings of the International Clay Conference, Tokyo, Japan, 5–10 September 1969; pp. 173–187.
31. Olis, A.C.; Malla, P.B.; Douglas, L.A. The rapid estimation of layer charges of 2:1 expanding clays from a single alkylammonium ion expansion. *Clay Miner.* **1990**, *25*, 39–50. [CrossRef]
32. Sato, T.; Murakami, T.; Watanabe, T. Change in layer charge of smectites and smectite layers in illite/smectite during diagenetic alteration. *Clays Clay Miner.* **1996**, *44*, 460–469. [CrossRef]
33. Bernard, E.; Jenni, A.; Fisch, M.; Grolimund, D.; Mäder, U. Micro-X-ray diffraction and chemical mapping of aged interfaces between cement pastes and Opalinus Clay. *Appl. Geochem.* **2020**, *115*, 1–17. [CrossRef]
34. Mazurek, M.; Haller, A. Pore-water evolution and solute-transport mechanisms in Opalinus Clay at Mont Terri and Mont Russelin (Canton Jura, Switzerland). *Swiss J. Geosci.* **2017**, *110*, 129–149. [CrossRef]
35. Sato, T.; Kuroda, M.; Yokoyama, S.; Tsutsui, M.; Fukushi, K.; Tanaka, T.; Nakayama, S. *Dissolution Mechanism and Kinetics of Smectite under Alkaline Conditions*; NUMO-TR-04-05; NUMO: Tokyo, Japan, 2004; pp. 38–41.
36. Cama, J.; Ganor, J.; Ayora, C.; Lasaga, C.A. Smectite dissolution kinetics at 80 °C and pH 8.8. *Geochim. Cosmochim. Acta* **2000**, *64*, 2701–2717. [CrossRef]

37. Cappelli, C.; Yokoyama, S.; Cama, J.; Huertas, F.J. Montmorillonite dissolution kinetics: Experimental and reactive transport modeling interpretation. *Geochim. Cosmochim. Acta* **2018**, *227*, 96–122. [CrossRef]

38. Yokoyama, S.; Kuroda, M.; Sato, T. Atomic force microscopy study of montmorillonite dissolution under highly alkaline conditions. *Clays Clay Miner.* **2002**, *53*, 147–154. [CrossRef]

39. Reynolds, R.C. Interstratified clay minerals. In *Crystal Structures of Clay Minerals and Their X-ray Identification*; Brindley, G.W., Brown, G., Eds.; Mineralogical Society: London, UK, 1984; pp. 249–303.

40. MacEwan, D.M.C.; Wilson, M.J. Interlayer and intercalation complexes of clay minerals. In *Crystal Structures of Clay Minerals and Their X-ray Identification*; Brindley, G.W., Brown, G., Eds.; Mineralogical Society: London, UK, 1984; pp. 197–248.

41. Herbert, H.J.; Kasbohm, J.; Sprenger, H.; Fernández, A.M.; Reichelt, C. Swelling pressures of MX-80 bentonite in solutions of different ionic strength. *Phys. Chem. Earth* **2008**, *33*, S327–S342. [CrossRef]

42. Muurinen, A.; Carlsson, T. Experiences of pH and Eh measurements in compacted MX-80 bentonite. *Appl. Clay Sci.* **2010**, *47*, 23–27. [CrossRef]

*Article*

# Evolution of the Reaction and Alteration of Mudstone with Ordinary Portland Cement Leachates: Sequential Flow Experiments and Reactive-Transport Modelling

Keith Bateman [1,*], Shota Murayama [1], Yuji Hanamachi [2], James Wilson [3], Takamasa Seta [2], Yuki Amano [1], Mitsuru Kubota [1], Yuji Ohuchi [1] and Yukio Tachi [1]

[1] Nuclear Fuel Cycle Engineering Laboratories, Japan Atomic Energy Agency, Tokai 319-1194, Ibaraki, Japan; murayama.shota@jaea.go.jp (S.M.); amano.yuki@jaea.go.jp (Y.A.); kubota.mitsuru@jaea.go.jp (M.K.); ohuchi.yuji@jaea.go.jp (Y.O.); tachi.yukio@jaea.go.jp (Y.T.)

[2] QJ Science, 8-1 Sakaecho, Yokohama 221-0052, Kanagawa, Japan; hanamachi.yuji@jaea.go.jp (Y.H.); takamasa.seta@qjscience.co.jp (T.S.)

[3] Wilson Scientific Ltd., Birchwood, Warrington WA3 6TR, UK; jim@wilsonscientific.co.uk

* Correspondence: bateman.keith@jaea.go.jp

**Abstract:** The construction of a repository for geological disposal of radioactive waste will include the use of cement-based materials. Following closure, groundwater will saturate the repository and the extensive use of cement will result in the development of a highly alkaline porewater, pH > 12.5; this fluid will migrate into and react with the host rock. The chemistry of the fluid will evolve over time, initially high [Na] and [K], evolving to a Ca-rich fluid, and finally returning to the groundwater composition. This evolving chemistry will affect the long-term performance of the repository, altering the physical and chemical properties, including radionuclide behaviour. Understanding these changes forms the basis for predicting the long-term evolution of the repository. This study focused on the determination of the nature and extent of the chemical reaction, as well as the formation and persistence of secondary mineral phases within a mudstone, comparing data from sequential flow experiments with the results of reactive transport modelling. The reaction of the mudstone with the cement leachates resulted in small changes in pH with the precipitation of calcium aluminium silicate hydrate (C-(A-)S-H) phases of varying compositions. As the system evolves, secondary C-(A-)S-H phases re-dissolve and are replaced by secondary carbonates. This general sequence was successfully simulated using reactive transport modelling.

**Keywords:** ordinary Portland cement; mudstone; sequential flow experiment; reactive-transport modelling; radioactive waste disposal

**Citation:** Bateman, K.; Murayama, S.; Hanamachi, Y.;Wilson, J.; Seta, T.; Amano, Y.; Kubota, M.; Ohuchi, Y.; Tachi, Y. Evolution of the Reaction and Alteration of Mudstone with Ordinary Portland Cement Leachates: Sequential Flow Experiments and Reactive-Transport Modelling. *Minerals* **2021**, *11*, 1026. https://doi.org/10.3390/min11091026

Academic Editor: Hegoi Manzano

Received: 9 August 2021
Accepted: 16 September 2021
Published: 21 September 2021

**Publisher's Note:** MDPI stays neutral with regard to jurisdictional claims in published maps and institutional affiliations.



## 1. Introduction

The construction of a repository for geological disposal of radioactive waste will include the use of cement-based materials [1–4]. Following closure, groundwater will saturate the repository and the use of cement will result in the development of a highly alkaline porewater, pH > 12.5, in the case of ordinary Portland cement (OPC) [5,6]. The fluid will migrate into and react with the host rock. The chemistry of the migrating fluid will evolve over time, initially being high in Na and K with high pH ~13.5 (Stage I), evolving to a Ca rich fluid with pH ~12.5 (Stage II), followed by C-S-H buffering (Stage III), and finally returning to the groundwater composition [7]. This evolving fluid chemistry will affect the long-term performance of the repository, altering the physical and chemical properties of the host rock, including radionuclide behaviour ([7] and references within). Understanding these changes forms the basis for modelling the long-term evolution of the repository.

Flow-through or column experiments [8] are a useful technique with which to obtain experimental data on the spatial and temporal changes as the chemistry of the migrating fluid evolves. However, many previous studies have used only a single fluid at a time [9–11], and although the reactions of different OPC leachates (representing Stages I and II) have sometimes been examined, these studies generally compared the reaction of unaltered solids (either single minerals, synthetic mineral assemblages, or real rocks) and a single leachate.

It is known, for example, that the Ca/Si ratios of the secondary calcium silicate hydrate (C-S-H) and calcium aluminium silicate hydrate (C-A-S-H) phases may change by re-dissolution and precipitation as the chemistry of the reacting fluid evolves, i.e., reduction in Na and K and increases in Ca concentration [12–15]. Indeed, this has been observed in laboratory column experiments [15] in which early-formed C-S-H and C-A-S-H gels were replaced by ones with a lower Ca/Si ratio during successive reaction with fluids representing the evolution of a cement leachate over time. The initial Na-K-Ca-OH fluid [15] was followed by a $Ca(OH)_2$-saturated fluid and, finally, a $Ca-HCO_3$-type fluid (neutral pH), representing eventual re-saturation of by background bicarbonate groundwater [15]. However, this work [15] only used a 'generic' crystalline rock.

Some previous modelling studies [16–18] have been undertaken to consider the fluid evolution impacts on the host and secondary mineralogy, but without experimental data with which to validate the predictions. In addition, most of these modelling studies used a single input fluid and tracked evolution of this fluid's chemistry and change in mineralogy with time and/or distance. An understanding of how the evolution in chemistry of the migrating cement pore fluids from the repository and subsequent interaction with the altered the host rock is critical for the prediction of the long-term evolution of the mineralogy, and will be part of the safety assessment for radioactive waste disposal repository.

This study focused on the sequence of alteration owing to the evolution of OPC-type leachate chemistry on argillaceous mudstone from the Horonobe Underground Research Laboratory (URL), Hokkaido, Japan [19]. This study describes the use of sequential fluids to represent the evolution of the cement leachate fluid chemistry with time and how it interacts with the host rock, which has been identified as a key area of uncertainty, particularly with the modelling of such systems [7]. This was performed by setting up a series of identical flow experiments to provide (at least in a laboratory time frame) information on the sequence of reaction. The first flow experiment was stopped after the reaction with a fluid representative of an early OPC cement leachate. In the remaining experiments, the reactant fluid was then changed to one representative of an evolved leachate. Again, after a period of reaction, one of the experiments was terminated and the reactant fluid changed to a natural groundwater for the final stage of reaction. Fluid chemistry was monitored throughout the experiments, providing a near continuous record of the evolution of the reacted fluid chemistry, and with the three different fluid types, 'snapshots' of the mineralogical variations could be determined. The specific aim of this work was the determination of the nature and extent of the chemical reaction as well as the formation and persistence of secondary mineral phases within the mudstone, as the fluid chemistry evolved, by the comparison of experimental data with geochemical reactive transport modelling.

## 2. Materials and Methods

### 2.1. Horonobe Mudstone

Samples of mudstone, for use in this study, were collected from the Horonobe URL site, Hokkaido, Japan [19]. The samples were taken from gallery walls located in the Koetoi formation, which is a massive and lithologically homogeneous, diatomaceous mudstone that contains amorphous silica (40–50 wt%), clay (17–25 wt%), quartz (7–10 wt%), feldspar (5–10 wt%), and pyrite (<2 wt%) [20,21]. The mudstone samples were crushed to <500 μm prior to being used in the experiments. The use of crushed materials increases the surface area available for reaction and encourages a greater degree of reaction within the time constraints of a laboratory study. However, it is recognised that the crushing process could also result in the generation of highly reactive 'fines', resulting in experimental artefacts, i.e., increased initial dissolution.

### 2.2. Description of Fluids

Cement pore fluids representative of the alkaline leachates expected from a cementitious repository [22,23] were used for these sequential experiments. The first fluid represented a 'young' OPC leachate pH ~13.4 with high [Na] and [K]. A second fluid, an 'evolved' OPC leachate, was Ca-rich and saturated with respect to portlandite, pH ~12.5. The OPC leachates were prepared from analytical grade reagents; Na and K were added as hydroxides and Ca as CaO. The third fluid was Horonobe groundwater (HGW) collected from the 07-V140-M03 borehole (depth 140 m, sampled on 22 July 2020), located in the Koetoi Formation, which was stored in stainless-steel containers under nitrogen before use. Details of the initial concentrations of ions in the fluids are given in Table 1.

**Table 1.** Initial concentrations of major ions in the OPC leachates and Horonobe groundwater.

| Leachate | pH @ 24 °C | Components (mg/L) | | | | | | |
|---|---|---|---|---|---|---|---|---|
| | | Na | K | Ca | $SiO_2$ | Mg | Cl | $HCO_3^-$ |
| 'young' OPC leachate (Na-K-Ca-OH) | 13.4 | 1500 | 7300 | 60 | - | - | - | - |
| | | | | Ca | $SiO_2$ | Mg | Cl | $HCO_3^-$ |
| 'evolved' OPC leachate (saturated $Ca(OH)_2$) | 12.5 | | | 800 | - | - | - | - |
| | | Na | K | Ca | $SiO_2$ | Mg | Cl | $HCO_3^-$ |
| Horonobe groundwater 07-V140-M03 (sampled 22 July 2020) | 7.93 | 2580 | 74 | 69 | 68 | 97 | 2900 | 2200 |

Upon sub-sampling the fluid from the reservoir used in the flow experiment, it was noted that the pH of the HGW (pH ~8.65) had increased from that of the original HGW (pH ~7.93, Table 1) as sampled from the stainless-steel container. This was accompanied by a corresponding decrease in [$HCO_3^-$] to ~1900 mg/L (from ~2200 mg/L), indicating that there had been some degassing of the HGW during the set-up of the experiment. Typically, HGW from the same borehole [24] has a ~pH 7.4~7.8 and a [$HCO_3^-$] ranging from 2700–3100 mg/L, suggesting that, even in the stainless-steel container, there had been some degassing of the HGW.

### 2.3. Description of the Flow Experiments

A small flow cell (SFC) as used in previous studies, examining differences in alteration of Horonobe mudstone with ordinary Portland and low alkali cement leachates [14], was used to conduct the experiments. A schematic of the small flow cell set-up is shown in Figure 1. The SFC was constructed from three pieces of acrylic plastic, sealed by a combination of 'O-rings' and bolts (Figure 1). Both inlet and outlet sides for the cell were fitted with filters and porous polypropylene disks, which, on the inlet side, aided the distribution the incoming fluid across the whole face of the mudstone sample. Polypropylene fluid reservoirs and sample collection bottles were used, with flow through the cell controlled by a Cole-Parmer MASTERFLEX® peristaltic pump (Cole-Parmer, Vernon Hills, IL, USA) [14]. The dry density of the packed mudstone could be controlled during the initial set-up. The experiments conducted here used a dry density of 1 g/cm$^3$ (corresponding to a solid mass ~3.14 g); this was achieved using a small stainless-steel hand-press to mould the mudstone sample. The experiments were conducted inside a glove box continuously flushed with $N_2$. The primary aim of this flushing was to prevent carbonation of the alkaline leachates by atmospheric carbon dioxide.

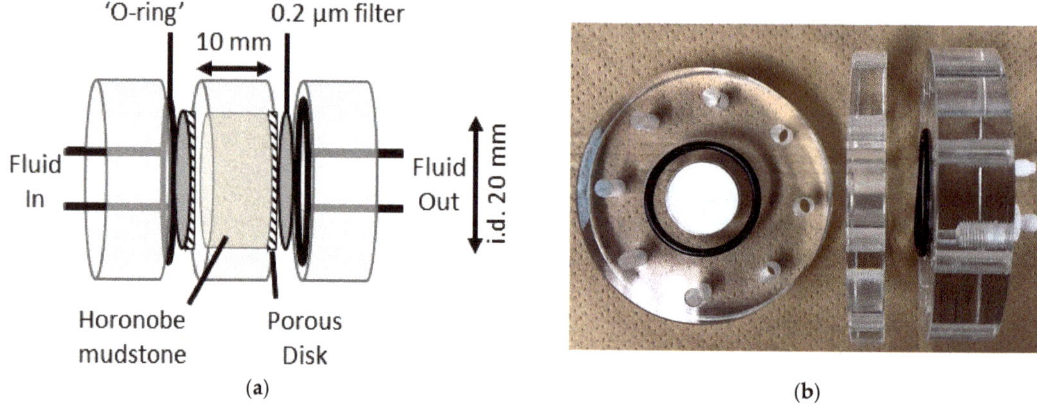



**Figure 1.** (**a**) Schematic (after [14]) and photograph (**b**) of small flow cell (SFC).

On completion of the experiments, the SFC was partly dismantled and the mudstone sample was vacuum dried, while still being held in the central section, before being carefully extruded and sectioned into ~1.5 mm thick slices, using a thin blade, before subsequent mineralogical analysis.

A series of three flow experiments were conducted to examine the sequential reaction of Horonobe mudstone with OPC leachates and groundwater. This was achieved by reacting mudstone samples with successive fluids starting with the 'young' OPC leachate (SFC-1), followed by the 'evolved' OPC leachate (SFC-2), and finally HGW (SFC-3). Figure 2 gives details of the experiments conducted, together with the reaction duration with each fluid.

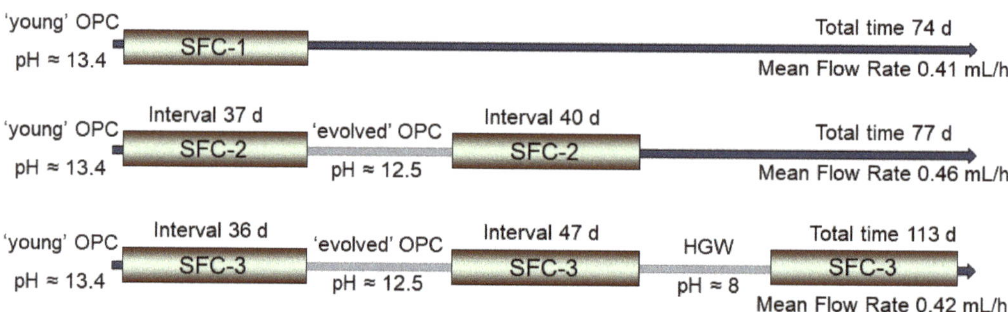



**Figure 2.** Details of the experiments, with interval and total times for the sequential experiments with the Horonobe mudstone reacting with OPC leachates and then ground water.

### 2.4. Fluid Sampling and Analysis

Sampling of the fluids was performed within the protective $N_2$ atmosphere of the glove box. All collected fluids were filtered using 0.2 µm syringe filters and then sub-sampled for determination of cations, and anions. Typically, a 4 mL sample of the fluid was diluted two-fold with 18 MΩ cm demineralised water (Millipore Simplicity® ultrapure water system) and then acidified with concentrated $HNO_3$ (1% *v/v*), in order to preserve the sample. This sample was used for the analysis of major cations, by a combination of ICP-OES (inductively coupled plasma—optical emission spectrometry) using a Shimadzu ICPE-9800 (Shimadzu Corporation, Kyoto, Japan), and ICP-MS (inductively coupled plasma—mass spectrometry) using a Perkin-Elmer NexION 300× (PerkinElmer, Inc., Waltham, MA, USA), both calibrated using matrix matched standards. A second subsample was taken for determination of major anions by IC (ion chromatography) using a Dionex ICS-5000 Ion Chromatograph system (Thermo Fisher Scientific, Waltham, MA, USA), calibrated using mixed anion standard solution (Kanto Chemical Co., Inc, Tokyo, Japan). All fluid samples were stored in a refrigerator at <5 °C until required for analysis. Considering instrumental and sample preparation errors (e.g., sample dilution), a total 5% error was assumed.

The pH of the experimental fluids was determined immediately upon sampling using a DKK-TOA corp., model HM-30P meter and combination electrode calibrated using DKK-TOA corp., buffers at 4.01, 6.86, and 9.18 pH (Japanese standard), pH accurate to $\pm 0.02$ pH.

### 2.5. Solids' Sampling and Analysis

On completion of the experiments, the mudstone samples were cut into ~1.5 mm thick slices, using a thin blade, with SEM stubs and powdered samples for XRD being prepared for subsequent mineralogical characterisation. Petrographic analysis of the solid samples was performed using a combination of both scanning electron microscopy (SEM) (JEOL JSM-6510 Series SEM, JEOL Ltd., Tokyo, Japan) and X-ray diffraction (XRD) analysis (RigaKu SmartLab XRD, Rigaku Corporation, Tokyo, Japan) with a 9 kW X-ray source. Sub-samples for SEM analysis were prepared as carbon coated, random mount stub samples. Techniques used included SEM using secondary electron imaging (SE) and backscattered electron (BSE) imaging and element distribution analysis using with energy-dispersive X-ray spectroscopy (EDS). Samples for XRD were prepared for analysis by taking a representative sub-sample and grinding to a fine powder

### 2.6. Mineral Saturation State Calculations

The saturation indices (SI = log (IAP/$K_s$)—where IAP: ion activity product and $K_s$: solubility constant) of the primary and potential secondary minerals—in the reacted fluids were calculated using the PHREEQC v3.6.3 geochemical code [25]. Calculations were performed using the JAEA thermodynamic database (JAEA-TDB) [26]. JAEA-TDB version PHREEQC20.dat (v1.2, 11 March 2021) was used for the calculations, being the latest version available at the time.

### 2.7. Reactive-Transport Modelling

#### 2.7.1. CABARET Model Concept

Fully-coupled 1D reactive transport models were constructed using the 'CABARET' (Cement And Bentonite Alteration due to REactive Transport) computer modelling code (Quintessa Limited, Henley-on-Thames, UK). CABARET was chosen for the modelling in preference to codes such as PHREEQC, as it allows coupling of porosity evolution (as minerals dissolve and/or precipitate) with diffusive and advective transport. CABARET uses the underlying QPAC Code, which has been used in previous reactive transport modelling studies [27,28]. CABARET uses an adaptive time-stepper to maximise the solver efficiency, which reduces the size of the time-step in response to external events (e.g., time-dependent inputs) or 'emergent events' (e.g., precipitation of secondary minerals or total dissolution of pre-existing minerals) and increases the time-step when the system is evolving less rapidly.

Supporting calculations (e.g., aqueous chemical speciation) were undertaken using PHREEQCv3 to generate the initial fluid compositions and to identify the key aqueous species specified in the CABARET models. The same JAEA-TDB [26] version was used for the CABARET calculations as used for the PHREEQC mineral saturation state calculations.

#### 2.7.2. CABARET Model Setup and Parameters

Figure 3 shows the geometry of the experimental set-up as represented in CABARET. As CABARET was designed to work with a single input fluid type, it was necessary to generate the fluids 'in situ', effectively reproducing the fluid reservoir used in the experimental set up. This reservoir contained defined 'hypothetical minerals' (Tables S1–S3) which, when reacted with the inflowing fluid (i.e., pure water), replicated the three different successive reacting fluids. The resulting chemistry matched the fluid compositions given in Table 1.

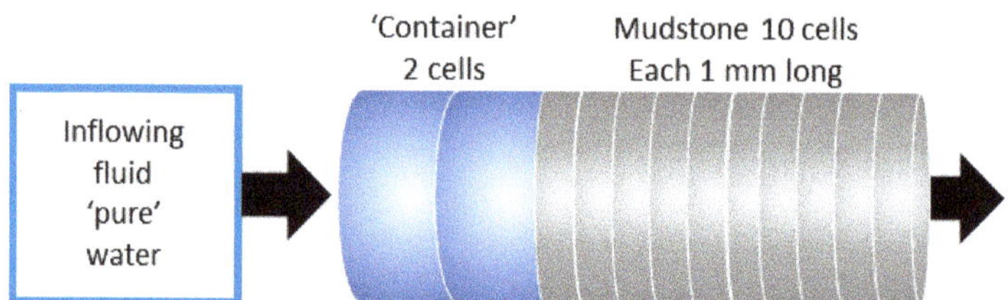


**Figure 3.** Geometry of the experimental set-up as represented in CABARET. 'Container' represents the fluid reservoir, with the mudstone divided into 10 sections, each 1 mm long and 20 mm in diameter (note: not to scale).

Although CABARET is an efficient model, as with other complex simulations, it is still desirable to minimise the number of active components (i.e., chemical species and minerals) in the system in order to obtain results in a reasonable computational time.

The choice of minerals considered in the model was determined by the analysis of the Horonobe mudstone and precipitates observed in this study together with information from previous studies of the Horonobe mudstone mineralogy [18,20,29]. In addition, previous geochemical modelling studies on various clayrock–cement fluid reactions were used to inform the choice of likely active minerals [15,30,31]. The active minerals are given in Table 2. Ten aqueous basis species ($H_2O$, $Al^{+++}$, $Ca^{++}$, $Cl^-$, $H^+$, $HCO_3^-$, $K^+$, $Mg^{++}$, $Na^+$, $Si(OH)_{4(aq)}$, $SO_4^-$) were active in the model, together with 35 related aqueous complex species. A full list of the active chemical species is available in Table S4. The choice of chemical species included was based on the chemical analysis of the reacted fluids and the compositions of the minerals included in the model.

**Table 2.** Details of the 'model' Horonobe mudstone composition and potentially active minerals.

| | Vol % | Formula in Thermodynamic Database (JAEA TDB, [26]) |
|---|---|---|
| Porosity | 56.4 | |
| $SiO_2$(am) | 27.5 | $SiO_2$ |
| Montmor | 4.18 | Montmor_Ca $Ca_{0.165}(Mg_{0.33}Al_{1.67})(Si_4)O_{10}(OH)_2$<br>Montmor_K $K_{0.33}(Mg_{0.33}Al_{1.67})(Si_4)O_{10}(OH)_2$<br>Montmor_Mg $Mg_{0.165}(Mg_{0.33}Al_{1.67})(Si_4)O_{10}(OH)_2$<br>Montmor_Na $Na_{0.33}(Mg_{0.33}Al_{1.67})(Si_4)O_{10}(OH)_2$ |
| Quartz | 4.15 | $SiO_2$ |
| Illite | 3.63 | $K_{0.6}(Mg_{0.25}Al_{1.8})(Al_{0.5}Si_{3.5})O_{10}(OH)_2$ |
| Albite | 1.91 | $NaAlSi_3O_8$ |
| K_Feldspar | 1.17 | $KAlSi_3O_8$ |
| Anorthite | 1.09 | $CaAl_2Si_2O_8$ |
| Portlandite | | $Ca(OH)_2$ |

**Table 2.** *Cont.*

| | Vol % | Formula in Thermodynamic Database (JAEA TDB, [26]) |
|---|---|---|
| CSH055<br>to<br>CSH165 | | $(CaO)_{1.65}(SiO_2)(H_2O)_{2.1167}$, $(CaO)_{1.55}(SiO_2)(H_2O)_{2.0167}$,<br>$(CaO)_{1.45}(SiO_2)(H_2O)_{1.9167}$,<br>$(CaO)_{1.35}(SiO_2)(H_2O)_{1.8167}$, $(CaO)_{1.25}(SiO_2)(H_2O)_{1.7167}$,<br>$(CaO)_{1.15}(SiO_2)(H_2O)_{1.6167}$<br>$(CaO)_{1.05}(SiO_2)(H_2O)_{1.5167}$,<br>$(CaO)_{1.00}(SiO_2)(H_2O)_{1.4667}$, $(CaO)_{0.95}(SiO_2)(H_2O)_{1.4167}$,<br>$(CaO)_{0.90}(SiO_2)(H_2O)_{1.3667}$, $(CaO)_{0.85}((SiO_2)(H_2O)_{1.3167}$,<br>$(CaO)_{0.80}(SiO_2)(H_2O)_{1.248}$,<br>$(CaO)_{0.75}(SiO_2)(H_2O)_{1.17}$, $(CaO)_{0.65}(SiO_2)(H_2O)_{1.014}$, $(CaO)_{0.55}(SiO_2)(H_2O)_{0.858}$ |
| Stratlingite_Al | | $(Ca_2Al(OH)_6)(AlSiO_2(OH)_4)(H_2O)_3$ |
| Analcime<br>(Analcite) | | $NaAlSi_2O_6(H_2O)$ |
| Clinoptilolite_alk,<br>Clinoptilolite_Ca<br>Clinoptilolite_K<br>Clinoptilolite_Na | | $K_{2.3}Na_{1.7}Ca_{1.4}(Al_{6.8}Si_{29.2}O_{72})(H_2O)_{26}$<br>$Ca_3(Al_6Si_{30}O_{72})(H_2O)_{20}$<br>$K_6(Al_6Si_{30}O_{72})(H_2O)_{20}$<br>$Na_6(Al_6Si_{30}O_{72})(H_2O)_{20}$ |
| Phillipsite_alk<br>Phillipsite_Ca<br>Phillipsite_K<br>Phillipsite_Na | | $K_{1.4}Na_{1.6}Ca_{0.4}(Al_{3.8}Si_{12.2}O_{32})(H_2O)_{12}$<br>$Ca_3(Al_6Si_{10}O_{32})(H_2O)_{12}$<br>$K_6(Al_6Si_{10}O_{32})(H_2O)_{12}$<br>$Na_6(Al_6Si_{10}O_{32})(H_2O)_{12}$ |
| Brucite | | $Mg(OH)_2$ |
| MSH06<br>to<br>MSH15 | | $(MgO)_{0.6}(SiO_2)(H_2O)_{1.08}$, $(MgO)_{0.7}(SiO_2)(H_2O)_{1.2}$, $(MgO)_{0.8}(SiO_2)(H_2O)_{1.32}$<br>$(MgO)_{0.9}(SiO_2)(H_2O)_{1.44}$, $(MgO)_1(SiO_2)(H_2O)_{1.56}$, $(MgO)_{1.1}(SiO_2)(H_2O)_{1.68}$<br>$(MgO)_{1.2}(SiO_2)(H_2O)_{1.8}$, $(MgO)_{1.3}(SiO_2)(H_2O)_{1.92}$, $(MgO)_{1.4}(SiO_2)(H_2O)_{2.04}$<br>$(MgO)_{1.5}(SiO_2)(H_2O)_{2.16}$ |
| Monocarbonate_Al | | $(Ca_2Al(OH)_6)_2(CO_3)(H_2O)_5$ |
| Monosulfate_Al | | $(Ca_2Al(OH)_6)_2(SO_4)(H_2O)_8$ |
| Magnesite | | $MgCO_3$ |
| Thaumasite | | $Ca_3Si(OH)_6(SO_4)(CO_3)(H_2O)_{12}$ |
| Calcite | | $Ca(CO_3)_2$ |
| Dolomite | | $CaMg(CO_3)_2$ |
| Gypsum | | $CaSO_4(H_2O)_2$ |
| Ettringite_Al | | $Ca_6(Al(OH)_6)_2(SO_4)_3(H_2O)_{26}$ |
| Friedel_Salt_Al | | $(Ca_2Al(OH)_6)_2(Cl)_2(H_2O)_4$ |
| Kuzel_Salt_Al | | $(Ca_2Al(OH)_6)_2((SO_4)_{0.5}Cl)(H_2O)_6$ |

Thermodynamic data (equilibrium constants and standard molal volume for minerals) were taken from the JAEA-TDB [26] version PHREEQC20.dat (v1.2, 11 March 2021). Kinetic rates for primary minerals were taken from [32,33] and the references within, those for montmorillonite from [34] and C-S-H from [35]. Data for surface areas were from [34,36]. Ion exchange of Na, K, and Ca from the reactant fluids with the clays in the mudstone was included, and the ion exchange parameters for montmorillonite were taken from [37].

In terms of the description of diffusion, including the effect of tortuosity, within CABARET, this was derived from [38] with dispersion and advection as described in [39].

## 3. Results

### 3.1. Aqueous Chemistry

The chemical evolution of the three flow-through experiments was nearly identical when comparing the same fluid type. That is, the fluid analytical data for the 'young' OPC leachate were the same for SFC-1, -2, and -3 experiments, and the data for the 'evolved' OPC leachate from the SFC-2 and 3 experiments were identical. The results of the SFC experiments with the 'young' OPC leachate have previously been discussed in detail in [14].

### 3.1.1. Changes in pH

Figure 4a shows the evolution of pH with time for the experiment with the 'young' OPC leachate, followed by the 'evolved' OPC leachate and then HGW (SFC-3). In all three experiments, there was an initial decrease to pH ~4 due to the presence of dissolution of sulphate phases, the discussion of which can be found in [14]. The pH recovered from the low values within the first 24 h to pH ~13.2. The pH reduction with the 'young' OPC leachate was ~0.2 pH units lower than the original leachate, and a similar drop was seen with the 'evolved' OPC leachate (pH ~12.4). With the change to HGW, the pH decreased, but even after ~36 d of reaction, the measured pH ~9.25 was still higher than that of the inflowing groundwater (pH ~8.6).

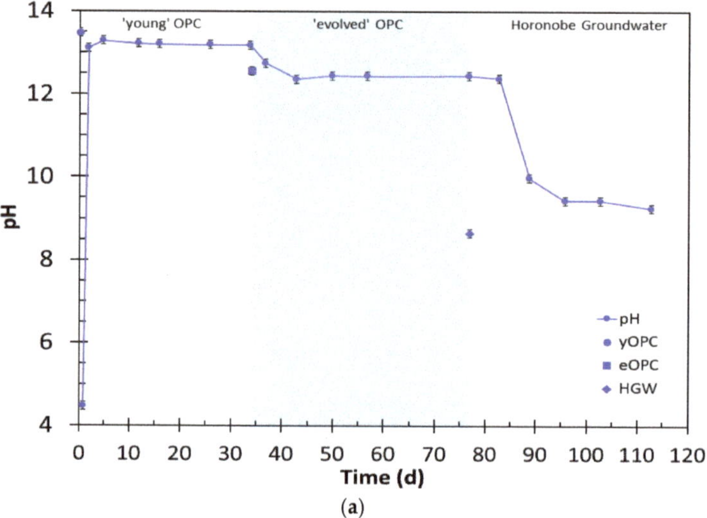


**Figure 4.** *Cont.*

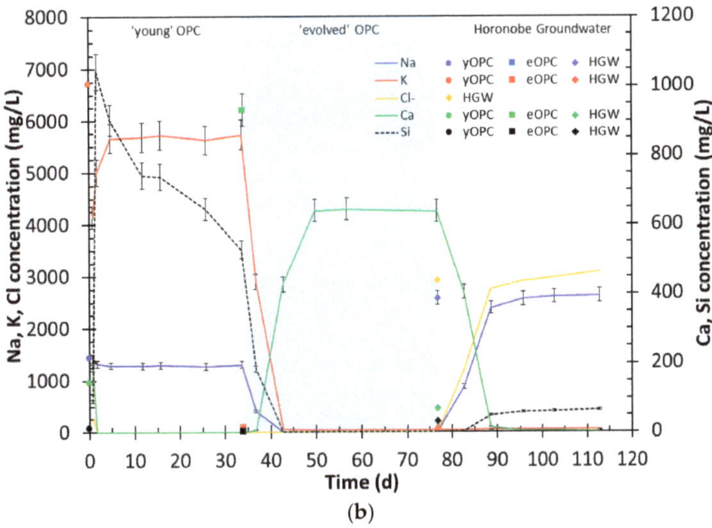



**Figure 4.** Major changes in fluid chemistry with time. Horonobe mudstone with 'young' OPC, and then the 'evolved OPC leachate followed by HGW. (**a**) pH; (**b**) [Na], [K], [Ca], [SiO$_2$], and [Cl]. Legend text: yOPC—'young' OPC leachate; eOPC—'evolved' OPC leachate; HGW—Horonobe groundwater. Lines indicate concentrations in reacted samples; single points the original concentration in the reacting fluids.

### 3.1.2. Changes in Na Concentration

The changes in Na concentrations in the reacted fluids are shown in Figure 4b. In all three experiments, [Na] decreased from the concentration in 'young' OPC leachate to ~1300 mg/L and remained at close to this value until the inflowing fluid was changed to the 'evolved' OPC leachate (Figures 4b and S1) when, as expected, [Na], reflecting the lack of Na in the 'evolved' OPC leachate, decreased to ~10 mg/L (analytical detection limit) (Figures 4b and S2). With the change to HGW, [Na] slowly rose over the next 25 days (Figure 4b) to close to that of the HGW (~2600 mg/L). This time to match the incoming fluid concentration was slower than the time taken for [Na] to reach steady state with the change from the 'young' to the 'evolved' OPC leachate, which was only ~10 d.

### 3.1.3. Changes in K Concentration

With the 'young' OPC leachate, [K] initially decreased to ~4100 mg/L, possibly owing to ion exchange reactions with the clays in the mudstone [14], before increasing to ~5700 mg/L (Figures 4b and S1) and remained, within analytical error, at this concentration until the fluid was changed to the 'evolved' OPC leachate (SFC-2 and -3) when the [K] dropped to <40 mg/L (Figures 4b and S2). As with [Na], with the change to HGW (SFC-3), [K] recovered to close to the concentration present in the groundwater.

### 3.1.4. Changes in Ca Concentration

With both of the OPC leachate types, [Ca] in the reacted fluids was significantly lower than of the original OPC leachates (<25 mg/L for the 'young' OPC leachate and ~640 mg/L for the 'evolved' OPC leachate (Figure 4b, Figures S1 and S2)). When HGW replaced the OPC leachate, [Ca] decreased over the next 25 days to <5 mg/L, mirroring the changes seen in [Na], and remained at this value for the remaining duration of the experiment (Figure 4b).

### 3.1.5. Changes in Silica Concentration

In all the experiments (Figure 4b, Figures S1 and S2), the silica concentrations increased rapidly in the first 6 days of the reaction to high levels ~1000 mg/L, and then slowly decreased to ~500 mg/L at the time of the change to the evolved' OPC leachate (~40 d). With the 'evolved' OPC leachate, silica decreased further, so that, after 7 d of flow with the evolved' OPC leachate, silica concentrations were <3 mg/L (Figures 4b and S2). When the inflowing fluid was changed to HGW, silica concentrations in the reacted fluids increased over 14 d to approximately twice that of the HGW, i.e., ~130 mg/L.

### 3.1.6. Changes in Other Ions

With the exception of Rb and Sr, which tracked the concentrations of Na and Ca, respectively, the concentrations of the remaining cations analysed (e.g., Al, Ba, Cs, Cu, Fe, Li, Mg, Mn) showed no significant change from the concentrations in original fluids.

In general, the concentrations of anions (Br, $NO_2^-$, $NO_3^-$, and F) reflected the respective concentrations present in the OPC leachates and groundwater. With the 'young' OPC leachate, sulphate concentrations in the first fluid samples collected showed an early large peak (up to ~800 mg/L), but, after 7 d, sulphate concentrations were <5 mg/L and showed no significant change during the subsequent reaction with neither the 'evolved' OPC leachate nor HGW.

In all three experiments, phosphate showed a small peak (~10 days reaction) before decreasing to <1 mg/L. Previous studies [29] reported rare apatite in some samples of Horonobe mudstone, the dissolution of which may have been the source of this phosphate peak.

The behaviour of Cl (Figure 4b) with the change to HGW was similar to that of Na, and took ~25 d to reach the concentration of the original HGW (~2900 mg/L).

Bicarbonate concentrations in the collected fluids from the reaction with both OPC leachates were below detection, but once the fluid was changed to HGW, $[HCO_3^-]$ increased to ~1300 mg/L over 14 d, but did not reach the concentration of the HGW (~1900 mg/L).

### 3.2. Mineralogical Analysis

### 3.2.1. 'Young' OPC Leachate Experiment (SFC-1)

A summary of the evolution of the mineralogy for the SFC-1 is shown in Figure 5 and discussed in greater detail in [14]. In the first section (~0–1.5 mm), there was evidence for the dissolution of the primary material, and no secondary phases were observed. However, in the next section, ~1.5–3 mm, a wide variety of secondary C-S-H and C-A-S-H phases were observed (Figure 5 and [14]). By section 3, ~3–5 mm, there was no further evidence for additional secondary phase precipitation, and mineral surfaces showed little sign of reaction.

### 3.2.2. 'Young' and Then 'Evolved' OPC Leachate Experiment (SFC-2)

A summary of the evolution of the mineralogy for the experiment with 'young' OPC leachate, followed by the 'evolved' OPC leachate (SFC-2), is shown in Figure 6. Again, the mudstone samples were cut into ~1.5 mm thick slices before being prepared for subsequent mineralogical characterisation.

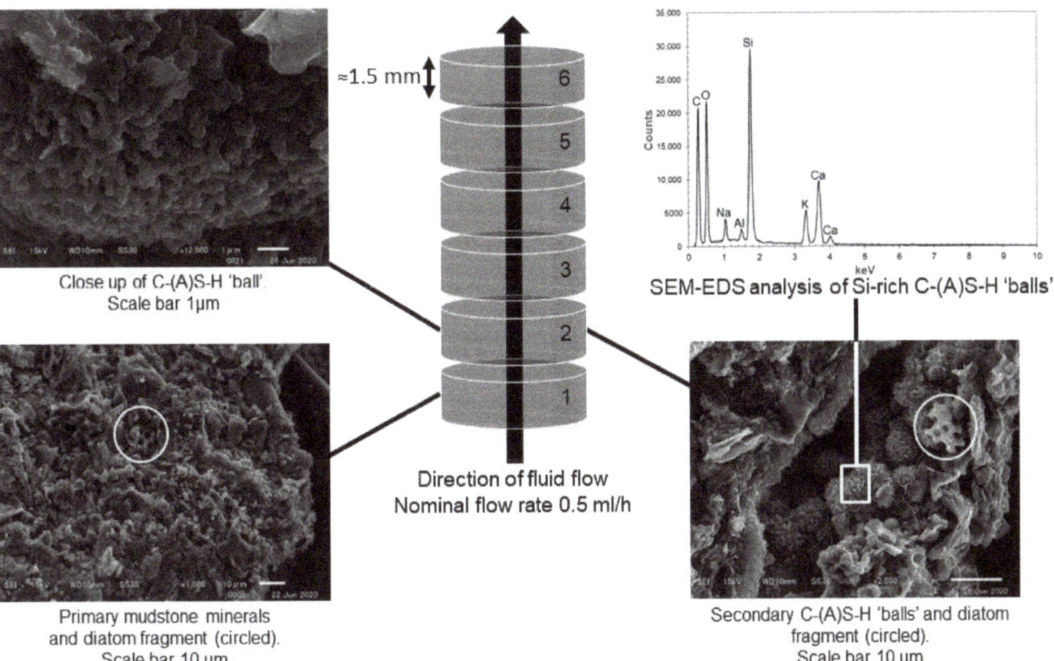



**Figure 5.** Summary of mineralogy of Horonobe mudstone with 'young' OPC leachate. All SEM photos are secondary electron (SE) images. The white square in the SEM image from Section 2 indicates the area used for SEM-EDS analysis of C(A)–S–H phases (reproduced from [14]).

In the first section (~0–1.5 mm), a variety of C-(A)-S-H phases were observed (Figure 6), identified by semi-quantitative SEM-EDS analysis. These phases varied in Ca/Si, Al content, and morphology. By section 3 (~3–4.5 mm), as with section 1, a variety of C-(A)-S-H phases were observed, but also present were 'needle-like' crystals. SEM-EDS analysis (Figure 6) suggested that they had a high S content, possibly representing an ettringite or monosulphate-like composition. Both ettringite and monosulphate are often found in hydrated Portland cement pastes [40]. XRD analysis (Figure S2) proved inclusive as to which phase(s) was present, but the SEM-EDS analysis suggested Si was also present. Ettringite has been shown to be able to accept the replacement of up to half of its Al by Si [41], which suggests ettringite as the likely phase, though monosulphate is still a possibility if the Si detected was due to the presence of the underlying primary mineral. An examination of section 6 (~8.5–10 mm) by SEM again showed the presence of a needle-like sulphur bearing phase and a variety of C-S-H and C-A-S-H phases (Figure 6).

PHREEQC was used to calculate mineral SI in the reacted fluids based on the analysed fluid chemistry (Figure S3). The calculations showed that, for most primary mineral phases, the degree of undersaturation increased in the 'evolved' OPC leachate, suggesting more dissolution compared with the 'young' OPC leachate. In addition, the modelling suggested that, as well as C-S-H phases being saturated or close to saturation, both ettringite and monosulphate would also be saturated in the collected fluids (Figure S3) when reacted with the 'evolved' OPC leachate.

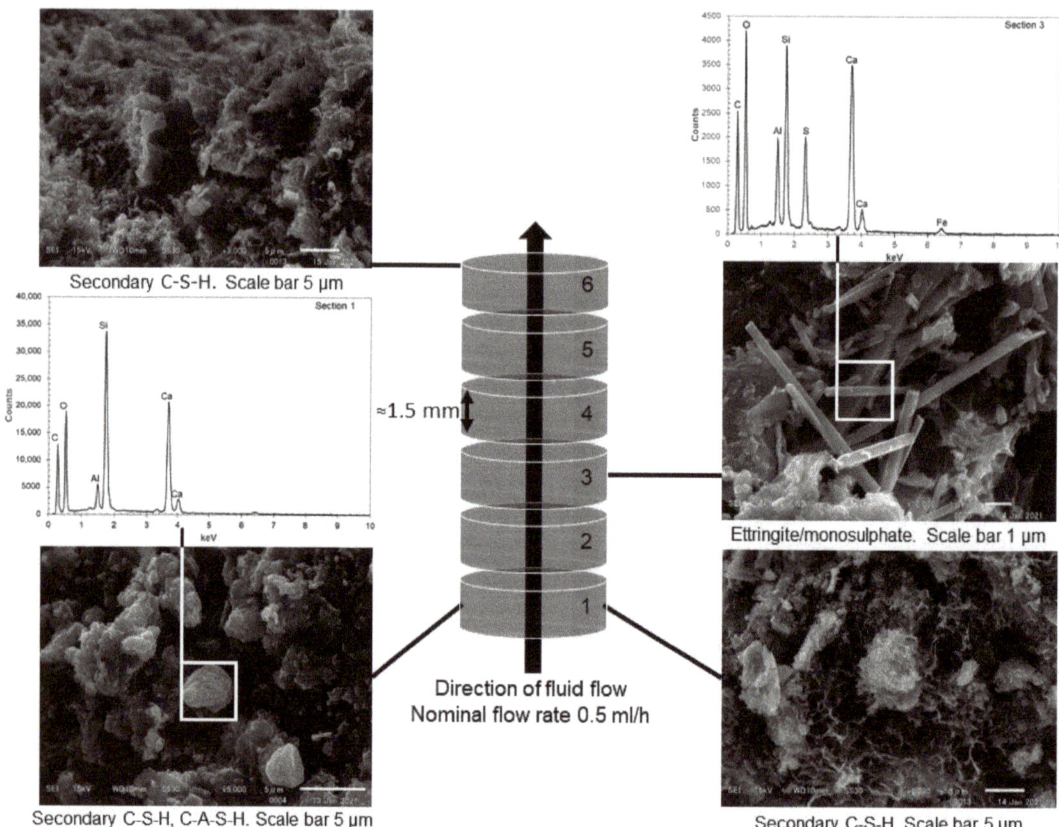

**Figure 6.** Summary of mineralogy of Horonobe mudstone with 'young' OPC (Na-K-Ca-OH) and then 'evolved OPC leachate (Ca(OH)$_2$) (SFC-2). All SEM photos are secondary electron (SE) images. The white square in the SEM images from Sections 1 and 3 indicate the areas used for SEM-EDS analysis of the secondary phases.

### 3.2.3. OPC Leachates Then HGW (SFC-3)

A summary of the evolution of the mineralogy for the experiment with the 'young' OPC and then the 'evolved' OPC leachate followed by HGW (SFC-3) is shown in Figure 7. Unlike the experiments with the OPC leachates (SFC-1,2), it was difficult, owing to partial cementation of the sample, to prepare equally sliced samples for mineralogical characterisation.

In the first section (~0–1.0 mm), many surfaces showed signs of significant secondary precipitation. SEM-EDS analysis (Figure 7) of these precipitates showed them to be aragonite/calcite, sometimes containing small amounts of Mg, of varying morphologies. Most of these crystals were small, ranging from around ~2 to ~10 μm, suggesting fairly rapid formation. As the HGW contained significant carbonate and the mudstone was saturated with the 'evolved' OPC leachate, it was expected that carbonate precipitation would occur, especially as, in the HGW itself, calcite was close to saturation. Indeed, calculation of mineral states, using PHREEQC, indicated that carbonates were close to saturation (Figure 8) in the collected reacted fluids. Apart from aragonite/calcite (Figures 7 and S4), no other secondary phases were found, and where visible, the primary mineral surfaces appeared to be clean of 'fines'. The secondary C-S-H phases, and ettringite/monosulphate, observed in the experiments with the OPC leachates (SFC-2) appeared to have re-dissolved in this section.

In Section 2 (~1.0–2.5 mm), again, 'clean' diatoms were visible. Secondary precipitates of differing compositions were also found. SEM-EDS analysis of these precipitates was sometimes inconclusive, with some rare precipitates having compositions that could be attributed to high Ca/Si, C-S-H phases, and/or calcite overlying primary silicate minerals. The examination of the saturation state of various C-S-H phases in the reacted fluids suggested that all C-S-H phases were undersaturated (Figure 8), and thus should have re-dissolved. Elsewhere in this section, secondary calcites of different morphologies, sometimes containing Mg, were found. Some larger carbonate crystals (Figure 7) with diameters of 40–50 µm were also observed, suggesting that these had formed more slowly than those observed in the first section of the SFC.

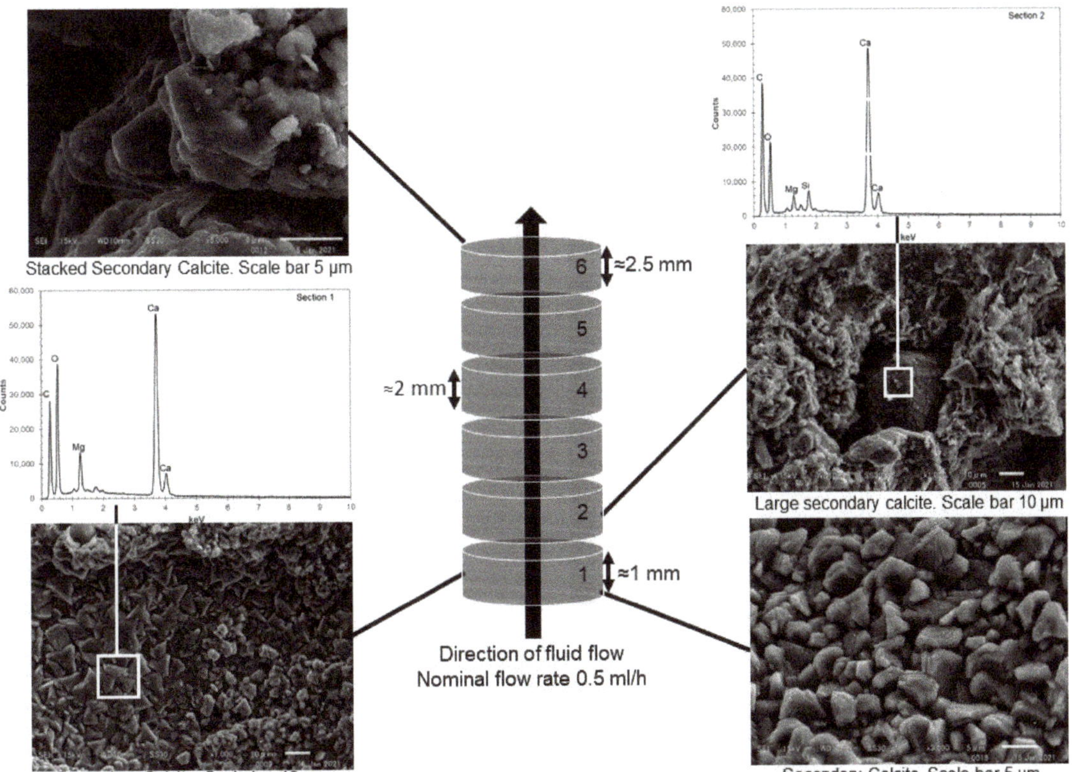


**Figure 7.** Summary of mineralogy of Horonobe mudstone with 'young' OPC (Na-K-Ca-OH), then 'evolved OPC leachate (Ca(OH)$_2$), and then HGW (SFC-3). Except where indicated, each section was ~1.5 mm. All SEM photos are secondary electron (SE) images. The white square in the SEM images from Sections 1 and 2 indicate the areas used for SEM-EDS analysis of the secondary carbonates.

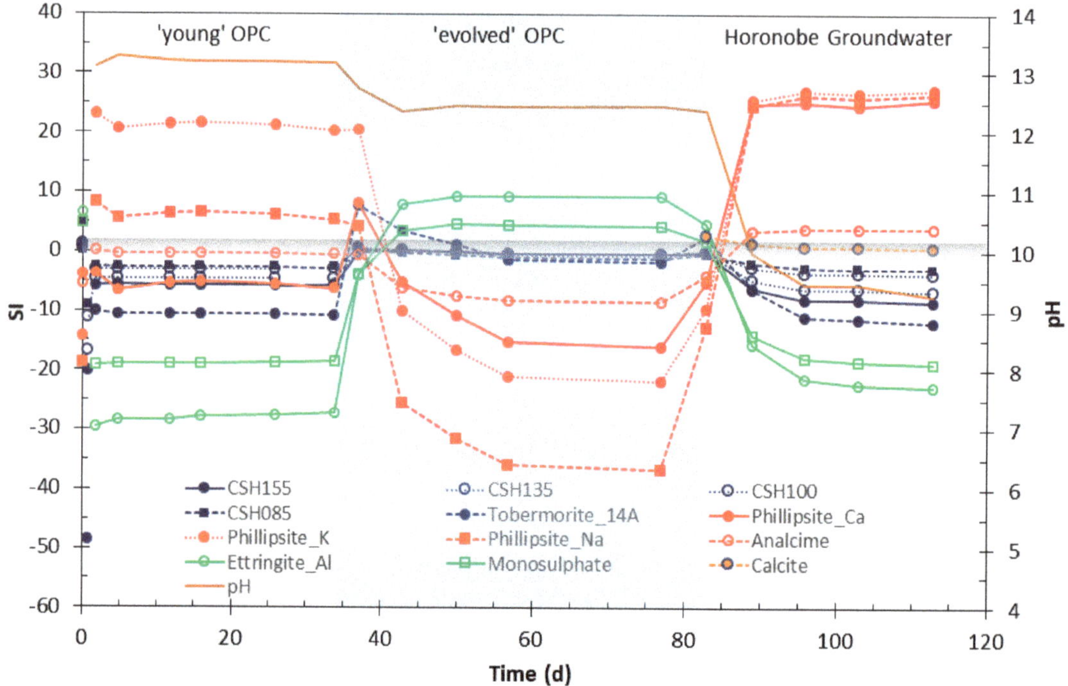


**Figure 8.** Selected secondary mineral saturation states in reacted fluids, experiment with 'young' OPC (Na-K-Ca-OH), and then the 'evolved OPC leachate (Ca(OH)$_2$), followed by HGW; all data from SFC-3. Representative C-S-H phases in blue; zeolites in red; and S-bearing cement phases in green.

By section 6 (~7.5–10 mm), 'stacked' carbonate grains (calcite) were frequently observed; again, some of these crystals were large (typical diameters of 30–40 µm), suggesting slow sustained growth. Although zeolites were identified as possible secondary minerals from the calculations for the mineral saturation states (Figure 8), no evidence was found for their formation, though it should be noted that the available thermodynamic data for many zeolites remain poorly known [17]. However, given the much faster precipitation rate of both aragonite and calcite relative to zeolites, carbonates would be highly favoured over zeolites as secondary phases over experimental timescales. In addition, other studies have suggested that higher temperatures (>60 °C) may be required for zeolite formation [42].

## 4. Results of Reactive Transport Modelling

To further examine the reaction sequence seen in these experiments and to extend the timescale beyond that possible in the laboratory experiments, coupled 1D reactive transport models were constructed using the 'CABARET' software.

*Model Predictions vs. Experimental Data*

Figure 9 shows the comparison of the results of the CABARET model compared with the experimental data for the major aqueous components. The model predictions for the evolution of pH (Figure 9a) were a good match for the experimental data. Both Na and K predicted concentrations were also a good match, especially for the OPC leachates (Figure 9a). However, Na with the HGW was predicted to increase faster to the concentration of the HGW than the experimental data. Calcium concentrations (Figure 9b) were also a reasonable match to the experimental data for the OPC leachates, although predicted [Ca] with the 'evolved' OPC leachate were slightly lower than in the experiments.

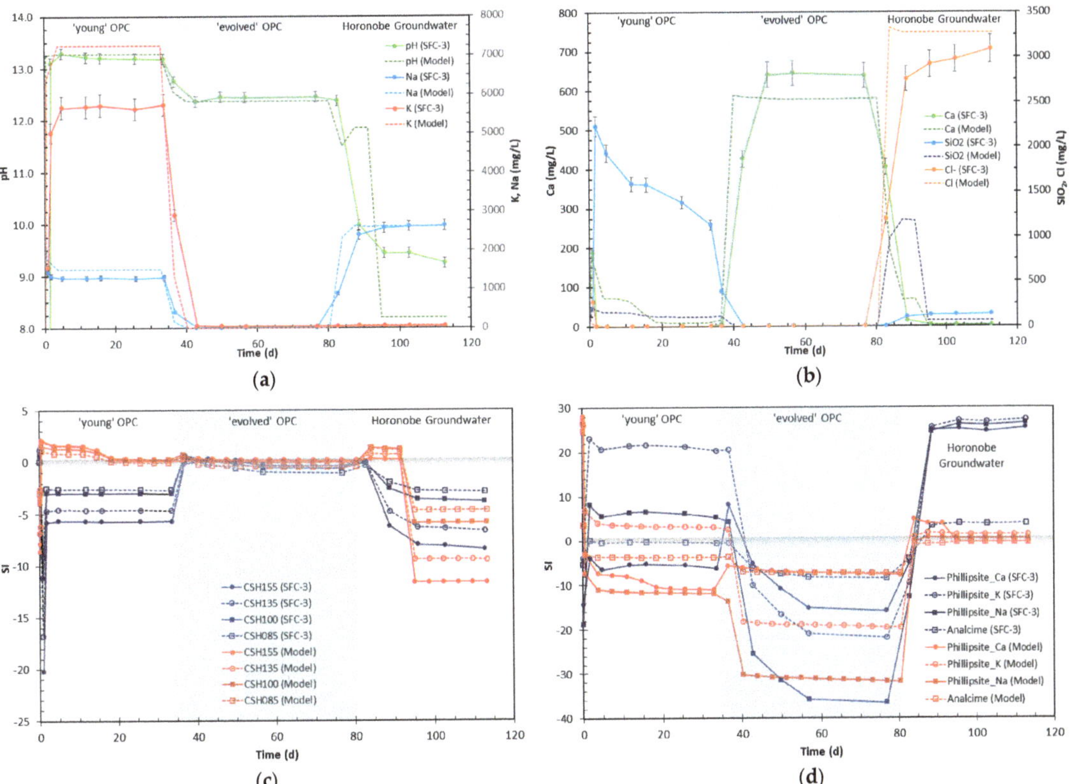

**Figure 9.** Summary of the model predictions compared with the experiment data (**a**) pH, [Na], and [K]; (**b**) [Ca], [SiO$_2$], and [Cl]; (**c**) selected C-S-H saturation indices; (**d**) zeolite saturation indices.

Silica concentrations were more difficult to match with the experimental data (Figure 9b). Although an initial increase in silica concentration was predicted by the CABARET model, this was significantly lower than the concentration observed in the experiments; the reason(s) for this mismatch are not clear, but one possible explanation could be that, in the experiments, the crushing process produced many highly reactive 'fines' that would have reacted before the bulk of the mudstone.

In addition, with the HGW, silica concentrations in the CABARET model initially rose to ~1200 mg/L (~2 × 10$^{-2}$ mol/kg) compared with ~140 mg/L (~2.3 × 10$^{-3}$) in the experiments, before decreasing after 100 d to ~63 mg/L (~7.65 × 10$^{-4}$ mol/kg), which was about half the concentration observed in the experiments.

Chloride concentrations with the HGW (Figure 9b) showed an almost immediate increase from <1 to ≈3200 mg/L in the model rather than the delayed increase seen in the SFC-3 experiment, showing a similar behaviour to sodium.

Comparison of the saturation indices (SI) for selected C-S-H and zeolite phases derived from the fluid chemistry (see Figure 8) and CABARET are shown in Figure 9c,d. Although the absolute numerical values for the mineral SI are often different, the trend in the model is a reasonable match for the experiment derived SI data, with the major difference being that, in the experiments, zeolites were oversaturated compared with CABARET model predictions.

Figure 10 shows the predicted variation of porosity and selected minerals with distance, at defined time steps. The variation in porosity with time and distance (Figure 10a) predicts that, for the 'young' OPC leachate (0 to 20 d) from 0 to 2.5 mm, the porosity would increase, i.e., dissolution of the primary minerals after 3 mm, the porosity decreases, indicating formation of a secondary phases(s). From ~7 to 10 mm, the porosity remains close to but slightly higher than the starting value (56.4%), suggesting continued partial dissolution of the mudstone minerals.

The predicted changes in porosity mirrored the prediction of C-S-H phase precipitation (Figure 10b), which was predicted at time step of 40.2 (d), i.e., the change from 'young' OPC to 'evolved' OPC to be the greatest, ~3–4.5 mm, which matches well with the mineralogical observations from the flow experiment with the 'young' OPC leachate (SFC-1) (Figure 5).

With the 'evolved' OPC leachate, the model suggests continued dissolution of the mudstone and increased precipitation of C-S-H phases (Figure 10a,b; time steps of 40.2, 58.4, and 76.7 (d)) with C-S-H phase precipitation extending throughout the flow cell (Figure 6), but still with its maximum between 2.5 and 4.5 mm.

With the change to HGW (Figure 10a; time steps of 94.9 and 113.2 (d)), the porosity close to the inlet (0–1 mm) decreases in direct response to the predicted formation of secondary calcite (Figure 10c), and because calcite is predicted to form throughout the cell, porosity is also reduced along the whole flow path. Again, this was in good agreement with the experimental observations of calcite precipitation (Figure 7). Also shown in Figure 10 are exemplar data for the primary minerals, which show continued dissolution with time and distance (Figure 10d,e).

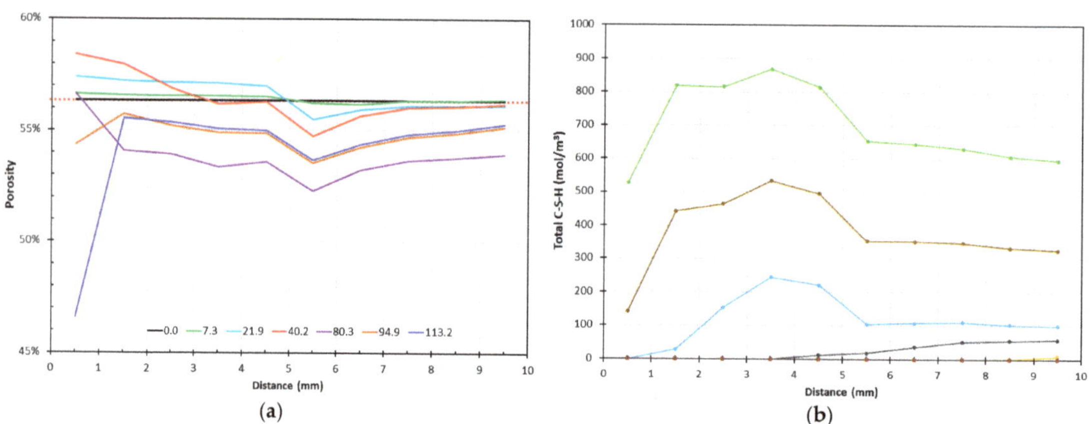

**Figure 10.** *Cont.*

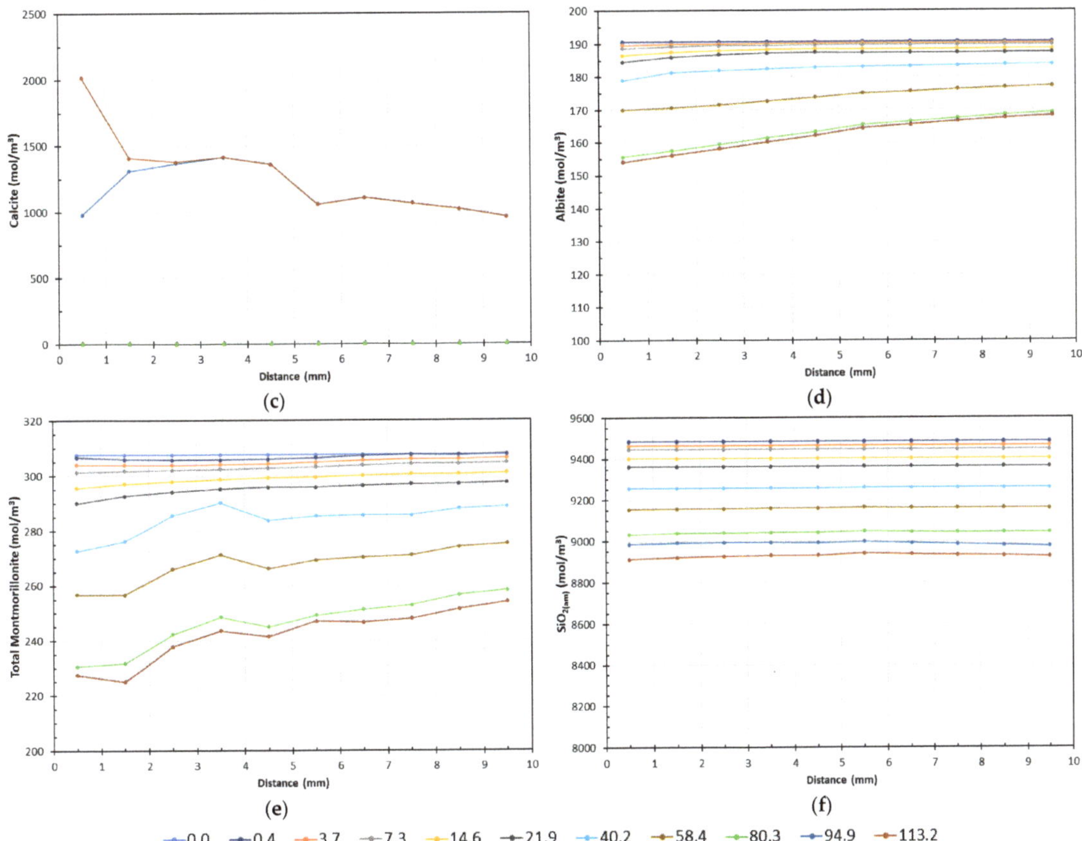



**Figure 10.** Legend for all plots, except porosity; (**a**) time step (days), selected model data showing variation along the length of SFC with time; (**a**) porosity, red dashed line indicates initial porosity; (**b**) total C-S-H present; (**c**) calcite; (**d**) albite (as representative of the feldspars); (**e**) total montmorillonite; and (**f**) amorphous silica. Time step of 40.2, (**d**) change from 'young' to 'evolved' OPC leachate; time step of 80.3, (**d**) change from "evolved' OPC leachate to HGW.

Figure 11 shows summary plots of the changes in composition of modelled mineral assemblage for the unreacted mudstone and after reaction with each fluid (i.e., time 0, ~40, ~80, ~120 d). This illustrates the variation in the relative proportions of the primary and secondary minerals with time and distance. The increase in volume of C-S-H phases precipitated with the 'evolved' OPC leachate compared with the 'young' leachate is evident, as is the complete removal of C-S-H phases and replacement by calcite following reaction with HGW, which is the same sequence observed in the experiments (see Figures 5–7). In addition, primary minerals, though showing slight dissolution, dominate the mineral assemblage throughout.

Zeolites were predicted to form, but, as noted above (Section 3.2.3), carbonates would be favoured over zeolites as potential secondary phases.

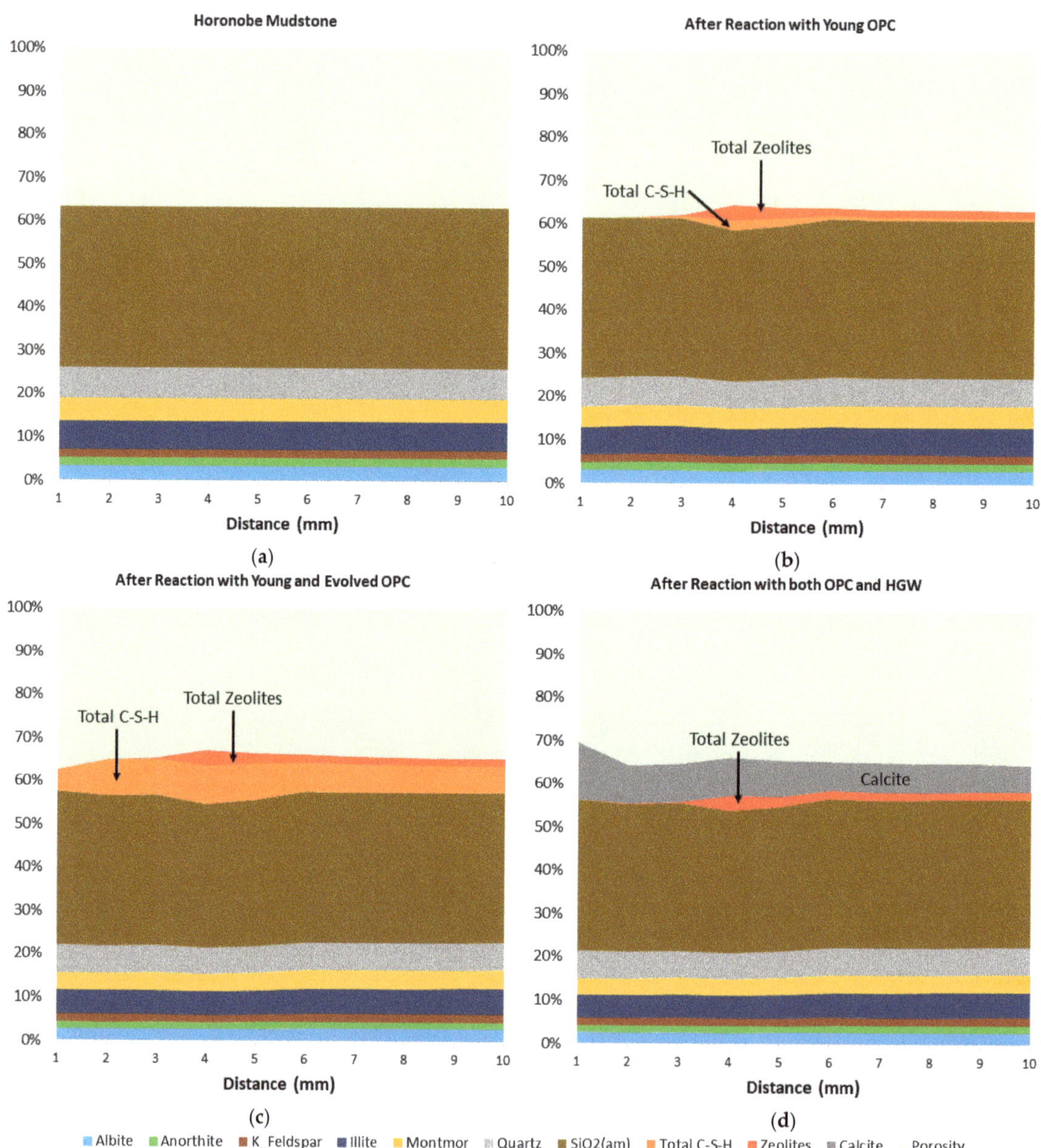



**Figure 11.** Legend for all plots. Summary of the composition of modelled mineral assemblage: (**a**) Horonobe mudstone before reaction; (**b**) after reaction with young for 40 d; (**c**) after successive reaction with both OPC leachates; and (**d**) after successive reaction with both OPC leachates and Horonobe groundwater.

## 5. Discussion

In summary, the experimental data obtained from this study on the reaction of mudstone with highly alkaline fluids are consistent with previous modelling and experimental studies on other host rocks [9–11,15,39], although the Horonobe mudstone is less reactive than some of the mineral assemblages previously investigated, i.e., Borrowdale Volcanic Group; Äspö Granite; Wellenberg marl [10], Opalinus Shale [11]; and a generic crystalline rock [15].

### 5.1. Chemistry and Mineralogy

The pH with the OPC leachates was buffered to <0.2 pH units below that of the unreacted leachates, and similar changes in pH have been observed in other studies with OPC leachates [9–11,14,15]. However, with the change to Horonobe groundwater, the pH of the reacted fluids did not, within the timeframe of this study, return to that of the HGW (pH~8.6), but remined slightly higher at pH~9.4 (Figure 4a), suggesting that there was some longer-term buffering of the pH by the remnant secondary mineral assemblage present after reaction with the successive OPC leachates, i.e., the presence of C-(A-)S-H phases.

It was also noticeable that, for [Na] and [Cl] (Figure 4b), there was a delay before the dissolved concentrations increased towards the levels present in the HGW (Figure 4b). As chloride is normally considered to be a conservative species, this suggests that the lag in the experiments is due to a physical/transport process rather than a chemical one. A possible explanation for this is that there was a slow mixing of the HGW with OPC leachate(s) trapped within the pore spaces by the precipitated C-(A-)S-H phases that continued until the phases completely dissolved, thereby releasing the trapped fluid; this would also result in a slightly higher pH than expected with the HGW. Previously, it has been reported that the presence of C-A-S-H phases caused fluid stagnation and poor flow in some pores/voids, resulting in some cases in the persistence of secondary precipitates [15]. The model included porosity in solute transport processes, but did not consider heterogeneity in the porosity distribution.

The sequential reaction of the mudstone with the OPC leachates was dominated by the initial precipitation of C-(A-)S-H phases of varying compositions (SEM-EDS, semi-quantitate analysis), with accompanying variations in the concentration of Ca and Si in the reacted fluids. This general sequence was also predicted by the modelling, allowing direct comparison of the experimental system evolution with the reactive transport simulations. With the change to the groundwater and the consequent reduction in pH, the secondary C-(A-)S-H phases re-dissolved, confirming model predictions that they would dissolve, owing to the lower pH of the groundwater, and be replaced by secondary carbonates.

In the experiments, [Ca] remained below that of the HGW owing to the formation of these carbonate minerals, and [Si] increased as the C-(A-)S-H phases dissolved. However, the simulated magnitude of the [Si] increase and the subsequent decrease was greater than that seen in the experiments, suggesting that there was either a kinetic inhibition to the dissolution of C-(A-)S-H phase(s) in the experiments or that a physical mechanism [15] not simulated by the model, as discussed above for [Na] and [Cl], was controlling the dissolution, and hence [Ca] and [Si]. This clearly illustrates the importance of being able to compare model predictions with analytical data.

### 5.2. Extent of Reaction

While, with the 'young' OPC leachate, the extent of reaction in the experiments was limited to only the first few millimetres (Figure 5), there was greater reaction with the 'evolved' OPC leachate, and thus greater secondary C-(A-)S-H precipitation (Figure 6), reflecting the higher [Ca] of that leachate. Although observed throughout, the secondary C-(A-)S-H phases were still, at least subjectivity, most prevalent in the first few millimetres. The model simulations showed a similar trend with regard to both the extent and degree of reaction (precipitation) to the experimental data.

In a previous study working with Opalinus clay (OPA) and a similar 'young' OPC leachate as used in this study [11], it was found that, even after 18 months, the precipitation zone within the OPA was limited to <2 cm, with C-A-S-H phases together with Ca-carbonate, portlandite, and brucite identified. Similarly, it has been previously observed during the reaction of Na-montmorillonite with simulated OPC and OPA porewaters [43] that porosity changes were limited to ~2 mm penetration into the clay, and comparable observations have also been made in related studies [43–46]. In a review of cement-clay modelling with high pH fluids [47], it was noted that many reactive transport models predicted limited zones of alteration similar to the extent seen in the experiments undertaken in this study.

However, the higher than expected pH in the reacted fluids and the trend in Ca and silica concentrations following the change to groundwater suggest that some C-(A-)S-H phases may persist for a longer time than predicted by the modelling as a result of a physical mechanism. A probable explanation for this is that the secondary phases will have formed in the pore spaces, blocking some fluid flow paths, thus reducing the impact of the inflowing groundwater, and furthermore that formation in voids will reduce the reactive surface area of the secondary phases, further slowing their dissolution [15]. Again, this illustrates the importance of having analytical data with which to validate model predictions.

The experiments provided mineralogical data only at four time steps (i.e., at the start and at ~40, ~80, and ~120 d) and the outflow fluid chemistry was only sampled intermittently. As the CABARET model predictions were in general agreement with the experimental data at these time steps, the model can then be used to examine how the chemical and spatial changes may have evolved between these data points, as well as to explore the changes in fluid chemistry within the flow cell, which are difficult (at least at the millimetre scale of the flow cell used here) to determine experimentally.

### 5.3. Recommendations for Model Improvements and Validation

The CABARET reactive transport model has been demonstrated to be in general agreement with the experimental data, which gives greater confidence in the potential use of this code to make predictions at greater scale. However, common with other reactive transport models, e.g., PHREEQC [25] and TOUGHREACT [48] and others, there is scope for improving the agreement with experimental data.

For example, in CABARET, the surface area for each mineral is assigned a fixed value in the initial set-up file, and this does not subsequently vary with dissolution or precipitation. Reactive-transport models are able to couple diffusive and advective flow to porosity, but processes such as heterogenous pore occlusion and secondary minerals coating primary solids are not considered, with only bulk porosity being able to evolve in each model cell.

The timing of the sequence of host rock alteration resulting in the formation of C-(A-)S-H phases, which later re-dissolve with the change to groundwater and the release of any sorbed radionuclides, requires further investigation, as the experimental data from this study suggest that this process may occur over a somewhat longer time than predicted by the reactive transport modelling.

To fully understand the temporal and spatial extent of the reactions on mineral evolution (precipitation and dissolution) and the subsequent effect on radionuclide migration requires longer duration experiments with intact samples, under carefully controlled laboratory conditions, and more realistic 'in situ' experiments to provide data for model validation. Further improvements to the available simulation software are also required to analyse and extend the results from such experimental programmes to repository scales.

**Supplementary Materials:** The following are available online at https://www.mdpi.com/article/10.3390/min11091026/s1, Figure S1: Major changes in fluid chemistry with time. Horonobe mudstone with 'young' OPC leachate (SFC-1) and Horonobe mudstone with 'young' OPC, and then 'evolved' OPC leachate (SFC-2). Figure S2: XRD analysis of unreacted and reacted mudstone samples experiment with 'young' OPC, and then the 'evolved' OPC leachate (SFC-2). Figure S3: Selected primary mineral and C-S-H phase saturation states in reacted fluids, experiment with 'young' OPC, and then the 'evolved' OPC leachate (SFC-2). Figure S4: XRD analysis of unreacted and reacted mudstone samples experiment with 'young' OPC, and then the 'evolved' OPC leachate, followed by HGW (SFC-3). Table S1: Calculated Log K for the hypothetical minerals; 'young' OPC leachate. Table S2:

Calculated Log K for the hypothetical minerals; 'evolved' OPC leachate. Table S3: Calculated Log K (for the hypothetical minerals); Horonobe groundwater. Table S4: Details of the dissolved chemical species included in the CABARET reactive transport model.

**Author Contributions:** The individual contributions to this paper are as follows: Conceptualization and methodology, K.B., Y.A., Y.T., Y.H., M.K., J.W. and T.S.; validation, K.B., S.M., M.K. and Y.O.; investigation and formal analysis, K.B., S.M., Y.H., M.K., J.W., T.S. and Y.O.; resources, Y.T. and Y.A.; data curation, K.B., S.M., Y.H., J.W., T.S. and Y.O.; writing—original draft preparation, K.B.; writing—review and editing, all authors. All authors have read and agreed to the published version of the manuscript.

**Funding:** This study was partly performed as a part of "The project for validating near-field assessment methodology in geological disposal (FY2020)" supported by the Ministry of Economy, Trade, and Industry of Japan.

**Data Availability Statement:** The data used in this study are available from the authors upon request.

**Acknowledgments:** We thank Hikari Beppu and Takashi Endo for assistance with fluid chemical analysis. This study was supported by the JAEA Horonobe Underground Research Center, Hokkaido, Japan, providing the rock samples and background information.

**Conflicts of Interest:** The authors declare no conflict of interest.

# References


1. JNC. H12: Project to Establish the Scientific and Technical Basis for HLW Disposal in Japan, Project Overview Report and three Supporting Reports. JNC TN1410 2000–001~004. 2000. Available online: https://jopss.jaea.go.jp/pdfdata/JNC-TN1410-2000-003 .pdf (accessed on 30 July 2021).
2. Baker, A.J.; Bateman, K.; Hyslop, E.K.; Ilett, D.J.; Linklater, C.M.; Milodowski, A.E.; Noy, D.J.; Rochelle, C.A.; Tweed, C.J. *Research on the Alkaline Disturbed Zone Resulting from Cement–Water–Rock Reactions around a Cementitious Repository*; UK Nirex Ltd.: Harwell, UK, 2002.
3. JAEA. Second Progress Report on Research and Development for TRU Waste Disposal in Japan; Repository Design, Safety Assessment and Means of Implementation in the Generic Phase (TRU–2). JAEA–Review 2007–010, 2007/03. 2007. Available online: https://www.jaea.go.jp/04/be/documents/doc_02.html (accessed on 30 July 2021).
4. Falck, W.E.; Nilsson, K.-F. Geological Disposal of Radioactive Waste–Moving Towards Implementation. In *JRC Reference Reports*, JRC45385 (EUR 23925 E); European Commission: Brussels, Belgium, 2009. Available online: http://ec.europa.eu/dgs/jrc/ downloads/jrc_reference_report_2009_10_geol_disposal.pdf (accessed on 30 July 2021).
5. Atkinson, A. *The time dependence of pH within a repository for radioactive waste disposal. UKAEA Report AERE–R 11777*; UKAEA: Harwell, UK, 1985.
6. Berner, U. Evolution of pore water chemistry during degradation of cement in a radioactive waste repository environment. *Waste Manag.* **1992**, *12*, 201–219. [CrossRef]
7. Wilson, J.; Bateman, K.; Tachi, Y. The impact of cement on argillaceous rocks in radioactive waste disposal systems: A review focusing on key processes and remaining issues. *Appl. Geochem.* **2021**, *130*, 104979. [CrossRef]
8. Levenspiel, O. *Chemical Reaction Engineering*, 3rd ed.; Wiley: Hoboken, NJ, USA, 1998; ISBN 978-0-471-25424-9.
9. Bateman, K.; Coombs, P.; Pearce, J.M.; Wetton, P.D. Nagra/Nirex/SKB Column Experiments; Fluid Chemical and Mineralogical Studies. In *British Geological Survey Report WE/95/26*; British Geological Survey: Nottingham, UK, 2001.
10. Bateman, K.; Coombs, P.; Pearce, J.M.; Noy, D.J.; Wetton, P.D. Fluid Rock Interactions in the Disturbed Zone: Nagra/Nirex/SKB Column Experiments—Phase II. In *British Geological Survey Report WE/99/5*; British Geological Survey: Nottingham, UK, 2001.
11. Taubald, H.; Bauer, A.; Schafer, T.; Geckeis, H.; Satir, M.; Kim, J.I. Experimental investigation of the effect of high–pH solutions on the Opalinus Shale and the Hammerschmiede Smectite. *Clay Miner.* **2000**, *35*, 515–524. [CrossRef]
12. Ochs, M.; Mallants, D.; Wang, L. *Radionuclide and Metal Sorption on Cement and Concrete*; Springer International Publishing: Cham, Switzerland, 2016.
13. Rochelle, C.A.; Milodowski, A.E.; Bateman, K.; Moyce, E.B.A. A long–term experimental study of the reactivity of basement rock with highly alkaline cement waters: Reactions over the first 15 months. *Mineral. Mag.* **2016**, *80*, 1089–1113. [CrossRef]
14. Bateman, K.; Amano, Y.; Kubota, M.; Ohuchi, Y.; Tachi, Y. Reaction and Alteration of Mudstone with Ordinary Portland Cement and Low Alkali Cement Pore Fluids. *Minerals* **2021**, *11*, 588. [CrossRef]
15. Small, J.S.; Byran, N.; Lloyd, J.R.; Milodowski, A.E.; Shaw, S.; Morris, K. Summary of the BIGRAD project and its implications for a geological disposal facility. In *Report NNL*; Radioactive Waste Management: Didcot OX11GD, UK, 2016. Available online: https:// rwm.nda.gov.uk/publication/summary-of-the-bigrad-project-and-its-implications-for-a-geological-disposal-facility/ (accessed on 30 July 2021).
16. Marty, N.; Bildstein, O.; Blanc, P.; Claret, F.; Cochepin, B.; Gaucher, E.C.; Jacques, D.; Lartigue, J.E.; Liu, S.; Mayer, K.U.; et al. Benchmarks for multicomponent reactive transport across a cement/clay interface. *Comput. Geosci.* **2015**, *19*, 635–653. [CrossRef]

17. Savage, D.; Noy, D.; Mihara, M. Modelling the Interaction of Bentonite with Hyperalkaline Fluids. *Appl. Geochem.* **2002**, *17*, 207–223. [CrossRef]

18. Watson, C.; Hane, K.; Savage, D.; Benbow, S.; Cuevas, J.; Fernandez, R. Reaction and diffusion of cementitious water in bentonite: Results of 'blind' modelling. *Appl. Clay Sci.* **2009**, *45*, 54–69. [CrossRef]

19. Hama, K.; Kunimaru, T.; Metcalfe, R.; Martin, A.J. The hydrogeochemistry of argillaceous rocks formations at the Horonobe URL site, Japan. *Phys. Chem. Earth* **2007**, *32*, 170–180. [CrossRef]

20. Mazurek, M.; Eggenberger, U. Mineralogical analysis of core samples from the Horonobe area. In *RWI Technical Report 05–01*; Institute of Geological Sciences, University of Bern: Bern, Switzerland, 2005.

21. Hiraga, N.; Ishii, E. Mineral and chemical composition of rock core and surface gas composition in Horonobe Underground Research Laboratory Project (Phase 1). In *JAEA Technical Report JAEA–Data/Code 2007–022*; Japan Atomic Energy Agency: Tokai–Mura, Japan, 2008. Available online: https://jopss.jaea.go.jp/pdfdata/JAEA-Data-Code-2007-022.pdf (accessed on 30 July 2021).

22. Savage, D.; Hughes, C.R.; Milodowski, A.E.; Bateman, K.; Pearce, J.; Rae, E.; Rochelle, C.A. The evaluation of chemical mass transfer in the disturbed zone of a deep geological disposal facility for radioactive wastes. I. Reaction of silicates with Calcium hydroxide fluids. In *Nirex Report NSS/R244*; UK Nirex Ltd.: Harwell, UK, 1998.

23. Savage, D.; Bateman, K.; Hill, P.; Hughes, C.R.; Milodowski, A.E.; Pearce, J.; Rochelle, C.A. The evaluation of chemical mass transfer in the disturbed zone of a deep geological disposal facility for radioactive wastes. II. Reaction of silicates with Na–K–Ca–hydroxide fluids. In *Nirex Report NSS/R283*; UK Nirex Ltd.: Harwell, UK, 1998.

24. Miyakawa, K.; Mezawa, T.; Mochizuki, A.; Sasamoto, H. Data of groundwater chemistry obtained in the Horonobe Underground Research Laboratory Project (FY2014-2016). In *JAEA Technical Report JAEA-Data/Code 2017-012*; Japan Atomic Energy Agency: Tokai, Japan, 2017. Available online: https://jopss.jaea.go.jp/pdfdata/JAEA-Data-Code-2017-012.pdf. (accessed on 30 July 2021).

25. Parkhurst, D.L.; Appelo, C.A.J. Description of Input and Examples for PHREEQC Version 3–A Computer Program for Speciation, Batch–Reaction, One–Dimensional Transport, and Inverse Geochemical Calculations. 2013. Available online: https://pubs.usgs.gov/tm/06/a43/ (accessed on 30 July 2021).

26. JAEA. *The Project for Validating Assessment Methodology in Geological Disposal System*; Annual Report for JFY2016, JAEA Technical Report; Japan Atomic Energy Agency (JAEA): Tokai, Ibaraki, Japan, 2017. (In Japanese)

27. Savage, D.; Wilson, J.; Benbow, S.; Sasamoto, H.; Oda, C.; Walker, C.; Kawama, D.; Tachi, Y. Using natural systems evidence to test models of transformation of montmorillonite. *Appl. Clay Sci.* **2020**, *195*, 105741. [CrossRef]

28. Wilson, J.; Benbow, S.; Metcalfe, R. Reactive transport modelling of a cement backfill for radioactive waste disposal. *Cem. Concr. Res.* **2018**, *111*, 81–93. [CrossRef]

29. Kemp, S.J.; Cave, M.R.; Hodgkinson, E.; Milodowski, A.E.; Kunimaru, T. Mineralogical Observations and Interpretation of Porewater Chemistry from the Horonobe Deep Boreholes HDB-1 and HDB-2, Hokkaido, Japan. In *British Geological Survey Report*; British Geological Survey: Nottingham, UK, 2002.

30. Bernard, E.; Jenni, A.; Fisch, M.; Grolimund, D.; Mäder, U. Micro-X-ray diffraction and chemical mapping of aged interfaces between cement pastes and Opalinus Clay. *Appl. Geochem.* **2020**, *115*, 104538. [CrossRef]

31. Dauzeres, A.; Le Bescop, P.; Sardini, P.; Cau Dit Coumes, C.; Brunet, F.; Bourbon, X.; Timonen, J.; Voutilainen, M.; Chomat, L.; Sardini, P. On the physico–chemical evolution of low–pH and CEM I cement pastes interacting with Callovo–Oxfordian pore water under its in situ $CO_2$ partial pressure. *Cem. Concr. Res.* **2014**, *58*, 76–88. [CrossRef]

32. JAEA 2014: H25 Project for Technical Study on Geological Disposal of High-Level Radioactive Waste. In *Report on Development of Advanced Cement Material Impact Assessment Technology*; Japan Atomic Energy Agency (JAEA): Tokai, Ibaraki, Japan, (In Japanese). Available online: https://www.enecho.meti.go.jp/category/electricity_and_gas/nuclear/rw/library/2013/25-13-1.pdf. (accessed on 7 September 2021).

33. JAEA 2019: H29 Project for Geological Disposal of High-Level Radioactive Waste. Technology Development Project for Geological Disposal of High-Level Radioactive Waste. Development of Technology for Confirmation of Evaluation of Disposal System. (In Japanese). Available online: https://www.enecho.meti.go.jp/category/electricity_and_gas/nuclear/rw/library/2017/29fy_hyoukakakushou.pdf (accessed on 7 September 2021).

34. Oda, C.; Walker, C.; Chino, D.; Ichige, S.; Honda, A.; Sato, T.; Yoneda, T. Na-montmorillonite dissolution rate determined by varying the Gibbs free energy of reaction in a dispersed system and its application to a coagulated system in 0.3M NaOH solution at 70 °C. *Appl. Clay Sci.* **2014**, *93–94*, 62–71. [CrossRef]

35. Trapote-Barreira, A.; Cama, J.; Soler, J.M. Dissolution kinetics of C-S-H gel: Flow-through experiments. *Phys. Chem. Earth Parts A/B/C* **2014**, *70–71*, 17–31. [CrossRef]

36. Steefel, C.I.; Lichtner, P.C. 1998 Multicomponent reactive transport in discrete fractures: II: Infiltration of hyperalkaline groundwater at Maqarin, Jordan, a natural analogue site. *J. Hydrol.* **1998**, *209*, 200–224. [CrossRef]

37. Tachi, Y.; Shibutani, T.; Sato, H.; Yui, M. Experimental and modeling studies on sorption and diffusion of radium in bentonite. *J. Contam. Hydrol.* **2001**, *47*, 171–186. [CrossRef]

38. Mihara, M.; Sasaki, R. RAdio-nuclides Migration DAtasets (RAMDA) on Cement, Bentonite and Rock for TRU Waste Repository in Japan. JNC TN1410 2000-001~004. 2005. Available online: https://jopss.jaea.go.jp/pdfdata/JNC-TN8400-2005-027.pdf (accessed on 7 September 2021).

39. Benbow, S.J.; Watson, C.E. 2014 QPAC Reactive Transport Module: Theory and Testing. Quintessa Report QRS-QPAC-RTM-2 v1.1.

40. Baur, I.; Keller, P.; Mavrocordatos, D.; Wehrli, B.; Johnson, C.A. Dissolution-precipitation behaviour of ettringite, monosulfate, and calcium silicate hydrate. *Cem. Concr. Res.* **2004**, *34*, 341–348. [CrossRef]

41. Barnett, S.J.; Adam, C.D.; Jackson, A.R.W. Solid solutions between ettringite, $Ca_6Al_2(SO_4)_3(OH)_{12}\cdot26H_2O$, and thaumasite, $Ca_3SiSO_4CO_3(OH)_6\cdot12H_2O$. *J. Mater. Sci.* **2000**, *35*, 4109–4114. [CrossRef]

42. Lalan, P.; Dauzères, A.; De Windt, L.; Bartier, D.; Sammaljärvi, J.; Barnichon, J.-D.; Techer, I.; Detilleux, V. Impact of a 70°C temperature on an ordinary Portland cement paste/claystone interface: An in situ experiment. *Cem. Concr. Res.* **2016**, *83*, 164–178. [CrossRef]

43. Shafizadeh, A.; Gimmi, T.; Van Loon, L.R.; Kaestner, A.P.; Mäder, U.K.; Churakov, S.V. Time-resolved porosity changes at cement-clay interfaces derived from neutron imaging. *Cem. Concr. Res.* **2020**, *127*, 105924. [CrossRef]

44. Read, D.; Glasser, F.P.; Ayora, C.; Guardiola, M.T.; Sneyers, A. Mineralogical and microstructural changes accompanying the interaction of Boom Clay with ordinary Portland cement. *Adv. Cem. Res.* **2001**, *13*, 175–183. [CrossRef]

45. Bartier, D.; Techer, I.; Dauzères, A.; Boulvais, P.; Blanc-Valleron, M.-M.; Cabrera, J. In situ investigations and reactive transport modelling of cement paste/argillite interactions in a saturated context and outside an excavated disturbed zone. *Appl. Geochem.* **2013**, *31*, 94–108. [CrossRef]

46. Gaboreau, S.; Prèt, D.; Tinseau, E.; Claret, F.; Pellegrine, D.; Stammose, D. 15 Years of in situ cement–argillite interaction from Tournemire URL characterisation of the multi-scale spatial heterogeneities of pore space evolution. *Appl. Geochem.* **2011**, *26*, 2159–2171. [CrossRef]

47. Savage, D.; Cloet, V. *A Review of Cement-Clay Modelling*; Nagra Report NAB 18–24; National Cooperative for the Disposal of Radioactive Waste: Wettingen, Switzerland, 2018. [CrossRef]

48. Xu, T.; Sonnenthal, E.; Spycher, N.; Pruess, K. *TOUGHREACT User's Guide: A Simulation Program for Non-Isothermal Multiphase Reactive Geochemical Transport in Variably Saturated Geologic Media (No. LBNL-55460)*; Lawrence Berkeley National Lab. (LBNL): Berkeley, CA, USA, 2004. [CrossRef]

*minerals*

*Article*

# Reactive Transport Simulation of Low-pH Cement Interacting with Opalinus Clay Using a Dual Porosity Electrostatic Model

Andreas Jenni * and Urs Mäder

Institute of Geological Sciences, University of Bern, CH-3012 Bern, Switzerland; urs.maeder@geo.unibe.ch
* Correspondence: andreas.jenni@geo.unibe.ch; Tel.: +41-31-684-87-65

**Abstract:** Strong chemical gradients between clay and concrete porewater lead to diffusive transport across the interface and subsequent mineral reactions in both materials. These reactions may influence clay properties such as swelling behaviour, permeability or radionuclide retention, which are relevant for the safety of a radioactive waste repository. Different cement types lead to different interactions with Opalinus Clay (OPA), which must be understood to choose the most suitable material. The consideration of anion-depleted porosity due to electrostatic repulsion in clay modelling substantially influences overall diffusive transport and pore clogging at interfaces. The identical dual porosity model approach previously used to predict interaction between Portland cement and OPA is now applied to low-alkali cement—OPA interaction. The predictions are compared with corresponding samples from the cement-clay interaction (CI) experiment in the Mont Terri underground rock laboratory (Switzerland). Predicted decalcification of the cement at the interface (depletion of C–S–H and absence of ettringite within 1 mm from the interface), the Mg enrichment in clay and cement close to the interface (neoformation of up to 17 vol% Mg hydroxides in concrete, and up to 6 vol% in OPA within 0.6 mm at the interface), and the slightly increased S content in the cement 3–4 mm away from the interface qualitatively match the sample characterisation. Simulations of Portland cement—OPA interaction indicate a weaker chemical disturbance over a larger distance compared with low-pH cement—OPA. In the latter case, local changes in porosity are stronger and lead to predicted pore clogging.

**Keywords:** cement—clay interaction; diffusion; dual porosity; electrostatic effects; reactive transport modelling; near field; radioactive waste repository; low-pH cement

**Citation:** Jenni, A.; Mäder, U. Reactive Transport Simulation of Low-pH Cement Interacting with Opalinus Clay Using a Dual Porosity Electrostatic Model. *Minerals* **2021**, *11*, 664. https://doi.org/10.3390/min11070664

Academic Editors: Ana María Fernández, Stephan Kaufhold, Markus Olin, Lian-Ge Zheng, Paul Wersin and James Wilson

Received: 27 April 2021
Accepted: 18 June 2021
Published: 22 June 2021

**Publisher's Note:** MDPI stays neutral with regard to jurisdictional claims in published maps and institutional affiliations.

## 1. Introduction

Deep geological repository designs include cementitious materials for structural elements, backfill or waste matrix. While mineralogy in Opalinus Clay (OPA) has proven stability for the last 170 million years, cementitious minerals are subject to deterioration; hence, concrete infrastructure has a typical service life between 50 and 100 years.

Chemical gradients between porewaters of cement and contrasting materials cause diffusive fluxes of dissolved species. In case of cement—clay interfaces, this may lead to mineralogical alterations, which in turn are expected to locally influence important barrier properties like permeability, swelling pressure or specific retention.

Experimental literature documents chemical and mineralogical changes at clay—cement interfaces [1–10]. Local decalcification (instability of portlandite, calcium silicate hydrate (C–S–H), and ettringite) is the main alteration of the cement. The resulting Ca depletion within a thin layer down to 0.2 mm can be measured by element mapping with various techniques, but mineralogical characterisation with such high spatial resolutions is still demanding [1]. Even more challenging is the characterisation of the measured Mg enrichments at the interface. The most likely neoformation, nano crystalline magnesium silicate hydrate (M–S–H), shows no distinct reflections in X-ray diffraction [11]. Mg X-ray absorption near-edge spectroscopy reveal a crystallographic structure similar to certain clay minerals [12,13], which impedes M–S–H detection within OPA by this technique. The location and Mg content of the Mg-enriched layer in OPA depend on the nature of the cement [14]. Other interactions between cement and OPA result in carbonation in the cement and a sulphate enrichment detached from the interface, and the change of the exchanger population in OPA [6]. Reactive transport model approaches predict such changes [11,14–19], but studies rarely compare modelling results with experimental data. Of value are modelling studies comparing different cement types interacting with clay under similar conditions to contribute to the engineering design of a repository.

Changes in porosity, its distribution and connectivity—and therefore also permeability—near such interfaces are important processes governing the long-term physicochemical evolution of the engineered barrier and its geological near-field [17]. Increasing porosity in the Portland cement close to the interface, and clogging in the claystone adjacent to it are experimentally observed and commonly predicted by reactive transport modelling [15–17]. These studies all use a simplified porosity concept and do not account for the clay-specific influence of clay layer charge on the porewater solute distribution and on diffusive transport.

Dual porosity reactive transport modelling can take into account such clay-specific features, and this may be crucial when modelled porosity clogging approaches zero-porosity. Jenni, Gimmi, Alt–Epping, Mäder and Cloet [14] describe a novel dual porosity approach, simulate the short-term evolution of Portland cement concrete (OPC) in contact with OPA, and compare the results with experimental data from the CI experiment (Cement-clay Interaction) carried out in the Mont Terri rock laboratory (www.mont-terri.ch). The current study repeated the simulation but exchanged the OPC concrete with a low-alkali cement concrete (ESDRED, named after the European project "Engineering Studies and Demonstrations of Repository Designs" 2004–2008, which included the development of this concrete formulation). Both types of concrete were installed in the CI experiment, and their interfaces were sampled and characterised several times after emplacement. The simulation outcome was compared with element maps of the appropriate interface sample, and with the outcome of the previous OPC-OPA model.

## 2. Materials and Methods

The multicomponent reactive transport modelling was performed in 1D with the FLOTRAN code [20] in the same way the OPC—OPA interaction was modelled [14,19]. These publications contain further details, especially on the dual porosity approach used for the OPA part of the model, which is summarised in Section 2.2. Advection is negligible in these types of clay host systems. Site-specific transport studies on undisturbed OPA state diffusion-dominated transport [21]. Small hydraulic heads in the OPA close to the repository are expected, which lead to Peclet numbers in the range of $10^{-2}$ to $10^{-3}$ [22]. Glaciations or tunnel convergence might increase the importance of advection only temporarily. Therefore, only diffusion and electrochemical migration were included in this model approach.

Comparable reactive transport simulations considering electrostatic effects on the solutes were carried out by using CrunchFlowMC and PhreeqC. Alt–Epping, et al. [23] compare the capabilities of these codes with respect to dual porosity. Results of OPC—OPA interface simulations with these codes have been presented in Jenni, et al. [24].

### 2.1. Modelling of the Concrete Domain

Diffusive transport followed by reactions between concrete and clay rock start when the freshly-mixed cement slurry is applied to a clay rock surface. Substantial interactions between the materials are likely to occur during the early stage of cement hydration. Therefore, the initial condition of the simulation should consist of unhydrated cement clinker and water. Initial conditions with hardened cement are appropriate in case of pre-cast elements brought in contact with clay, and should not be used to simulate shotcrete applications.

Concrete properties used in the model (e.g., aggregate content, ESDRED cement composition) matched the concrete used in the Mont Terri CI experiment (Table 1, taken from Jenni, Mäder, Lerouge, Gaboreau and Schwyn [6]). The concrete aggregates were considered as non-porous and were included in the initial condition because they lowered total diffusive transport. The low specific surface areas and reaction rates of the aggregate minerals suggested treating them as inert for the limited time-scale of the model, in contrast to reactive clinker phases and nanosilica.

**Table 1.** Characteristics of concrete and cement. Binder represents all reactive components (cement, silica). Cement composition was determined by XRD Riedtveld analysis using an internal standard.

| Concrete Composition | | |
|---|---|---|
| CEM I 42.5 N | $[kg/m^3]$ | 210 |
| silica fume | $[kg/m^3]$ | 140 |
| superplasticiser | $[kg/m^3]$ | 4.2 |
| accelerator | $[kg/m^3]$ | 16.8 |
| water | $[kg/m^3]$ | 175 |
| water/binder weight ratio | – | 0.5 |
| sand, gravel | $[kg/m^3]$ | 1800 |
| **Cement Composition** | | |
| alite | [wt%] | 52.8 |
| belite | [wt%] | 22.8 |
| aluminate | [wt%] | 5.0 |
| ferrite | [wt%] | 7.9 |
| periclase | [wt%] | 1.0 |
| calcite | [wt%] | 2.9 |
| quartz | [wt%] | 0.5 |
| anhydrite | [wt%] | 1.9 |
| hemihydrate | [wt%] | 1.7 |
| gypsum | [wt%] | 0.0 |
| syngenite | [wt%] | 2.0 |
| dolomite | [wt%] | 1.5 |

Within the cement, a geometric factor (ratio of constrictivity/tortuosity) of G = 0.1 was chosen, following arguments in Jenni, Gimmi, Alt-Epping, Mäder and Cloet [14].

The clinker dissolution rates mainly control the overall cement hydration kinetics, and the formation of hydrates is generally considered to be instantaneous [25,26]. Here, all cement hydrates formed with a fast rate constant of $10^{-3}$ mol/m$^3$/s and a generic surface area of 0.01 m$^2$/g (rate law described in [27,28]). Two combined dissolution rates per mineral (Table 2) were derived by a best-fit of clinker and silica fume dissolution data, measured by quantitative XRD on reference samples of the same cement [29]. The two rates were implemented by incorporating two reacting versions of the same mineral. Reaction rates of potentially forming non-cement hydrates were taken from Palandri and Kharaka [30] (generic surface area of 0.1 m$^2$/g). Thermodynamic data was taken from the Cemdata07.2 [26,31–36] and EQ3/6 databases. More details about clinker composition, solid solutions of hydrates, and phase reactions are given in [14].

The set accelerator added to the ESDRED concrete leads to a formate content of more than 0.2 M in the early porewater, slightly decreasing with time [29]. In the initial condition of the model, 0.2 M formate, charge-balanced by Ca, was added to the pure water to account for this substantial anionic charge.

### 2.2. Dual Porosity Modelling of Opalinus Clay

Opalinus Clay was modeled in the exact same manner as described in Jenni, Gimmi, Alt-Epping, Mäder and Cloet [14], that contains further explanations and figures. Rock composition (Table 3) and porewater chemistry data were taken from Berner, et al. [38], based on measured and modelled data [39–41]. Thermodynamic data was taken from

EQ3/6, except for chlorite (7 Å chamosite), K-feldspar, and albite, which were treated as inert. Too many reactive minerals in the simulation violate the phase rule, and no equilibrium initial condition can be achieved. Pyrophyllite represented the TOT layer of smectite, considered as inert due to the small extent of reaction to be expected within the time-scale of the model. Reaction rates for all reactive minerals were taken from Palandri and Kharaka [30].

**Table 2.** Stoichiometric clinker composition calculated from measured cement bulk chemical composition after Taylor [37], measured silica fume composition, implemented dissolution rates, and surface areas.

| | Alite $C_3S$ | Belite $C_2S$ | Aluminate $C_3A$ | Ferrite $C_4AF$ | Si Fume |
|---|---|---|---|---|---|
| CaO [formula units] | 2.911 | 1.975 | 2.793 | 4.202 | 0.024 |
| $SiO_2$ [formula units] | 0.956 | 0.914 | 0.17 | 0.293 | 0.981 |
| $Al_2O_3$ [formula units] | 0.027 | 0.041 | 0.85 | 1.066 | 0.001 |
| $Fe_2O_3$ [formula units] | 0.01 | 0.01 | 0.088 | 0.665 | 0 |
| $Na_2O$ [formula units] | 0.002 | 0.002 | 0.036 | 0.005 | 0 |
| MgO [formula units] | 0.054 | 0.015 | 0 | 0 | 0.003 |
| $K_2O$ [formula units] | 0.002 | 0.014 | 0.016 | 0.007 | 0.003 |
| log rate constant1 [mol/m$^2$/s] | −5.06374 | −5.34159 | −4.06095 | −4.69578 | −9.24229 |
| log rate constant2 [mol/m$^2$/s] | −7.27421 | −8.48021 | −8.21623 | −8.2 | −11.39708 |
| surface area for 1 and 2 [m$^2$/g] | 0.00319732 | 0.00301906 | 0.00329394 | 0.0026751 | 21 |

**Table 3.** Mineral composition of OPA in wt% of dry rock.

| | [wt%] |
|---|---|
| illite | 23.9 |
| kaolinite | 17.7 |
| smectite | 12.2 |
| calcite | 13.7 |
| dolomite | 0.5 |
| quartz | 17.1 |
| siderite | 3.3 |
| pyrite | 1.0 |
| kalifeldspar | 1.3 |
| albite | 1.0 |
| chlorite | 8.3 |

In OPA, 6 vol% of 12 vol% total pores contains porewater with cations balancing the negative charge of the clay layers. This porosity consists of the water in the clay interlayer plus the water on clay outer particle surfaces (often called diffuse double layer). Because both of these porewater domains are similarly influenced by the negative charge of the clay layers, the sum of the domains was implemented in the model, and is called *Donnan porewater* (and Donnan porosity). Its homogeneous chemical composition was calculated by assuming Donnan equilibrium with the remaining fraction of the porewater, which is here referred to as freely accessible porewater, *free porewater* in short (contained in the free porosity). Donnan equilibrium assumes a homogeneous composition of the Donnan porewater (averaging a Poisson-Boltzmann distribution), and equal activity coefficients in both porosity domains [14]. The code inhibited mineral precipitation in the Donnan porosity, and all minerals kinetically equilibrated with the free porewater. Thermodynamically, mineral phases at saturation had equal chemical potentials in both domains, despite the depletion of anions (compensated by an electrostatic potential term). The restricted space in the interlayer was assumed to inhibit nucleation. Chagneau, et al. [42] give experimental evidence for mineral precipitation in the free porosity only in compacted clay. This implies that the Donnan porosity was not clogged and always provides a pathway for mainly neutral species and cations that diffuse while maintaining charge balance (by considering streaming potentials in multi-component transport). The model approach presented here can handle this type of transport [14].

In a Donnan porosity concept, Donnan equilibrium between the Donnan porewater and the free porewater replaces conventional cation exchange. An electrostatic term dependent on ionic strength and the mean potential governs the Donnan equilibrium. In CrunchflowMC or PhreeqC [23,43–47], this is implemented by an explicit partitioning function that distributes aqueous species between Donnan and free porewater. In the present FLOTRAN approach, the ion partitioning resulted from rapid diffusion and electrochemical migration between the free and Donnan porewater. The cation exchange capacity (CEC) was represented by immobile anions in the Donnan porewater. Technical implementation details are given in [14]. The geometric factor (G = constrictivity/tortuosity) in the free porosity was set to 0.023, and to 0.006 in the Donnan porosity (derived from diffusion experiments in OPA [14]).

The Donnan porosity was not linked to a specific clay mineral in the current approach, and the total ion charge in the Donnan porewater balanced the bulk CEC of OPA. Dissolution or precipitation of clay minerals did not decrease or increase the CEC or the Donnan porosity as they should. This simplification is acceptable, because the clay minerals are virtually stable on the time-scale of the model presented here (rate constant $<10^{-13}$ mol/m$^2$/s after [30]).

The arithmetic averaging of concentrations in the electromigration term between adjacent cells in FLOTRAN [20] was changed to harmonic averaging, as argued in [48]. More details are given in [14].

### 3. Results and Discussion

*3.1. Model Predictions of Low-pH Cement—Opalinus Clay Interaction*

The initial conditions of the unhydrated ESDRED cement matrix consisted of pure water with Ca-formate (accelerator) in the free porewater, which was undersaturated with respect to the cement clinkers (Figure 1). The OPA free porewater was in chemical equilibrium with the reactive OPA minerals as well as in Donnan equilibrium with the Donnan porewater. This initial condition describes the fresh (unhydrated), well-mixed concrete right after casting into the borehole drilled into OPA (start of the Mont Terri CI experiment), or by analogy a shotcrete just applied to a tunnel wall for stabilisation within a repository.

Comparably fast dissolution of clinkers and silica fume led to the precipitation of mainly C–S–H and ettringite (Figure 2). The high Si content and lower pH in the porewater compared to OPC suppressed portlandite formation and favoured low Ca/Si C–S–H, which is the main difference compared with Portland cements [49]. During the dissolution of the clinker phases, Ca and OH species diffused directly into the OPA due to the concentration gradient. At the same time, some OPA porewater species diffused into the concrete. At the interface, the pore solution in ESDRED was slightly lower in pH and dissolved Ca, which led to the instability of C–S–H and ettringite. In turn, Mg diffusing in from OPA porewater led to precipitation of Mg-phases.

Over time, the model predicted the precipitation of hydrotalcite in the unaffected ESDRED due to a slow dissolution of MgO associated with clinker, as predicted in [29]. However, hydrotalcite was not observed experimentally as a hydration product of ESDRED (same reference). In the present approach, hydrotalcite might have been more stable due to a overpredicted availability of Al in the pore solution that might be consumed in reality by Al uptake by C–S–H [50] and by M–S–H [51].

At the interface, hydrotalcite, M–S–H and/or talc precipitated. M–S–H is described in literature as a hydrated nano-crystalline precursor of phyllosilicates like talc or Mg-smectite [52], and therefore the presence of talc in the database inhibited the precipitation of early M–S–H due to its precursor character, i.e., M–S–H is a less stable phase compared to talc. Predictions of the nature of the Mg precipitates are uncertain due to uncertainties in thermodynamic data for Mg (Si-) hydroxides. The formation of a specific Mg (Si-)

hydroxide depends on local pH and Mg, Ca, Al, and Si concentrations. Regardless of its nature, the Mg (Si-) hydroxides precipitated within the Ca-depleted zone of the cement, and could partly compensate the additional pore space available due to C–S–H dissolution. Availability of Mg from OPA, the Mg content in the cement, and the capability of the OPA to lower the pH in the cement determined the extent to which the porosity in the cement at the interface could be clogged. The model prediction clearly shows that Mg (Si-) hydroxides have the highest potential to clog the porosity in the cement at the interface.

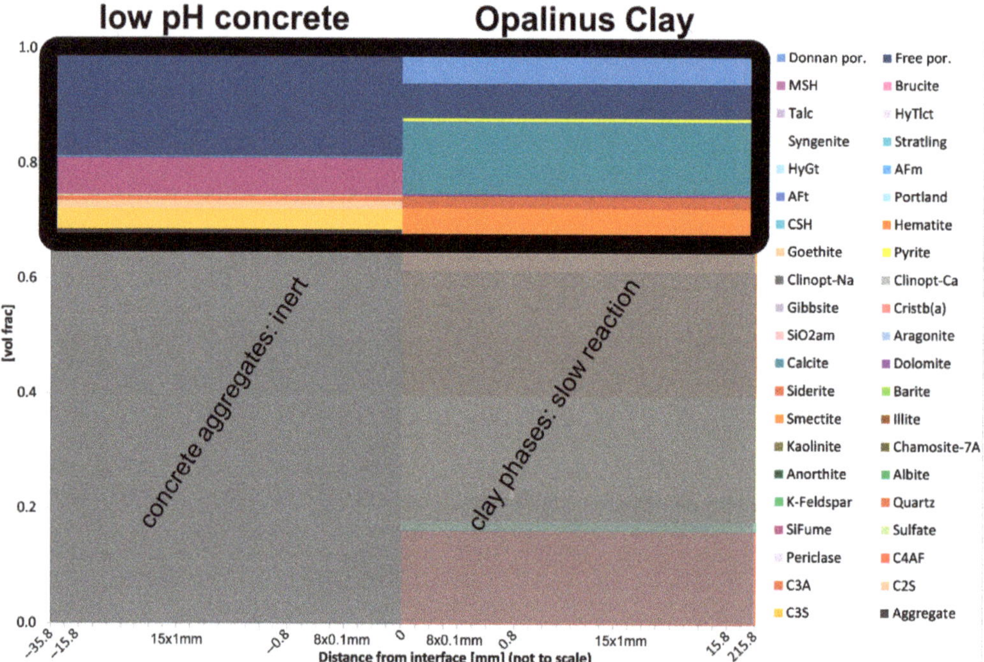


**Figure 1.** Phase volume fractions of initial condition of the concrete—OPA interface. The legend shows all phases considered in the model (por.: porosity; HyTlct: hydrotalcite; Stratling: strätlingite; HyGt: hydrogarnet; AFm: monosulphate; AFt: ettringite; Portland: portlandite; Clinopt: clinoptilolyte; Crist (a): alpha-cristobalite; SiO2am: amorphous silica; SiFume: silica fume; Sulfate: sum of gypsum, hemihydrate, anhydrite; C4AF: ferrite; C3A: aluminate; C2S: belite; C3S: alite). Only reactive phases ondergoing visible volume changes are shown in subsequent figures (top 30 vol%, black frame). The entire domain was discretised into $16 \times 100$ μm-cells close to the interface, $30 \times 1$ mm-cells further away, and two large cells at the outer concrete and OPA boundaries.

Calcite formed in the cement with ongoing in-diffusion of inorganic carbon species from OPA. Transport was fast enough to carbonate the entire cement domain, but only resulted in calcite contents below 1 vol% due to the limited reservoir in the OPA porewater. Therefore, carbonation reduced porosity and transport insignificantly according to this model prediction.

From the very beginning of interaction, the high-pH front entered the OPA, where it was buffered by dissolution of clay minerals. Only negligible volumes of dissolving clay were required to efficiently buffer pH, but the limiting factor was the slow reaction kinetics relative to the OH diffusion. Two scenarios were calculated using the slow reaction rates in Palandri and Kharaka [30], and the increased rates (four orders of magnitude) proposed by Marty, et al. [53]. Figure 3 shows a more advanced high-pH front in case of slower reaction rates, whereas the faster clay reactivity hindered the pH front more efficiently from entering the OPA.

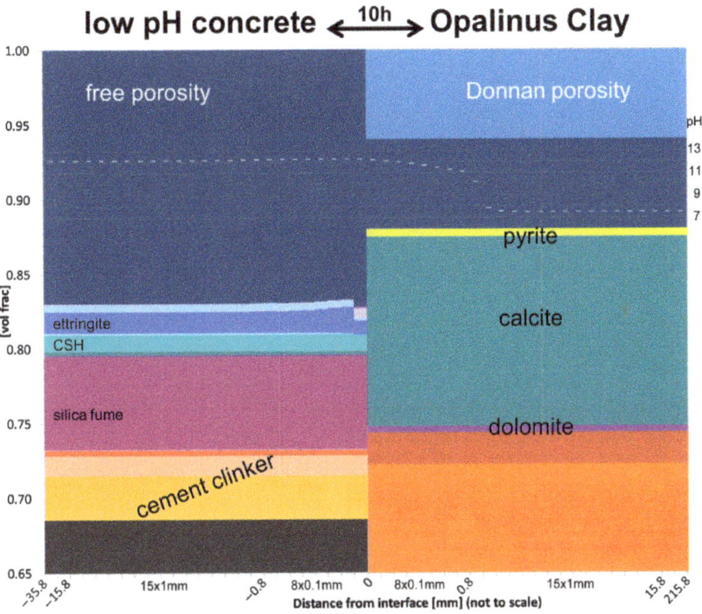


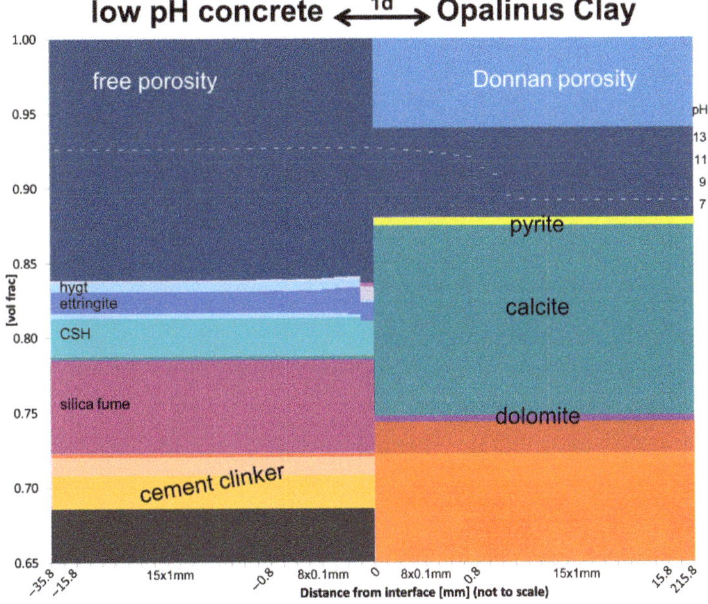


**Figure 2.** *Cont.*

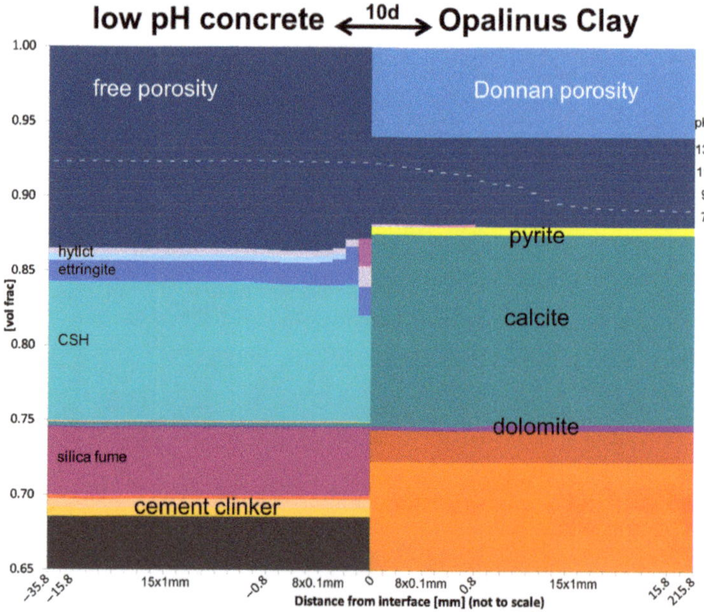


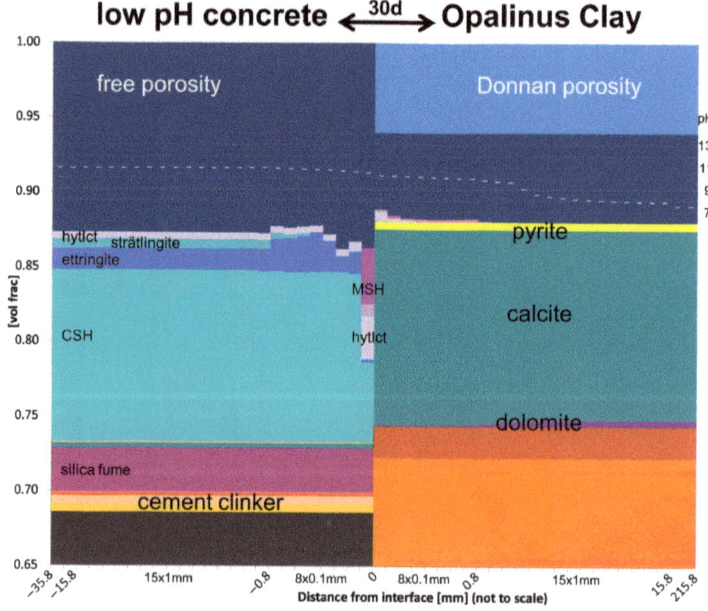


**Figure 2.** *Cont.*

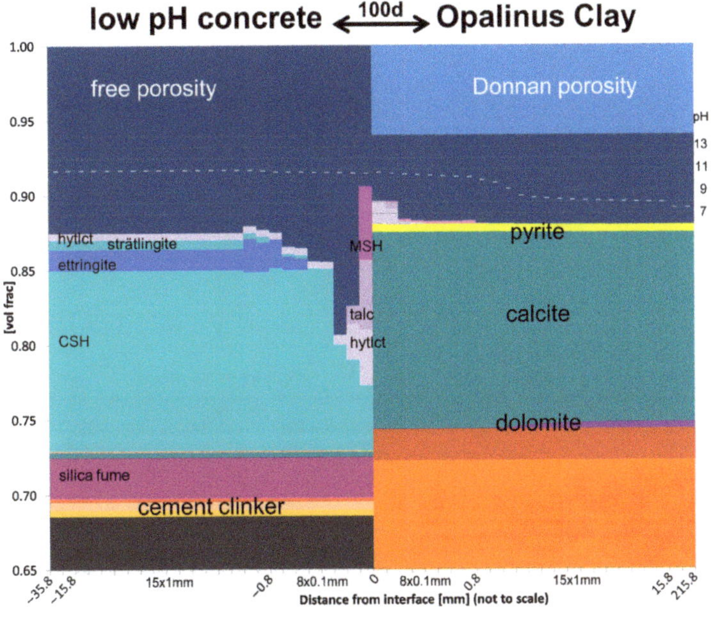


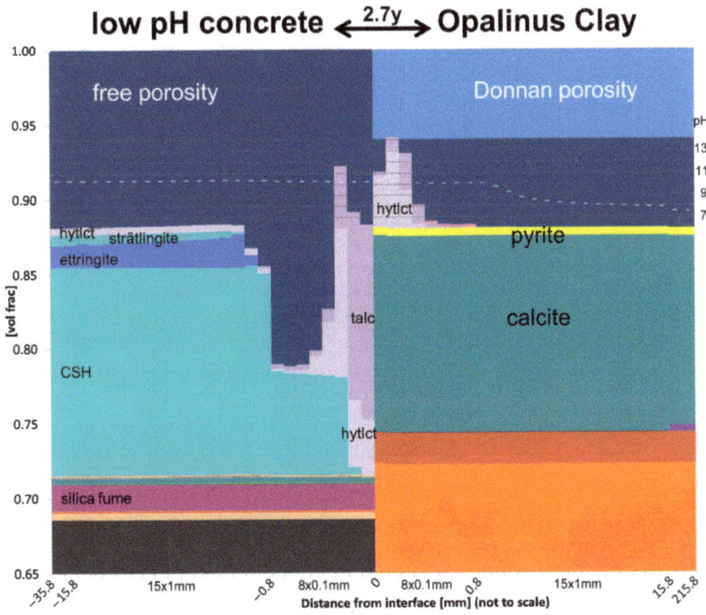


**Figure 2.** Predicted volume evolution of the reactive phases (legend and explanation of horizontal axis in Figure 1 and captions). The pH of the free porewater is indicated in the dark blue free porosity area.

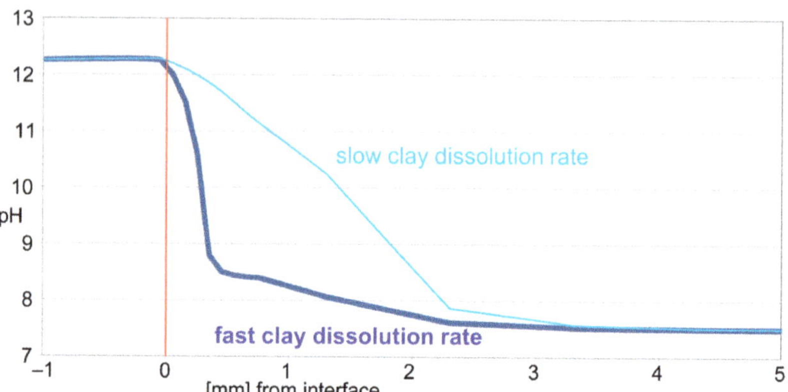


**Figure 3.** pH profiles across the OPA free porewater after 10 h of interaction with ESDRED, modelled with slow and increased reaction rates of illite and kaolinite. Interface at 0 mm, ESDRED on the left, OPA on the right.

Following a strong concentration gradient, Mg migrated into the cement. The decreasing Mg in the OPA free porewater is buffered by the large Mg content in the Donnan porosity (equivalent to the clay exchangeable cations). The high mobility of the cations in the Donnan porosity led to a depletion of Mg up to 15 mm into the OPA within 2.7 y, compensated mainly by Ca (Figure 4), originating from the instability of mainly C–S–H at the interface (decalcification of the cement). The predicted dissolution of the minor dolomite content in the OPA contributed to a smaller extent to the Mg supply. The freed Ca and carbonate species were bound in newly formed calcite, but were also released to the OPA porewater.

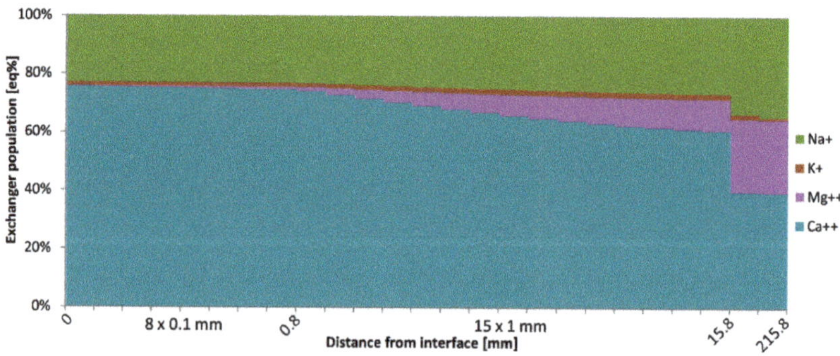


**Figure 4.** The cation concentration profile in the Donnan porewater (in charge equivalent fractions) across the OPA after 2.7 y of interaction with ESDRED. In the current modelling approach, the Donnan porewater concentrations replaced the cation occupancies of clay models with explicit ion-exchange. Explanation of horizontal axis in Figure 1 captions.

Mg-hydroxides also formed in the OPA at the interface within a layer of increased pH. Between 100 d and 2.7 y, the free porosity became completely clogged, but the model was set up in a way that transport between the free porewater in the cement and the Donnan porewater in the OPA could continue (Section 2.2). However, anion mobility across the interface was extremely low due to the low anion concentration in the Donnan porosity. In turn, neutral species (e.g., water tracer) and cations still migrated considerably across the interface, as proposed in Jenni, Gimmi, Alt-Epping, Mäder and Cloet [14], and by a diffusion experiment with a fully clogged illite core [42]. A single porosity model approach cannot predict this ongoing interaction across an interface clogged on the clay side of an interface.

### 3.2. Comparison of Low-pH and Portland Cement Interactions with Opalinus Clay

The modelling of the ESDRED low-pH cement interaction with Opalinus Clay predicted an early clogging of the free porosity on the clay-side, which led to a substantial decrease of the interaction rates: phase distributions after 4.8 y of interaction resemble the 2.7 y old interface, and the penetration of the pH front virtually stopped (Figure 5). In contrast, the porosity at the OPC-OPA interface was only predicted to minimally clog after 4.8 y [14]. In both predictions, OPA initial conditions and transport models are equal. The main difference is the presence of rapidly dissolving silica fume and the formate as additional anion charge carrier in the ESDRED, leading to a lower pH and an increased Si content in the porewater compared to OPC.

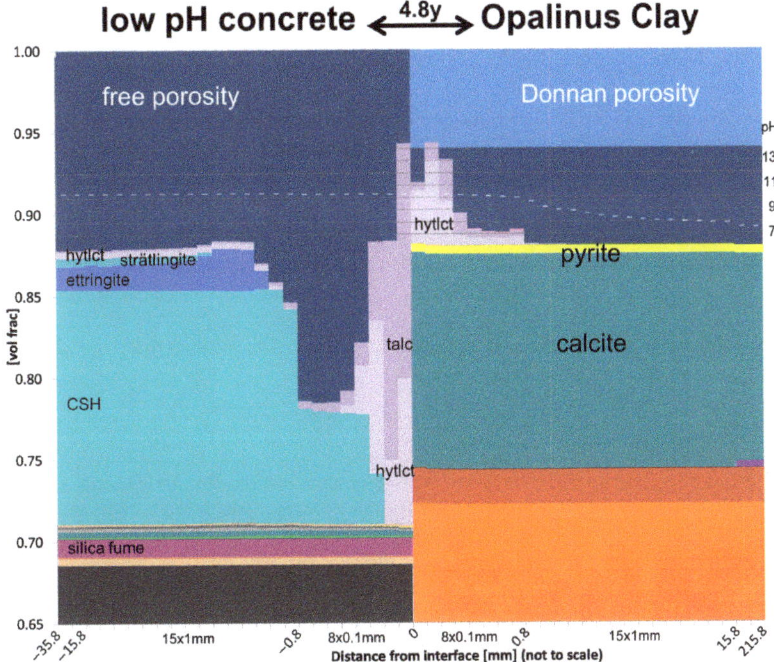


**Figure 5.** *Cont.*

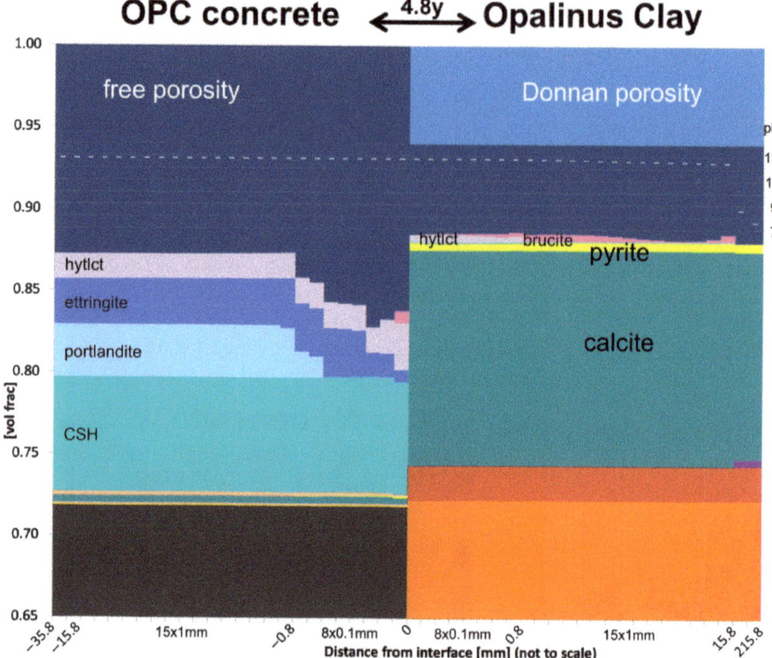



**Figure 5.** Predicted volume evolution of the reactive phases after 4.8 y of interaction of ESDRED (top), and OPC (bottom, modified from [14]) with OPA. The phase legend and explanation of horizontal axis is shown in Figure 1 and captions, the pH of the free porewater is indicated in the dark blue free porosity area.

### 3.3. Comparison of Model Predictions with Measurements

Quantitative agreement of the model predictions with measurements is not perfect but shares essential communalities. It must be kept in mind that significant local variability exists across samples of equivalent interfaces in terms of amount of neoformations and extent of interaction zones. The qualitative agreement of measurements on both ESDRED—OPA and OPC—OPA interfaces with predicted interaction is striking. Both simulations predicted decalcification and depletion in sulphur in the cement at the interface (portlandite, C–S–H, and ettringite dissolution) in agreement with chemical maps (Figure 6 and [14]). Whereas a substantial Mg enrichment was measured in ESDRED and in adjacent OPA only close to the interface, OPC—OPA interfaces showed no or very limited Mg increase at the interface, but a small detached enrichment at approximately 7 mm distance from the interface in the OPA (Figure 7). These cement-specific features were predicted by both simulations: the Mg-mineral hydrotalcite occured in both ESDRED and OPA right at the interface, but in OPC—OPA, brucite (very high in Mg) occurred distant from the interface (further details given in [14]). Both models also suggested that the Mg enrichments in the OPA traced the high-pH fronts, which reached different positions in the two interfaces (Figure 5). The exact identity of the Mg-phase is under investigation [1,12].

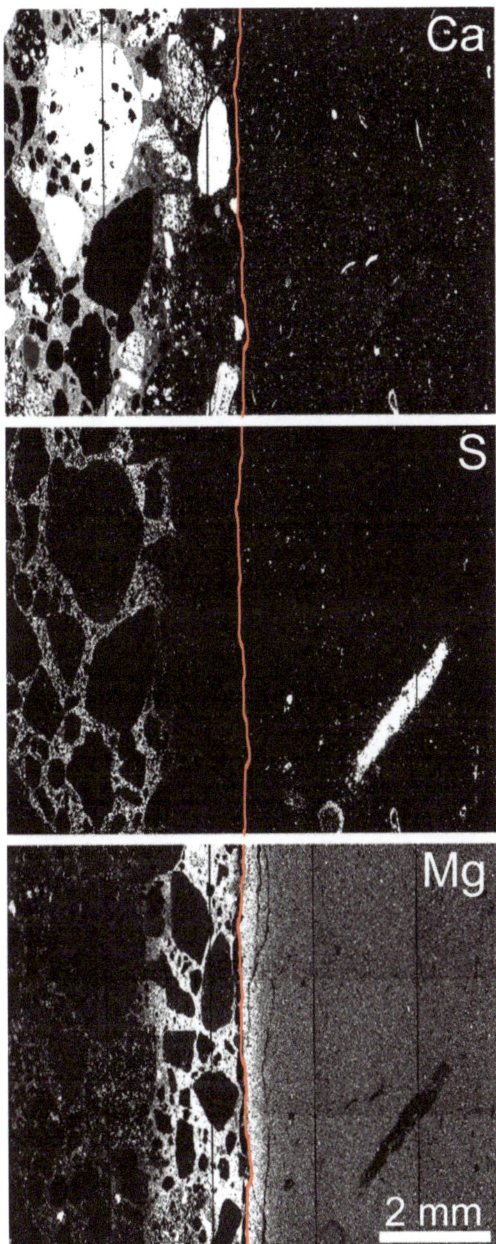

**Figure 6.** SEM EDX element maps of 4.8 y ESDRED-OPA interfaces, concrete on the left, OPA on the right of the interface marked in red. Bright areas represent high concentrations of the element indicated. On the concrete side, only the cement matrix between the aggregates is relevant. OPA contains pyrite (high S areas) and calcite (high Ca spots).

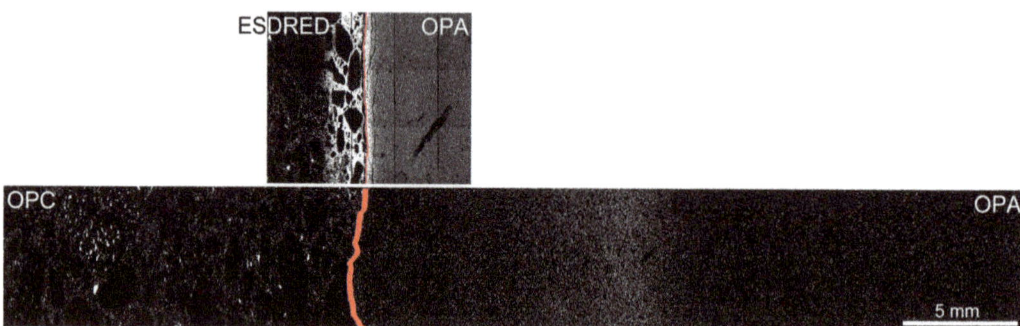



**Figure 7.** Comparison of SEM EDX Mg maps of interfaces (4.8 y), concrete on the left, OPA on the right of the interface marked in red. Bright areas represent high Mg concentrations. On the concrete side, only the cement matrix between the aggregates is relevant.

### 4. Conclusions

Several uncertainties of this modelling approach limit a quantitative prediction of the evolution at the cement—clay interface, especially for extended interaction times to hundreds or thousands of years. Transport parameters, especially geometric factors in the free and Donnan porosities, are not well known and large errors may arise from a simple estimation from OPA diffusion experiments. Clay dissolution rates given in literature differ significantly, but strongly influence the interaction model outcome: the relationship between transport, e.g., migration of the pH-front, and rates of mineral dissolution, e.g., buffering pH, determine the extent of cement—clay interaction. An additional crucial uncertainty is the degree to which the free porosity can be clogged. Further, assuming a minimal remaining free porosity, as strongly suspected in the concrete, leads to considerable interaction within thousands of years. In contrast, full clogging of the total porosity in the concrete at the interface (implemented in the model by allowing precipitations up to zero porosity) would completely stop cement-clay interaction. Such conceptual questions cannot be answered by modelling studies, but require long-term transport experiments, whose outcome can then be used to conceptualise the models.

It is striking that the predicted extent of alteration in terms of overall mass transfer in case of the ESDRED ("low-pH cement") was in fact larger than that in the OPC case after a given time, e.g., 4.7 y (Figure 5). This was predicted despite that the mobility of solutes across the interface became restricted due to clogging of free porosity in case of the low-alkali cement. Only the penetration of the high-pH front into the claystone was more extensive in the OPC case, but it was associated with smaller mineral mass transfers. In the ESDRED, the decrease of Ca-containing hydrates at the interface was higher than in OPC. More dissolved Ca may have diffused from ESDRED into the OPA and exchanged with Mg, which precipitated within the narrow high-pH zone. The clogging of OPA's free porosity could not stop this cation exchange, because it could continue via the Donnan porosity. In contrast, $OH^-$ was mainly present in the free porosity and its diffusion was therefore affected by free porosity clogging. In the OPA in contact with OPC, the Mg neoformations distributed over a far wider high-pH zone and clogged only a small porosity fraction.

A supposedly higher inherent reactivity of OPC compared to low-alkali cementitious products due to its contrasting initial pH has motivated implementers to reduce, substitute or omit OPC in some repository designs for radioactive waste. If our model indeed captured the key reactive processes, a more extensive alteration at OPC-claystone interfaces may not be expected, and thus this a priori assumption is incorrect and should not be a decisive issue. A similar conclusion was reached [2] simply by comparing the extents of alteration based on highly-resolved element maps and other methods (similar to Figures 6 and 7), and this despite a significantly larger water/cement ratio of 0.8 in OPC compared to 0.5 in the ESDRED cement. Based on the extensive characterisation and modelling of interfaces from the Mont Terri CI experiment reacted for up to 10 years, the high-pH front coming from OPC enters deeper into the OPA compared with the ESDRED high-pH front. In turn, the volumes of dissolved cement and neoformations are larger at the ESDRED-OPA interface. Simplified numerical simulations of Portland cement—clay interaction run for longer times [17,38,54] predicted clogging after considerably longer interactions than 10 years. This suggests that both high-pH and low-pH cements will clog eventually. The expected difference in clogging time might be crucial if the buffer resaturation requires water transport across a clay-cement interface.

**Author Contributions:** Conceptualization, A.J. and U.M.; methodology, A.J.; software, A.J.; validation, A.J. and U.M.; formal analysis, A.J. and U.M.; investigation, A.J. and U.M.; resources, A.J. and U.M.; data curation, A.J.; writing—original draft preparation, A.J.; writing—review and editing, A.J. and U.M.; visualization, A.J.; supervision, U.M.; project administration, A.J. and U.M.; funding acquisition, A.J. and U.M. All authors have read and agreed to the published version of the manuscript.

**Funding:** This research was partially funded by the Mont Terri Consortium.

**Data Availability Statement:** The data presented in this study are available on request from the corresponding author. The data are not publicly available due to technical reasons.

**Acknowledgments:** Peter Alt-Epping and Thomas Gimmi are acknowledged for their modifications of FLOTRAN. The Mont Terri Consortium via the CI Experiment group (ANDRA, CRIEPI, FANC, IRSN, Nagra, NWMO, Obayashi, RWM, SCK-CEN) provided funding for sampling campaigns and sample characterisation, as well as partial funding for this modeling study. The paper was improved by the internal review of Ellina Bernard.

**Conflicts of Interest:** The authors declare no conflict of interest.

## References

1. Bernard, E.; Jenni, A.; Fisch, M.; Grolimund, D.; Mäder, U. Micro-X-ray diffraction and chemical mapping of aged interfaces between cement pastes and Opalinus Clay. *Appl. Geochem.* **2020**, *115*, 104538. [CrossRef]
2. Mäder, U.; Jenni, A.; Lerouge, C.; Gaboreau, S.; Miyoshi, S.; Kimura, Y.; Cloet, V.; Fukaya, M.; Claret, F.; Otake, T.; et al. 5-year chemico-physical evolution of concrete–claystone interfaces, Mont Terri rock laboratory (Switzerland). *Swiss J. Geosci.* **2017**, 1–21. [CrossRef]
3. Idiart, A.; Laviña, M.; Kosakowski, G.; Cochepin, B.; Meeussen, J.C.L.; Samper, J.; Mon, A.; Montoya, V.; Munier, I.; Poonoosamy, J.; et al. Reactive transport modelling of a low-pH concrete/clay interface. *Appl. Geochem.* **2020**, *115*, 104562. [CrossRef]
4. Lerouge, C.; Gaboreau, S.; Grangeon, S.; Claret, F.; Warmont, F.; Jenni, A.; Cloet, V.; Mäder, U. In situ interactions between Opalinus Clay and Low Alkali Concrete. *Phys. Chem. Earth Parts A/B/C* **2017**, *99*, 3–21. [CrossRef]
5. Fisch, M.; Jenni, A.; Mäder, U.; Grolimund, D.; Cloet, V. *CI Experiment: Micro-XRD Studies on 3.2-Year-Old Interfaces between Opalinus Clay and OPC and ESDRED Mortar (3rd Sampling Campaign 2015)*; TN 2016-73; Mont Terri Project: St. Ursanne, Switzerland, 2017.
6. Jenni, A.; Mäder, U.; Lerouge, C.; Gaboreau, S.; Schwyn, B. In situ interaction between different concretes and Opalinus Clay. *Phys. Chem. Earth Parts A/B/C* **2014**, *70–71*, 71–83. [CrossRef]
7. Dauzères, A.; Le Bescop, P.; Sardini, P.; Coumes, C.C.D. Physico-chemical investigation of clayey/cement-based materials interaction in the context of geological waste disposal: Experimental approach and results. *Cem. Concr. Res.* **2010**, *40*, 1327–1340. [CrossRef]
8. Read, D.; Glasser, F.P.; Ayora, C.; Guardiola, M.T.; Sneyers, A. Mineralogical and microstructural changes accompanying the interaction of Boom Clay with ordinary Portland cement. *Adv. Cem. Res.* **2001**, *13*, 175–183. [CrossRef]
9. Tinseau, E.; Bartier, D.; Hassouta, L.; Devol-Brown, I.; Stammose, D. Mineralogical characterization of the Tournemire argillite after in situ interaction with concretes. *Waste Manag.* **2006**, *26*, 789–800. [CrossRef] [PubMed]
10. Gaboreau, S.; Prêt, D.; Tinseau, E.; Claret, F.; Pellegrini, D.; Stammose, D. 15 years of in situ cement-argillite interaction from Tournemire URL: Characterisation of the multi-scale spatial heterogeneities of pore space evolution. *Appl. Geochem.* **2011**, *26*, 2159–2171. [CrossRef]
11. Dauzères, A.; Achiedo, G.; Nied, D.; Bernard, E.; Alahrache, S.; Lothenbach, B. Magnesium perturbation in low-pH concretes placed in clayey environment—solid characterizations and modeling. *Cem. Concr. Res.* **2016**, *79*, 137–150. [CrossRef]
12. Vespa, M.; Borca, C.; Huthwelker, T.; Lothenbach, B.; Dähn, R.; Wieland, E. Structural characterisation of magnesium (sodium) aluminium silicate hydrate (M-(N)-A-S-H) phases by X-ray absorption near-edge spectroscopy. *Appl. Geochem.* **2020**, *123*, 104750. [CrossRef]

13. Vespa, M.; Lothenbach, B.; Dähn, R.; Huthwelker, T.; Wieland, E. Characterisation of magnesium silicate hydrate phases (M–S–H): A combined approach using synchrotron-based absorption-spectroscopy and ab initio calculations. *Cem. Concr. Res.* **2018**, *109*, 175–183. [CrossRef]

14. Jenni, A.; Gimmi, T.; Alt-Epping, P.; Mäder, U.; Cloet, V. Interaction of ordinary Portland cement and Opalinus Clay: Dual porosity modelling compared to experimental data. *Phys. Chem. Earth Parts A/B/C* **2017**, *99*, 22–37. [CrossRef]

15. Marty, N.C.M.; Tournassat, C.; Burnol, A.; Giffaut, E.; Gaucher, E.C. Influence of reaction kinetics and mesh refinement on the numerical modelling of concrete/clay interactions. *J. Hydrol.* **2009**, *364*, 58–72. [CrossRef]

16. De Windt, L.; Marsal, F.; Tinseau, E.; Pellegrini, D. Reactive transport modeling of geochemical interactions at a concrete/argillite interface, Tournemire site (France). *Phys. Chem. Earth* **2008**, *33*, S295–S305. [CrossRef]

17. Kosakowski, G.; Berner, U. The evolution of clay rock/cement interfaces in a cementitious repository for low- and intermediate level radioactive waste. *Phys. Chem. Earth* **2013**, *64*, 65–86. [CrossRef]

18. Jenni, A.; Mäder, U. Reactive transport modelling of cement-clay interaction accounting for electrostatic effects in the clay. In Proceedings of the Mechanisms and Modelling of Waste/Cement Interactions, Karlsruhe, Germany, 25 March 2019.

19. Jenni, A.; Gimmi, T.; Alt-Epping, P.; Mäder, U. *CI Experiment: Interaction of Ordinary Portland Cement and Opalinus Clay: Dual Porosity Modelling Compared to Experimental Data*; TN 2015-99; Mont Terri Project: St. Ursanne, Switzerland, 2016.

20. Lichtner, P.C. *FLOTRAN User's Manual Version 2.0: LA-CC 02-036*; Los Alamos National Laboratory: Los Alamos, NM, USA, 2007.

21. Beauhheim, R.L. *Hydraulic Conductivity and Head Distributions in the Host Rock Formations of the Proposed Siting Regions*; Nagra Arbeitsbericht NAB: Wettingen, Switzerland, 2013.

22. Kosakowski, G. *Time-Dependent Flow and Transport Calculations for Project Opalinus Clay*; Nagra Arbeitsbericht NAB: Wettingen, Switzerland, 2004.

23. Alt-Epping, P.; Tournassat, C.; Rasouli, P.; Steefel, C.I.; Mayer, K.U.; Jenni, A.; Mäder, U.; Sengor, S.S.; Fernández, R. Benchmark reactive transport simulations of a column experiment in compacted bentonite with multispecies diffusion and explicit treatment of electrostatic effects. *Comput. Geosci.* **2015**, 1–16. [CrossRef]

24. Jenni, A.; Mäder, U.; Wieland, E.; Lerouge, C.; Gaboreau, S. Concrete-clay interaction: Give-and-take without a loser? In Proceedings of the 6th Meeting on Clays in Natural and Engineered Barrier for Radioactive Waste Confinement, Brussels, Belgium, 25 March 2015.

25. Parrot, L.J.; Killoh, D.C. *Prediction of Cement Hydration*; British Ceramic Society: Stoke-on-Trent, UK, 1984; pp. 41–53.

26. Lothenbach, B.; Winnefeld, F. Thermodynamic modelling of the hydration of Portland cement. *Cem. Concr. Res.* **2006**, *36*, 209–226. [CrossRef]

27. Lasaga, A.C. Transition state theory. *Rev. Mineral. Geochem.* **1981**, *8*, 135–168.

28. Lasaga, A.C. Chemical kinetics of water-rock interactions. *J. Geophys. Res.* **1984**, *89*, 4009–4025. [CrossRef]

29. Lothenbach, B.; Rentsch, D.; Wieland, E. Hydration of a silica fume blended low-alkali shotcrete cement. *Phys. Chem. Earth Parts A/B/C* **2014**, *70–71*, 3–16. [CrossRef]

30. Palandri, J.L.; Kharaka, Y.K. *A Compilation of Rate Parameters of Water-Mineral Interaction Kinetics for Application to Geochemical Modeling*; U.S. Geological Survey: Washington, WA, USA, 2004.

31. Babushkin, V.I.; Matveyev, G.M.; Mchedlov-Petrossyan, O.P. *Thermodynamics of Silicates*; Springer: Berlin, Germany, 1985.

32. Hummel, W.; Berner, U.; Curti, E.; Pearson, F.J.; Thoenen, T. Nagra/PSI chemical thermodynamic data base 01/01. *Radiochim. Acta* **2002**, *90*, 805–813. [CrossRef]

33. Matschei, T.; Lothenbach, B.; Glasser, F.P. Thermodynamic properties of Portland cement hydrates in the system $CaO$-$Al_2O_3$-$SiO_2$-$CaSO_4$-$CaCO_3$-$H_2O$. *Cem. Concr. Res.* **2007**, *37*, 1379–1410. [CrossRef]

34. Moeschner, G.; Lothenbach, B.; Rose, J.; Ulrich, A.; Figi, R.; Kretzschmar, R. Solubility of Fe-ettringite ($Ca_6[Fe(OH)_6]_2(SO_4)_3$ $26H_2O$). *Geochim. Cosmochim. Acta* **2008**, *72*, 1–18. [CrossRef]

35. Moeschner, G.; Lothenbach, B.; Winnefeld, F.; Ulrich, A.; Figi, R.; Kretzschmar, R. Solid solution between Al-ettringite and Fe-ettringite ($Ca_6[Al_{1-x}Fe_x(OH)_6]_2(SO_4)_3$ $26H_2O$). *Cem. Concr. Res.* **2009**, *39*, 482–489. [CrossRef]

36. Schmidt, T.; Lothenbach, B.; Romer, M.; Scrivener, K.; Rentsch, D.; Figi, R. A thermodynamic and experimental study of the conditions of thaumasite formation. *Cem. Concr. Res.* **2008**, *38*, 337–349. [CrossRef]

37. Taylor, H.F.W. *Cement Chemistry*, 2nd ed.; Thomas Telford Publishing: London, UK, 1997.

38. Berner, U.; Kulik, D.A.; Kosakowski, G. Geochemical impact of a low-pH cement liner on the near field of a repository for spent fuel and high-level radioactive waste. *Phys. Chem. Earth* **2013**, *64*, 46–56. [CrossRef]

39. Mäder, U. *Reference Pore Water for the Opalinus Clay and "Brown Dogger" for the Provisional Safety-Analysis in the Framework of the Sectorial Plan–Interim Results (SGT-ZE)*; NAB 09-14; Nagra: Wettingen, Switzerland, 2009.

40. Mazurek, M. *Aufbau und Auswertung der Gesteinsparameter-Datenbank für Opalinuston, den 'Braunen Dogger', Effinger Schichten und Mergel-Formationen des Helvetikums*; NAB 11-020; Nagra: Wettingen, Switzerland, 2011.

41. Pearson, F.J.; Tournassat, C.; Gaucher, E.C. Biogeochemical processes in a clay formation in situ experiment: Part E - Equilibrium controls on chemistry of pore water from the Opalinus Clay, Mont Terri Underground Research Laboratory, Switzerland. *Appl. Geochem.* **2011**, *26*, 990–1008. [CrossRef]

42. Chagneau, A.; Tournassat, C.; Steefel, C.I.; Bourg, I.C.; Gaboreau, S.; Esteve, I.; Kupcik, T.; Claret, F.; Schaefer, T. Complete Restriction of $^{36}Cl^-$ Diffusion by Celestite Precipitation in Densely Compacted Illite. *Environ. Sci. Technol. Lett.* **2015**, *2*, 139–143. [CrossRef]

43. Appelo, C.A.J.; Wersin, P. Multicomponent diffusion modeling in clay systems with application to the diffusion of tritium, iodide, and sodium in Opalinus Clay. *Environ. Sci. Technol.* **2007**, *41*, 5002–5007. [CrossRef]

44. Steefel, C.I. CrunchFlow: Software for Modeling Multicomponent Reactive Flow and Transport. 2009. Available online: https://bitbucket.org/crunchflow/crunchtope-dev/wiki/Home (accessed on 23 February 2021).

45. Tournassat, C.; Appelo, C.A.J. Modelling approaches for anion-exclusion in compacted Na-bentonite. *Geochim. Cosmochim. Acta* **2011**, *75*, 3698–3710. [CrossRef]

46. Tournassat, C.; Steefel, C.I. Ionic Transport in Nano-Porous Clays with Consideration of Electrostatic Effects. *Rev. Mineral. Geochem.* **2015**, *80*, 287–329. [CrossRef]

47. Tournassat, C.; Steefel, C.I. Reactive Transport Modeling of Coupled Processes in Nanoporous Media. *Rev. Mineral. Geochem.* **2019**, *85*, 75–109. [CrossRef]

48. Gimmi, T.; Alt-Epping, P. Simulating Donnan equilibria based on the Nernst-Planck equation. *Geochim. Cosmochim. Acta* **2018**, *232*, 1–13. [CrossRef]

49. Lothenbach, B. Hydration of Blended Cements. In *Cement-Based Materials for Nuclear Waste Storage*; Bart, F., Cau-di-Coumes, C., Frizon, F., Lorente, S., Eds.; Springer: New York, NY, USA, 2013; pp. 33–41. [CrossRef]

50. Myers, R.J.; L'Hôpital, E.; Provis, J.L.; Lothenbach, B. Effect of temperature and aluminium on calcium (alumino)silicate hydrate chemistry under equilibrium conditions. *Cem. Concr. Res.* **2015**, *68*, 83–93. [CrossRef]

51. Bernard, E.; Lothenbach, B.; Cau-Dit-Coumes, C.; Pochard, I.; Rentsch, D. Aluminum incorporation into magnesium silicate hydrate (M–S–H). *Cem. Concr. Res.* **2020**, *128*, 105931. [CrossRef]

52. Roosz, C.; Grangeon, S.; Blanc, P.; Montouillout, V.; Lothenbach, B.; Henocq, P.; Giffaut, E.; Vieillard, P.; Gaboreau, S. Crystal structure of magnesium silicate hydrates (M–S–H): The relation with 2:1 Mg–Si phyllosilicates. *Cem. Concr. Res.* **2015**, *73*, 228–237. [CrossRef]

53. Marty, N.C.M.; Claret, F.; Lassin, A.; Tremosa, J.; Blanc, P.; Madé, B.; Giffaut, E.; Cochepin, B.; Tournassat, C. A database of dissolution and precipitation rates for clay-rocks minerals. *Appl. Geochem.* **2015**, *55*, 108–118. [CrossRef]

54. Marty, N.M.; Bildstein, O.; Blanc, P.; Claret, F.; Cochepin, B.; Gaucher, E.; Jacques, D.; Lartigue, J.-E.; Liu, S.; Mayer, K.U.; et al. Benchmarks for multicomponent reactive transport across a cement/clay interface. *Comput. Geosci.* **2015**, *19*, 1–19. [CrossRef]

*Article*

# Reactive Transport Modelling of the Long-Term Interaction between Carbon Steel and MX-80 Bentonite at 25 °C

**M. Carme Chaparro *, Nicolas Finck, Volker Metz and Horst Geckeis**

Karlsruhe Institute of Technology (KIT), Institute for Nuclear Waste Disposal (INE), P.O. Box 3640, 76021 Karlsruhe, Germany; nicolas.finck@kit.edu (N.F.); Volker.Metz@kit.edu (V.M.); horst.geckeis@kit.edu (H.G.)
* Correspondence: carme.chaparro@kit.edu

**Abstract:** The geological disposal in deep bedrock repositories is the preferred option for the management of high-level radioactive waste (HLW). In some of these concepts, carbon steel is considered as a potential canister material and bentonites are planned as backfill material to protect metallic waste containers. Therefore, a 1D radial reactive transport model has been developed in order to better understand the processes occurring during the long-term iron-bentonite interaction. The numerical model accounts for diffusion, aqueous complexation reactions, mineral dissolution/precipitation and cation exchange at a constant temperature of 25 °C under anoxic conditions. Our results suggest that Fe is sorbed at the montmorillonite surface via cation exchange in the short-term, and it is consumed by formation of the secondary phases in the long-term. The numerical model predicts precipitation of nontronite, magnetite and greenalite as corrosion products. Calcite precipitates due to cation exchange in the short-term and due to montmorillonite dissolution in the long-term. Results further reveal a significant increase in pH in the long-term, while dissolution/precipitation reactions result in limited variations of the porosity. A sensitivity analysis has also been performed to test the effect of selected parameters, such as corrosion rate, diffusion coefficient and composition of the bentonite porewater, on the corrosion processes. Overall, outcomes suggest that the predicted main corrosion products in the long-term are Fe-silicate minerals, such phases thus should deserve further attention as a chemical barrier in the diffusion of radionuclides to the repository far field.

**Keywords:** radioactive waste disposal; iron–bentonite interaction; reactive transport; numerical model

**Citation:** Chaparro, M.C.; Finck, N.; Metz, V.; Geckeis, H. Reactive Transport Modelling of the Long-Term Interaction between Carbon Steel and MX-80 Bentonite at 25 °C. *Minerals* **2021**, *11*, 1272. https://doi.org/10.3390/min11111272

Academic Editors: Ana María Fernández, Stephan Kaufhold, Markus Olin, Paul Wersin and James Wilson

Received: 31 August 2021
Accepted: 7 November 2021
Published: 16 November 2021

**Publisher's Note:** MDPI stays neutral with regard to jurisdictional claims in published maps and institutional affiliations.

## 1. Introduction

Burial in a stable deep geological formation is the generally accepted strategy for the safe management of high-level radioactive waste. In some repository concepts, the waste is foreseen to be emplaced in metallic canisters (e.g., made of carbon steel), which would be surrounded by bentonite as buffer material (engineered barrier system or EBS). During the geochemical evolution of such a repository system, groundwater is expected to move through the barriers and contact the metallic canisters, which will start corroding. In case of canister failure and subsequent waste matrix alteration, bentonite, consisting mainly of montmorillonite, would be able to retain the released radionuclides. However, during container corrosion, long-term geochemical processes at the corroding container/bentonite interface will proceed and have an impact on the properties of the geotechnical barrier. Such information may be of relevance for the Safety Case of the repository. To gain insights into container corrosion and bentonite alteration, several experiments and numerical models studying the iron-bentonite interactions have been reported in literature (see, e.g., in [1–5]).

In laboratory experiments at 90 °C studying the interaction between iron and argillite [6,7], or iron and bentonite [8], the formation of a layer of secondary phases, made of magnetite, siderite and Fe-silicates such as nontronite and cronstedtite, was found at the reacting interface. More recently, Kaufhold et al. [9] reported that the presence of highly reactive silica results in the predominant formation of Fe-silicates as corrosion products rather than magnetite. These findings suggest that the nature of formed corrosion products depends on the composition of the infiltrating porewater and thus on the mineralogical composition of the bentonite and argillite but also on the coupling of reaction kinetics and transport. Some of these reactions take a long time and cannot be reproduced over laboratory time scales [10]. Due to the strong coupling of kinetically controlled dissolution/precipitation/transport processes, reactive transport models play an important role for studying the long-term iron-bentonite interaction. Some of the reactive transport calculations considered magnetite and siderite as principal corrosion products [11,12], and most of them also considered the formation of Fe-silicates such as berthierine, greenalite or cronstedtite [10,13–19]. The predicted results suggested that not only magnetite but also Fe-silicates can form during the long-term iron-bentonite interaction. Hence, the conceptual model considered has an effect on the predicted results. For instance, some reported reactive transport models considered a gap between the canister and the bentonite (see, e.g., in [15,18]), but this approach enhances the alteration of both canister and bentonite because the reactive surface area is much larger than the one considering that canister and bentonite are in contact without any gap in between. In the real system, any gap would be closed due to bentonite swelling upon hydration. Thus, the reactive surface area of the bentonite minerals is only related to the porosity of the bentonite. Other authors considered the canister as porous material (see, e.g., in [13,19]) allowing dissolution of the canister and precipitation of corrosion products within the canister pores. However, in the real system, the canister can only react at the interface with bentonite due to the presence of porewater. The temperature of the system is under discussion in the literature, some reported reactive transport models of EBS in clays considered a fully water-saturated isothermal system with a temperature of 50 °C [13], 70 °C [18] or 100 °C [17,20] conditioning the nature of the secondary phases that can form during the long-term interaction. However, according to Finsterle et al. [21], the heat pulse could last between a few decades to a few hundred years, then the ambient temperature prior to waste emplacement will prevail in the long-term anaerobic conditions. Then, a temperature around 25 °C would be expected from the depth taking into account the geothermal gradient. Indeed, isothermal reactive transport models at 25 °C have been reported [19], however, for an EBS in granite.

The objective of this study is, therefore, to better understand the processes that will occur in the long-term iron-bentonite (i.e., MX-80) interaction and the nature of the secondary phases for an EBS in clay rock by means of reactive transport models. The conceptual model assumes that bentonite is fully water-saturated and that heat pulse is already dissipated, therefore, anaerobic conditions at a constant temperature of 25 °C are considered. Reported corrosion products that were found experimentally [6,8,22,23] are considered as potential secondary phases. Due to the lack of long-term experimental data, our results are compared with other published numerical studies on iron–clay/bentonite interaction [13,18,19]. The conceptual model and the results of the reference model are presented in Sections 2 and 3 respectively. In addition, a sensitivity analysis, Section 4, has been performed in order to study the effect of selected parameters, such as corrosion rate, diffusion coefficient, reactive surface area and the porewater composition, on the corrosion products.

## 2. Reactive Transport Model

### 2.1. Processes and Governing Equations

The simulations were carried out using Retraso-CodeBright [24]. This software package couples CodeBright [25], a finite element computer code that can handle multiphase flow and heat transfer to a module for reactive transport, which includes aqueous complexation reactions, precipitation/dissolution of minerals and adsorption. For more details about the code we refer to Olivella et al. [25] and Saaltink et al. [24]. Only an outline of the main processes taken into account in the present work is given here.

The conceptual model considers that the system is fully water-saturated, then the mass balance of reactive transport can be written as

$$\mathbf{U_a}\frac{\delta\phi\rho_l\mathbf{c_a}}{\delta t} + \mathbf{U_d}\frac{\delta\phi\rho_l\mathbf{c_d}}{\delta t} + \mathbf{U_m}\frac{\delta(1-\phi)\rho_s\mathbf{c_m}}{\delta t} = \mathbf{U_a}i(\mathbf{c_a}) + \mathbf{U}\mathbf{S_k^t}\mathbf{r_m}(\mathbf{c_a}) \tag{1}$$

where $\phi$ is the porosity, $\rho_l$ and $\rho_s$ are the liquid and solid densities, respectively (kg m$^{-3}$). Vectors $\mathbf{c_a}$, $\mathbf{c_d}$ and $\mathbf{c_m}$ (mol kg$^{-1}$) are the concentrations of aqueous species, adsorbed species and minerals respectively. Matrix $\mathbf{S_k^t}$ and vector $\mathbf{r_m}$ contain the stoichiometric coefficients and the rates of the kinetic reactions, which can be considered as functions of all aqueous concentrations. Matrices $\mathbf{U_a}$, $\mathbf{U_d}$ and $\mathbf{U_m}$ are called the component matrices for aqueous, adsorbed and mineral species and relate the concentrations of the species with the total concentrations of the components. The matrix $\mathbf{U}$ is the component matrix for all species. These matrices can be computed from the stoichiometric coefficient of the chemical reactions. $i$ is the diffusion term, and according to Fick's law [25] can be written as

$$i(\mathbf{c_a}) = \nabla \cdot (\mathbf{D_p}\phi\rho_l\nabla(\mathbf{c_a})) \tag{2}$$

where $\mathbf{D_p}$ is the molecular diffusion coefficient in porous media (m$^2$ s$^{-1}$):

$$D_p = \tau 1.1 \times 10^{-4}\exp\left(\frac{-24530}{R(273.15+T)}\right) \tag{3}$$

$$D_e = \phi D_p \tag{4}$$

where $\tau$ is the tortuosity factor, $R = 8.31$ J mol$^{-1}$ K$^{-1}$, $T$ is the temperature (°C) and $D_e$ is the effective diffusion coefficient (m$^2$ s$^{-1}$).

### 2.2. Conceptual Model

Our conceptual model considers a cylindrical carbon steel canister surrounded by bentonite (Figure 1). It takes into account the reactions that could occur during the interaction between steel, bentonite and the porewater. We assume anaerobic conditions, when the bentonite is fully water-saturated, and therefore interaction between canister, bentonite and porewater can occur. This also implies that the thermal pulse has dissipated [19,26], thus a temperature of 25 °C is considered. No gas phase has been considered in the numerical model due to a water-saturated system, and the negligible presence of some type of dissolved gases such as hydrogen or $CO_2$. These assumptions are similar to that reported in literature by other authors (see, e.g., in [19]) implying that the model does not consider the potential impact of $H_2$ evolution on the integrity of the bentonite barrier. However, in the real system aerobic conditions will prevail first and temperature will decrease with time, and during the hydration process the bentonite will be altered. According to Bossart et al. [26], consumption of oxygen in bentonite pores will be consumed within decades. During this phase, a mixture of crystalline and amorphous Fe(II) and Fe(III) corrosion products will be generated [22]. Thermal equilibration usually is assumed to take decades up to some hundreds of years [26]. Recent simulations [27], however, result in much longer thermal equilibration times. This will have a certain influence on corrosion and alteration kinetics. However, the temperature dependence of iron corrosion rates appears to be moderate (see, e.g., in [28]). Full bentonite saturation is established after some decades [26,27,29]. As long as the packing density of the bentonite does not reach threshold values of 1250–1450 kg m$^{-3}$ [30] significant microbial impact cannot be excluded. In experimental studies complex corrosion product mixtures are found, predominantly magnetite [30] but also sulphide minerals such a pyrite [31]. However, the anoxic steel corrosion phase under saturated bentonite conditions and at relatively low temperature appears to dominate thereafter for thousands of years. The main fraction of secondary iron corrosion phases will, thus, form under those conditions investigated in the present study and finally will represent the corrosion phase layer which will interact with radionuclides released from waste forms first upon loss of container integrity.

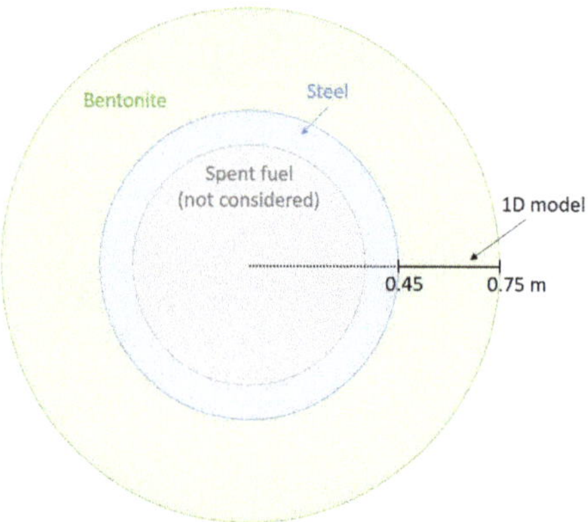


**Figure 1.** Geometry and materials of the 1D radial model. The axisymmetric model only considers the canister-bentonite interface, at x = 0.45 m, and a thickness of 0.3 m of bentonite, from x = 0.45 to x = 0.75 m.

The geometry of the 1D axisymmetric model reflects the canister–bentonite interface and 0.3 m of bentonite (Figure 1). The canister is considered only as a node in the boundary. The bentonite is discretised into 45 unidimensional finite elements with sizes ranging from 0.5 cm near to the interface to 1 cm near to the boundary. The total iron–bentonite interaction time is 10,000 years. The model considers only diffusion as transport mechanism (Equation (2)). We assume a pore diffusion coefficient of $1 \times 10^{-10}$ m$^2$ s$^{-1}$ (Equation (3)) [19]. The porosity of bentonite is 0.4 [18,19], yielding an effective diffusion coefficient of $4 \times 10^{-11}$ m$^2$ s$^{-1}$ (Equation (4)). Table 1 summarises our conceptual model, which is also compared with other reported numerical models.

**Table 1.** Comparison of the main differences in conceptual models between the present study and other reported models in the literature.

| | Present Study | Bildstein et al. [13] | Wilson et al. [18] | Samper et al. [19] |
|---|---|---|---|---|
| Geometry | Canister as boundary | Porous canister | Canister as boundary and porous canister | Porous canister |
| Temperature | 25 °C | 50 °C | 70 °C | 25 °C |
| Bentonite | MX-80 | MX-80 | MX-80 | FEBEX |
| Secondary phases | Magnetite, nontronite greenalite, siderite cronstedtite | Magnetite, siderite, chamosite, saponite cronstedtite | Magnetite, siderite amesite, berthierine chlorite, greenalite, lizardite, analcime, cronstedtite, saponite | Magnetite, siderite analcime, cronstedtite |
| Corrosion rate | 2 µm y$^{-1}$ | 4.3 µm y$^{-1}$ | 1 µm y$^{-1}$ | 2 µm y$^{-1}$ |
| Interaction time | 10,000 y | 16,250 y | 10,000 y | $1 \times 10^6$ y |

### 2.3. Geochemical System

The geochemical model accounts for steel corrosion, aqueous complexation reactions, mineral dissolution/precipitation and cation exchange reactions.

### 2.3.1. Solid Composition

The mineral composition of MX-80 bentonite and the volumetric fractions of each mineral were based on that given by Karnland [32], which is comparable to the characterisation made by Kiviranta et al. [33]. The primary phases considered are montmorillonite, quartz, muscovite, albite, illite, pyrite and calcite. The volumetric fractions of each phase are listed in Table 2. We assume that carbon steel is composed only of iron, Fe(s), which is the main phase of carbon steel. The potential corrosion products considered in the numerical model have been selected according to published experiments [6–8,23] and from the Iron Corrosion in Bentonite (IC-A) experiment in the Mont-Terri rock laboratory (Switzerland) [22,31,34], which are magnetite, greenalite, cronstedtite, siderite, nontronite-Na and nontronite-Mg. Other modelling works reported in literature (see, e.g., in [18,19]) considered zeolites as secondary phases such as analcime; however, we did not consider them because they were not reported by Morelová et al. [31].

**Table 2.** Bentonite mineralogical composition considered in the numerical model. The volumetric fractions are also listed for each phase.

| Mineral | Volumetric Fraction ($m^3 m^{-3}$) |
|---|---|
| Montmorillonite | $5.01 \times 10^{-1}$ |
| Quartz | $4.70 \times 10^{-2}$ |
| Albite | $2.10 \times 10^{-2}$ |
| Muscovite | $2.10 \times 10^{-2}$ |
| Illite | $5.00 \times 10^{-3}$ |
| Pyrite | $4.00 \times 10^{-3}$ |
| Calcite | $1.00 \times 10^{-3}$ |

2.3.2. Solution Composition

Table 3 shows the chemical composition of the MX-80 bentonite initial porewater used in the numerical model. In the real system, the bentonite will be hydrated by the clay groundwater reacted with the concrete liner. Therefore, the composition of the MX-80 bentonite porewater will depend not only on the reactions with the primary phases of the materials, but also on the secondary phases formed during these reactions. In order to simplify the problem, the majority of authors in literature calculate the porewater as a result of the equilibrium of bentonite with a groundwater (see, e.g., in [35]). Other authors calculate the initial water composition of the bentonite using mineral constraints (see, e.g., in [13]). In this work, a clay groundwater that was not oversaturated with the minerals used in the model has been selected. The initial porewater calculated, first using PhreeqC, corresponds to the composition of the porewater used by Martin et al. [8,36] equilibrated with montmorillonite, because it is the main phase of the bentonite. Ca, Na, K, Fe, Mg, Al and Si are thus at equilibrium with the montmorillonite used in the numerical model (Table 4), and $SO_4$ corresponds to the value used by Martin et al. [8]. Then, the obtained values from the PhreeqC calculation have been fixed in Retraso-CodeBright as well as the pH, where Cl has been used for the charge balance and $HCO_3^-$ is at equilibrium with calcite. The resulting porewater composition is a simplification, therefore a sensitivity analysis studying the effect of the variation of the initial porewater has been done (Section 4.3). The primary and secondary species taken into account in the numerical model have been selected according to the speciation made using PhreeqC with the clay porewater [8,36] equilibrated with montmorillonite at 25 °C, which pH is 7.30. The resulting primary and secondary species have been compared with those reported by Samper et al. [19]. The redox is described by the redox pairs $O_2(aq)/H_2O$ and $Fe^{3+}/Fe^{2+}$.

**Table 3.** MX-80 bentonite initial porewater composition used in the numerical model. Imposed constraints to calculate some of these values (equilibrium with solids and charge balance) are also indicated.

| Component | Water Composition (mol L$^{-1}$) |
| --- | --- |
| $Al(OH)_4^-$ | $2.00 \times 10^{-7}$ |
| $Ca^{2+}$ | $8.60 \times 10^{-3}$ |
| $Cl^-$ | $3.54 \times 10^{-2}$ charge balance |
| $Fe^{+2}$ | $7.19 \times 10^{-8}$ |
| $HCO_3^-$ | $1.31 \times 10^{-3}$ calcite |
| $K^+$ | $2.50 \times 10^{-3}$ |
| $Mg^{2+}$ | $6.50 \times 10^{-3}$ |
| $Na^+$ | $7.30 \times 10^{-3}$ |
| $SiO_2(aq)$ | $1.00 \times 10^{-4}$ |
| $SO_4^{2-}$ | $1.70 \times 10^{-3}$ |
| $O_2(aq)$ | $1.00 \times 10^{-15}$ |
| pH | 7.30 |

### 2.3.3. Thermodynamic Data

The used thermodynamic data at 25 °C for all mineral and aqueous complexation reactions are given in Tables 4 and 5, respectively (also in Tables S1–S3 from the supplementary material). The equilibrium constants (log K) are taken from the Geochemist's Workbench database *thermo.com.V8.R6.full* (generated by GEMBOCHS.V2-Jewel.src.R6 [37]).

**Table 4.** Equilibrium constants (log $K_{eq}$ at 25 °C) of the minerals taken into account in the numerical model. Reactions are written as the dissolution of 1 mol of mineral in terms of the primary species $Ca^{2+}$, $Fe^{2+}$, $SiO_2(aq)$, $Al(OH)_4^-$, $H^+$, $SO_4^{2-}$, $HCO_3^-$, $K^+$, $Na^+$, $Mg^{2+}$, $Cl^-$ and $O_2(aq)$.

| Mineral | Formula/ Reaction | log $K_{eq}$ |
| --- | --- | --- |
| Iron | $Fe(s) + 2H^+ + 0.5O_2(aq) = Fe^{2+} + H_2O$ | 59.01 |
| Montmorillonite | $Ca_{0.02}Na_{0.15}K_{0.2}Fe_{0.45}Mg_{0.9}Al_{1.25}Si_{3.75}O_{10}(OH)_2$ | −18.02 |
| Quartz | $SiO_2$ | −4.00 |
| Albite | $NaAlSi_3O_8$ | −19.39 |
| Muscovite | $KAl_3Si_3O_{10}(OH)_2$ | −52.86 |
| Illite | $K_{0.6}Mg_{0.25}Al_{2.3}Si_{3.5}O_{10}(OH)_2$ | −41.92 |
| Pyrite | $FeS_2$ | 217.34 |
| Calcite | $CaCO_3$ | 1.85 |
| Magnetite | $Fe_3O_4$ | −6.51 |
| Greenalite | $Fe_3Si_2O_5(OH)_4$ | 22.65 |
| Cronstedtite | $Fe_4SiO_5(OH)_4$ | −0.73 |
| Nontronite-Na | $Na_{0.33}Fe_2Al_{0.33}Si_{3.67}H_2O_{12}$ | −35.82 |
| Nontronite-Mg | $Mg_{0.165}Fe_2Al_{0.33}Si_{3.67}H_2O_{12}$ | −35.92 |
| Siderite | $FeCO_3$ | −0.19 |

### 2.3.4. Cation Exchange

The MX-80 bentonite cation exchange reactions and log K at 25 °C taken into account in the numerical model are listed in Table 6 (also in Table S4 from the supplementary material). The Gaines–Thomas convention is used for cation exchange reactions based on Appelo and Postma [38]. A cation exchange capacity (CEC) of 750 meq kg$^{-1}$ is used for the bentonite [32].

**Table 5.** Secondary species with their equilibrium constants (log $K_{eq}$ at 25 °C) taken into account in the numerical model. Reactions are written as the dissolution of 1 mol of mineral in terms of the primary species $Ca^{2+}$, $Fe^{2+}$, $SiO_2(aq)$, $Al(OH)_4^-$, $H^+$, $SO_4^{2-}$, $HCO_3^-$, $K^+$, $Na^+$, $Mg^{2+}$, $Cl^-$ and $O_2(aq)$.

| Formula | log $K_{eq}$ | Formula | log $K_{eq}$ |
|---|---|---|---|
| $Al(OH)_3(aq)$ | −5.99 | $Fe(OH)_4^-$ | 13.12 |
| $Al(OH)_2^+$ | −11.55 | $Fe(OH)_2^+$ | −2.59 |
| $AlOH^{2+}$ | −17.18 | $Fe(SO_4)_2^-$ | −11.73 |
| $Al^{3+}$ | −22.14 | $FeSO_4(aq)$ | −2.19 |
| $CaCO_3(aq)$ | 7.01 | $Fe_2(OH)_2^{4+}$ | −14.02 |
| $CaSO_4(aq)$ | −2.10 | $HSO_4^-$ | −1.98 |
| $CaOH^+$ | 12.85 | $HS^-$ | 138.27 |
| $CaCl^+$ | 0.70 | $OH^-$ | 13.99 |
| $CaHCO_3^+$ | −1.04 | $MgCO_3(aq)$ | 7.35 |
| $CaCl_2(aq)$ | 0.65 | $MgHCO_3^+$ | −1.03 |
| $CaH_3SiO_4^+$ | 8.79 | $MgSO_4(aq)$ | −2.38 |
| $CO_3^{2-}$ | 10.32 | $MgOH^+$ | 11.78 |
| $CO_2(aq)$ | −6.34 | $MgCl^+$ | 0.14 |
| $Fe^{3+}$ | −8.48 | $MgH_3SiO_4^+$ | 8.54 |
| $FeHCO^{+3}$ | −2.04 | $NaOH(aq)$ | 14.79 |
| $FeCO_3(aq)$ | 5.53 | $NaCO_3^-$ | 9.82 |
| $FeCl^+$ | −0.16 | $NaHCO_3(aq)$ | −0.15 |
| $FeCl^{2+}$ | −7.67 | $NaCl(aq)$ | 0.78 |
| $FeCl_2(aq)$ | 2.46 | $NaSO_4^-$ | −0.81 |
| $FeOH^+$ | 9.63 | $KCl(aq)$ | 1.50 |
| $FeOH^{2+}$ | −6.29 | $KOH(aq)$ | 14.46 |
| $Fe(OH)_2(aq)$ | 20.59 | $KSO_4^-$ | −0.87 |
| $Fe(OH)_3(aq)$ | 3.64 | | |

**Table 6.** MX-80 bentonite cation exchange reactions and log K at 25 °C [38].

| Exchange Reaction | log K |
|---|---|
| $2NaX + Ca^{2+} = 2Na^+ + CaX_2$ | 0.80 |
| $2NaX + Fe^{2+} = 2Na^+ + FeX_2$ | 0.50 |
| $2NaX + Mg^{2+} = 2Na^+ + MgX_2$ | 0.60 |
| $NaX + K^+ = Na^+ + KX$ | 0.70 |

### 2.3.5. Kinetic Data

A kinetic approach is used for the dissolution-precipitation of mineral phases (Equation (5)). The form of the rate law is from [39]

$$R_m = \sigma_m k_m (\Omega_m - 1)^\eta \tag{5}$$

where $R_m$ is the mineral dissolution–precipitation rate (mol m$^{-3}$ s$^{-1}$), $\sigma_m$ is the surface area (m$^2$m$^{-3}$), $k_m$ is the kinetic rate constant (mol m$^{-2}$ s$^{-1}$), $\Omega_m$ is $\frac{IAP}{K}$ where $IAP$ is the ion activity product and $K$ is the equilibrium constant (ion activity product at equilibrium), and $\eta$ is an empirical parameter of the kinetic law, which is 1 for all the minerals except the iron in which is 0 in order to get a constant iron corrosion rate.

The reactive surface areas and kinetic constants used for the primary minerals considered in the bentonite and corrosion products, based on those used by Bildstein et al. [13], are given in Table 7. For the secondary phases, large values of the surface areas are used, leading to precipitation at equilibrium.

**Table 7.** Reactive surface areas and kinetic constants (at 25 °C) of the primary minerals considered in the bentonite and the possible corrosion products. (a) Cama et al. [40] (b) Bildstein et al. [13] (c) Dove [41] (d) Knauss and Wolery [42] (e) Kamei and Ohmoto [43] (f) Plummer et al. [44] (g) Wilson et al. [18] .

| Mineral | $\sigma_m$ (m$^2$/m$^3$) | $k_m$ (mol/m$^2$s) | References ($k_m$) |
|---|---|---|---|
| Montmorillonite | $1 \times 10^3$ | $1.58 \times 10^{-13}$ | (a,b) |
| Quartz | 0.01 | $1.00 \times 10^{-14}$ | (c,b) |
| Albite | 0.01 | $1.00 \times 10^{-13}$ | (d,b) |
| Muscovite | 0.01 | $1.58 \times 10^{-13}$ | Montmorillonite analogue |
| Illite | 0.01 | $1.58 \times 10^{-13}$ | Montmorillonite analogue |
| Pyrite | 0.01 | $5.01 \times 10^{-6}$ | (e,b) |
| Calcite | 0.01 | $6.31 \times 10^{-7}$ | (f,b) |
| Magnetite | $1 \times 10^3$ | $4.47 \times 10^{-11}$ | (g) |
| Greenalite | $1 \times 10^3$ | $4.90 \times 10^{-11}$ | (g) |
| Cronstedtite | $1 \times 10^3$ | $1.58 \times 10^{-13}$ | (b) |
| Nontronite-Na | $1 \times 10^3$ | $1.58 \times 10^{-13}$ | Montmorillonite analogue |
| Nontronite-Mg | $1 \times 10^3$ | $1.58 \times 10^{-13}$ | Montmorillonite analogue |
| Siderite | $1 \times 10^3$ | $1.00 \times 10^{-9}$ | (b) |

After the closure of the repository the available oxygen will be consumed, then, in the long-term, anaerobic conditions will prevail. Thus, iron will corrode according to

$$Fe(s) + 2H_2O = Fe^{2+} + 2OH^- + H_2(g) \tag{6}$$

The following reaction is obtained by rewriting reaction (6) in terms of the primary species used in the numerical model:

$$Fe(s) + 2H^+ + 0.5O_2(aq) = Fe^{2+} + H_2O \tag{7}$$

Following Samper et al. [19] the iron corrosion rate ($r_c$) is calculated as

$$r_c = \frac{k_c M_w}{\rho} \tag{8}$$

where $k_c$ is the kinetic constant corrosion rate (mol m$^{-2}$s$^{-1}$), $M_w$ is the molecular weight (55.85 g mol$^{-1}$) and $\rho$ is the density of carbon steel (7860 kg m$^{-3}$).

A constant corrosion rate of 2 µm y$^{-1}$ [22], which gives $k_c = 8.93 \times 10^{-9}$ mol m$^{-2}$s$^{-1}$, is used in the reference model.

## 3. Results of the Reference Model

### 3.1. Solution Composition

Figure 2 shows the calculated concentration of dissolved aqueous species and adsorbed species against distance. During the first 100 years, the Fe concentration in the porewater increases due to iron dissolution. However, Fe is partly adsorbed by the montmorillonite and partly consumed by the secondary phases (e.g., nontronite). At 1000 years, it decreases considerably in the porewater (below $1.6 \times 10^{-9}$ mol L$^{-1}$) because it is mainly consumed by the formation of secondary phases (e.g., magnetite, nontronite and greenalite) rather than being adsorbed, and also because of the high pH. The Si concentration increases in the first 5 years due to dissolution of bentonite minerals, after that, it decreases because of the relatively rapid precipitation rate for Fe-silicates and the slow dissolution rate of montmorillonite. The Ca concentration decreases in the first 1000 years due to calcite precipitation, afterwards it increases due to dissolution of montmorillonite. The HCO$_3$ concentration decreases with time due to calcite precipitation. The pH of the porewater increases progressively from 7.5 to 10.9 because of the dissolution of iron (Equation (7)) and some bentonite minerals, such as montmorillonite, albite and illite consume protons (Table 4). This increase in pH has also been reported by Bildstein et al. [13].

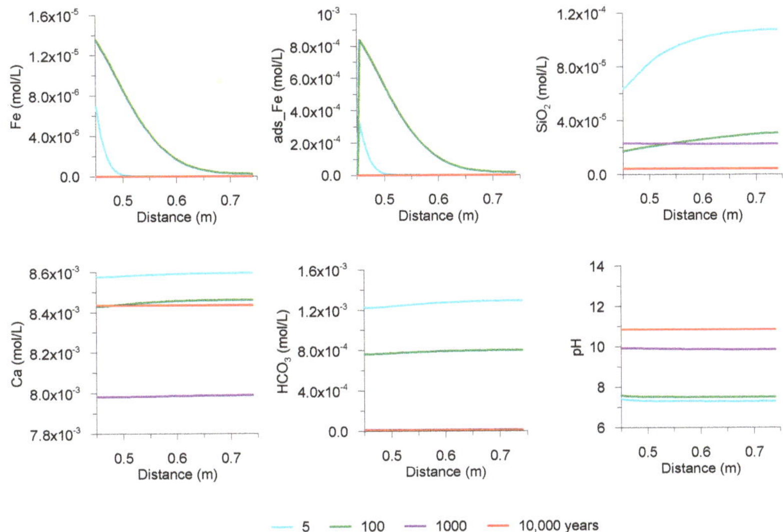


**Figure 2.** Concentration of dissolved species in the porewater and of adsorbed species against distance for the reference model ($r_c = 2$ $\mu$m y$^{-1}$). The interface between steel and bentonite is at 0.45 m. Results after 5, 100, 1000 and 10,000 years of interaction are compared.

### 3.2. Mineral Composition of the Bentonite

Figure 3 displays the calculated variation of the volumetric fraction of the minerals against distance. Bentonite minerals dissolve because the dissolution of iron provokes an increase in pH that makes silicates unstable. Montmorillonite dissolution releases Ca yielding calcite precipitation, however, specially in the first 100 years, Ca in the montmorillonite interlayer can also be exchanged by Fe and thereby likewise induce calcite precipitation ($7.5 \times 10^{-6}$ m$^3$m$^{-3}$). The volumetric fraction of calcite remains constant from 1000 to 10,000 years indicating no precipitation/dissolution. Dissolution of the primary phases of the bentonite are more important after 10,000 years when the pH is higher. Results from the numerical model show precipitation of nontronite, magnetite and greenalite as secondary phases. At 5 years, specially nontronite-Mg but also nontronite-Na precipitate because there is available Fe and Si in the porewater due to iron and bentonite dissolution, respectively. Magnetite starts precipitating at 100 years, when enough Fe is available. At 1000 years, pH is increasing that provokes the dissolution of nontronite-Na and nontronite-Mg, which release Si and Fe. At the same time, the bentonite minerals and the canister are still dissolving. These conditions lead to greenalite formation. Greenalite is more stable than magnetite under these conditions, therefore, magnetite dissolves. At 10,000 years, greenalite is the main corrosion product. Corrosion products (nontronite, magnetite and greenalite) precipitate quite close to the iron-bentonite interface, and calcite also precipitates along the first 5 cm right after the interface. The reference model predicts neither precipitation of cronstedtite nor that of siderite.

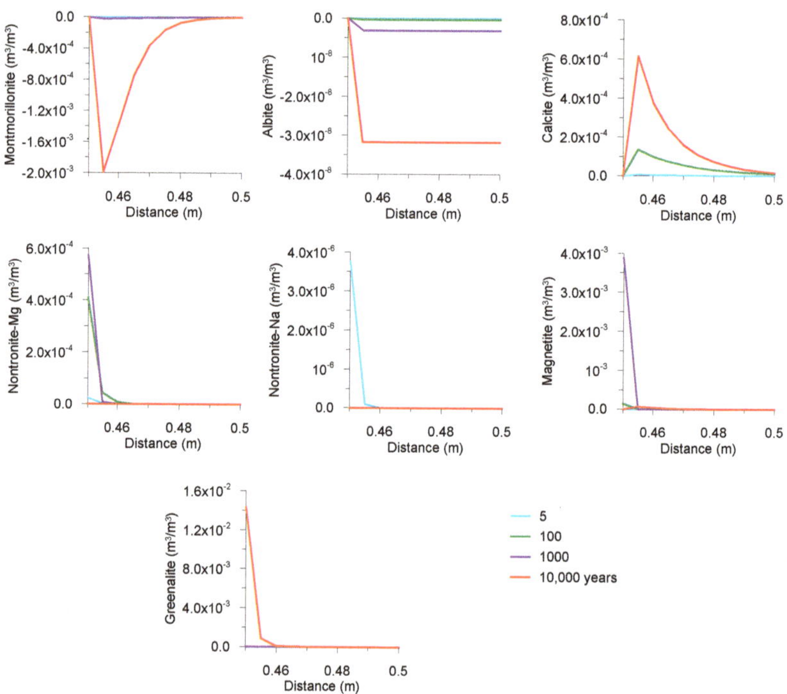

**Figure 3.** Volumetric fraction of minerals versus distance for the reference model ($r_c$ = 2 μm y$^{-1}$). For the primary phases (montmorillonite, albite and calcite) the variation of the volumetric fraction is plotted. Positive values mean precipitation and negative values mean dissolution. The interface between steel and bentonite is at 0.45 m. Results at 5, 100, 1000 and 10,000 years of interaction are compared.

Precipitation of magnetite, calcite and greenalite has been recently found in the Iron Corrosion in Bentonite (IC-A) experiment in the Mont-Terri rock laboratory (Switzerland) [31], which supports our model results. Experiments performed by Schlegel et al. [6,7] and Martin et al. [8] reported the presence of magnetite, cronstedtite and siderite, but our reference model does not predict the formation of siderite or cronstedtite. These reported experiments were performed at 90 °C with Callovo-Oxfordian argilite, which contains about 30 wt% calcite. Thus, the temperature, bentonite and porewater compositions could have a significant impact on the nature of formed corrosion products.

Other modelling works reported in literature, such as Bildstein et al. [13] and Samper et al. [19], predicted the precipitation of cronstedtite and siderite. However, they did not consider greenalite as corrosion product in their numerical models. Therefore, the potential secondary phases selected in the numerical model have an impact on the predicted results. On the other hand, the porewater composition considered had higher $HCO_3^-$ concentration in solution allowing siderite precipitation, and the differences in the bentonite composition and temperature of the system could have an effect as well. Although small differences, in general, our results are consistent with other reported models in the literature [13,18,19], where a carbonate phase (calcite or siderite) and Fe-silicate phases (greenalite, nontronite or cronstedtite) will precipitate as corrosion products together with magnetite. The other reported numerical models [13,18,19] consider the canister as porous media, then magnetite prevails as corrosion product precipitating in the canister. However, when iron is considered as boundary [18], Fe-silicates are dominating, which is in agreement with our findings, indicating that bentonite is mainly altered to Fe-silicates rather than to magnetite.

Simulation results are always influenced by certain assumptions and simplifications in the conceptual model. For instance, pyrite forming due to the presence of sulphate-reducing bacteria has been observed in experiments [45]. However, our conceptual model does not consider biological processes, therefore pyrite formation cannot be reproduced. The relevance of microbially induced iron corrosion may, however, be relevant only temporarily as long as the bentonite packing density is not high enough to suppress microbial activity significantly [30,46].

### 3.3. Porosity

Figure 4 shows the porosity variation at 5 and 10,000 years. The numerical model calculates the total mineral volumetric fraction in each node ($Vf$), ranging from 0 to 1, and the porosity is calculated with the following relationship:

$$\phi = 1 - Vf \tag{9}$$

In general, the model does not predict important changes in porosity, being unnoticeable at 5 years. At 1000 years, the porosity decreases due to the precipitation of magnetite, calcite and nontronite, but especially magnetite forming in larger amounts. At the end of the calculations, at 10,000 years, there is a decrease in porosity right after the interface due to precipitation of greenalite. Although there is also dissolution of bentonite near the interface, precipitation is higher than dissolution thus porosity decreases. Slightly further away from the interface, there is an increase of porosity mainly due to montmorillonite dissolution. Numerical models reported in literature, such as that by Bildstein et al. [13] and Samper et al. [19] predicted pore clogging due to the formation of magnetite followed by an increase of porosity due to dissolution of bentonite. They considered the canister as porous material, thus magnetite, which has larger molar volume (44.52 cm$^3$mol$^{-1}$) than iron (7.09 cm$^3$mol$^{-1}$), could precipitate clogging the pores. However, our conceptual model considers the canister as boundary, hence, the secondary phases precipitate in the porosity of the bentonite. Dissolution of montmorillonite and precipitation of greenalite take place at the same time, both with similar pore volumes, 133.96 cm$^3$mol$^{-1}$ and 115 cm$^3$mol$^{-1}$, respectively, therefore, not much changes on porosity are predicted.

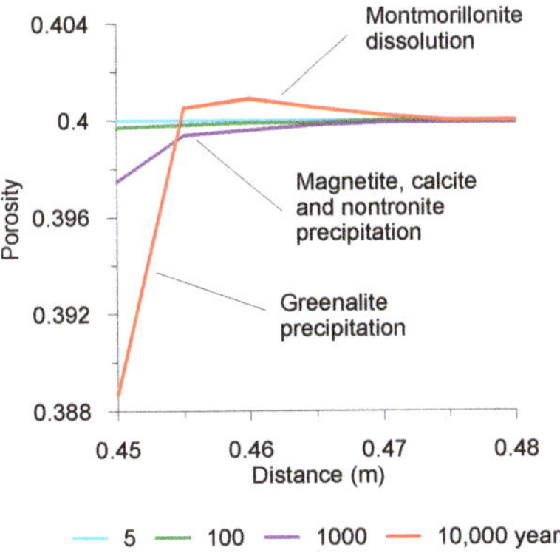


**Figure 4.** Porosity against length for the reference model ($r_c$ = 2 μm y$^{-1}$). The interface between steel and bentonite is at 0.45 m. Results at 5, 100, 1000, and 10,000 years of interaction are compared.

## 4. Sensitivity Analysis

### 4.1. Corrosion Rate of the Canister

In the real system, it is expected to have larger corrosion rates at the beginning that will decrease with time [22]. The formation of corrosion products at the surface of the canister will result in the formation of a protective layer, which would likely decrease the corrosion rate. Therefore, larger and lower corrosion rates with respect to the reference model have been considered in order to study the effect of the corrosion rate on the formation of secondary phases. In the reference model we assume a constant rate of 2 $\mu m\ y^{-1}$, which is replaced with a rate of either 10 or 0.1 $\mu m\ y^{-1}$. Our numerical results are comparable with those reported by Samper et al. [19].

Figure 5 shows the model results considering a constant corrosion rate of 10 $\mu m\ y^{-1}$. The higher corrosion rate yields a rapid increase in pH, which is then almost constant at 1000 years. This can be explained by the higher dissolution of iron and montmorillonite, consuming protons. In comparison with the reference model, in this model, at 5 years, the Fe concentration in the porewater is higher, yielding a rapid precipitation of secondary phases, and then it decreases with time because of increasing pH, which is linked to secondary phases formation. At 100 years, calcite, nontronite-Mg and magnetite are the main secondary phases, in volumetric fractions larger than in the reference model. The precipitation of greenalite starts earlier and is the principal consumer of Si at 1000 years, and therefore nontronite does not continue forming. At 10,000 years, considerable amount of greenalite forms.

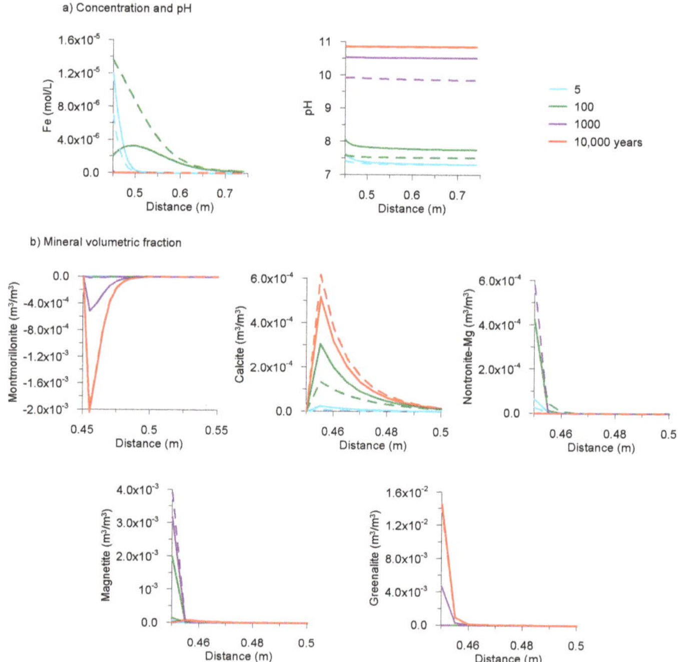



**Figure 5.** Results for the numerical model with a corrosion rate of 10 $\mu m\ y^{-1}$ (continuous lines) are compared with those from the reference model (r = 2 $\mu m\ y^{-1}$, dashed lines). The interface between steel and bentonite is at 0.45 m. Results at 5, 100, 1000 and 10,000 years of interaction are given. (**a**) Fe concentration and pH against distance and (**b**) volumetric fraction of minerals versus distance. For the primary phases (montmorillonite and calcite) the variation of the volumetric fraction is plotted. Positive values mean precipitation and negative values mean dissolution.

Figure 6 displays the model results considering a constant corrosion rate of 0.1 $\mu$m y$^{-1}$. In the first 1000 years, the Fe concentration in the porewater increases with time. During this time, Fe(II) species mainly diffuses through the bentonite and adsorbs via cation exchange. However, at 1000 years, it decreases due to precipitation of corrosion products. The pH is barely increasing with time reaching a maximum value of 8. Therefore, bentonite does not dissolve much. Calcite precipitates more noticeably at 1000 years due to Fe(II)/Ca(II) cation exchange within the montmorillonite interlayer, and at 10,000 years, calcite precipitates due to the Ca coming from the montmorillonite dissolution. Nontronite precipitates due to the available Fe and Si in the porewater, however, at 10,000 years, Fe is mainly consumed by magnetite precipitation. The numerical model does not predict precipitation of other corrosion products such as greenalite, because there is not enough Fe and Si in the system to let greenalite precipitate due to the low dissolution of bentonite and low corrosion rate. In this model, the predicted alteration thickness is larger because diffusion dominates rather than mineral dissolution–precipitation processes. The cation exchange is relevant for the first 1000 years when there is not much precipitation of magnetite.

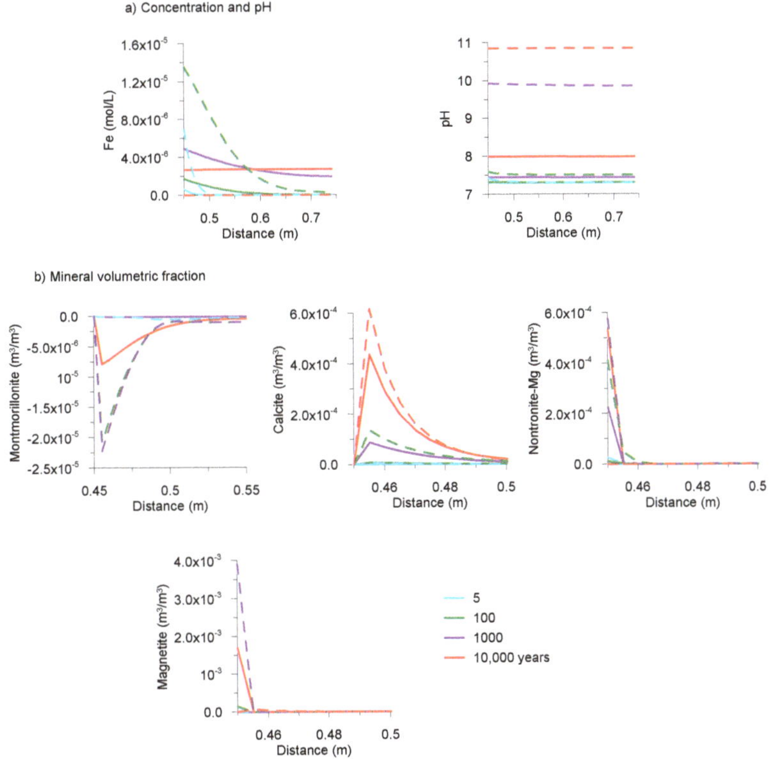


**Figure 6.** Results of the numerical model with a corrosion rate of 0.1 $\mu$m y$^{-1}$ (continuous lines) are compared with those from the reference model (r = 2 $\mu$m y$^{-1}$, dashed lines). The interface between steel and bentonite is at 0.45 m. Results at 5, 100, 1000 and 10,000 years of interaction are compared. (**a**) Fe concentration and pH against distance. (**b**) Volumetric fraction of minerals versus distance. For the primary phases (montmorillonite and calcite) a variation of the volumetric fraction is plotted. Positive values mean precipitation and negative values mean dissolution. The reference model results for montmorillonite at 10,000 are not plotted (see Figure 3 for that detail).

### 4.2. Diffusion Coefficient

Other authors working on similar systems used a larger pore diffusion coefficient for the bentonite [18]. In order to study the impact of the diffusion coefficient on the formation of corrosion products, we assumed a pore diffusion coefficient being six times higher ($6 \times 10^{-10}$ m$^2$ s$^{-1}$), which is the same as used by Wilson et al. [18]. This model considers the same corrosion rate as in the reference model (2 μm y$^{-1}$). Model results are shown in Figure 7. In comparison with the reference model, a higher diffusion coefficient gives a larger thickness of the altered bentonite, which has also been reported by Samper et al. [19]. Ca released from bentonite dissolution is transported away and therefore gives a larger thickness of calcite precipitation. The Si supply from bentonite also gives a larger precipitation area for nontronite and greenalite. The thickness of the porosity alteration is larger, however, there is less variation in porosity because there is higher diffusion of dissolved species away from the corroding iron.

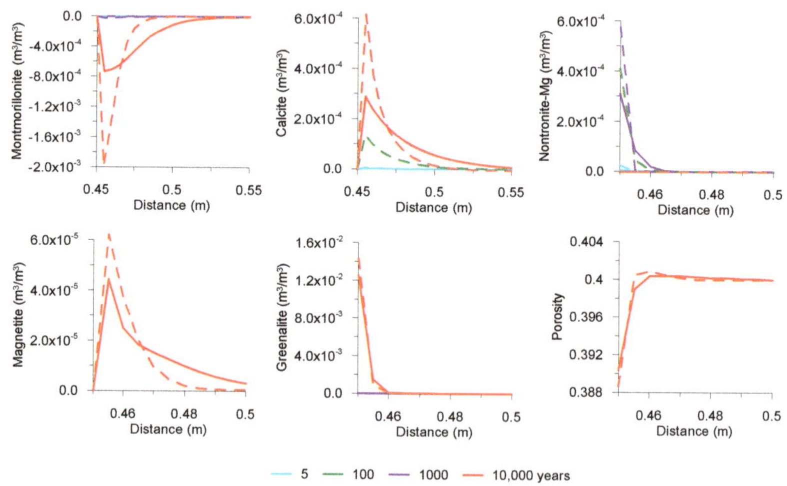


**Figure 7.** Results of the numerical model with a diffusion coefficient of $6 \times 10^{-10}$ m$^2$ s$^{-1}$ ($r_c$ = 2 μm y$^{-1}$) (continuous line) are compared with those from the reference model (dashed line). Volumetric fraction of minerals and porosity versus distance. For the primary phases (montmorillonite and calcite) the variation of the volumetric fraction is plotted. Positive values mean precipitation and negative values mean dissolution. The interface between steel and bentonite is at 0.45 m. Results at 5, 100, 1000 and 10,000 years of interaction are compared. For magnetite and porosity only results at 10,000 years are plotted.

### 4.3. Porewater

There is disagreement in literature about how to determine the initial composition of the porewater of the bentonite, because it is not easy to obtain a reliable composition. Thus, we study the effect of the porewater composition on the nature of formed corrosion products. The composition considered (Table 8) corresponds to a granitic porewater reported by Samper et al. [19]. This porewater composition has been selected because it contains the concentration of all primary species used in our model and higher Si content. In this case, the porewater is not at equilibrium with the montmorillonite. The interaction of low mineralized groundwater with bentonite is investigated in scenarios simulating the intrusion of glacial melt water into emplacement caverns of a repository in crystalline rock (see, e.g., Bouby et al. [47]).

**Table 8.** Initial porewater composition used in the sensitivity analysis.

| Component | Water Composition (mol L$^{-1}$) |
| --- | --- |
| $Al(OH)_4^-$ | $1.00 \times 10^{-8}$ |
| $Ca^{2+}$ | $1.52 \times 10^{-4}$ |
| $Cl^-$ | $3.95 \times 10^{-4}$ |
| $Fe^{+2}$ | $1.79 \times 10^{-8}$ |
| $HCO_3^-$ | $5.05 \times 10^{-3}$ |
| $K^+$ | $5.37 \times 10^{-5}$ |
| $Mg^{2+}$ | $1.60 \times 10^{-4}$ |
| $Na^+$ | $4.35 \times 10^{-3}$ |
| $SiO_2(aq)$ | $3.76 \times 10^{-4}$ |
| $SO_4^{2-}$ | $1.56 \times 10^{-5}$ |
| $O_2(aq)$ | $1.00 \times 10^{-15}$ |
| pH | 7.83 |

Figure 8 shows the results calculated for the numerical model. The pH remains constant around a value of 8. In this system, the Fe-silicates are the main corrosion products and no precipitation of magnetite is predicted. Calcite precipitates during the 10,000 years due to Ca/Fe(II) exchange. Nontronite-Na, cronstedtite and greenalite precipitate during the first 10 years. However, afterwards they dissolve and the Fe and Si are consumed by precipitation of nontronite-Mg, which remains until the end of the run (10,000 years). Thus, our model results suggest that the porewater composition can affect the nature of corrosion products, being Fe-silicates as main corrosion products. Our numerical model results differ from that from Samper et al. [19] because of different assumptions in the conceptual model, such as the selected secondary phases, considering the canister as porous material and a boundary flow.

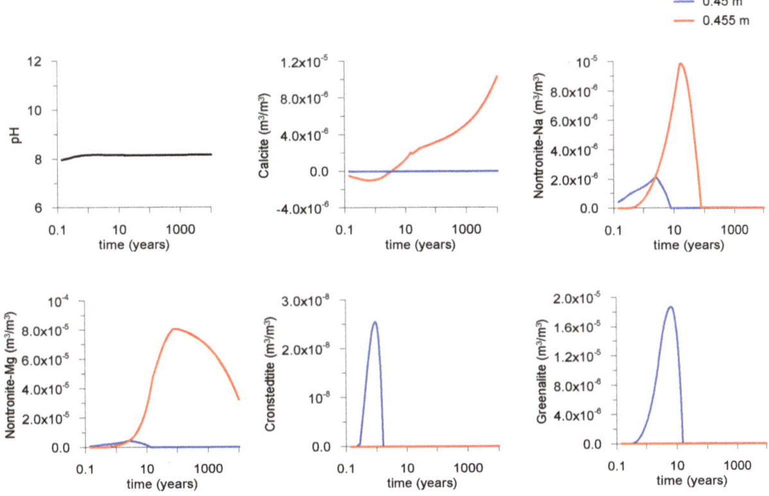


**Figure 8.** Evolution of pH and volumetric fraction of secondary phases at two distances from the steel/bentonite interface considering a granitic porewater ($r_c$ = 2 µm y$^{-1}$).

### 4.4. Reactive Surface Area

In the reference model, large reactive surface areas ($\sigma$ = 1000 m$^2$/m$^3$) have been used for the secondary phases in order to let them precipitate at equilibrium. However, a sensitivity analysis on the surface area has been done, using a much lower value ($\sigma$ = 0.1

m$^2$/m$^3$). Figure 9 shows the calculated evolution of volumetric fraction of the secondary phases. Calcite, magnetite, nontronite-Mg and nontronite-Na precipitate within the first 1000 years. However, afterwards, magnetite and nontronites dissolve, while the volumetric fraction of calcite remains constant. Magnetite dissolves slowly within 3000 years, after that no magnetite remains in the first node (interface), and only $6 \times 10^{-5}$ m$^3$m$^{-3}$ remains in the following nodes (see Figure 3). Nontronites dissolve rapidly, not only the first node, but also in the following nodes. At the same time, greenalite starts precipitating consuming the Fe and Si of the system. In general, a lower reactive surface area does not give important changes compared to the reference model for calcite, magnetite and greenalite, because although the low kinetics they reach the equilibrium. However, the lower surface area provokes a delay on time on nontronite precipitation, because more time is needed to reach the equilibrium.

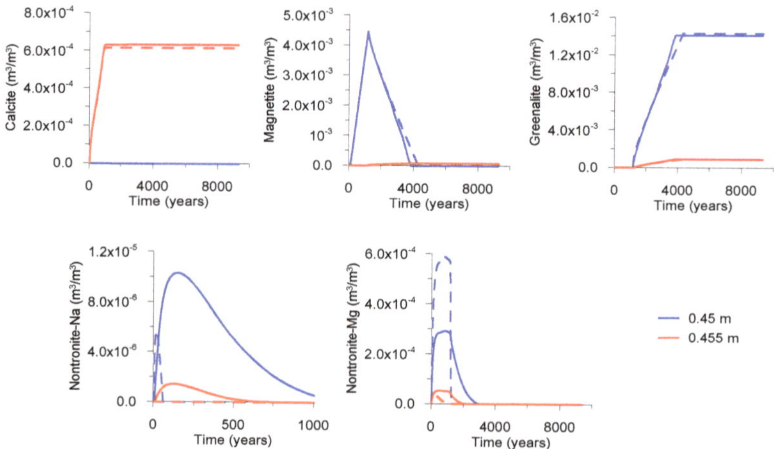

**Figure 9.** Evolution of the volumetric fraction of the secondary phases of the model with a low reactive surface area ($\sigma$ = 0.1 m$^2$/m$^3$, continuous lines) compared with the reference model ($\sigma$ = 1000 m$^2$/m$^3$, dashed lines). Results of the node at the interface (0.45 m) and the node right after the interface (0.455 m) are plotted.

*4.5. Longer Interaction Time*

The reference model considers an iron-bentonite interaction of 10,000 years because it is the time expected that the canister should resist failure [48]. However, due to the slow kinetics of the primary minerals dissolution, longer interaction times can cause more dissolution of them, and therefore, more precipitation of secondary phases. Thus, the effect of a 100,000 years interaction is also studied, taking into account the same corrosion rate as in the reference model (2 μm y$^{-1}$). Figure 10 displays the calculated volumetric fractions only for the minerals that present some variations with respect to the reference model (albite, magnetite and cronstedtite). From the period of time between 10,000 and 100,000 years, the volumetric fraction of montmorillonite remains constant, therefore the volumetric fraction of calcite is also constant. However, the dissolution of albite and quartz increase with time. This Si supply could inhibit montmorillonite dissolution [49,50], and is consumed by precipitation of cronstedtite ($1.8 \times 10^{-5}$ m$^3$ m$^{-3}$), which precipitates not only near to the interface, but extends into the bentonite. Magnetite slightly dissolves, at 10,000 years the predicted volumetric fraction is $6 \times 10^{-5}$ m$^3$ m$^{-3}$ and at 100,000 years is $5 \times 10^{-5}$ m$^3$ m$^{-3}$. Even for longer interaction times and with the formation of cronstedtite, the main corrosion product of the system at the end of the run (100,000 years) is greenalite, being the same as that in the reference model.

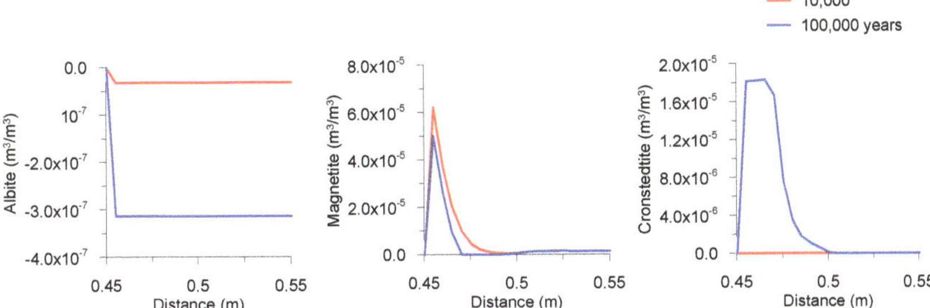



**Figure 10.** Volumetric fraction of minerals of the model with 100,000 years interaction ($r_c = 2~\mu\text{m}~\text{y}^{-1}$). For albite the variation of the volumetric fraction is plotted. Positive values mean precipitation and negative values mean dissolution. The interface between steel and bentonite is at 0.45 m. Results at 10,000 and 100,000 years of interaction are compared.

## 5. Conclusions

A 1D radial reactive transport model has been developed in order to better understand the processes occurring in the long-term iron–bentonite interaction. Results suggest that Fe partly diffuses and is partly adsorbed to montmorillonite by cation exchange, in the short-term. In the long-term, Fe is mainly consumed by corrosion products formation. Calcite precipitates due to the release of Ca into the porewater, in the short-term because of cation exchange, and in the long-term because of montmorillonite dissolution. The reference model predicts precipitation of nontronite, magnetite and greenalite as corrosion products. In the short-term, nontronite and magnetite form consuming the Fe and Si of the system, however, in the long-term, greenalite is predicted as main corrosion product forming upon alteration of the bentonite. Due to the slow kinetics of the minerals, not much mineral dissolution/precipitation is predicted, and therefore, the variation of porosity is not relevant. However, porosity slightly decreases near the interface due to precipitation of mainly greenalite, and increases thereafter because of montmorillonite dissolution.

Results of the sensitivity analysis show that a larger corrosion rate would provoke a rapid precipitation of greenalite due to the rapid Si supply from the dissolution of bentonite. On the contrary, when a lower corrosion rate is considered, the model predicts slower dissolution of bentonite, hence, only nontronite and magnetite precipitate. When a larger diffusion coefficient is considered, the transport of dissolved Fe(II) species dominates with less mineral dissolution/precipitation occurring. The porewater composition can affect the nature of corrosion products formed, Fe-silicates being more relevant than magnetite. A lower reactive surface area for the secondary phases does not seem to have an important impact. Longer iron–bentonite interaction time allows higher dissolution of bentonite, thus cronstedtite forms coexisting with calcite, magnetite and greenalite.

Overall, outcomes suggest that the predicted main corrosion products on the long-term are Fe-silicates minerals. Such phases thus should deserve further attention as chemical barrier in the diffusion of radionuclides to the repository far field. Our results can differ from other reactive transport calculations reported in the literature because of the assumptions considered in the conceptual model, such as the selected secondary phases, considering the canister as porous material or the temperature of the system.

**Supplementary Materials:** The following are available online at https://www.mdpi.com/article/10.3390/min11111272/s1, Tables S1 and S2: Log K at 25 °C and stoichiometric coefficients of homogeneous reactions, Table S3: Log K at 25 °C and stoichiometric coefficients of mineral reactions, Table S4: Log K at 25 °C and stoichiometric coefficients of cation exchange reactions.

**Author Contributions:** Conceptualization, M.C.C., N.F., V.M. and H.G.; investigation, M.C.C.; methodology, M.C.C.; validation, M.C.C., N.F., V.M. and H.G.; visualization, M.C.C.; writing—original draft preparation, M.C.C.; writing—review and editing, M.C.C., N.F., V.M. and H.G.; funding acquisition, H.G. All authors have read and agreed to the published version of the manuscript.

**Funding:** This research was funded by the Helmholtz Association (Grant SO-093) and the German Federal Ministry of Education and Research (Grant 02NUK053).

**Data Availability Statement:** Not applicable.

**Acknowledgments:** This work was funded by the German Federal Ministry of Education and Research (BMBF, Grant 02NUK053) and the Helmholtz Association (Grant SO-093). We acknowledge support by the KIT-Publication Fund of the Karlsruhe Institute of Technology.

**Conflicts of Interest:** The authors declare no conflict of interest.

## References


1. Savage, D. *Prospects for Coupled Modelling*; Technical Report, STUK-TR 13; Radiation and Nuclear Safety Authority: Helsinki, Finland, 2012.
2. Bildstein, O.; Claret, F. Chapter 5—Stability of Clay Barriers Under Chemical Perturbations. In *Natural and Engineered Clay Barriers*; Developments in Clay Science; Tournassat, C., Steefel, C.I., Bourg, I.C., Bergaya, F., Eds.; Elsevier: Amsterdam, The Netherlands, 2015; Volume 6, pp. 155–188. [CrossRef]
3. Claret, F.; Marty, N.; Tournassat, C. Modeling the Long-Term Stability of Multibarrier Systems for Nuclear Waste Disposal in Geological Clay Formations. In *Reactive Transport Modeling*; Xiao, Y.; Whitaker, F.; Xu, T.; Steefel, C., Eds.; John Wiley & Sons, Ltd.: Chichester, UK, 2018; pp. 395–436.
4. Bildstein, O.; Claret, F.; Frugier, P. RTM for Waste Repositories. *Rev. Mineral. Geochem.* **2019**, *85*, 419–457. [CrossRef]
5. Deissmann, G.; Mouheb, N.A.; Martin, C.; Turrero, M.; Torres, E.; Kursten, B.; Weetjens, E.; Jacques, D.; Cuevas, J.; Samper, J.; et al. Experiments and Numerical Model Studies on Interfaces; Final version as of 11.05.2021 of Deliverable D2.5 of the HORIZON 2020 Project EURAD. EC Grant Agreement no: 847593; 2021.
6. Schlegel, M.L.; Bataillon, C.; Benhamida, K.; Blanc, C.; Menut, D.; Lacour, J.L. Metal corrosion and argillite transformation at the water-saturated, high-temperature iron-clay interface: A microscopic-scale study. *Appl. Geochem.* **2008**, *23*, 2619–2633. [CrossRef]
7. Schlegel, M.L.; Bataillon, C.; Brucker, F.; Blanc, C.; Prêt, D.; Foy, E.; Chorro, M. Corrosion of metal iron in contact with anoxic clay at 90 °C: Characterization of the corrosion products after two years of interaction. *Appl. Geochem.* **2014**, *51*, 1–4. [CrossRef]
8. Martin, F.; Bataillon, C.; Schlegel, M. Corrosion of iron and low alloyed steel within a water saturated brick of clay under anaerobic deep geological disposal conditions: An integrated experiment. *J. Nucl. Mater.* **2008**, *379*, 80–90. [CrossRef]
9. Kaufhold, S.; Schippers, A.; Marx, A.; Dohrmann, R. SEM study of the early stages of Fe-bentonite corrosion—The role of naturally present reactive silica. *Corros. Sci.* **2020**, *171*, 108716. [CrossRef]
10. Savage, D.; Watson, C.; Benbow, S.; Wilson, J. Modelling iron-bentonite interactions. *Appl. Clay Sci.* **2010**, *47*, 91–98. [CrossRef]
11. Samper, J.; Lu, C.; Montenegro, L. Reactive transport model of interactions of corrosion products and bentonite. *Phys. Chem. Earth Parts A/B/C* **2008**, *33*, S306–S316. [CrossRef]
12. Lu, C.; Samper, J.; Fritz, B.; Clement, A.; Montenegro, L. Interactions of corrosion products and bentonite: An extended multicomponent reactive transport model. *Phys. Chem. Earth Parts A/B/C* **2011**, *36*, 1661–1668. [CrossRef]
13. Bildstein, O.; Trotignon, L.; Perronnet, M.; Jullien, M. Modelling iron-clay interactions in deep geological disposal conditions. *Phys. Chem. Earth Parts A/B/C* **2006**, *31*, 618–625. [CrossRef]
14. Wersin, P.; Birgersson, M.; Olsson, O.; Karnland, O.; Snellman, M. *Impact of Corrosion-Derived Iron on the Bentonite Buffer within the KBS-3H Disposal Concept The Olkiluoto Site as Case Study*; Technical Report SKB-TR-08-34; Swedish Nuclear Fuel and Waste Management Co.: Stockholm, Sweden, 2008.
15. Marty, N.C.; Fritz, B.; Clément, A.; Michau, N. Modelling the long term alteration of the engineered bentonite barrier in an underground radioactive waste repository. *Appl. Clay Sci.* **2010**, *47*, 82–90. [CrossRef]
16. Wersin, P.; Birgersson, M. Reactive transport modelling of iron-bentonite interaction within the KBS-3H disposal concept—The Olkiluoto site as case study. *Geol. Soc. Spec. Publ.* **2014**, *400*, 237–250. [CrossRef]
17. Ngo, V.V.; Delalande, M.; Clément, A.; Michau, N.; Fritz, B. Coupled transport-reaction modeling of the long-term interaction between iron, bentonite and Callovo-Oxfordian claystone in radioactive waste confinement systems. *Appl. Clay Sci.* **2014**, *101*, 430–443. [CrossRef]
18. Wilson, J.C.; Benbow, S.; Sasamoto, H.; Savage, D.; Watson, C. Thermodynamic and fully-coupled reactive transport models of a steel-bentonite interface. *Appl. Geochem.* **2015**, *61*, 10–28. [CrossRef]
19. Samper, J.; Naves, A.; Montenegro, L.; Mon, A. Reactive transport modelling of the long-term interactions of corrosion products and compacted bentonite in a HLW repository in granite: Uncertainties and relevance for performance assessment. *Appl. Geochem.* **2016**, *67*, 42–51. [CrossRef]
20. Montes-H, G.; Marty, N.; Fritz, B.; Clement, A.; Michau, N. Modelling of long-term diffusion–reaction in a bentonite barrier for radioactive waste confinement. *Appl. Clay Sci.* **2005**, *30*, 181–198. [CrossRef]
21. Finsterle, S.; Muller, R.A.; Baltzer, R.; Payer, J.; Rector, J.W. Thermal Evolution near Heat-Generating Nuclear Waste Canisters Disposed in Horizontal Drillholes. *Energies* **2019**, *12*, 596. [CrossRef]

22. Necib, S.; Diomidis, N.; Keech, P.; Nakayama, M. *Corrosion of Carbon Steel in Clay Environments Relevant to Radioactive Waste Geological Disposals, Mont Terri Rock Laboratory (Switzerland)*; Springer International Publishing: Cham, Switzerland, 2018; pp. 331–344. [CrossRef]
23. Carrière, C.; Dillmann, P.; Foy, E.; Neff, D.; Dynes, J.; Linard, Y.; Michau, N.; Martin, C. Use of nanoprobes to identify iron-silicates in a glass/iron/argillite system in deep geological disposal. *Corros. Sci.* **2019**, *158*, 108104. [CrossRef]
24. Saaltink, M.W.; Batlle, F.; Ayora, C.; Carrera, J.; Olivella, S. RETRASO, a code for modeling reactive transport in saturated and unsaturated porous media. *Geol. Acta* **2004**, *2*, 235–251.
25. Olivella, S.; Gens, A.; Carrera, J.; Alonso, E.E. Numerical formulation for a simulator (CODE_BRIGHT) for the coupled analysis of saline media. *Eng. Comput.* **1996**, *13*, 87–112. [CrossRef]
26. Bossart, P.; Bernier, F.; Birkholzer, J.; Bruggeman, C.; Connolly, P.; Dewonck, S.; Fukaya, M.; Herfort, M.; Jensen, M.; Matray, J.; et al. *Mont Terri Rock Laboratory, 20 Years of Research: Introduction, Site Characteristics and Overview of Experiments*; Birkhäuser: Cham, Swizerland, 2018; Volume 5. [CrossRef]
27. Zheng, L.; Rutqvist, J.; Birkholzer, J.T.; Liu, H.H. On the impact of temperatures up to 200 °C in clay repositories with bentonite engineer barrier systems: A study with coupled thermal, hydrological, chemical, and mechanical modeling. *Eng. Geol.* **2015**, *197*, 278–295. [CrossRef]
28. King, F. *Corrosion of Carbon Steel under Anaerobic Conditions in a Repository for SF and HLW in Opalinus Clay*; Technical Report NAGRA-NTB–08-12; National Cooperative for the Disposal of Radioactive Waste: Wettingen, Switzerland, 2008.
29. Zheng, L.; Rutqvist, J.; Xu, H.; Birkholzer, J.T. Coupled THMC models for bentonite in an argillite repository for nuclear waste: Illitization and its effect on swelling stress under high temperature. *Eng. Geol.* **2017**, *230*, 118–129. [CrossRef]
30. Smart, N.R.; Reddy, B.; Rance, A.P.; Nixon, D.J.; Frutschi, M.; Bernier-Latmani, R.; Diomidis, N. The anaerobic corrosion of carbon steel in compacted bentonite exposed to natural Opalinus Clay porewater containing native microbial populations. *Corros. Eng. Sci. Technol.* **2017**, *52*, 101–112. [CrossRef]
31. Morelová, N.; Schild, D.; Heberling, F.; Finck, N.; Dardenne, K.; Metz, V.; Geckeis, H. Anaerobic Corrosion of Carbon Steel in Compacted Bentonite Exposed to Natural Opalinus Clay Porewater: Bentonite Alteration Study. *Goldschmidt Virtual*, 4–9 July 2021.
32. Karnland, O. *Chemical and Mineralogical Characterization of the Bentonite Buffer for the Acceptance Control Procedure in a KBS-3 Repository*; Technical Report SKB-TR-10-60; Department of Energy, Office of Scientific and Technical Information: Oak Ridge, TN, USA, 2010.
33. Kiviranta, L.; Kumpulainen, S.; Pintado, X.; Karttuten, P.; Schatz, T. *Characterization of Bentonite and Clay Materials 2012–2015*; Working Report 2016-05; POSIVA OY: Eurajoki, Finland, 2018.
34. Reddy, B.; Padovani, C.; Smart, N.R.; Rance, A.P.; Cook, A.; Milodowski, A.; Field, L.; Kemp, S.; Diomidis, N. Further results on the in situ anaerobic corrosion of carbon steel and copper in compacted bentonite exposed to natural Opalinus Clay porewater containing native microbial populations. *Mater. Corros.* **2021**, *72*, 268–281. [CrossRef]
35. Curti, E. *Modelling Bentonite Pore Waters for the Swiss High-Level Radioactive Waste Repository*; Technical Report, PSI Report No. 93-05; PSI: Villigen, Switzerland, 1993.
36. Melkior, T.; Mourzagh, D.; Yahiaoui, S.; Thoby, D.; Alberto, J.; Brouard, C.; Michau, N. Diffusion of an alkaline fluid through clayey barriers and its effect on the diffusion properties of some chemical species. *Appl. Clay Sci.* **2004**, *26*, 99 – 107. [CrossRef]
37. Johnson, J.; Lundeen, S. *GEMBOCHS Thermodynamic Datafiles for Use with the EQ3/6 Software Package*; Technical Report, Yucca Mountain Laboratory, Milestone Report M0L72; Lawrence Livermore National Laboratory: Livermore, CA, USA, 1994.
38. Appelo, C.; Postma, D. *Geochemistry, Groundwater and Pollution*; A.A. Balkema: Rotterdam, The Netherland, 2005.
39. Lasaga, A.C. Chemical kinetics of water-rock interactions. *J. Geophys. Res. Solid Earth* **1984**, *89*, 4009–4025. [CrossRef]
40. Cama, J.; Ganor, J.; Ayora, C.; Lasaga, C. Smectite dissolution kinetics at 80 °C and pH 8.8. *Geochim. Cosmochim. Acta* **2000**, *64*, 2701–2717. [CrossRef]
41. Dove, P.M. The dissolution kinetics of quartz in aqueous mixed cation solutions. *Geochim. Cosmochim. Acta* **1999**, *63*, 3715–3727. [CrossRef]
42. Knauss, K.G.; Wolery, T.J. Dependence of albite dissolution kinetics on ph and time at 25 °C and 70 °C. *Geochim. Cosmochim. Acta* **1986**, *50*, 2481–2497. [CrossRef]
43. Kamei, G.; Ohmoto, H. The kinetics of reactions between pyrite and O2-bearing water revealed from in situ monitoring of DO, Oh and pH in a closed system. *Geochim. Cosmochim. Acta* **2000**, *64*, 2585–2601. [CrossRef]
44. Plummer, L.; Parkhurst, D.; Wigley, T. Critical review of the kinetics of calcite dissolution and precipitation. In *Chemical Modeling of Aqueous Systems*; ACS Symposium Series: Washington, DC, USA, 1979; pp. 537–573; doi:10.1021/bk-1979-0093.ch025 [CrossRef]
45. Necib, S.; Schlegel, M.L.; Bataillon, C.; Daumas, S.; Diomidis, N.; Keech, P.; Crusset, D. Long-term corrosion behaviour of carbon steel andstainless steel in Opalinus clay: influence of stepwise temperature increase. *Corros. Eng. Sci. Technol.* **2019**, *54*, 516–528. [CrossRef]
46. King, F.; Padovani, C. Review of the corrosion performance of selected canister materials for disposal of UK HLW and/or spent fuel. *Corros. Eng. Sci. Technol.* **2011**, *46*, 82–90. [CrossRef]
47. Bouby, M.; Kraft, S.; Kuschel, S.; Geyer, F.; Moisei-Rabung, S.; Schäfer, T.; Geckeis, H. Erosion dynamics of compacted raw or homoionic MX80 bentonite in a low ionic strength synthetic water under quasi-stagnant flow conditions. *Appl. Clay Sci.* **2020**, *198*, 105797. [CrossRef]

48. National Cooperative for the Disposal of Radioactive Waste (NAGRA). *W.S. Project Opalinus Clay—Safety Report—Demonstration of Disposal Feasibility for Spent Fuel, Vitrified High-Level Waste and Long-Lived Intermediate-Level Waste (Entsorgungsnachweis)*; Technical Report NTB-02-05; Nagra: Wettingen, Switzerland, 2002.
49. Arthur, R.; Savage, D. *Long-Term Stability of Clay Minerals in the Buffer and Backfill*; Technical Report, STUK Report STUK-TR22; Finnish Radiation and Nuclear Safety Authority (STUK): Helsinki, Finland, 2016.
50. Savage, D.; Cloet, V. *A Review of Cement-Clay Modelling*; Working Report NAB-18-24; Nagra: Wettingen, Switzerland, 2018.

MDPI

St. Alban-Anlage 66

4052 Basel

Switzerland

Tel. +41 61 683 77 34

Fax +41 61 302 89 18

www.mdpi.com

*Minerals* Editorial Office

E-mail: minerals@mdpi.com

www.mdpi.com/journal/minerals